Graduate Texts in Mathematics 262

Graduate Texts in Mathematics

Graduate Texts in Mathematics bridge the gap between passive study and creative understanding, offering graduate-level introductions to advanced topics in mathematics. The volumes are carefully written as teaching aids and highlight characteristic features of the theory. Although these books are frequently used as textbooks in graduate courses, they are also suitable for individual study.

More information about this series at http://www.springer.com/series/136

Daniel W. Stroock

Essentials of Integration Theory for Analysis

Second Edition

Springer

Daniel W. Stroock
Department of Mathematics
Massachusetts Institute of Technology
Cambridge, MA, USA

ISSN 0072-5285 ISSN 2197-5612 (electronic)
Graduate Texts in Mathematics
ISBN 978-3-030-58477-1 ISBN 978-3-030-58478-8 (eBook)
https://doi.org/10.1007/978-3-030-58478-8

Mathematics Subject Classification: 28-00, 26A42

This Springer imprint is published by the registered company Springer Nature Switzerland AG
The registered company address is: Gewerbestrasse 11, 6330 Cham, Switzerland

Preface

From the first edition: With some justification, measure theory has a bad reputation. It is regarded by most students as a subject that has little aesthetic appeal and lots of fussy details. In order to make the subject more palatable, many authors have chosen to add spice by embedding measure theory inside one of the many topics in which measure theory plays a central role. In the past, Fourier analysis was usually the topic chosen, but in recent years Fourier analysis has been frequently displaced by probability theory. There is a lot to be said for the idea of introducing a running metaphor with which to motivate the technical definitions and minutiae with which measure theory is riddled. However, I[1] have not adopted this pedagogic device. Instead, I have attempted to present measure theory as an essential branch of analysis, one that has the merit of its own. Thus, although I often digress to demonstrate how measure theory answers questions whose origins are in other branches of analysis, this book is about measure theory, unadorned.

In the first chapter, I give a resume of Riemann's theory of integration, including Stieltjes' extension of that theory. My reason for including Riemann's theory is twofold. In the first place, when I turn to Lebesgue's theory, I want Riemann's theory available for comparison purposes. Secondly, and perhaps more importantly, I believe that Riemann's theory provides many of the basic tools with which one does actual computations. Lebesgue's theory enables one to prove equalities between abstract quantities, but evaluation of those quantities usually requires Riemann's theory. The final section of Chapter 1 contains an analysis of the rate at which Riemann sums approximate his integral. In no sense is this section serious numerical analysis. On the other hand, it gives an amusing introduction to the Euler–Maclaurin formula.

[1] Contrary to the convention in most modern mathematical exposition, I often use the first person singular rather than the "royal we" when I expect the reader to be playing a passive role. I restrict the use of "we" to places, like proofs, where I expect the active participation of my readers.

Modern (i.e., *après* Lebesgue) measure theory is introduced in Chapter 2. I begin by trying to explain why countable additivity is the *sine qua non* in Lebesgue's theory of integration. This explanation is followed by the derivation of a few elementary properties possessed by countably additive measures. In the second section of the chapter, I develop a somewhat primitive procedure for constructing measures on metric spaces and then apply this procedure to the construction of Lebesgue measure $\lambda_{\mathbb{R}^N}$ on $\mathbb{R}^N$, the measure μ_F on $\mathbb{R}$ determined by a distribution function F, and the Bernoulli measures β_p on $\{0,1\}^{\mathbb{Z}^+}$. Included here is a proof of the way in which Lebesgue measure transforms under linear maps and of the relationship between $\lambda_{\mathbb{R}^N}$ and $\beta_{\frac{1}{2}}$.

Lebesgue integration theory is taken up in Chapter 3. The basic theory is covered in the first section, and its miraculous stability (i.e., the monotone convergence and Lebesgue's dominated convergence theorems as well as Fatou's lemma) is demonstrated in the next section. The third section is a bit of a digression. There I give a proof, based on Riesz's Sunrise Lemma, of Lebesgue's Differentiation Theorem for increasing functions.

The first section of Chapter 4 is devoted to the construction of product measures and the proof of Fubini's Theorem. As an application, in the second section, I describe Steiner's symmetrization procedure and use it to prove the isodiametric inequality, which I then apply to show that N-dimensional Hausdorff measure in $\mathbb{R}^N$ is Lebesgue measure there.

In Chapter 5, I discuss several topics that are tied together by the fact that they all involve changes of variables. The first of these is the application of distribution functions to show that Lebesgue integrals can be represented as Riemann integrals, and the second topic is polar coordinates. Both of these are in § 5.1. The second section contains a proof of Jacobi's transformation formula and an application of his formula to the construction of surface measure for hypersurfaces in $\mathbb{R}^N$. My treatment of these is, from a differential geometric perspective, extremely inelegant: there are no differential forms here. In particular, my construction of surface measure is concertedly non-intrinsic. Instead, I have adopted a more geometric measure-theoretic point of view and constructed surface measure by "differentiating" Lebesgue measure. Similarly, my derivation in § 5.3 of the Divergence Theorem is devoutly extrinsic and devoid of differential form technology.

Some of the bread and butter inequalities (specifically, Jensen's, Hölder's, and Minkowski's) of integration theory are derived in the first section of Chapter 6. In the second section, these inequalities are used to study some elementary geometric facts about the Lebesgue spaces L^p as well as the mixed Lebesgue spaces $L^{(p,q)}$. The results obtained in § 6.2 are applied in § 6.3 to the analysis of boundedness properties for transformations defined by kernels on the Lebesgue space. Particular emphasis is placed on transformations given by convolution, for which Young's inequality is proved. The chapter ends with a brief discussion of Friedrichs mollifiers.

In the preparation for Fourier analysis, Chapter 7 begins with a cursory introduction to Hilbert spaces. The basic L^2-theory of Fourier series is given in § 7.2 and is applied there to complete the program, started in § 1.3 of Chapter 1, of understanding the Euler–Maclaurin formula. The elementary theory of the Fourier transform is developed in § 7.3, where I first give the L^1-theory and then the L^2-theory. My approach to the latter is via Hermite functions.

The concluding chapter contains several vital topics that were either given short shrift or entirely neglected earlier. The first of these is the Radon–Nikodym Theorem, which I, following von Neumann, prove as an application of Riesz's Representation Theorem for Hilbert space. The second topic is Daniell's theory of integration, which I use first to derive the standard criterion that says when a finite, finitely additive measure admits a countably additive extension and second to derive Riesz's Representation Theorem for non-negative linear functionals on continuous functions. The final topic is Carathéodory's method for constructing measures from subadditive functions and its application to the construction of Hausdorff measures on $\mathbb{R}^N$. Although my treatment of Hausdorff measures barely touches on the many beautiful and deep aspects of this subject, I do show that the restriction of $(N-1)$-dimensional Hausdorff measure to a hypersurface in $\mathbb{R}^N$ coincides with the surface measure constructed in § 5.2.2.

It is my hope that this book will be useful both as a resource for students who are studying measure theory on their own and as a text for a course. I have used it at MIT as the text for a one-semester course. However, MIT students are accustomed to abuse,[2] and it is likely that as a text for a one-semester course elsewhere some picking and choosing will be necessary. My suggestion is that one be sure to cover the first four chapters and the first sections of Chapters 7 and 8, perhaps skipping § 1.3, § 3.3, and § 4.2. Depending on the interests of the students, one can supplement this basic material with selections from Chapters 5, 6, as well as from material that one skipped earlier.

There are exercises at the end of each section. Some of these are quite trivial and others are quite challenging. Especially for those attempting to learn the subject on their own, I strongly recommend that, at the very least, my readers look at all the exercises and solve enough of them to become facile with the techniques they wish to master. At least for me, it is not possible to learn mathematics as a spectator.

[2] It has been said that getting an education at M.I.T. is like taking a drink from a fire hydrant.

What is new: Although the core of the material in the previous edition is present in this one, there have been quite a few changes. For one thing, I have incorporated minor changes as well as corrections to the errors I and others (particularly Robert Koirala) found. Secondly, § 2.2.5 and § 8.3 have been substantially revised. Thirdly, I have added a few new topics. As an application of the material about Hermite functions in subsection § 7.3.2, I have added a brief introduction to L. Schwartz's theory of tempered distributions in § 7.3.4. Section § 7.4 is entirely new and contains applications, including the Central Limit Theorem, of Fourier analysis to measures. Related to this are subsections § 8.2.5 and § 8.2.6, where I prove Lévy's Continuity Theorem and Bochner's characterization of the Fourier transforms of Borel probability on $\mathbb{R}^N$. Also, subsection § 8.1.2 is new and contains a proof of the Hahn Decomposition Theorem. Finally, there are quite a few new exercises, some covering material from the original edition and others based on newly added material.

Nederland, Colorado Daniel W. Stroock

Contents

Notation

General

$\alpha^+ \,\&\, \alpha^-$	The positive and negative parts of $\alpha \in \mathbb{R}$ (§ 1.2).
$\lfloor t \rfloor$	The integer part of $t \in \mathbb{R}$.
$f \restriction S$	The restriction of the function f to the set S.
$\\|\cdot\\|_{\mathrm{u}}$	The uniform (supremum) norm.
$\mathrm{Var}(\psi; J)$	Variation norm of function ψ on interval J (§ 1.2.2).
i	The imaginary number $\sqrt{-1}$.
ω_{N-1}	The surface area of unit sphere $\mathbb{S}^{N-1}$ in $\mathbb{R}^N$ (Exercise 5.2.3).
Ω_N	The volume, $N^{-1}\omega_{N-1}$, of the unit ball $B(\mathbf{0};1)$ in $\mathbb{R}^N$
$\mathcal{O}(g(t))$	A function f for which $\frac{f(t)}{g(t)}$ is bounded as t tends to a limit.
$o(g(t))$	A function f for which $\frac{f(t)}{g(t)}$ tends to 0 as t tends to a limit.
$(\mathbf{v},\mathbf{w})_{\mathbb{R}^N}$	The inner product of $\mathbf{v}$ with $\mathbf{w}$ in $\mathbb{R}^N$.
$\overline{\lim}_{n\to\infty} A_n, \underline{\lim}_{n\to\infty} A_n$	The limit superior $\bigcap_{m=1}^{\infty}\bigcup_{n\geq m} A_n$ and limit inferior $\bigcup_{m=1}^{\infty}\bigcap_{n\geq m} A_n$ of the sets $\{A_n\}_1^{\infty}$ (Exercise 2.1.6).
$B_E(a,r)$	The ball of radius r around a in the metric space E.
$\mathring{\Gamma}, \bar{\Gamma}, \partial\Gamma$	The interior, closure, and boundary of a subset in a topological space.

Sets, Functions, and Spaces

$\mathbb{C}$	The complex numbers.
$\mathbb{Z}$ & $\mathbb{Z}^+$	The integers and the positive integers.
$\mathbb{N}$	The non-negative integers: $\mathbb{N} = \{0\} \cup \mathbb{Z}^+$.
$\mathbb{Q}$	The set of rational numbers.
$\mathcal{P}(E)$	The power set of E, consisting of all subsets of E (§ 2.1.2).
$A^\complement$	The complement of the set A.
$A^{(\delta)}$	The δ-hull around the set A (§ 5.2.1).
$\mathfrak{G}(E)$ & $\mathfrak{G}_\delta(E)$	The set of open subsets and the set of countable intersections of open subsets of E.
$\mathfrak{F}(E)$ & $\mathfrak{F}_\sigma(E)$	The set of closed subsets and the set of countable unions of closed subsets of E.
$\mathbb{S}^{N-1}$	The unit sphere in $\mathbb{R}^N$.
$\Gamma(\rho)$	The tubular neighborhood of Γ (§ 5.3.2).
$B_E(a,r)$	The ball of radius r around a in E. When E is omitted, it is assumed to be $\mathbb{R}^N$ for some $N \in \mathbb{Z}^+$.
$K \subset\subset E$	To be read: *K is a compact subset of E.*
$\lVert\mathcal{C}\rVert$	The mesh size of the collection $\mathcal{C}$ (§ 1.1).
$\mathbf{1}_A$	The indicator function of the set A (§ 3.1.1).
$f(x+)$ & $f(x-)$	The right and left limits of f at $x \in \mathbb{R}$.
$\mathbf{e}_\xi(x)$	The imaginary exponential function $e^{i2\pi(\xi,x)_{\mathbb{R}^N}}$.
$\operatorname{sgn}(x)$	The signum function: equal to 1 if $x \geq 0$ and -1 if $x<0$.
$\Gamma(t)$	Euler's gamma function (Exercise 5.1.3).
$\mathbf{T}_p(M)$	The tangent space to M at p (§ 5.2.2).
T_x	Translation by x: $T_x(y) = x+y$ (§ 2.2.1).
T_A	The linear transformation determined by the matrix A (§ 2.2.1).
$J\Phi$	The Jacobian of Φ: $J\Phi$ is the absolute value of the determinant of the Jacobian matrix $\frac{\partial \Phi}{\partial x}$ (§ 5.2.1).
$C_b(E;\mathbb{R})$ or $C_b(E;\mathbb{C})$	The space of bounded continuous functions from a topological space E into $\mathbb{R}$ or $\mathbb{C}$.

Sets, Functions, and Spaces Continued

$L^p(\mu;\mathbb{R})$ or $L^p(\mu;\mathbb{C})$	The Lebesgue space of $\mathbb{R}$-valued or $\mathbb{C}$-valued functions f for which $\|f\|^p$ is μ-integrable.
$\ell^2(\mathbb{N};\mathbb{R})$	The space $L^2(\mu;\mathbb{R})$ when μ is counting measure on $\mathbb{N}$ (§ 7.1.2).
$C_b(E;\mathbb{R})$	The space of bounded, continuous, $\mathbb{R}$-valued functions on the topological space E.
$C^n(E;\mathbb{R}^N)$	The space of $f : G \longrightarrow \mathbb{R}^N$ with $n \in \mathbb{N}\cup\{\infty\}$ continuous derivatives.
$C^n_c(E;\mathbb{R}^N)$	The space of $f \in C^n(G;\mathbb{R}^N)$ that vanish off of a compact set.

Measure Theoretic

(a.e., μ)	To be read *almost everywhere with respect to* μ (§ 3.3).
$\sigma(\mathcal{C})$	The σ-algebra generated by $\mathcal{C}$ (§ 2.1.2).
$\mathcal{B}_E$	The Borel σ-algebra $\sigma(\mathfrak{G}(E))$ over E.
$\mathcal{B}[E']$	The restriction of the σ-algebra $\mathcal{B}$ to the subset E' (§ 3.1).
$\bar{\mathcal{B}}^\mu$	The completion of the σ-algebra $\mathcal{B}$ with respect to the measure μ (§ 3.1).
$\mathcal{B}_1 \times \mathcal{B}_2$	The product σ-algebra generated by sets of the form $\Gamma_1 \times \Gamma_2$ for $\Gamma_i \in \mathcal{B}_i$ (§ 3.2).
$\bar{\mathcal{B}}^\mu$	The completion of the σ-algebra $\mathcal{B}$ with respect to μ (§ 2.1.2).
δ_a	The unit point mass at a.
$\mathbf{H}^s$	Hausdorff measure (§ 8.3.4).
λ_S	Lebesgue measure on the set $S \in \mathcal{B}_{\mathbb{R}^N}$.
$\mu \ll \nu$	μ is absolutely continuous with respect to ν (Exercise 2.1.8)
$\mu \perp \nu$	μ is singular to ν (Exercise 2.1.9).
$\Phi_*\mu$	The pushforward (image) of μ under Φ (Exercise 2.1.3).
$\int_\Gamma f(x)\,dx$	Equivalent to the Lebesgue integral $\int_\Gamma f\,d\lambda_{\mathbb{R}^N}$ of f on Γ.

Measure Theoretic Continued

$\fint_\Gamma f\,d\mu$	The average value $\frac{1}{\mu(\Gamma)}\int_\Gamma f\,d\mu$ of f on Γ with respect to the measure μ.
$(\mathrm{R})\int_{[a,b]}\varphi(t)\,d\psi(t)$	The Riemann–Stieltjes integral of φ on $[a,b]$ with respect to ψ (§1.2).
$\mathcal{C}_1 \leq \mathcal{C}_2$	To be read $\mathcal{C}_2$ *is a refinement of* $\mathcal{C}_1$ (§ 1.1).
$\mathcal{C}_1 \vee \mathcal{C}_2$	The least common refinement of $\mathcal{C}_1$ and $\mathcal{C}_2$ (§ 1.1).
$\Delta_I\psi$	Change $\psi(I^+) - \psi(I^-)$ of ψ over the interval I (1.2.2).

Chapter 1
The Classical Theory

I begin by recalling a few basic facts about the integration theory that is usually introduced in advanced calculus. I do so not only for purposes of later comparison with the modern theory but also because it is the theory with which most computations are eventually performed.

1.1 Riemann Integration

Let $N \in \mathbb{Z}^+$ (throughout $\mathbb{Z}^+$ will denote the positive integers). A **rectangle** in $\mathbb{R}^N$ is a subset I of $\mathbb{R}^N$ that can be written as the Cartesian product $\prod_{k=1}^N [a_k, b_k]$ of compact intervals $[a_k, b_k]$, where it is assumed that $-\infty < a_k \le b_k < \infty$ for each $1 \le k \le N$. If I is such a rectangle, its **diameter** and **volume** are, respectively,

$$\operatorname{diam}(I) \equiv \sup\{|y-x| : x, y \in I\} = \sqrt{\sum_{k=1}^N (b_k - a_k)^2} \text{ and } \operatorname{vol}(I) \equiv \prod_{k=1}^N (b_k - a_k).$$

For the purposes of this exposition, it will be convenient to also take the empty set to be a rectangle with diameter and volume 0.

Given a collection $\mathcal{C}$,[1] I will say that $\mathcal{C}$ is **non-overlapping** if distinct elements of $\mathcal{C}$ have disjoint interiors. In that its conclusions seem *obvious*, the following lemma is surprisingly difficult to prove.

Lemma 1.1.1 *If $\mathcal{C}$ is a non-overlapping, finite collection of rectangles each of which is contained in the rectangle J, then* $\operatorname{vol}(J) \ge \sum_{I \in \mathcal{C}} \operatorname{vol}(I)$. *On the other hand, if $\mathcal{C}$ is any finite collection of rectangles and J is a rectangle that is covered by $\mathcal{C}$ (i.e., $J \subseteq \bigcup \mathcal{C}$), then* $\operatorname{vol}(J) \le \sum_{I \in \mathcal{C}} \operatorname{vol}(I)$.

[1] Throughout this chapter, $\mathcal{C}$ will denote a collection of rectangles.

D. W. Stroock, *Essentials of Integration Theory for Analysis*, Graduate Texts in Mathematics 262, https://doi.org/10.1007/978-3-030-58478-8_1

Proof. Since $\mathrm{vol}(I \cap J) \le \mathrm{vol}(I)$, assume throughout that $J \supseteq \bigcup_{I \in \mathcal{C}} I$. Also, without loss in generality, we will assume that $\mathring{J} \neq \emptyset$.

The proof is by induction on N. Thus, suppose that $N = 1$. Given a closed interval, use a_I and b_I to denote its left and right endpoints. Choose $a_J \le c_0 < \cdots < c_\ell \le b_J$ so that

$$\{c_k : 0 \le k \le \ell\} = \{a_I : I \in \mathcal{C}\} \cup \{b_I : I \in \mathcal{C}\},$$

and set $\mathcal{C}_k = \{I \in \mathcal{C} : [c_{k-1}, c_k] \subseteq I\}$. Clearly, for each $I \in \mathcal{C}$, $\mathrm{vol}(I) = \sum_{\{k:\, I \in \mathcal{C}_k\}} (c_k - c_{k-1})$.[2]

When $\mathcal{C}$ is non-overlapping, no $\mathcal{C}_k$ contains more than one $I \in \mathcal{C}$, and so

$$\begin{aligned}\sum_{I \in \mathcal{C}} \mathrm{vol}(I) &= \sum_{I \in \mathcal{C}} \sum_{\{k:I \in \mathcal{C}_k\}} (c_k - c_{k-1}) = \sum_{k=1}^{\ell} \mathrm{card}(\mathcal{C}_k)(c_k - c_{k-1}) \\ &\le \sum_{k=1}^{\ell} (c_k - c_{k-1}) \le (b_J - a_J) = \mathrm{vol}(J).\end{aligned}$$

If $J = \bigcup \mathcal{C}$, then $c_0 = a_J$, $c_\ell = b_J$, and, for each $1 \le k \le \ell$, there is an $I \in \mathcal{C}$ for which $I \in \mathcal{C}_k$. To prove this last assertion, simply note that if $x \in (c_{k-1}, c_k)$ and $x \in I \in \mathcal{C}$, then $[c_{k-1}, c_k] \subseteq I$ and therefore $I \in \mathcal{C}_k$. Knowing this, we have

$$\begin{aligned}\sum_{I \in \mathcal{C}} \mathrm{vol}(I) &= \sum_{I \in \mathcal{C}} \sum_{\{k:I \in \mathcal{C}_k\}} (c_k - c_{k-1}) = \sum_{k=1}^{\ell} \mathrm{card}(\mathcal{C}_k)(c_k - c_{k-1}) \\ &\ge \sum_{k=1}^{\ell} (c_k - c_{k-1}) = (b_J - a_J) = \mathrm{vol}(J).\end{aligned}$$

Now assume the result for N. Given a rectangle I in $\mathbb{R}^{N+1}$, determine $a_I, b_I \in \mathbb{R}$ and the rectangle R_I in $\mathbb{R}^N$ so that $I = R_I \times [a_I, b_I]$. Choose $a_J \le c_0 < \cdots < c_\ell \le b_J$ as before, and define $\mathcal{C}_k$ accordingly. Then, for each $I \in \mathcal{C}$,

$$\mathrm{vol}(I) = \mathrm{vol}(R_I)(b_I - a_I) = \mathrm{vol}(R_I) \sum_{\{k:I \in \mathcal{C}_k\}} (c_k - c_{k-1}).$$

If $\mathcal{C}$ is non-overlapping, then $\{R_I : I \in \mathcal{C}_k\}$ is non-overlapping for each k. Hence, since $\bigcup_{I \in \mathcal{C}_k} R_I \subseteq R_J$, the induction hypothesis implies $\sum_{I \in \mathcal{C}_k} \mathrm{vol}(R_I) \le \mathrm{vol}(R_J)$ for each $1 \le k \le \ell$, and therefore

[2] Here, and elsewhere, the sum over the empty set is taken to be 0.

$$\sum_{I\in\mathcal{C}} \mathrm{vol}(I) = \sum_{I\in\mathcal{C}} \mathrm{vol}(R_I) \sum_{\{k:\, I\in\mathcal{C}_k\}} (c_k - c_{k-1})$$
$$\le \sum_{k=1}^{\ell} (c_k - c_{k-1}) \sum_{I\in\mathcal{C}_k} \mathrm{vol}(R_I) \le (b_J - a_J)\mathrm{vol}(R_J) = \mathrm{vol}(J).$$

Finally, assume that $J = \bigcup_{I\in\mathcal{C}}$. In this case, $c_0 = a_J$ and $c_\ell = b_J$. In addition, for each $1 \le k \le \ell$, $R_J = \bigcup_{I\in\mathcal{C}_k} R_I$. To see this, note that if $\mathbf{x} = (x_1, \dots, x_{N+1}) \in J$ and $x_{N+1} \in (c_{k-1}, c_k)$, then $I \ni \mathbf{x} \implies [c_{k-1}, c_k] \subseteq [a_I, b_I]$ and therefore that $I \in \mathcal{C}_k$. Hence, by the induction hypothesis, $\mathrm{vol}(R_J) \le \sum_{I\in\mathcal{C}_k} \mathrm{vol}(R_I)$ for each $1 \le k \le \ell$, and therefore

$$\sum_{I\in\mathcal{C}} \mathrm{vol}(I) = \sum_{I\in\mathcal{C}} \mathrm{vol}(R_I) \sum_{\{k:I\in\mathcal{C}_k\}} (c_k - c_{k-1})$$
$$= \sum_{k=1}^{\ell} (c_k - c_{k-1}) \sum_{I\in\mathcal{C}_k} \mathrm{vol}(R_I) \ge (b_J - a_J)\mathrm{vol}(R_J) = \mathrm{vol}(J).$$

□

Given a collection $\mathcal{C}$ of rectangles I, say that $\xi : \mathcal{C} \longrightarrow \bigcup\mathcal{C}$ is a **choice map** for $\mathcal{C}$ if $\xi(I) \in I$ for each $I \in \mathcal{C}$, and use $\Xi(\mathcal{C})$ to denote the set of all such maps. Given a finite collection $\mathcal{C}$, a choice map $\xi \in \Xi(\mathcal{C})$, and a function $f : \bigcup\mathcal{C} \longrightarrow \mathbb{R}$, define the **Riemann sum of f over $\mathcal{C}$ relative to ξ** to be

$$\mathcal{R}(f;\mathcal{C},\xi) \equiv \sum_{I\in\mathcal{C}} f(\xi(I))\mathrm{vol}\,(I). \tag{1.1.1}$$

Finally, if J is a rectangle and $f : J \longrightarrow \mathbb{R}$ is a function, f is said to be **Riemann integrable** on J if there is a number $A \in \mathbb{R}$ with the property that, for all $\epsilon > 0$, there is a $\delta > 0$ such that

$$|\mathcal{R}(f;\mathcal{C},\xi) - A| < \epsilon$$

whenever $\xi \in \Xi(\mathcal{C})$ and $\mathcal{C}$ is a non-overlapping, finite, **exact cover** of J (i.e., $J = \bigcup\mathcal{C}$) whose **mesh size** $\|\mathcal{C}\|$, given by $\|\mathcal{C}\| \equiv \max\{\mathrm{diam}(I) : I \in \mathcal{C}\}$, is less than δ. When f is Riemann integrable on J, the associated number A is called the **Riemann integral of f on** J, and I will use

$$(\mathrm{R})\int_J f(x)\,dx$$

to denote A.

It is a relatively simple matter to see that any $f \in C(J;\mathbb{R})$ (the space of continuous real-valued functions on J) is Riemann integrable on J. However, in order to determine when more general bounded functions are Riemann integrable, it is useful to introduce the **upper Riemann sum**

$$\mathcal{U}(f;\mathcal{C}) \equiv \sum_{I\in\mathcal{C}} \sup_{x\in I} f(x)\mathrm{vol}\,(I)$$

and the **lower Riemann sum**

$$\mathcal{L}(f;\mathcal{C}) \equiv \sum_{I\in\mathcal{C}} \inf_{x\in I} f(x)\mathrm{vol}\,(I).$$

Clearly, one always has

$$\mathcal{L}(f;\mathcal{C}) \le \mathcal{R}(f;\mathcal{C},\xi) \le \mathcal{U}(f;\mathcal{C})$$

for any $\mathcal{C}$ and $\xi \in \Xi(\mathcal{C})$. Also, by Cauchy's convergence criterion, it is clear that a bounded f is Riemann integrable if and only if

$$\varliminf_{\|\mathcal{C}\|\to 0} \mathcal{L}(f;\mathcal{C}) \ge \varlimsup_{\|\mathcal{C}\|\to 0} \mathcal{U}(f;\mathcal{C}), \tag{1.1.2}$$

where the limits are taken over non-overlapping, finite, exact covers of J. My goal now is to show that the preceding can be replaced by the condition[3]

$$\sup_{\mathcal{C}} \mathcal{L}(f;\mathcal{C}) \ge \inf_{\mathcal{C}} \mathcal{U}(f;\mathcal{C}) \tag{1.1.3}$$

where the $\mathcal{C}$'s run over all non-overlapping, finite, exact covers of J.

To this end, partially order the covers $\mathcal{C}$ by *refinement*. That is, say that $\mathcal{C}_2$ is **more refined** than $\mathcal{C}_1$ and write $\mathcal{C}_1 \le \mathcal{C}_2$, if, for every $I_2 \in \mathcal{C}_2$, there is an $I_1 \in \mathcal{C}_1$ such that $I_2 \subseteq I_1$. Note that, for every pair $\mathcal{C}_1$ and $\mathcal{C}_2$, the *least* common refinement $\mathcal{C}_1 \vee \mathcal{C}_2$ is given by $\mathcal{C}_1 \vee \mathcal{C}_2 = \{I_1 \cap I_2 : I_1 \in \mathcal{C}_1 \text{ and } I_2 \in \mathcal{C}_2\}$.

Lemma 1.1.2 *For any pair of non-overlapping, finite, exact covers $\mathcal{C}_1$ and $\mathcal{C}_2$ of J and any bounded function $f : J \longrightarrow \mathbb{R}$, $\mathcal{L}(f;\mathcal{C}_1) \le \mathcal{U}(f;\mathcal{C}_2)$. Moreover, if $\mathcal{C}_1 \le \mathcal{C}_2$, then $\mathcal{L}(f;\mathcal{C}_1) \le \mathcal{L}(f;\mathcal{C}_2)$ and $\mathcal{U}(f;\mathcal{C}_1) \ge \mathcal{U}(f;\mathcal{C}_2)$. Finally, if f is bounded, then* (1.1.3) *holds if and only if for every $\epsilon > 0$ there exists a $\mathcal{C}$ such that*[4]

$$\mathcal{U}(f;\mathcal{C}) - \mathcal{L}(f;\mathcal{C}) = \sum_{I\in\mathcal{C}} \left(\sup_I f - \inf_I f\right) \mathrm{vol}(I) < \epsilon. \tag{1.1.4}$$

Proof. We will begin by proving the second statement. Noting that

$$\mathcal{L}(f;\mathcal{C}) = -\mathcal{U}(-f;\mathcal{C}), \tag{1.1.5}$$

[3] In many texts, this condition is adopted as the definition of Riemann integrability. Obviously, since, as is about to be shown, it is equivalent to the definition that was given earlier, there is no harm in doing so. However, when working in the more general setting studied in § 1.2, the distinction between these two definitions does make a difference.

[4] Here, and elsewhere, $\sup_I f = \sup_{x\in I} f(x)$ and $\inf_I f = \inf_{x\in I} f(x)$.

one sees that it suffices to check that $\mathcal{U}(f;\mathcal{C}_1) \geq \mathcal{U}(f;\mathcal{C}_2)$ if $\mathcal{C}_1 \leq \mathcal{C}_2$. But, for each $I_1 \in \mathcal{C}_1$,

$$\sup_{x\in I_1} f(x)\text{vol}\,(I_1) \geq \sum_{\{I_2\in\mathcal{C}_2: I_2\subseteq I_1\}} \sup_{x\in I_2} f(x)\text{vol}\,(I_2),$$

where Lemma 1.1.1 was used to see that

$$\text{vol}\,(I_1) = \sum_{\{I_2\in\mathcal{C}_2:\, I_2\subseteq I_1\}} \text{vol}\,(I_2).$$

After summing the above over $I_1 \in \mathcal{C}_1$, one arrives at the required result.

Given the preceding, the first assertion is immediate. Namely, for any $\mathcal{C}_1$ and $\mathcal{C}_2$,

$$\mathcal{L}(f;\mathcal{C}_1) \leq \mathcal{L}(f;\mathcal{C}_1 \vee \mathcal{C}_2) \leq \mathcal{U}(f;\mathcal{C}_1 \vee \mathcal{C}_2) \leq \mathcal{U}(f;\mathcal{C}_2).$$

Finally, if for each $\epsilon > 0$ (1.1.4) holds for some $\mathcal{C}_\epsilon$, then, for each $\epsilon > 0$,

$$\inf_{\mathcal{C}} \mathcal{U}(f;\mathcal{C}) - \sup_{\mathcal{C}} \mathcal{L}(f;\mathcal{C}) \leq \mathcal{U}(f;\mathcal{C}_\epsilon) - \mathcal{L}(f;\mathcal{C}_\epsilon) < \epsilon,$$

and so (1.1.3) holds. Conversely, if (1.1.3) holds and $\epsilon > 0$, choose $\mathcal{C}_1$ and $\mathcal{C}_2$ for which $\sup_{\mathcal{C}} \mathcal{L}(f;\mathcal{C}) \leq \mathcal{L}(f;\mathcal{C}_1) + \frac{\epsilon}{2}$ and $\inf_{\mathcal{C}} \mathcal{U}(f;\mathcal{C}) \geq \mathcal{U}(f;\mathcal{C}_2) - \frac{\epsilon}{2}$. Then (1.1.4) holds with $\mathcal{C}_\epsilon = \mathcal{C}_1 \vee \mathcal{C}_2$. □

Lemma 1.1.2 really depends only on properties of our order relation for covers and not on the properties of vol(I). On the other hand, the next lemma depends on the continuity of volume with respect to the side-lengths of rectangles.

Lemma 1.1.3 *Assume that $\mathring{J} \neq \emptyset$, and let $\mathcal{C}$ be a non-overlapping, finite, exact cover of the rectangle J. If $f : J \longrightarrow \mathbb{R}$ is a bounded function, then, for each $\epsilon > 0$, there is a $\delta > 0$ such that*

$$\mathcal{U}(f;\mathcal{C}') \leq \mathcal{U}(f;\mathcal{C}) + \epsilon \quad \textit{and} \quad \mathcal{L}(f;\mathcal{C}') \geq \mathcal{L}(f;\mathcal{C}) - \epsilon$$

whenever $\mathcal{C}'$ is a non-overlapping, finite, exact cover of J with the property that $\|\mathcal{C}'\| < \delta$.

Proof. In view of (1.1.5), we need consider only the upper Riemann sums.

Let $J = \prod_1^N [c_k, d_k]$. Given a $\delta > 0$, a rectangle $I = \prod_1^N [a_k, b_k]$ and $1 \leq k \leq N$, define $I_k^-(\delta)$ and $I_k^+(\delta)$ to be the rectangles

$$J \cap \left[\left(\prod_{1\leq j<k} [c_j, d_j]\right) \times [a_k - \delta, a_k + \delta] \times \left(\prod_{k<j\leq N} [c_j.d_j]\right)\right]$$

and

$$J \cap \left[\left(\prod_{1 \le j < k} [c_j, d_j] \right) \times [b_k - \delta, b_k + \delta] \times \left(\prod_{k < j \le N} [c_j . d_j] \right) \right]$$

respectively. Then, for any rectangle $I' \subseteq J$ with $\operatorname{diam}(I') < \delta$, either $I' \subseteq I$ for some $I \in \mathcal{C}$ or, for some $I \in \mathcal{C}$ and $1 \le k \le N$, $I' \subseteq I_k^+(\delta)$ or $I' \subseteq I_k^-(\delta)$.

Now let $\mathcal{C}'$ with $\|\mathcal{C}'\| < \delta$ be given. Then, by an application of Lemma 1.1.1, we can write

$$\begin{aligned} \mathcal{U}(f; \mathcal{C}') &= \sum_{I' \in \mathcal{C}'} \sup_{I'} f \operatorname{vol}(I') = \sum_{I \in \mathcal{C}} \sum_{I' \in \mathcal{C}'} \sup_{I'} f \operatorname{vol}(I \cap I') \\ &= \sum_{I \in \mathcal{C}} \sum_{I' \in \mathcal{C}'} \sup_{I \cap I'} f \operatorname{vol}(I \cap I') + \sum_{I \in \mathcal{C}} \sum_{I' \in \mathcal{C}'} \left(\sup_{I'} f - \sup_{I \cap I'} f \right) \operatorname{vol}(I \cap I'). \end{aligned}$$

But clearly

$$\sum_{I \in \mathcal{C}} \sum_{I' \in \mathcal{C}'} \sup_{I \cap I'} f \operatorname{vol}(I \cap I') \le \sum_{I \in \mathcal{C}} \sum_{I' \in \mathcal{C}'} \sup_{I} f \operatorname{vol}(I \cap I') = \mathcal{U}(f; \mathcal{C}),$$

where the final step is another application of Lemma 1.1.1. Thus, it remains to estimate

$$\sum_{I \in \mathcal{C}} \sum_{I' \in \mathcal{C}'} \left(\sup_{I'} f - \sup_{I \cap I'} f \right) \operatorname{vol}(I \cap I').$$

However, by the discussion in the preceding paragraph, for each $I' \in \mathcal{C}'$, either $I' \subseteq I$ for some $I \in \mathcal{C}$, in which case

$$\sum_{I \in \mathcal{C}} \left(\sup_{I'} f - \sup_{I \cap I'} f \right) \operatorname{vol}(I \cap I') = 0,$$

or, for some $I \in \mathcal{C}$ and $1 \le k \le N$, $I' \subseteq I_k^+(\delta)$ or $I' \subseteq I_k^-(\delta)$. Thus, if

$$\mathcal{B}(k, I)^{\pm} = \left\{ I' \in \mathcal{C}' : I' \subseteq I_k^{\pm}(\delta) \right\},$$

then

$$\begin{aligned} &\sum_{I \in \mathcal{C}} \sum_{I' \in \mathcal{C}'} \left(\sup_{I'} f - \sup_{I \cap I'} f \right) \operatorname{vol}(I \cap I') \\ &\qquad \le \|f\|_{\mathrm{u}} \sum_{k=1}^{N} \sum_{I \in \mathcal{C}} \left(\sum_{I' \in \mathcal{B}(k,I)^+} \operatorname{vol}(I \cap I') + \sum_{I' \in \mathcal{B}(k,I)^-} \operatorname{vol}(I \cap I') \right). \end{aligned}$$

(In the preceding, I have introduced the notation, to be used throughout, that $\|f\|_{\mathrm{u}}$ denotes the **uniform norm of** f: the supremum of $|f|$ over the set on which f is defined.) Finally, by Lemma 1.1.1, for each $1 \le k \le N$ and $I \in \mathcal{C}$,

$$\sum_{I'\in\mathcal{B}(k,I)^{\pm}} \mathrm{vol}(I\cap I') \le \mathrm{vol}\big(I_k^{\pm}(\delta)\big) \le 2\delta\frac{\mathrm{vol}(J)}{d_k-c_k},$$

and so we have now proved that, whenever $\|\mathcal{C}'\|\le\delta$,

$$\mathcal{U}(f;\mathcal{C}')-\mathcal{U}(f;\mathcal{C}) \le \sum_{I\in\mathcal{C}}\sum_{I'\in\mathcal{C}'}\left(\sup_{I'} f - \sup_{I\cap I'} f\right)\mathrm{vol}(I\cap I') \le K\|f\|_{\mathrm{u}}\delta,$$

where

$$K \equiv 4N\mathrm{card}(\mathcal{C}) \max_{1\le k\le N}\frac{\mathrm{vol}(J)}{d_k-c_k}.$$

□

As an essentially immediate consequence of Lemma 1.1.3, we have the following theorem.

Theorem 1.1.4 *Let $f: J\longrightarrow\mathbb{R}$ be a bounded function on the rectangle J. Then*

$$\lim_{\|\mathcal{C}\|\to 0}\mathcal{L}(f;\mathcal{C}) = \sup_{\mathcal{C}}\mathcal{L}(f;\mathcal{C}) \quad \text{and} \quad \lim_{\|\mathcal{C}\|\to 0}\mathcal{U}(f;\mathcal{C}) = \inf_{\mathcal{C}}\mathcal{U}(f;\mathcal{C}),$$

where $\mathcal{C}$ runs over non-overlapping, finite, exact covers of J. In particular, (1.1.3) *is a necessary and sufficient condition that a bounded f on J be Riemann integrable, and so every $f\in C(J;\mathbb{R})$ is Riemann integrable.*

Proof. First note that, by Lemma 1.1.3, for every $\mathcal{C}$ and $\epsilon>0$ there exists a $\delta>0$ for which

$$\|\mathcal{C}'\|<\delta \implies \mathcal{U}(f;\mathcal{C}')\le\inf_{\mathcal{C}}\mathcal{U}(f;\mathcal{C})+\epsilon \text{ and } \mathcal{L}(f;\mathcal{C}')\ge\sup_{\mathcal{C}}\mathcal{L}(f;\mathcal{C})-\epsilon.$$

Hence,

$$\varlimsup_{\|\mathcal{C}\|\searrow 0}\mathcal{U}(f;\mathcal{C})\le\inf_{\mathcal{C}}\mathcal{U}(f;\mathcal{C}) \text{ and } \varliminf_{\|\mathcal{C}\|\searrow 0}\mathcal{L}(f;\mathcal{C})\ge\sup_{\mathcal{C}}\mathcal{L}(f;\mathcal{C}).$$

Since

$$\varliminf_{\|\mathcal{C}\|\searrow 0}\mathcal{U}(f;\mathcal{C})\ge\inf_{\mathcal{C}}\mathcal{U}(f;\mathcal{C}) \text{ and } \varlimsup_{\|\mathcal{C}\|\searrow 0}\mathcal{L}(f;\mathcal{C})\le\sup_{\mathcal{C}}\mathcal{L}(f;\mathcal{C})$$

trivially, the first assertion follows. Given this, it is obvious that (1.1.2) is equivalent to (1.1.3) and therefore that f is Riemann integrable if and only if (1.1.3) holds. Finally, if $f\in C(J;\mathbb{R})$, then for each $\epsilon>0$ there is a $\delta>0$ such that[5] $\max_{I\in\mathcal{C}}(\sup_I f-\inf_I f)<\epsilon$ and therefore $\mathcal{U}(f;\mathcal{C})-\mathcal{L}(f;\mathcal{C})<\epsilon\mathrm{vol}(J)$ for any $\mathcal{C}$ with $\|\mathcal{C}\|<\delta$. □

[5] Recall that a continuous function on a compact set is uniformly continuous there.

1.1.1 Exercises for § 1.1

Exercise 1.1.1 Suppose that f and g are bounded, Riemann integrable functions on J. Show that $f \vee g \equiv \max\{f, g\}$, $f \wedge g \equiv \min\{f, g\}$, and, for any α, $\beta \in \mathbb{R}$, $\alpha f + \beta g$ are all Riemann integrable on J. In addition, check that

$$(\mathrm{R})\int_J (f \vee g)(x)\,dx \geq \left((\mathrm{R})\int_J f(x)\,dx\right) \vee \left((\mathrm{R})\int_J g(x)\,dx\right),$$

$$(\mathrm{R})\int_J (f \wedge g)(x)\,dx \leq \left((\mathrm{R})\int_J f(x)\,dx\right) \wedge \left((\mathrm{R})\int_J g(x)\,dx\right),$$

and

$$(\mathrm{R})\int_J (\alpha f + \beta g)(x)\,dx = \alpha\left((\mathrm{R})\int_J f(x)\,dx\right) + \beta\left((\mathrm{R})\int_J g(x)\,dx\right).$$

Conclude, in particular, that if f and g are Riemann integrable on J and $f \leq g$, then $(\mathrm{R})\int_J f(x)\,dx \leq (\mathrm{R})\int_J g(x)\,dx$.

Exercise 1.1.2 Show that if f is a bounded real-valued function on the rectangle J, then f is Riemann integrable if and only if, for each $\epsilon > 0$, there is a $\delta > 0$ such that

$$\|\mathcal{C}\| < \delta \implies \sum_{\{I\in\mathcal{C}:\sup_I f - \inf_I f > \epsilon\}} \mathrm{vol}\,(I) < \epsilon.$$

In fact, show that f will be Riemann integrable if, for each $\epsilon > 0$, there exists some $\mathcal{C}$ for which $\sum_{\{I\in\mathcal{C}:\sup_I f - \inf_I f > \epsilon\}} \mathrm{vol}(I) < \epsilon$.

Exercise 1.1.3 Show that a bounded f on J is Riemann integrable if it is continuous on J at all but a finite number of points. See Theorem 5.1.2 for more information.

1.2 Riemann–Stieltjes Integration

In Section 1.1 I developed the classical integration theory with respect to the standard notion of Euclidean volume. In the present section, I will extend the classical theory, at least for integrals in one dimension, to cover more general notions of volume.

Let $J = [a, b]$ be a compact interval in $\mathbb{R}$ and φ and ψ a pair of bounded, real-valued functions on J. Given a non-overlapping, finite, exact cover $\mathcal{C}$ of J by closed intervals I and a choice map $\xi \in \Xi(\mathcal{C})$, define the **Riemann sum of φ over $\mathcal{C}$ with respect to ψ relative to ξ** to be

$$\mathcal{R}(\varphi|\psi;\mathcal{C},\xi) = \sum_{I\in\mathcal{C}} \varphi(\xi(I))\Delta_I\psi,$$

where $\Delta_I\psi \equiv \psi(b_I) - \psi(a_I)$ and b_I and a_I denote, respectively, the right- and left-hand endpoints of the interval I. Obviously, when $\psi(x) = x, x \in J$, $\mathcal{R}(\varphi|\psi;\mathcal{C},\xi) = \mathcal{R}(\varphi;\mathcal{C},\xi)$. Thus, it is consistent to say that φ is **Riemann integrable on J with respect to ψ**, or, more simply, **ψ-Riemann integrable on J**, if there is a number A with the property that, for each $\epsilon > 0$, there is a $\delta > 0$ such that

$$\sup_{\xi\in\Xi(\mathcal{C})} |\mathcal{R}(\varphi|\psi;\mathcal{C},\xi) - A| < \epsilon \tag{1.2.1}$$

whenever $\mathcal{C}$ is a non-overlapping, finite, exact cover of J satisfying $\|\mathcal{C}\| < \delta$. Assuming that φ is ψ-Riemann integrable on J, the number A in (1.2.1) is called the **Riemann–Stieltjes integral**, and I will use

$$(\mathrm{R})\int_J \varphi(x)\,d\psi(x)$$

to denote A.

Examples 1.2.1 The following examples may help to explain what is going on here. Throughout, $J = [a,b]$ is a compact interval.

(i) If $\varphi \in C(J;\mathbb{R})$ and $\psi \in C^1(J;\mathbb{R})$ (i.e., ψ is continuously differentiable on J), then one can use the Mean Value Theorem to check that φ is ψ-Riemann integrable on J and that

$$(\mathrm{R})\int_J \varphi(x)\,d\psi(x) = (\mathrm{R})\int_J \varphi(x)\psi'(x)\,dx.$$

(ii) If there exist $a = a_0 < a_1 < \cdots < a_n = b$ such that ψ is constant on each of the intervals (a_{m-1}, a_m), then every $\varphi \in C\big([a,b];\mathbb{R}\big)$ is ψ-Riemann integrable on $[a,b]$, and

$$(\mathrm{R})\int_{[a,b]} \varphi(x)\,d\psi(x) = \sum_{m=0}^{n} \varphi(a_m)d_m,$$

where $d_0 = \psi(a+) - \psi(a)$, $d_m = \psi(a_m+) - \psi(a_m-)$ for $1 \le m \le n-1$, and $d_n = \psi(b) - \psi(b-)$. (I use $f(x+)$ and $f(x-)$ to denote the right and left limits of a function f at x.)

(iii) If both $(\mathrm{R})\int_J \varphi_1(x)\,d\psi(x)$ and $(\mathrm{R})\int_J \varphi_2(x)\,d\psi(x)$ exist (i.e., φ_1 and φ_2 are both ψ-Riemann integrable on J), then, for all real numbers α and β, $(\alpha\varphi_1 + \beta\varphi_2)$ is ψ-Riemann integrable on J and

$$
\begin{aligned}
&(\mathrm{R})\int_J (\alpha\varphi_1+\beta\varphi_2)(x)\,d\psi(x)\\
&\quad=\alpha\left((\mathrm{R})\int_J \varphi_1(x)\,d\psi(x)\right)+\beta\left((\mathrm{R})\int_J \varphi_2(x)\,d\psi(x)\right).
\end{aligned}
$$

(iv) If $J = J_1 \cup J_2$ where $\mathring{J}_1 \cap \mathring{J}_2 = \emptyset$ and if φ is ψ-Riemann integrable on J, then φ is ψ-Riemann integrable on both J_1 and J_2, and

$$
(\mathrm{R})\int_J \varphi(x)\,d\psi(x) = (\mathrm{R})\int_{J_1} \varphi(x)\,d\psi(x) + (\mathrm{R})\int_{J_2} \varphi(x)\,d\psi(x).
$$

All the assertions made in Example 1.2.1 are reasonably straightforward consequences of the definition of Riemann–Stieltjes integrability, and the reader will be asked to verify them in Exercise 1.2.1 below.

1.2.1 Riemann–Stieltjes Integrability

Perhaps the most important reason for introducing the Riemann–Stieltjes integral is the following theorem, which shows that the notion of Riemann–Stieltjes integrability possesses a remarkable *symmetry.*

Theorem 1.2.1 (Integration by Parts) *If φ is ψ-Riemann integrable on $J = [a, b]$, then ψ is φ-Riemann integrable on J and*

$$
(\mathrm{R})\int_J \psi(x)\,d\varphi(x) = \psi(b)\varphi(b) - \psi(a)\varphi(a) - (\mathrm{R})\int_J \varphi(x)\,d\psi(x). \tag{1.2.2}
$$

Proof. Let $\mathcal{C} = \{[\alpha_{m-1}, \alpha_m] : 1 \le m \le n\}$, where $a = \alpha_0 \le \cdots \le \alpha_n = b$; and let $\xi \in \Xi(\mathcal{C})$ with $\xi([\alpha_{m-1}, \alpha_m]) = \beta_m \in [\alpha_{m-1}, \alpha_m]$. Set $\beta_0 = a$ and $\beta_{n+1} = b$. Then

$$
\begin{aligned}
\mathcal{R}(\psi|\varphi;\mathcal{C},\xi) &= \sum_{m=1}^{n} \psi(\beta_m)\big(\varphi(\alpha_m)-\varphi(\alpha_{m-1})\big)\\
&= \sum_{m=1}^{n} \psi(\beta_m)\varphi(\alpha_m) - \sum_{m=0}^{n-1} \psi(\beta_{m+1})\varphi(\alpha_m)\\
&= \psi(\beta_n)\varphi(\alpha_n) - \sum_{m=1}^{n-1} \varphi(\alpha_m)\big(\psi(\beta_{m+1})-\psi(\beta_m)\big) - \psi(\beta_1)\varphi(\alpha_0)\\
&= \psi(b)\varphi(b) - \psi(a)\varphi(a) - \sum_{m=0}^{n} \varphi(\alpha_m)\big(\psi(\beta_{m+1})-\psi(\beta_m)\big)\\
&= \psi(b)\varphi(b) - \psi(a)\varphi(a) - \mathcal{R}(\varphi|\psi;\mathcal{C}',\xi'),
\end{aligned}
$$

where $\mathcal{C}' = \{[\beta_{m-1}, \beta_m] : 1 \le m \le n+1\}$ and $\xi' \in \Xi(\mathcal{C}')$ is defined by $\xi'([\beta_m, \beta_{m+1}]) = \alpha_m$ for $0 \le m \le n$. Noting that $\|\mathcal{C}'\| \le 2\|\mathcal{C}\|$, one sees that if φ is ψ-Riemann integrable then ψ is φ-Riemann integrable and that (1.2.2) holds. □

It is hardly necessary to point out, but notice that when $\psi \equiv 1$ and φ is continuously differentiable, then, by **(i)** in Examples 1.2.1, (1.2.2) becomes *the Fundamental Theorem of Calculus.*

Although the preceding theorem indicates that it is natural to consider φ and ψ as playing symmetric roles in the theory of Riemann–Stieltjes integration, it turns out that, in practice, one wants to impose a condition on ψ that will guarantee that every $\varphi \in C(J;\mathbb{R})$ is Riemann integrable with respect to ψ and that, in addition (recall that $\|\varphi\|_{\mathrm{u}}$ is the uniform norm of φ),

$$\left|(\mathrm{R}) \int_J \varphi(x)\, d\psi(x)\right| \le K_\psi \|\varphi\|_{\mathrm{u}} \tag{1.2.3}$$

for some $K_\psi < \infty$ and all $\varphi \in C(J;\mathbb{R})$. Example **(i)** in Examples 1.2.1 tells us that one condition on ψ that guarantees the ψ-Riemann integrability of every continuous φ is that $\psi \in C^1(J;\mathbb{R})$. Moreover, from the expression given there, it is an easy matter to check that in this case (1.2.3) holds with $K_\psi = \|\psi'\|_{\mathrm{u}}(b-a)$. On the other hand, example **(ii)** makes it clear that ψ need not be even continuous, much less differentiable, in order that Riemann integration with respect to ψ have the above properties. The following result emphasizes this same point.

Theorem 1.2.2 *Let ψ be non-decreasing on J. Then every $\varphi \in C(J;\mathbb{R})$ is ψ-Riemann integrable on J. In addition, if φ is non-negative and ψ-Riemann integrable on J, then* $(\mathrm{R})\int_J \varphi(x)\, d\psi(x) \ge 0$. *In particular,* (1.2.3) *holds with $K_\psi = \Delta_J\psi$.*

Proof. The fact that $(\mathrm{R})\int_J \varphi(x)\, d\psi(x) \ge 0$ if φ is a non-negative function that is ψ-Riemann integrable on J follows immediately from the fact that $\mathcal{R}(\varphi|\psi;\mathcal{C},\xi) \ge 0$ for any $\mathcal{C}$ and $\xi \in \Xi(\mathcal{C})$. After applying this to the functions $\|\varphi\|_{\mathrm{u}} \pm \varphi$ and using the linearity property in **(iii)** of Examples 1.2.1, one concludes that (1.2.3) holds with $K_\psi = \Delta_J\psi$. Thus, all that we have to do is check that every $\varphi \in C(J;\mathbb{R})$ is ψ-Riemann integrable on J.

Let $\varphi \in C(J;\mathbb{R})$ be given and define

$$\mathcal{U}(\varphi|\psi;\mathcal{C}) = \sum_{I\in\mathcal{C}} \big(\sup_I \varphi\big)\Delta_I\psi \quad \text{and} \quad \mathcal{L}(\varphi|\psi;\mathcal{C}) = \sum_{I\in\mathcal{C}} \big(\inf_I \varphi\big)\Delta_I\psi.$$

Then, just as in § 1.1,

$$\mathcal{L}(\varphi|\psi;\mathcal{C}) \le \mathcal{R}(\varphi|\psi;\mathcal{C},\xi) \le \mathcal{U}(\varphi|\psi;\mathcal{C})$$

for any $\xi \in \Xi(\mathcal{C})$. In addition (cf. Lemma 1.1.2), for any pair $\mathcal{C}_1$ and $\mathcal{C}_2$, one has that $\mathcal{L}(\varphi|\psi;\mathcal{C}_1) \le \mathcal{U}(\varphi|\psi;\mathcal{C}_2)$. Finally, for any $\mathcal{C}$,

$$\mathcal{U}(\varphi|\psi;\mathcal{C}) - \mathcal{L}(\varphi|\psi;\mathcal{C}) \le \omega(\|\mathcal{C}\|)\Delta_J\psi,$$

where

$$\omega(\delta) \equiv \sup\left\{|\varphi(y) - \varphi(x)| : x, y \in J \text{ and } |y - x| \le \delta\right\}$$

is the **modulus of continuity** of φ. Hence, since, by uniform continuity, $\lim_{\delta\searrow 0}\omega(\delta) = 0$,

$$\lim_{\|\mathcal{C}\|\to 0}\big(\mathcal{U}(\varphi|\psi;\mathcal{C}) - \mathcal{L}(\varphi|\psi;\mathcal{C})\big) = 0.$$

But this means that, for every $\epsilon > 0$, there is a $\delta > 0$ for which

$$\mathcal{U}(\varphi|\psi;\mathcal{C}) - \mathcal{U}(\varphi|\psi;\mathcal{C}') \le \mathcal{U}(\varphi|\psi;\mathcal{C}) - \mathcal{L}(\varphi|\psi;\mathcal{C}) < \epsilon$$

no matter what $\mathcal{C}'$ is chosen as long as $\|\mathcal{C}\| < \delta$. From the above it is clear that

$$\inf_{\mathcal{C}}\mathcal{U}(\psi;\mathcal{C}) = \lim_{\|\mathcal{C}\|\to 0}\mathcal{U}(\varphi|\psi;\mathcal{C}) = \lim_{\|\mathcal{C}\|\to 0}\mathcal{L}(\varphi|\psi;\mathcal{C}) = \sup_{\mathcal{C}}\mathcal{L}(\psi;\mathcal{C})$$

and therefore that φ is ψ-Riemann integrable on J and (R) $\int_J \varphi(x)\,d\psi(x) = \inf_{\mathcal{C}}\mathcal{U}(\varphi;\mathcal{C})$. □

One obvious way to extend the preceding result is to note that if φ is Riemann integrable on J with respect to both ψ_1 and ψ_2, then it is Riemann integrable on J with respect to $\psi \equiv \psi_2 - \psi_1$ and

$$(\mathrm{R})\int_J \varphi(x)\,d\psi(x) = (\mathrm{R})\int_J \varphi(x)\,d\psi_2(x) - (\mathrm{R})\int_J \varphi(x)\,d\psi_1(x).$$

(This can be seen directly or as a consequence of Theorem 1.2.1 combined with **(iii)** in Examples 1.2.1.) In particular, we have the following corollary to Theorem 1.2.2.

Corollary 1.2.3 *If* $\psi = \psi_2 - \psi_1$, *where* ψ_1 *and* ψ_2 *are non-decreasing functions on* J, *then every* $\varphi \in C(J;\mathbb{R})$ *is Riemann integrable with respect to* ψ *and* (1.2.3) *holds with* $K_\psi = \Delta_J\psi_1 + \Delta_J\psi_2$.

1.2.2 Functions of Bounded Variation

Let $J = [a, b]$ be a compact interval. In this subsection I will carry out a program that will show that, at least among ψ's that are right-continuous on $J\setminus\{b\}$ and have left limits at each point in $J\setminus\{a\}$, the ψ's in Corollary (1.2.3) are the only ones with the properties that every $\varphi \in C(J;\mathbb{R})$ is ψ-Riemann integrable on J and (1.2.3) holds for some $K_\psi < \infty$.

The first step is to provide an alternative description of those ψ 's that can be expressed as the difference between two non-decreasing functions. To this end, let ψ be a real-valued function on J, and define

$$\mathcal{S}(\psi;\mathcal{C}) = \sum_{I\in\mathcal{C}} |\Delta_I\psi|$$

for any non-overlapping, finite, exact cover $\mathcal{C}$ of J. Clearly

$$\mathcal{S}(\alpha\psi;\mathcal{C}) = |\alpha|\mathcal{S}(\psi;\mathcal{C}) \quad \text{for all } \alpha\in\mathbb{R},$$

$$\mathcal{S}(\psi_1+\psi_2;\mathcal{C}) \le \mathcal{S}(\psi_1;\mathcal{C}) + \mathcal{S}(\psi_2;\mathcal{C}) \quad \text{for all } \psi_1 \text{ and } \psi_2,$$

and

$$\mathcal{S}(\psi;\mathcal{C}) = |\Delta_J\psi|$$

if ψ is monotone on J. Moreover, if $\mathcal{C}$ is given and $\mathcal{C}'$ is obtained from $\mathcal{C}$ by replacing one of the I 's in $\mathcal{C}$ by a pair $\{I_1, I_2\}$, where $I = I_1\cup I_2$ and $\mathring{I}_1\cap\mathring{I}_2 = \emptyset$, then, by the triangle inequality,

$$\begin{aligned}&\mathcal{S}(\psi;\mathcal{C}') - \mathcal{S}(\psi;\mathcal{C})\\ &\quad = |\psi(b_{I_1}) - \psi(a_{I_1})| + |\psi(b_{I_2}) - \psi(a_{I_2})| - |\psi(b_I) - \psi(a_I)| \ge 0.\end{aligned}$$

Hence, $\mathcal{C}\le\mathcal{C}' \implies \mathcal{S}(\psi;\mathcal{C}) \le \mathcal{S}(\psi;\mathcal{C}')$.

Define the **variation** of ψ on J to be the number (possibly infinite)

$$\mathrm{Var}(\psi;J) \equiv \sup_{\mathcal{C}} \mathcal{S}(\psi;\mathcal{C}),$$

where the $\mathcal{C}$'s run over all non-overlapping, finite, exact covers of J. Also, say that ψ has **bounded variation on** J if $\mathrm{Var}(\psi;J)<\infty$. It should be clear that if $\psi=\psi_2-\psi_1$ for non-decreasing ψ_1 and ψ_2 on J, then ψ has bounded variation on J and $\mathrm{Var}(\psi;J)\le\Delta_J\psi_1+\Delta_J\psi_2$. What is less obvious is that every ψ having bounded variation on J can be expressed as the difference of two non-decreasing functions. In order to prove this, introduce the quantities

$$\mathrm{Var}_\pm(\psi;J) \equiv \frac{\mathrm{Var}(\psi;J)\pm\Delta_J\psi}{2} = \sup_{\mathcal{C}}\sum_{I\in\mathcal{C}}(\Delta_I\psi)^\pm$$

where $\alpha^+\equiv\alpha\vee 0$ and $\alpha^-\equiv-(\alpha\wedge 0)$ for $\alpha\in\mathbb{R}$, and call $\mathrm{Var}_+(\psi;J)$ and $\mathrm{Var}_-(\psi;J)$ the **positive variation** and the **negative variation** of ψ on J. After noting that

$$\begin{aligned}2\mathrm{Var}_\pm(\psi;J) &= \mathrm{Var}(\psi;\mathcal{C})\pm\Delta_J\psi\\ V_+(\psi;J) - \mathrm{Var}_-(\psi;J) &= \Delta_J\psi\\ \mathrm{Var}_+(\psi;J) + \mathrm{Var}_-(\psi;J) &= \mathrm{Var}(\psi;J),\end{aligned} \tag{1.2.4}$$

one sees that

$$\mathrm{Var}_+(\psi; J) < \infty \iff \mathrm{Var}(\psi; J) < \infty \iff \mathrm{Var}_-(\psi; J) < \infty.$$

Lemma 1.2.4 *If ψ has bounded variation on $[a, b]$ and $a < c < b$, then*

$$\mathrm{Var}_\pm(\psi; [a, b]) = \mathrm{Var}_\pm(\psi; [a, c]) + \mathrm{Var}_\pm(\psi; [c, b]),$$

and therefore also $\mathrm{Var}(\psi; [a, b]) = \mathrm{Var}(\psi; [a, c]) + \mathrm{Var}(\psi; [c, b])$.

Proof. Because of (1.2.4), it suffices to check the equality only for "Var" itself. But if $\mathcal{C}_1$ and $\mathcal{C}_2$ are non-overlapping, finite, exact covers of $[a, c]$ and $[c, b]$, then $\mathcal{C} = \mathcal{C}_1 \cup \mathcal{C}_2$ is a non-overlapping, finite, exact cover of $[a, b]$; and so

$$\mathcal{S}(\psi; \mathcal{C}_1) + \mathcal{S}(\psi; \mathcal{C}_2) = \mathcal{S}(\psi; \mathcal{C}) \le \mathrm{Var}(\psi; [a, b]).$$

Hence $\mathrm{Var}(\psi; [a, c]) + \mathrm{Var}(\psi; [c, b]) \le \mathrm{Var}(\psi; [a, b])$. On the other hand, if $\mathcal{C}$ is a non-overlapping, finite, exact cover of $[a, b]$, then $\mathcal{C}_1 = \{[a, c] \cap I : I \in \mathcal{C}\}$ and $\mathcal{C}_2 = \{[c, b] \cap I : I \in \mathcal{C}\}$ are non-overlapping, finite, exact covers of $[a, c]$ and $[c, b]$ such that $\mathcal{C} \le \mathcal{C}_1 \cup \mathcal{C}_2$. Hence,

$$\mathcal{S}(\psi; \mathcal{C}) \le \mathcal{S}(\psi; \mathcal{C}_1 \cup \mathcal{C}_2) = \mathcal{S}(\psi; \mathcal{C}_1) + \mathcal{S}(\psi; \mathcal{C}_2) \le \mathrm{Var}(\psi; [a, c]) + \mathrm{Var}(\psi; [c, b]).$$

Since this is true for every $\mathcal{C}$, the asserted equality is now proved. □

We have now proved the following decomposition theorem for functions having bounded variation.

Theorem 1.2.5 *Let $\psi : J \longrightarrow \mathbb{R}$ be given. Then ψ has bounded variation on J if and only if there exist non-decreasing functions ψ_1 and ψ_2 on J such that $\psi = \psi_2 - \psi_1$. In fact, if ψ has bounded variation on $J = [a, b]$ and we define $\psi_\pm(x) = \mathrm{Var}_\pm(\psi; [a, x])$ for $x \in J$, then ψ_+ and ψ_- are non-decreasing and $\psi(x) = \psi(a) + \psi_+(x) - \psi_-(x)$, $x \in J$. Finally, if ψ has bounded variation on J, then every $\varphi \in C(J; \mathbb{R})$ is Riemann integrable on J with respect to ψ and*

$$\left| (\mathrm{R}) \int_J \varphi(x)\, d\psi(x) \right| \le \mathrm{Var}(\psi; J) \|\varphi\|_{\mathrm{u}}.$$

In order to complete our program, we need to establish a couple of elementary facts about functions $\psi : \mathbb{R} \longrightarrow \mathbb{R}$ with discontinuities. Discontinuities of functions on $\mathbb{R}$ can be of two types. The first type is a **jump discontinuity**: one that arises because both the right limit $\lim_{t \searrow s} \psi(t)$ and the left limit $\lim_{t \nearrow s} \psi(t)$ exist but one or both are not equal to $\psi(s)$. The second type of discontinuity is an oscillatory discontinuity: one that occurs because either or both the left and right limits fail to exist. All discontinuities of monotone functions, and therefore also of all functions having bounded variation, are jump discontinuities. An example of a ψ with an oscillatory discontinuity is the one for which $\psi(t)$ equals $\sin \frac{1}{t}$ when $t \ne 0$ and 0 when $t = 0$. The following lemma contains evidence that functions that have only jump discontinuities cannot be too wild.

Lemma 1.2.6 *If $\psi : J = [a,b] \longrightarrow \mathbb{R}$ has a right limit in $\mathbb{R}$ at every $x \in J \setminus \{b\}$ and a left limit in $\mathbb{R}$ at every $x \in J \setminus \{a\}$, then ψ is bounded and*

$$\operatorname{card}\big(\{x \in \mathring{J} : |\psi(x) - \psi(x+)| \vee |\psi(x) - \psi(x-)| \geq \epsilon\}\big) < \infty \quad \text{for each } \epsilon > 0.$$

In particular, ψ has at most countably many discontinuities. Also, if $\tilde{\psi}(x) \equiv \psi(x+)$ for $x \in \mathring{J}$ and $\tilde{\psi}(x) = \psi(x)$ for $x \in \{a,b\}$, then $\tilde{\psi}$ is right-continuous on $\mathring{J}$, has a left limit in $\mathbb{R}$ at every $x \in J \setminus \{a\}$, and coincides with ψ at all points where ψ is continuous. Thus, if φ is Riemann integrable on J with respect to both ψ and $\tilde{\psi}$, then $(\mathrm{R}) \int_J \varphi(x)\, d\tilde{\psi}(x) = (\mathrm{R}) \int_J \varphi(x)\, d\psi(x)$. Finally, if $\varphi \in C(J;\mathbb{R})$ is Riemann integrable on J with respect to ψ, then it is also Riemann integrable on J with respect to $\tilde{\psi}$.

Proof. Suppose that ψ were unbounded. Then we could find a sequence $\{x_n : n \geq 1\} \subseteq J$ for which $|\psi(x_n)| \longrightarrow \infty$ as $n \to \infty$; and clearly there is no loss of generality to assume that either $x_{n+1} < x_n$ or $x_{n+1} > x_n$ for all $n \geq 1$. But, in the first case, this would mean that $|\psi(x+)| = \infty$, where $x = \lim_{n\to\infty} x_n$, and so no such sequence can exist. Similarly, the second case would lead to the non-existence of $\psi(x-)$ in $\mathbb{R}$. Thus ψ must be bounded.

The proof that $\operatorname{card}\big(\{x \in \mathring{J} : |\psi(x) - \psi(x+)| \vee |\psi(x) - \psi(x-)| \geq \epsilon\}\big) < \infty$ is very much the same. Namely, if not, we could assume that there exists a sequence $\{x_n : n \geq 1\} \subseteq \mathring{J}$ that converges to some $x \in J$, is either strictly decreasing or strictly increasing, and has the property that

$$|\psi(x_n) - \psi(x_n+)| \vee |\psi(x_n) - \psi(x_n-)| \geq \epsilon \quad \text{for each } n \geq 1.$$

But, in the decreasing case, for each $n \geq 1$, we could find $x_n' \in (x, x_n)$ and $x_n'' \in (x_n, x_n + \frac{1}{n}) \cap \mathring{J}$ so that

$$|\psi(x_n) - \psi(x_n')| \vee |\psi(x_n) - \psi(x_n'')| \geq \frac{\epsilon}{2},$$

and clearly this would contradict the existence in $\mathbb{R}$ of $\psi(x+)$. Essentially the same argument shows that $\psi(x-)$ cannot exist in $\mathbb{R}$ in the increasing case.

The preceding makes it obvious that ψ can be discontinuous at only countably many points, from which it follows that $\tilde{\psi}(x\pm) = \psi(x\pm)$ for all $x \in \mathring{J}$. To prove the equality of Riemann integrals with respect to ψ and $\tilde{\psi}$ of φ's that are Riemann integrable with respect to both, note that, because ψ coincides with $\tilde{\psi}$ on $\{a,b\}$ as well as on a dense subset of $\mathring{J}$, we can always evaluate these integrals using Riemann sums that are the same whether they are computed with respect to ψ or to $\tilde{\psi}$.

Finally, we must show that if $J = [a,b]$ and $\varphi \in C(J;\mathbb{R})$ is Riemann integrable with respect to ψ, then it also is with respect to $\tilde{\psi}$. To do this, it clearly suffices to show that for any $\mathcal{C}$, choice map $\xi \in \Xi(\mathcal{C})$, and $\epsilon > 0$, there exit a $\mathcal{C}'$ and a $\xi' \in \Xi(\mathcal{C}')$ such that $\|\mathcal{C}'\| \leq 2\|\mathcal{C}\|$ and

$$|\mathcal{R}(\varphi|\tilde{\psi};\mathcal{C},\xi) - \mathcal{R}(\varphi|\psi;\mathcal{C}',\xi')| < \epsilon.$$

To this end, assume, without loss in generality, that

$$\mathcal{C} = \{[c_k, c_{k+1}] : 0 \le k \le n\} \text{ where } a = c_0 < \cdots < c_{n+1} = b.$$

For $0 < \alpha < \min_{0 \le k \le n}(c_{k+1} - c_k)$, set

$$c_{k,\alpha} = \begin{cases} c_k & \text{if} \quad k \in \{0, n+1\} \\ c_k + \alpha & \text{if} \quad 1 \le k \le n, \end{cases}$$

and let $\mathcal{C}_\alpha = \{[c_{k,\alpha}, c_{k+1,\alpha}] : 0 \le k \le n\}$. Clearly $\|\mathcal{C}_\alpha\| \le 2\|\mathcal{C}\|$. Moreover, $c_{k,\alpha} \le c_{k+1} \le c_{k+1,\alpha}$ for each $0 \le k \le n$, and so we can take

$$\xi_\alpha([c_{k,\alpha}, c_{k+1,\alpha}]) = c_{k,\alpha} \vee \xi([c_k, c_{k+1}]).$$

Then, because φ is continuous and $\tilde{\psi}$ is right-continuous on $J \setminus \{b\}$,

$$\mathcal{R}(\varphi|\tilde{\psi}; \mathcal{C}, \xi) = \lim_{\alpha \searrow 0} \mathcal{R}(\varphi|\tilde{\psi}; \mathcal{C}_\alpha, \xi_\alpha).$$

At the same time,

$$\mathcal{R}(\varphi|\tilde{\psi}; \mathcal{C}_\alpha, \xi_\alpha) = \mathcal{R}(\varphi|\psi; \mathcal{C}_\alpha, \xi_\alpha)$$

for all but a countable number of α's. Thus, for any $\epsilon > 0$, there is an $\alpha > 0$ for which $|\mathcal{R}(\varphi|\tilde{\psi}; \mathcal{C}, \xi) - \mathcal{R}(\varphi|\tilde{\psi}; \mathcal{C}_\alpha, \xi_\alpha)| < \epsilon$ and $\mathcal{R}(\varphi|\tilde{\psi}, \mathcal{C}_\alpha, \xi_\alpha) = \mathcal{R}(\varphi|\psi, \mathcal{C}_\alpha, \xi_\alpha)$. □

Theorem 1.2.7 *Let ψ be a function on $J = [a, b]$ that satisfies the hypotheses of Lemma 1.2.6, and define $\tilde{\psi}$ accordingly. If every $\varphi \in C(J; \mathbb{R})$ is Riemann integrable on J with respect to ψ, and if there is a $K < \infty$ such that*

$$\left|(\mathrm{R}) \int_J \varphi(x)\, d\psi(x)\right| \le K\|\varphi\|_{\mathrm{u}}, \qquad \varphi \in C(J; \mathbb{R}), \tag{1.2.5}$$

then $\tilde{\psi}$ has bounded variation on J and

$$\begin{aligned} \mathrm{Var}(\tilde{\psi}; J) &= \sup\left\{(\mathrm{R}) \int_J \varphi(x)\, d\psi(x) : \varphi \in C(J; \mathbb{R}) \text{ and } \|\varphi\|_{\mathrm{u}} = 1\right\} \\ &= \sup\left\{(\mathrm{R}) \int_J \varphi(x)\, d\tilde{\psi}(x) : \varphi \in C(J; \mathbb{R}) \text{ and } \|\varphi\|_{\mathrm{u}} = 1\right\}. \end{aligned}$$

In particular, if ψ itself is right-continuous on $\mathring{J}$, then ψ has bounded variation on J if and only if every $\varphi \in C(J; \mathbb{R})$ is ψ-Riemann integrable on J and (1.2.5) *holds for some $K < \infty$, in which case* $\mathrm{Var}(\psi; J)$ *is the optimal choice of K.*

Proof. In view of what we already know, all that we have to do is check that for each $\mathcal{C}$ and $\epsilon > 0$ there is a $\varphi \in C(J; \mathbb{R})$ such that $\|\varphi\|_{\mathrm{u}} = 1$ and

$\mathcal{S}(\tilde{\psi};\mathcal{C}) \leq (\mathrm{R})\int_J \varphi(x)\,d\psi(x) + \epsilon$. Moreover, because $\tilde{\psi}$ is right continuous and continuous at a dense set of points, we may and will assume that $\mathcal{C} = \{[c_k, c_{k+1}] : 0 \leq k \leq n\}$ where $a = c_0 < \cdots < c_{n+1} = b$ and c_k is a point of continuity of ψ for each $1 \leq k \leq n$.

Given $0 < \alpha < \min_{0\leq k\leq n} \frac{c_{k+1}-c_k}{2}$, define $\varphi_\alpha \in C(J;\mathbb{R})$ so that

$$\varphi_\alpha(x) = \begin{cases} \operatorname{sgn}\big(\Delta_{[c_0,c_1]}\psi\big) & \text{for } x \in [c_0, c_1 - \alpha], \\ \operatorname{sgn}\big(\Delta_{[c_k,c_{k+1}]}\psi\big) & \text{for } x \in [c_k + \alpha, c_{k+1} - \alpha] \text{ and } 1 \leq k < n, \\ \operatorname{sgn}\big(\Delta_{[c_n,c_{n+1}]}\psi\big) & \text{for } x \in [c_n + \alpha, c_{n+1}], \end{cases}$$

and φ_α is linear on each of the intervals $[c_k - \alpha, c_k + \alpha]$, $1 \leq k \leq n$. (The **signum function** $t \in \mathbb{R} \longmapsto \operatorname{sgn}(t)$ is defined so that $\operatorname{sgn}(t)$ is -1 or 1 according to whether $t < 0$ or $t \geq 0$.) Then, by **(iv)** in Examples 1.2.1,

$$\begin{aligned} &(\mathrm{R})\int_J \varphi_\alpha(x)\,d\psi(x) - \mathcal{S}(\tilde{\psi};\mathcal{C}) \\ &\quad= \sum_{k=0}^{n} (\mathrm{R})\int_{[c_k,c_{k+1}]} \big(\varphi_\alpha(x) - \operatorname{sgn}(\Delta_{[c_k,c_{k+1}]}\psi)\big)\,d\psi(x) \\ &\quad= \sum_{k=1}^{n} \Bigg[(\mathrm{R})\int_{[c_k-\alpha,c_k]} \big(\varphi_\alpha(x) - \varphi_\alpha(c_k - \alpha)\big)\,d\psi(x) \\ &\qquad\qquad + (\mathrm{R})\int_{[c_k,c_k+\alpha]} \big(\varphi_\alpha(x) - \varphi_\alpha(c_k + \alpha)\big)\,d\psi(x)\Bigg]. \end{aligned}$$

For each $1 \leq k \leq n$, either $\varphi_\alpha \equiv \varphi_\alpha(c_k - \alpha)$ on $[c_k - \alpha, c_k + \alpha]$, in which case the integrals over $[c_k - \alpha, c_k]$ and $[c_k, c_k + \alpha]$ do not contribute to the preceding sum, or $\varphi_\alpha(c_k) = 0$ and $\varphi'_\alpha \equiv \big(\varphi_\alpha(c_k + \alpha) - \varphi_\alpha(c_k - \alpha)\big)/2\alpha$ on $[c_k - \alpha, c_k + \alpha]$. In the latter case, we apply Theorem 1.2.1 and the equation in part **(i)** of Examples 1.2.1 to show that

$$\begin{aligned} &(\mathrm{R})\int_{[c_k-\alpha,c_k]} \big(\varphi_\alpha(x) - \varphi_\alpha(c_k - \alpha)\big)\,d\psi(x) \\ &\qquad\qquad + (\mathrm{R})\int_{[c_k,c_k+\alpha]} \big(\varphi_\alpha(x) - \varphi_\alpha(c_k + \alpha)\big)\,d\psi(x) \\ &\quad= \big[\varphi_\alpha(c_k + \alpha) - \varphi_\alpha(c_k - \alpha)\big]\psi(c_k) \\ &\qquad\qquad - \frac{\varphi_\alpha(c_k + \alpha) - \varphi_\alpha(c_k - \alpha)}{2\alpha}(\mathrm{R})\int_{[c_k-\alpha,c_k+\alpha]} \psi(x)\,dx, \end{aligned}$$

which, since φ is continuous at c_k, clearly tends to 0 as $\alpha \searrow 0$. In other words, we now see that

$$\mathcal{S}(\tilde{\psi};\mathcal{C}) = \lim_{\alpha\searrow 0} (\mathrm{R})\int_J \varphi_\alpha(x)\,d\psi(x),$$

which is all that we had to prove. □

Theorem 1.2.5 can be viewed as a very special case of the Riesz Representation Theorem for continuous linear functionals on spaces of continuous functions. See Theorem 8.2.9 and Remark 8.2.1 for further information.

1.2.3 Exercises for § 1.2

Exercise 1.2.1 Check all of the assertions in Examples 1.2.1. The only one that presents a real challenge is the assertion in **(iv)** that φ is Riemann integrable on both J_1 and J_2 with respect to ψ.

Exercise 1.2.2 The goal of this exercise is to develop and apply a formula discovered by Euler. In what follows, for any $t \in \mathbb{R}$, $\lfloor t \rfloor$ is **integer part** and $\{t\} = t - \lfloor t \rfloor$ is the **fractional part** of t.

(i) Assume that $f : [0, \infty) \longrightarrow \mathbb{C}$ is a continuously differentiable function. Show that for integers $m < n$

$$\sum_{k=m}^{n} f(k) = f(m) + (R)\int_{[m,n]} f(t)d\lfloor t \rfloor = f(m) + (R)\int_{[m,n]} f(t)\, dt - (R)\int_{[m,n]} f(t)\, d\{t\},$$

and use this and integration by parts to conclude that

$$\sum_{k=m}^{n} f(k) = f(m) + (R)\int_{[m,n]} f(t)\, dt + (R)\int_{[m,n]} f'(t)\{t\}\, dt, \tag{1.2.6}$$

which is one of Euler's many formulas.

(ii) Next set $\rho(t) = \{t\} - \frac{1}{2}$ and $R(t) = -(R)\int_{[s,t]} \rho(s)\, ds$. Show that

$$R(t+1) = R(t) \text{ and } 0 \le R(t) \le \frac{1}{8} \text{ for all } t \ge 0.$$

In particular, $R(n) = 0$ for all $n \in \mathbb{N}$. Assuming that f is twice continuously differentiable, use (1.2.6) and integration by parts to show that

$$\sum_{k=m}^{n} f(k) = (R)\int_{[m,n]} f(t)\, dt + \frac{f(n)+f(m)}{2} + (R)\int_{[m,n]} f''(t)R(t)\, dt. \tag{1.2.7}$$

(iii) Apply (1.2.7) to show that

$$\sum_{k=1}^{n} \frac{1}{k} = \log n + \frac{n+1}{2n} + 2(R)\int_{[1,n]} \frac{R(t)}{t^3}\, dt,$$

and use this to conclude that

$$\sum_{k=1}^{n}\frac{1}{k}=\log n+\gamma+\frac{1}{2n}-E(n),$$

where

$$\gamma\equiv\frac{1}{2}+2(R)\int_{[1,\infty)}\frac{R(t)}{t^3}\,dt$$

is called **Euler's constant** and $0\le E(n)\le\frac{1}{8n^2}$.

(**iv**) Now use (1.2.7) to show that

$$\sum_{k=1}^{n}\log k=n\log n-n+1+\frac{\log n}{2}-(R)\int_{[1,n]}\frac{R(t)}{t^2}\,dt.$$

Finally, show that

$$0\le(R)\int_{[n,\infty)}\frac{R(t)}{t^2}\,dt\le\frac{1}{8n},$$

for all $n\ge 1$, and thereby conclude that

$$1\le\frac{n!}{\sqrt{Cn}\left(\frac{n}{e}\right)^n}\le e^{\frac{1}{8n}}$$

where

$$C=\exp\left(2(R)\int_{[1,\infty)}\frac{1-R(t)}{t^2}\,dt\right).$$

This result is known as **Stirling's formula** Although Stirling's name is attached to it, the result here is a somewhat more precise form of a result originally proved by A. de Moivre, a contemporary of Euler and Stirling. Shortly afterwards, Stirling showed that $C=2\pi$, and, in spite of his acknowledgment of de Moivre's seminal contribution, Stirling ended up getting the full credit. Stirling obtained his result as an application of Wallis's formula. A more transparent derivation can be obtained as an application of the Central Limit Theorem, which is a generalization of the result at which de Moivre was aiming. See Exercise 7.4.2 for such a proof.

Exercise 1.2.3 Define $\psi:[0,1]\longrightarrow\mathbb{R}$ so that $\psi(0)=0$ and $\psi(t)=t\cos\frac{1}{t}$ for $t\in(0,1]$. Show that ψ is a continuous function with unbounded variation. Also, give an example of a ψ having bounded variation on $[0,1]$ for which

$$\sup\left\{(\mathrm{R})\int_{[0,1]}\varphi(x)\,d\psi(x):\varphi\in C(J;\mathbb{R})\text{ and }\|\varphi\|_{\mathrm{u}}=1\right\}<\mathrm{Var}(\psi;J).$$

Exercise 1.2.4 This exercise is a variation on Exercise 1.1.2. If ψ is non-decreasing on J, show that a bounded function φ is Riemann–Stieltjes integrable on J with respect to ψ if and only if, for every $\epsilon>0$, there is a $\delta>0$

for which

$$\sum_{\{I \in \mathcal{C}: \sup_I \varphi - \inf_I \varphi \geq \epsilon\}} \Delta_I \psi < \epsilon \tag{1.2.8}$$

whenever $\mathcal{C}$ is a non-overlapping, finite, exact cover of J satisfying $\|\mathcal{C}\| < \delta$. Also, show that when, in addition, $\psi \in C(J;\mathbb{R})$, the preceding can be replaced by the condition that, for each $\epsilon > 0$, (1.2.8) holds for some $\mathcal{C}$.

Hint: For the last part, compare the situation here to the one handled in Lemma 1.1.3.

Exercise 1.2.5 If $\psi \in C(J;\mathbb{R})$, show that

$$\mathrm{Var}_\pm(\psi;J) = \lim_{\|\mathcal{C}\|\to 0} \mathcal{S}_\pm(\psi;\mathcal{C}) \; \big(\in [0,\infty]\big)$$

and conclude that $\mathrm{Var}(\psi,J) = \lim_{\|\mathcal{C}\|\to 0} \mathcal{S}(\psi;\mathcal{C})$. Also, show that if $\psi \in C^1(J;\mathbb{R})$, then

$$\mathrm{Var}_\pm(\psi;J) = (\mathrm{R}) \int_J \psi'(x)^\pm \, dx,$$

and therefore that $\mathrm{Var}(\psi;J) = (\mathrm{R}) \int_J |\psi'(x)| \, dx$.

Exercise 1.2.6 Let ψ be a function of bounded variation on the interval $J = [a,b]$, and define the non-decreasing functions ψ_+ and ψ_- accordingly, as in Theorem 1.2.5. Given any other pair of non-decreasing functions ψ_1 and ψ_2 on J satisfying $\psi = \psi_2 - \psi_1$, show that $\psi_2 - \psi_+$ and $\psi_1 - \psi_-$ are both non-decreasing functions. In particular, this means that $\psi_+ \leq \psi_2 - \psi_2(a)$ and $\psi_- \leq \psi_1 - \psi_1(a)$ whenever ψ_2 and ψ_1 are non-decreasing functions for which $\psi = \psi_2 - \psi_1$. Using Lemma 1.2.4 and the preceding, show that

$$\psi_\pm(x+) - \psi_\pm(x) = \big(\psi(x+) - \psi(x)\big)^\pm, \quad x \in [a,b),$$

and

$$\psi_\pm(x) - \psi_\pm(x-) = \big(\psi(x) - \psi(x-)\big)^\pm, \quad x \in (a,b].$$

Conclude, in particular, that the size of the jumps in $x \longmapsto \mathrm{Var}(\psi;[a,x])$, from both the right and left, coincide with the absolute value of the corresponding jumps in ψ. Hence, ψ is continuous if $x \in J \longmapsto \mathrm{Var}\big(\psi;[a,x]\big)$ is; and if ψ is continuous, then so are ψ_+, ψ_-, and therefore also $x \rightsquigarrow \mathrm{Var}\big(\psi;[a,\cdot\,]\big)$. See Exercise 8.2.1 for further information about the functions $\psi_\pm$.

Hint: In order to handle the last part, show that it is enough to check that $\psi_+(a+) - \psi_+(a) = \big(\psi(a+) - \psi(a)\big)^+$. Next, show that this comes down to checking that $\beta \equiv \big(\psi_+(a+) - \psi_+(a)\big) \wedge \big(\psi_-(a+) - \psi_-(a)\big) = 0$. Finally, define ψ_1 and ψ_2 on $[a,b]$ so that $\psi_1(a) = 0$, $\psi_2(a) = \psi(a)$, and, for $x \in (a,b]$, $\psi_1(x) = \psi_-(x) - \beta$ and $\psi_2(x) = \psi(a) + \psi_+(x) - \beta$, and apply the first part of this exercise to see that $\psi_- \leq \psi_1$.

Exercise 1.2.7
(**i**) Suppose that ψ is a right-continuous function of bounded variation on the finite interval $J = [c,d]$, and set $D(\psi) \equiv \{x \in (c,d] : \psi(x) \neq \psi(x-)\}$. As a consequence of Lemma 1.2.6, one knows that D is countable. Show that

$$\sum_{x \in D(\psi)} \big(\psi(x) - \psi(x-)\big)^{\pm} \leq \mathrm{Var}_{\pm}(\psi; J).$$

(**ii**) Say that $\psi : J \longrightarrow \mathbb{R}$ is a **pure jump function** if it is a right-continuous function of bounded variation with the property that

$$\psi(x) = \sum_{\xi \in D(\psi) \cap (c,x]} \big(\psi(\xi) - \psi(\xi-)\big) \quad \text{for all } x \in J,$$

in which case show that

$$\mathrm{Var}_{\pm}(\psi; J) = \sum_{x \in D(\psi)} \big(\psi(x) - \psi(x-)\big)^{\pm}.$$

(**iii**) Given a right-continuous function $\psi : J \longrightarrow \mathbb{R}$ of bounded variation, define $\psi_{\mathrm{d}} : J \longrightarrow \mathbb{R}$ by

$$\psi_{\mathrm{d}}(x) = \sum_{\{\xi \in D(\psi) \cap (c,x]\}} \big(\psi(\xi) - \psi(\xi-)\big).$$

Show that ψ_{d} is a pure jump function and that

$$\mathrm{Var}_{\pm}(\psi_d; J) = \sum_{x \in D(\psi)} \big(\psi(x) - \psi(x-)\big)^{\pm} \leq \mathrm{Var}_{\pm}(\psi; J).$$

(**iv**) Continuing (**iii**), show that ψ_{d} is the one and only pure jump function φ with the property that $\psi - \varphi$ is continuous. For this reason, $\psi_{\mathrm{c}} \equiv \psi - \psi_{\mathrm{d}}$ and ψ_{d} are called the **continuous** and **discontinuous** parts of ψ.

(**v**) Show that

$$\mathrm{Var}_{\pm}(\psi; J) = \mathrm{Var}_{\pm}(\psi_{\mathrm{c}}; J) + \mathrm{Var}_{\pm}(\psi_{\mathrm{d}}; J).$$

Exercise 1.2.8 Given $-\infty < c < d < \infty$ and a function $\psi : J = [c,d] \longrightarrow \mathbb{R}$ of bounded variation, define the **arc length** $\mathrm{Arc}\big(\psi, [c,d]\big)$ of the graph of ψ to be the supremum of

$$\sum_{I \in \mathcal{C}} \sqrt{\mathrm{vol}(I)^2 + (\Delta_I \psi)^2}$$

as $\mathcal{C}$ runs over all finite, exact covers of $[c,d]$ by non-overlapping intervals.

(**i**) Show that $\mathrm{Arc}\big(\psi; (c,d]\big)$ lies above $\sqrt{(d-c)^2 + \mathrm{Var}\big(\psi; [c,d]\big)^2}$ and below $(d-c) + \mathrm{Var}\big(\psi; [c,d]\big)$.

(ii) Show that the lower bound in **(i)** is achieved when ψ is linear and that the upper bound is achieved when ψ is pure jump. In Exercise 3.3.3 below, it will be shown that, counter to intuition, there are continuous, non-decreasing functions for which the upper bound is achieved.

(iii) When ψ is continuous, show that

$$\mathrm{Arc}\big(\psi;(c,d]\big) = \lim_{\|\mathcal{C}\|\searrow 0} \sum_{I\in\mathcal{C}} \sqrt{\mathrm{vol}(I)^2 + (\Delta_I\psi)^2}.$$

1.3 Rate of Convergence

This section[6] probably should be skipped by those readers whose primary interest is Lebesgue's theory and who would prefer not to waste any more time getting there. On the other hand, for those interested in knowing how fast Riemann approximations can converge, the contents of this section may come as something of a surprise.

For the purposes of this section, it is convenient to consider complex-valued integrands. Notice that there is no trouble doing so since, by applying the theory developed for real-valued integrands to the real and imaginary parts, no new ideas are involved. See Exercise 1.3.1.

1.3.1 Periodic Functions

Suppose that $f \in C^1\big([0,1];\mathbb{C}\big)$. Then, for any non-overlapping, finite, exact cover $\mathcal{C}$ of $[0,1]$ and any choice map $\xi \in \Xi(\mathcal{C})$,

$$\left|(\mathrm{R})\int_{[0,1]} f(x)\,dx - \mathcal{R}(f;\mathcal{C},\xi)\right| \le \|f'\|_{\mathrm{u}}\|\mathcal{C}\|. \tag{1.3.1}$$

Moreover, at least qualitatively, (1.3.1) is optimal. To see this, take $f(x) = x$, and observe that

$$\frac{1}{n}\sum_{m=1}^{n} f\left(\frac{m}{n}\right) = \frac{n(n+1)}{2n^2} = \frac{1}{2} + \frac{1}{2n} = (\mathrm{R})\int_{[0,1]} f(x)\,dx + \frac{1}{2n}.$$

Hence, in this case,

[6] The contents of this section are adapted from the article "Some Riemann sums are better than others," that I wrote with V. Guillemin and which appeared in the book *Representations, Wavelets, and Frames*, edited by P. Jorgensen, K. Merrill, and J. Packer and published in 2008 by Birkhäuser.

$$\left|(\mathrm{R})\int_{[0,1]} f(x)\,dx - \mathcal{R}(f;\mathcal{C}_n,\xi_n)\right| \geq \frac{\|f'\|_{\mathrm{u}}\|\mathcal{C}_n\|}{2},$$

where $\mathcal{C}_n = \left\{\left[\frac{m-1}{n},\frac{m}{n}\right] : 1 \leq m \leq n\right\}$ and $\xi_n\left(\left[\frac{m-1}{n},\frac{m}{n}\right]\right) = \frac{m}{n}$.

In spite of the preceding, in this subsection we will see that (1.3.1) is very far from optimal when f is a **smooth** (i.e., C^∞), periodic function. That is, when f is the restriction to $[0,1]$ of a smooth function on $\mathbb{R}$ that has period 1. In fact, I will show that if

$$\mathcal{R}_n(f) \equiv \frac{1}{n}\sum_{m=1}^{n} f\left(\tfrac{m}{n}\right) \tag{1.3.2}$$

and f on $[0,1]$ is a smooth, periodic function, then

$$\lim_{n\to\infty} n^\ell \left|\mathcal{R}_n(f) - (\mathrm{R})\int_{[0,1]} f(x)\,dx\right| = 0 \quad \text{for all } \ell \in \mathbb{Z}^+. \tag{1.3.3}$$

Before proceeding, one should recognize how essential both periodicity and the selection of the choice map are. Indeed, the preceding example shows that periodicity cannot be dispensed with. To see the importance of the choice map, take $f(x) = e^{i2\pi x}$, where $i = \sqrt{-1}$. This function is certainly smooth and periodic. Next, take $\alpha_n = 1 - \frac{1}{n}$. Then $\left[\frac{m-1}{n},\frac{m}{n}\right] \in \mathcal{C}_n \longmapsto \frac{m\alpha_n}{n} \in \mathbb{R}$ is an allowable choice map ξ_n and

$$\mathcal{R}(f;\mathcal{C}_n,\xi_n) = \frac{e^{i2\pi\frac{\alpha_n}{n}}}{n}\,\frac{1 - e^{i2\pi\alpha_n}}{1 - e^{i2\pi\frac{\alpha_n}{n}}}.$$

Thus, since

$$(\mathrm{R})\int_{[0,1]} f(x)\,dx = 0 \quad \text{and} \quad \lim_{n\to\infty} e^{i2\pi\frac{\alpha_n}{n}}\,\frac{1 - e^{i2\pi\alpha_n}}{1 - e^{i2\pi\frac{\alpha_n}{n}}} = -1,$$

we conclude that

$$\lim_{n\to\infty} n\left|(\mathrm{R})\int_{[0,1]} f(x)\,dx - \mathcal{R}(f;\mathcal{C}_n,\xi_n)\right| = 1.$$

Turning to the proof of (1.3.3), let f be a smooth, periodic function, and note that

$$(\mathrm{R})\int_{[0,1]} f(x)\,dx - \mathcal{R}_n(f) = \sum_{m=1}^{n}(\mathrm{R})\int_{I_{m,n}}\left(f(x) - f\left(\tfrac{m}{n}\right)\right)dx,$$

where $I_{m,n} = \left[\frac{m-1}{n},\frac{m}{n}\right]$. Integrating by parts, one finds that the mth summand equals

$$-(\mathrm{R})\int_{I_{m,n}} \left[x - \tfrac{m-1}{n}\right] f'(x)\,dx,$$

and so

$$(\mathrm{R})\int_{[0,1]} f(x)\,dx - \mathcal{R}_n(f) = -\sum_{m=1}^{n} (\mathrm{R})\int_{I_{m,n}} \left[x - \tfrac{m-1}{n}\right] f'(x)\,dx.$$

So far, we have used only smoothness but not periodicity. But we now use periodicity to see that the integral of f' over $[0,1]$ is 0 and therefore that the sum is unchanged when $\left[x - \frac{m-1}{n}\right]$ is replaced by $\left[x - \frac{m-1}{n} - c\right]$ for any $c \in \mathbb{R}$. In particular, by taking $c = \frac{1}{2n}$, which is the average value of $x - \frac{m-1}{n}$ on $I_{m,n}$, each of the summands can be replaced by

$$(\mathrm{R})\int_{I_{m,n}} \left[\left(x - \tfrac{m-1}{n}\right) - \tfrac{1}{2n}\right] f'(x)\,dx,$$

in which case $f'(x)$ can be replaced by $f'(x) - f'\left(\frac{m}{n}\right)$. After making these replacements, one arrives at

$$(\mathrm{R})\int_{[0,1]} f(x)\,dx - \mathcal{R}_n(f) = -\sum_{m=1}^{n} (\mathrm{R})\int_{I_{m,n}} \left(f'(x) - f'\left(\tfrac{m}{n}\right)\right)\left[x - \tfrac{m-1}{n} - \frac{1}{2n}\right] dx.$$

To see that we have already made progress toward (1.3.3), note that the absolute value of each summand in the preceding expression is dominated by $\frac{\|f''\|_{\mathrm{u}}}{4n^3}$ and therefore that we have shown that

$$\left|(\mathrm{R})\int_{[0,1]} f(x)\,dx - \mathcal{R}_n(f)\right| \le \frac{\|f''\|_{\mathrm{u}}}{4n^2}.$$

Before attempting to go further, it is best to introduce the following notation. For $\ell \ge 1$, let $C_1^\ell([0,1];\mathbb{C})$ be the space of $f \in C^\ell([0,1];\mathbb{C})$ with the property that $f^{(k)} \equiv \partial^k f$ takes the same value at 0 and 1 for each $0 \le k < \ell$, and set $C_1^\infty([0,1];\mathbb{C}) = \bigcap_{\ell=0}^\infty C_1^\ell([0,1];\mathbb{C})$. Next, given $k \in \mathbb{N}$, set

$$\Delta_n^{(k)}(f) = \frac{1}{k!}\sum_{m=1}^{n} (\mathrm{R})\int_{I_{m,n}} \left[x - \tfrac{m-1}{n}\right]^k \left(f(x) - f\left(\tfrac{m}{n}\right)\right) dx. \tag{1.3.4}$$

Then

$$\Delta_n^{(0)}(f) = (\mathrm{R})\int_{[0,1]} f(x)\,dx - \mathcal{R}_n(f),$$

and the preceding calculation shows that

$$\Delta_n^{(0)}(f) = \frac{1}{2n}\Delta^{(0)}(f') - \Delta^{(1)}(f').$$

More generally, integration by parts shows that the mth term in the expression for $\Delta^{(k)}(f)$ equals

$$-\frac{1}{(k+1)!}(\mathrm{R})\int_{I_{m,n}}\left[x-\tfrac{m-1}{n}\right]^{k+1} f'(x)\,dx.$$

Next, assuming that $f \in C^1_1([0,1];\mathbb{C})$ and using periodicity in the same way as we did above, one sees that the sum of these is the same as the sum of

$$\begin{aligned}
&-\frac{1}{(k+1)!}(\mathrm{R})\int_{I_{m,n}}\left[\left(x-\tfrac{m-1}{n}\right)^{k+1}-\tfrac{1}{(k+2)n^{k+1}}\right]\left(f'(x)-f'\left(\tfrac{m}{n}\right)\right)dx\\
&=\frac{1}{(k+2)!n^{k+1}}(\mathrm{R})\int_{I_{m,n}}\left(f'(x)-f'\left(\tfrac{m}{n}\right)\right)dx\\
&\qquad-\frac{1}{(k+1)!}(\mathrm{R})\int_{I_{m,n}}\left[x-\tfrac{m-1}{n}\right]^{k+1}\left(f'(x)-f'\left(\tfrac{m}{n}\right)\right)dx.
\end{aligned}$$

Hence, we have now shown that

$$\Delta_n^{(k)}(f)=\frac{1}{(k+2)!n^{k+1}}\Delta_n^{(0)}(f')-\Delta_n^{(k+1)}(f') \tag{1.3.5}$$

for any $f \in C^1_1([0,1];\mathbb{C})$.

Working by induction on $\ell \in \mathbb{N}$, one can use (1.3.5) to check that, for any $f \in C^\ell_1([0,1];\mathbb{C})$,

$$\Delta_n^{(0)}(f)=\frac{1}{n^{\ell+1}}\sum_{k=0}^{\ell}(-1)^k b_{\ell-k} n^{k+1}\Delta_n^{(k)}(f^{(\ell)}), \tag{1.3.6}$$

where $\{b_k : k \geq 0\}$ is determined inductively by

$$b_0 = 1 \quad\text{and}\quad b_{\ell+1}=\sum_{k=0}^{\ell}\frac{(-1)^k}{(k+2)!}b_{\ell-k}. \tag{1.3.7}$$

Noting that

$$n^{k+1}\left|\Delta_n^{(k)}(f)\right| \leq \frac{\|f'\|_{\mathrm{u}}}{(k+2)!}, \tag{1.3.8}$$

it follows from (1.3.6) that, for any $f \in C^{\ell+1}_1([0,1];\mathbb{C})$,

$$\left|(\mathrm{R})\int_{[0,1]} f(x)\,dx-\mathcal{R}_n(f)\right| \leq \frac{K_{\ell+1}}{n^{\ell+1}}\|f^{(\ell+1)}\|_{\mathrm{u}}, \tag{1.3.9}$$

where

$$K_{\ell+1}=\sum_{k=0}^{\ell}\frac{|b_{\ell-k}|}{(k+2)!}. \tag{1.3.10}$$

Obviously, (1.3.9) does more than just prove (1.3.3); it even gives a rate. To get a more explicit result, one has to get a handle on the numbers b_k and the associated quantities $K_{\ell+1}$. As we will see in § 7.2.2, the numbers b_k are intimately related to a famous sequence known as the Bernoulli numbers (so named for their discoverer, Jacob Bernoulli). However, here we will settle for less refined information and be content with knowing that

$$\lim_{\ell\to\infty} K_\ell^{\frac{1}{\ell}} = \frac{1}{2\pi}. \tag{1.3.11}$$

To prove (1.3.11), first observe if $f(x) = e^{i2\pi x}$, then $\Delta_1^{(0)}(f) = -1$, $\|f^{(\ell)}\|_{\mathrm{u}} = (2\pi)^\ell$, and so (1.3.9) shows that $K_{\ell+1} \geq (2\pi)^{-\ell-1}$ and therefore that $\underline{\lim}_{\ell\to\infty} K_\ell^{\frac{1}{\ell}} \geq (2\pi)^{-1}$. To prove the corresponding upper bound, begin by using (1.3.7) to inductively check that $|b_k| \leq \beta^{-k}$, where β is the element of $(0,\infty)$ that satisfies $e^\beta = 1 + 2\beta$. As a consequence, we know that the generating function $B(\lambda) \equiv \sum_{k=1}^\infty b_k \lambda^{k-1}$ is well defined for $\lambda \in \mathbb{C}$ with $|\lambda| < \beta$. Moreover, from (1.3.7), for $|\lambda| < \beta$,

$$B(\lambda) = \sum_{\ell=0}^\infty b_{\ell+1}\lambda^\ell = \sum_{k=0}^\infty \frac{(-\lambda)^k}{(k+2)!} \sum_{\ell=k}^\infty b_{\ell-k}\lambda^{\ell-k} = \big(1+\lambda B(\lambda)\big)\frac{e^{-\lambda}-1+\lambda}{\lambda^2},$$

and therefore

$$B(\lambda) = \frac{1 - e^\lambda + \lambda e^\lambda}{\lambda(e^\lambda - 1)}, \tag{1.3.12}$$

where it is to be understood that the expression on the right is defined at 0 by analytic continuation. In other words, it equals $\frac{1}{2}$ at $\lambda = 0$. Since $e^\lambda \neq 1$ for $0 < |\lambda| < 2\pi$ and $e^{i2\pi} = 1$, it follows that 2π is the radius of convergence for $B(\lambda)$ and therefore that

$$\overline{\lim_{k\to\infty}} |b_k|^{\frac{1}{k}} \leq \frac{1}{2\pi}. \tag{1.3.13}$$

Finally, plugging this into (1.3.10), we see that, for each $\theta \in (0,1)$, $1 > c > (2\pi)^{-1}$, and sufficiently large ℓ,

$$K_{\ell+1} \leq c^{(1-\theta)\ell} \sum_{0\leq k\leq \theta\ell} \frac{1}{(k+2)!} + M^\ell \sum_{\ell\geq k>\theta\ell} \frac{1}{(k+2)!} \leq ec^{(1-\theta)\ell} + \frac{eM^\ell}{(\lfloor\theta\ell\rfloor+1)!},$$

where $M \equiv \sup_{k\geq 0} |b_k|^{\frac{1}{k}} \in [1,\infty)$, $\lfloor t\rfloor$ denotes the integer part of $t \in \mathbb{R}$ (i.e., the largest integer dominated by t), and Taylor's remainder formula has been applied to see that

$$\sum_{k>n} \frac{1}{k!} = e - \sum_{k=0}^n \frac{1}{k!} = \frac{1}{n!}(\mathrm{R})\int_{[0,1]} (1-t)^n e^t\,dt \leq \frac{e}{(n+1)!}.$$

Using the trivial estimate $n! \geq \left(\frac{n}{2}\right)^{\frac{n}{2}}$, one can pass from the preceding to

$$K_{\ell+1} \leq e\left(1+\left(\left(\frac{2}{\theta\ell}\right)^{\frac{\theta}{2}}\frac{M}{c^{1-\theta}}\right)^{\ell}\right)c^{(1-\theta)\ell},$$

from which it is clear that $\overline{\lim}_{\ell\to\infty} K_\ell^{\frac{1}{\ell}} \leq c^{1-\theta}$ for every $\theta \in (0,1)$ and $1 > c > (2\pi)^{-1}$. Thus, $\overline{\lim}_{\ell\to\infty} K_\ell^{\frac{1}{\ell}} \leq (2\pi)^{-1}$ follows after one lets $\theta \searrow 0$ and $c \searrow (2\pi)^{-1}$.

These findings are summarized in the following.

Theorem 1.3.1 *If* $\ell \in \mathbb{N}$ *and* $f \in C_1^{\ell+1}([0,1];\mathbb{C})$, *then* (1.3.9) *holds. In particular, if* $f \in C_1^{\infty}([0,1];\mathbb{C})$ *and* $n \in \mathbb{Z}^+$, *then*

$$\overline{\lim_{\ell\to\infty}} \|f^{(\ell)}\|_{\mathrm{u}}^{\frac{1}{\ell}} < 2\pi n \implies (\mathrm{R})\int_{[0,1]} f(x)\,dx = \mathcal{R}_n(f).$$

Proof. The first assertion needs no further comment. As for the second, it is an immediate consequence of the first combined with (1.3.11). □

At first sight, the concluding part of Theorem 1.3.1 looks quite striking. However, as will be shown in § 7.2.2, it really only reflects the fact that there are relatively few smooth, periodic functions whose successive derivatives grow at most geometrically fast.

1.3.2 The Non-Periodic Case

It is interesting to see what can be said when the function f is not periodic. Thus, define $\Delta_n^{(k)}(f)$ as in (1.3.4). Then

$$\Delta_n^{(k)}(f) = -\frac{1}{(k+1)!}\sum_{k=1}^{n}(\mathrm{R})\int_{I_{m,n}}\left[x-\tfrac{m-1}{n}\right]^{k+1} f'(x)\,dx$$

still holds. Now add and subtract $\frac{f(1)-f(0)}{(k+2)!n^{k+1}}$ from the right-hand side to arrive first at

$$\Delta_n^{(k)}(f) = -\frac{1}{(k+1)!}\sum_{m=1}^{n}(\mathrm{R})\int_{I_{m,n}}\left[\left(x-\tfrac{m-1}{n}\right)^{k+1} - \frac{1}{(k+2)n^{k+1}}\right]f'(x)\,dx$$
$$-\frac{f(1)-f(0)}{(k+2)!n^{k+1}}$$

and then at

$$\Delta_n^{(k)}(f) = \frac{1}{(k+2)!n^{k+1}}\Big[\Delta_n^{(0)}(f') - \big(f(1)-f(0)\big)\Big] - \Delta_n^{(k+1)}(f').$$

Finally, proceed by induction on $\ell \in \mathbb{N}$ to show that, for any $f \in C^\ell\big([0,1];\mathbb{C}\big)$,

$$\begin{aligned}(\mathrm{R})\int_{[0,1]} f(x)\,dx - \mathcal{R}_n(f) =& \frac{1}{n^{\ell+1}}\sum_{k=0}^{\ell}(-1)^k b_{\ell-k} n^{k+1}\Delta_n^{(k)}(f^{(\ell)}) \\ &- \sum_{k=1}^{\ell}\frac{b_k}{n^k}\big(f^{(k-1)}(1) - f^{(k-1)}(0)\big),\end{aligned} \tag{1.3.14}$$

where $\{b_k : k \geq 0\}$ is the sequence described in (1.3.7). Since (1.3.8) did not require periodicity, (1.3.14) yields

$$\begin{aligned}&\left|(\mathrm{R})\int_{[0,1]} f(x)\,dx - \mathcal{R}_n(f) + \sum_{k=1}^{\ell}\frac{b_k}{n^k}\big(f^{(k-1)}(1) - f^{(k-1)}(0)\big)\right| \\ &\qquad\qquad \leq \frac{K_{\ell+1}\|f^{(\ell+1)}\|_{\mathrm{u}}}{n^{\ell+1}},\end{aligned} \tag{1.3.15}$$

where $K_{\ell+1}$ is the constant in (1.3.10). Just as before, (1.3.14) together with (1.3.11) implies that

$$\begin{aligned}&(\mathrm{R})\int_{[0,1]} f(x)\,dx = \mathcal{R}_n(f) - \sum_{k=1}^{\infty}\frac{b_k}{n^k}\big(f^{(k-1)}(1) - f^{(k-1)}(0)\big) \\ &\qquad \text{for } f \in C^\infty\big([0,1];\mathbb{C}\big) \text{ with } \varlimsup_{\ell\to\infty}\|f^{(\ell)}\|_{\mathrm{u}}^{\frac{1}{\ell}} < 2\pi n.\end{aligned} \tag{1.3.16}$$

In fact, by (1.3.13), the series in (1.3.16) will be absolutely convergent.

The formulas (1.3.14) and (1.3.16) are examples of what are called the **Euler–Maclaurin formula**. Although they are of theoretical interest, they do not, in practice, provide an efficient tool for computing integrals.

1.3.3 Exercises for § 1.3

Exercise 1.3.1 Given a compact rectangle J in $\mathbb{R}^N$ and an $f : J \longrightarrow \mathbb{C}$, define $\mathcal{R}(t;\mathcal{C},\xi)$ for a non-overlapping, exact cover $\mathcal{C}$ and a $\xi \in \Xi(\mathcal{C})$ as before, and say that f is Riemann integrable on J if $(\mathrm{R})\int_J f(x)\,dx = \lim_{\|\mathcal{C}\|\to 0}\mathcal{R}(f;\mathcal{C},\xi)$ exists in the same sense as when f is $\mathbb{R}$-valued. Show that f is Riemann integrable if and only if both $f_1 = \mathfrak{Re}(f)$ and $f_2 = \mathfrak{Im}(f)$ are, in which case $(\mathrm{R})\int_J f(x)\,dx = (\mathrm{R})\int_J f_1(x)\,dx + i(\mathrm{R})\int_J f_2(x)\,dx$.

Exercise 1.3.2 The usual statement of the Euler–Maclaurin formula is not the one in (1.3.14) but instead the formula

$$(\mathrm{R})\int_{[0,n]} f(x)\,dx - \sum_{m=1}^{n} f(m) + \sum_{k=1}^{\ell} b_k\big(f^{(k-1)}(n) - f^{(k-1)}(0)\big)$$
$$= (\mathrm{R})\int_{[0,n]} P_\ell(x - \lfloor x \rfloor)\Big(f^{(\ell)}(x) - f^{(\ell)}(\lceil x \rceil)\Big)\,dx, \tag{1.3.17}$$
$$\text{with } P_\ell(x) \equiv \sum_{k=0}^{\ell} \frac{(-1)^k b_{\ell-k}}{k!} x^k,$$

where $\lceil x \rceil$ is the smallest integer greater than or equal to x. Derive (1.3.17) from (1.3.14).

Exercise 1.3.3 Continuing Exercise 1.3.2, show that the sequence of polynomials P_ℓ is uniquely determined by the properties

$$P_0 \equiv 1, \quad P'_{\ell+1} = -P_\ell \quad \text{for } \ell \in \mathbb{N}, \quad \text{and } P_\ell(1) = P_\ell(0) \text{ for } \ell \geq 2. \tag{1.3.18}$$

Here are steps that you might want to take.

(i) First show that there is exactly one sequence of polynomials satisfying the properties in (1.3.18).

(ii) It is easy to check that $P_0 \equiv 1$ and that $P'_{\ell+1} = -P_\ell$. To see that $P_\ell(1) = P_\ell(0)$ when $\ell \geq 2$, use (1.3.7).

The periodicity property of the P_ℓ's for $\ell \geq 2$ is quite remarkable. Indeed, it is not immediately obvious that there exist ℓth order polynomials that, together with their derivatives up to order $\ell - 2$, are periodic on $[0, 1]$.

Exercise 1.3.4 Show that $b_{2\ell+1} = 0$ for $\ell \geq 1$.

Hint: Let $B(\lambda)$ be the function in (1.3.12), and show that it suffices to check that $B'(-\lambda) = B'(\lambda)$.

Chapter 2
Measures

In this chapter I will introduce the notion of a measure, give a procedure for constructing one, and apply that procedure to construct Lebesgue's measure on $\mathbb{R}^N$ as well as the Bernoulli measures for coin tossing.

2.1 Some Generalities

This section introduces a mathematically precise definition of what a measure is and then derives a few elementary properties that follow from the definition. However, to avoid getting lost in the formalities, it will be important to keep the ultimate goal in mind, and for this reason I will begin with a brief summary of what that goal is.

2.1.1 The Idea

The essence of any theory of integration is a *divide and conquer* strategy. That is, given a space E and a family $\mathcal{B}$ of subsets $\Gamma \subseteq E$ for which one has a *reasonable notion* of *measure* assignment $\Gamma \in \mathcal{B} \longmapsto \mu(\Gamma) \in [0, \infty]$, the *integral* of a function $f : E \longrightarrow \mathbb{R}$ with respect to μ is computed by a prescription that contains the following ingredients. First, one has to choose a partition $\mathcal{P}$ of the space E into subsets $\Gamma \in \mathcal{B}$. Second, having chosen $\mathcal{P}$, one has to select for each $\Gamma \in \mathcal{P}$ a *typical* value a_Γ of f on Γ. Third, given both the partition $\mathcal{P}$ and the selection

$$\Gamma \in \mathcal{P} \longmapsto a_\Gamma \in f(\Gamma) \equiv \operatorname{Range}(f \restriction \Gamma),$$

one forms the sum

D. W. Stroock, *Essentials of Integration Theory for Analysis*, Graduate Texts in Mathematics 262, https://doi.org/10.1007/978-3-030-58478-8_2

$$\sum_{\Gamma \in \mathcal{P}} a_\Gamma \, \mu(\Gamma). \tag{2.1.1}$$

Finally, using a limit procedure if necessary, one removes the ambiguity (inherent in the notion of *typical*) by choosing the partitions $\mathcal{P}$ in such a way that the restriction of f to each Γ is increasingly close to a constant.

Obviously, even if one ignores all questions of convergence, the only way in which one can make sense out of (2.1.1) is to restrict oneself to partitions $\mathcal{P}$ that are either finite or, at worst, countable. Hence, in general, the final limit procedure will be essential. Be that as it may, when E is itself countable and $\{x\} \in \mathcal{B}$ for every $x \in E$, there is an *obvious* way to avoid the limit step; namely, one chooses $\mathcal{P} = \big\{\{x\} : x \in E\big\}$ and takes

$$\sum_{x \in E} f(x)\mu(\{x\}) \tag{2.1.2}$$

to be the *integral.* (I continue, for the present, to systematically ignore all problems arising from questions of convergence.) Clearly, this is the idea on which Riemann based his theory of integration. On the other hand, Riemann's is not the only *obvious* way to proceed, even in the case of countable spaces E. For example, again assume that E is countable, and take $\mathcal{B}$ to be the set of all subsets of E. Given $f : E \longrightarrow \mathbb{R}$, set $\Gamma(a) = \{x \in E : f(x) = a\} \in \mathcal{B}$ for every a in the range of f. Then Lebesgue would say that

$$\sum_{a \in \mathbf{Range}(f)} a \, \mu\big(\Gamma(a)\big) \tag{2.1.3}$$

is an equally *obvious* candidate for the *integral* of f.

In order to reconcile these two *obvious* definitions, one has to examine the assignment $\Gamma \in \mathcal{B} \longmapsto \mu(\Gamma) \in [0, \infty]$ of *measure.* Indeed, even if E is countable and $\mathcal{B}$ contains every subset of E, (2.1.2) and (2.1.3) give the same answer only if one knows that, for any countable collection $\big\{\Gamma_n\big\} \subseteq \mathcal{B}$,

$$\mu\left(\bigcup_n \Gamma_n\right) = \sum_n \mu(\Gamma_n) \quad \text{when } \Gamma_m \cap \Gamma_n = \emptyset \text{ for } m \neq n. \tag{2.1.4}$$

The property in (2.1.4) is called **countable additivity**, and, as will become increasingly apparent, it is crucial. When E is countable, (2.1.4) is equivalent to taking

$$\mu(\Gamma) = \sum_{x \in \Gamma} \mu(\{x\}), \quad \Gamma \subseteq E.$$

However, when E is uncountable, the property in (2.1.4) becomes highly non-trivial. In fact, it is unquestionably Lebesgue's most significant achievement to have shown that there are non-trivial *assignments of measure* that possess this property.

Having compared Lebesgue's ideas to Riemann's in the countable setting, I close this introduction to Lebesgue's theory with a few words about the same comparison for uncountable spaces. For this purpose, suppose that $E = [0,1]$ and, without worrying about exactly which subsets of E are included in $\mathcal{B}$, assume that $\Gamma \in \mathcal{B} \longmapsto \mu(\Gamma) \in [0,1]$ is a mapping that satisfies (2.1.4).

Now let $f : [0,1] \longrightarrow \mathbb{R}$ be given. In order to integrate f, Riemann says that one should divide up $[0,1]$ into small intervals, choose a representative value of f from each interval, form the associated Riemann sum, and then take the limit as the mesh size of the division tends to 0. As we know, his procedure works beautifully as long as the function f respects the topology of the real line: that is, as long as f is sufficiently continuous. However, Riemann's procedure is doomed to failure when f does not respect the topology of $\mathbb{R}$. The problem is, of course, that Riemann's partitioning procedure is tied to the topology of the reals and is therefore too rigid to accommodate functions that pay little or no attention to that topology. To get around this problem, Lebesgue tailors his partitioning procedure to the particular function f under consideration. Thus, for a given function f, Lebesgue might consider the sequence of partitions $\mathcal{P}_n$, $n \in \mathbb{N}$, consisting of the sets

$$\Gamma_{n,k} = \left\{x \in E : f(x) \in \left[k2^{-n}, (k+1)2^{-n}\right)\right\}, \quad k \in \mathbb{Z}.$$

Obviously, all values of f restricted to any one of the $\Gamma_{n,k}$'s can differ from one another by at most $\frac{1}{2^n}$. Hence, assuming that $\Gamma_{n,k} \in \mathcal{B}$ for every $n \in \mathbb{N}$ and $k \in \mathbb{Z}$ and ignoring convergence problems,

$$\lim_{n\to\infty} \sum_{k\in\mathbb{Z}} \frac{k}{2^n}\, \mu\big(\Gamma_{n,k}\big)$$

simply must be the *integral* of f!

When one hears Lebesgue's ideas for the first time, one may well wonder what there is left to be done. On the other hand, after a little reflection, some doubts begin to emerge. For example, what is so sacrosanct about the partitioning suggested in the preceding paragraph and, for instance, why should one not have done the same thing relative to powers of 3 rather than 2? The answer is, of course, that there is nothing to recommend 2 over 3 and that it should make no difference which of them is used. Thus, one has to check that it really does not matter, and, once again, the verification entails repeated application of countable additivity. In fact, it will become increasingly evident that Lebesgue's entire program rests on countable additivity.

2.1.2 Measures and Measure Spaces

With the preceding discussion in mind, the following should seem quite natural.

Given a non-empty set E, the **power set** $\mathcal{P}(E)$ (denoted by 2^E by some authors) is the collection of all subsets of E, and a **σ-algebra** $\mathcal{B}$ over E is any subset of $\mathcal{P}(E)$ with the properties that $E \in \mathcal{B}$, $\mathcal{B}$ is **closed under countable unions** (i.e., $\{B_n : n \geq 1\} \subseteq \mathcal{B} \implies \bigcup_{n=1}^{\infty} B_n \in \mathcal{B}$), and $\mathcal{B}$ is **closed under complementation** (i.e., $B \in \mathcal{B} \implies B^{\complement} = E \setminus B \in \mathcal{B}$). Observe that if $\{B_n : n \geq 1\} \subseteq \mathcal{B}$, then

$$\bigcap_{n=1}^{\infty} B_n = \left(\bigcup_{n=1}^{\infty} B_n^{\complement}\right)^{\complement} \in \mathcal{B},$$

and so $\mathcal{B}$ is also **closed under countable intersections**. Given E and a σ-algebra $\mathcal{B}$ of its subsets, the pair $(E, \mathcal{B})$ is called a **measurable space**. Finally, if $(E, \mathcal{B})$ and $(E', \mathcal{B}')$ are measurable spaces, then a map $\Phi : E \longrightarrow E'$ is said to be **measurable** if (cf. Exercise 2.1.3 below) $\Phi^{-1}(B') \in \mathcal{B}$ for every $B' \in \mathcal{B}'$. Notice the analogy between the definitions of measurability and continuity. In particular, it is clear that if Φ is a measurable map on $(E_1, \mathcal{B}_1)$ into $(E_2, \mathcal{B}_2)$ and Ψ is a measurable map on $(E_2, \mathcal{B}_2)$ into $(E_3, \mathcal{B}_3)$, then $\Psi \circ \Phi$ is a measurable map on $(E_1, \mathcal{B}_1)$ into $(E_3, \mathcal{B}_3)$.

Obviously both $\{\emptyset, E\}$ and $\mathcal{P}(E)$ are σ-algebras over E. In fact, they are, respectively, the smallest and largest σ-algebras over E. More generally, given any[1] $\mathcal{C} \subseteq \mathcal{P}(E)$, there is a smallest σ-algebra over E, denoted by $\sigma(\mathcal{C})$ and known as the σ-algebra **generated** by $\mathcal{C}$, which contains $\mathcal{C}$. To construct $\sigma(\mathcal{C})$, note that there is at least one, namely $\mathcal{P}(E)$, σ-algebra containing $\mathcal{C}$, and check that the intersection of all the σ-algebras containing $\mathcal{C}$ is again a σ-algebra that contains $\mathcal{C}$. When E is a topological space, the σ-algebra generated by its open subsets is called the **Borel σ-algebra** and is denoted by $\mathcal{B}_E$.

Given a σ-algebra $\mathcal{B}$ over E, the reason why the pair $(E, \mathcal{B})$ is called a measurable space is that it is the natural structure on which measures are defined. Namely, a **measure** on $(E, \mathcal{B})$ is a map $\mu : \mathcal{B} \longrightarrow [0, \infty]$ that assigns 0 to $\emptyset$ and is countably additive in the sense that (2.1.4) holds whenever $\{\Gamma_n\}$ is a sequence of mutually disjoint elements of $\mathcal{B}$. If μ is a measure on $(E, \mathcal{B})$, then the triple $(E, \mathcal{B}, \mu)$ is called a **measure space**. A measure μ on $(E, \mathcal{B})$ is said to be **finite** if $\mu(E) < \infty$, and it is said to be a **probability measure** if $\mu(E) = 1$, in which case $(E, \mathcal{B}, \mu)$ is called a **probability space**.

Note that if $B,\, C \in \mathcal{B}$, then $B \cup C = B \cup \big(C \setminus (B \cap C)\big)$ and therefore

$$\mu(B) \leq \mu(B) + \mu(C \setminus B) = \mu(C) \quad \text{for all } B,\, C \in \mathcal{B} \text{ with } B \subseteq C. \tag{2.1.5}$$

In addition, because $C = (B \cap C) \cup \big(C \setminus (B \cap C)\big)$ and $B \cup C = B \cup \big(C \setminus (B \cap C)\big)$,

[1] Even if $E = \mathbb{R}^N$, the elements of $\mathcal{C}$ need not be rectangles.

$$\mu(B) + \mu(C) = \mu(B) + \mu\big(B \cap C\big) + \mu\big(C \setminus (B \cap C)\big) = \mu(B \cup C) + \mu(B \cap C),$$

and so

$$\begin{gathered} \mu(B \cup C) = \mu(C) + \mu(B) - \mu(B \cap C) \\ \text{for all } B,\, C \in \mathcal{B} \text{ with } \mu(B \cap C) < \infty \end{gathered} \tag{2.1.6}$$

and

$$\begin{gathered} \mu(C \setminus B) = \mu(C) - \mu(B) \\ \text{for all } B, C \in \mathcal{B} \text{ with } B \subseteq C \text{ and } \mu(B) < \infty. \end{gathered} \tag{2.1.7}$$

Finally, μ is **countably subadditive** in the sense that

$$\mu\left(\bigcup_{n=1}^{\infty} B_n\right) \leq \sum_{n=1}^{\infty} \mu(B_n) \quad \text{for any } \{B_n : n \geq 1\} \subseteq \mathcal{B}. \tag{2.1.8}$$

To check this, set $A_1 = B_1$ and $A_{n+1} = B_{n+1} \setminus \bigcup_{m=1}^{n} B_m$, note that the A_n's are mutually disjoint elements of $\mathcal{B}$ whose union is the same as that of the B_n's, and apply (2.1.4) and (2.1.5) to conclude that

$$\mu\left(\bigcup_{n=1}^{\infty} B_n\right) = \mu\left(\bigcup_{n=1}^{\infty} A_n\right) = \sum_{n=1}^{\infty} \mu(A_n) \leq \sum_{n=1}^{\infty} \mu(B_n).$$

As a consequence, the countable union of B_n's with μ-measure 0 again has μ-measure 0. More generally,

$$\mu\left(\bigcup_{n=1}^{\infty} B_n\right) = \sum_{n=1}^{\infty} \mu(B_n) \tag{2.1.9}$$

for any $\{B_n : n \geq 1\} \subseteq \mathcal{B}$ with $\mu(B_m \cap B_n) = 0$ when $m \neq n$.

To see this, take $C = \bigcup\{B_m \cap B_n : m \neq n\}$, use the preceding to see that $\mu(C) = 0$, set $B_n' = B_n \setminus C$, and apply (2.1.4) to $\{B_n' : n \geq 1\}$. Another important property of measures is that they are continuous under non-decreasing limits. To explain this property, say that $\{B_n : n \geq 1\}$ **increases** to B and write $B_n \nearrow B$ if $B_{n+1} \supseteq B_n$ for all $n \in \mathbb{Z}^+$ and $B = \bigcup_{n=1}^{\infty} B_n$. Then

$$\{B_n : n \geq 1\} \subseteq \mathcal{B} \text{ and } B_n \nearrow B \implies \mu(B_n) \nearrow \mu(B). \tag{2.1.10}$$

To check this, set $A_1 = B_1$ and $A_{n+1} = B_{n+1} \setminus B_n$, note that the A_n's are mutually disjoint and $B_n = \bigcup_{m=1}^{n} A_m$, and conclude that

$$\mu(B_n) = \sum_{m=1}^{n} \mu(A_m) \nearrow \sum_{m=1}^{\infty} \mu(A_m) = \mu(B).$$

Next say that $\{B_n : n \geq 1\}$ **decreases** to B and write $B_n \searrow B$ if $B_{n+1} \subseteq B_n$ for each $n \in \mathbb{Z}^+$ and $B = \bigcap_{n=1}^{\infty} B_n$. Obviously, $B_n \searrow B$ if and only if $B_1 \setminus B_n \nearrow B_1 \setminus B$. Hence, by combining (2.1.10) with (2.1.7), one finds that

$$\{B_n : n \geq 1\} \subseteq \mathcal{B},\ B_n \searrow B, \text{ and } \mu(B_1) < \infty \implies \mu(B_n) \searrow \mu(B). \quad (2.1.11)$$

To see that the condition $\mu(B_1) < \infty$ cannot be dispensed with in general, define μ on $\big(\mathbb{Z}^+, \mathcal{P}(\mathbb{Z}^+)\big)$ to be the counting measure (i.e., $\mu(B) = \text{card}(B)$ for $B \subseteq \mathbb{Z}^+$), and take $B_m = \{n \in \mathbb{Z}^+ : n \geq m\}$. Clearly $B_m \searrow \emptyset$, and yet $\mu(B_m) = \infty$ for all m.

Very often one encounters a situation in which two measures agree on a collection of sets and one wants to know that they agree on the σ-algebra generated by those sets. To handle such a situation, the following concepts are sometimes useful. A collection $\mathcal{C} \subseteq \mathcal{P}(E)$ is called a **Π-system** if it is closed under finite intersections. Given a Π-system $\mathcal{C}$, it is important to know what additional properties a Π-system must possess in order to be a σ-algebra. For this reason one introduces a notion that complements that of a Π-system. Namely, say that $\mathcal{H} \subseteq \mathcal{P}(E)$ is a **Λ-system over** E if

(**a**) $E \in \mathcal{H}$,
(**b**) $\Gamma, \Gamma' \in \mathcal{H}$ and $\Gamma \cap \Gamma' = \emptyset \implies \Gamma \cup \Gamma' \in \mathcal{H}$,
(**c**) $\Gamma, \Gamma' \in \mathcal{H}$ and $\Gamma \subseteq \Gamma' \implies \Gamma' \setminus \Gamma \in \mathcal{H}$,
(**d**) $\{\Gamma_n : n \geq 1\} \subseteq \mathcal{H}$ and $\Gamma_n \nearrow \Gamma \implies \Gamma \in \mathcal{H}$.

Notice that the collection of sets on which two finite measures agree satisfies (**b**), (**c**), and (**d**). Hence, if they agree on E, then they agree on a Λ-system.[2]

Lemma 2.1.1 *The intersection of an arbitrary collection of Π-systems or of Λ-systems is again a Π-system or a Λ-system. Moreover, $\mathcal{B} \subseteq \mathcal{P}(E)$ is a σ-algebra over E if and only if it is both a Π-system and a Λ-system over E. Finally, if $\mathcal{C} \subseteq \mathcal{P}(E)$ is a Π-system, then $\sigma(\mathcal{C})$ is the smallest Λ-system over E containing $\mathcal{C}$.*

Proof. The first assertion requires no comment. To prove the second one, it suffices to prove that if $\mathcal{B}$ is both a Π-system and a Λ-system over E, then it is a σ-algebra over E. To this end, first note that $A^{\complement} = E \setminus A \in \mathcal{B}$ for every $A \in \mathcal{B}$ and therefore that $\mathcal{B}$ is closed under complementation. Second, if $\Gamma_1, \Gamma_2 \in \mathcal{B}$, then $\Gamma_1 \cup \Gamma_2 = \Gamma_1 \cup (\Gamma_2 \setminus \Gamma_3)$ where $\Gamma_3 = \Gamma_1 \cap \Gamma_2$. Hence $\mathcal{B}$ is closed under finite unions. Finally, if $\{\Gamma_n : n \geq 1\} \subseteq \mathcal{B}$, set $A_n = \bigcup_1^n \Gamma_m$ for $n \geq 1$. Then $\{A_n : n \geq 1\} \subseteq \mathcal{B}$ and $A_n \nearrow \bigcup_1^\infty \Gamma_m$. Hence $\bigcup_1^\infty \Gamma_m \in \mathcal{B}$, and so $\mathcal{B}$ is a σ-algebra.

To prove the final assertion, let $\mathcal{C}$ be a Π-system and $\mathcal{H}$ the smallest Λ-system over E containing $\mathcal{C}$. Clearly $\sigma(\mathcal{C}) \supseteq \mathcal{H}$, and so all that we have to do is show that $\mathcal{H}$ is Π-system over E. To do this, first set

[2] I learned these ideas from E. B. Dynkin's treatise on Markov processes. In fact, the Λ- and Π-system scheme is often attributed to Dynkin, who certainly deserves the credit for its exploitation by a whole generation of probabilists. On the other hand, Richard Gill has informed me that, according to A.J. Lenstra, their origins go back to W. Sierpiński's article *Un théorème général sur les familles d'ensembles*, which appeared in *Fund. Math.* 12 (1928), pp. 206–210.

$$\mathcal{H}_1 = \{\Gamma \subseteq E : \Gamma \cap \Delta \in \mathcal{H} \text{ for all } \Delta \in \mathcal{C}\}.$$

It is then easy to check that $\mathcal{H}_1$ is a Λ -system over E. Moreover, since $\mathcal{C}$ is a Π-system, $\mathcal{C} \subseteq \mathcal{H}_1$, and therefore $\mathcal{H} \subseteq \mathcal{H}_1$. In other words, $\Gamma \cap \Delta \in \mathcal{H}$ for all $\Gamma \in \mathcal{H}$ and $\Delta \in \mathcal{C}$. Next set

$$\mathcal{H}_2 = \{\Gamma \subseteq E : \Gamma \cap \Delta \in \mathcal{H} \text{ for all } \Delta \in \mathcal{H}\}.$$

Again it is clear that $\mathcal{H}_2$ is a Λ-system. Also, by the preceding, $\mathcal{C} \subseteq \mathcal{H}_2$. Hence $\mathcal{H} \subseteq \mathcal{H}_2$, and so $\mathcal{H}$ is a Π-system. □

As a consequence of Lemma 2.1.1 and the remark preceding it, one has the following important result.

Theorem 2.1.2 *Let $(E, \mathcal{B})$ be a measurable space and $\mathcal{C}$ is Π-system that generates $\mathcal{B}$. If μ and ν are a pair of finite measures on $(E, \mathcal{B})$ and $\mu(\Gamma) = \nu(\Gamma)$ for all $\Gamma \in \{E\} \cup \mathcal{C}$, then $\mu = \nu$.*

Proof. As was remarked above, additivity, (2.1.7), and (2.1.10) imply that $\mathcal{H} = \{\Gamma \in \mathcal{B} : \mu(\Gamma) = \nu(\Gamma)\}$ is a Λ-system. Hence, since $\mathcal{B}$ is the smallest σ-algebra containing $\mathcal{C}$, it follows from Lemma 2.1.1 that $\mathcal{H} = \mathcal{B}$. □

A measure space $(E, \mathcal{B}, \mu)$ is said to be **complete** if $\Gamma \in \mathcal{B}$ whenever there exist C, $D \subset \mathcal{B}$ such that $C \subseteq \Gamma \subseteq D$ with $\mu(D \setminus C) = 0$. The following simple lemma shows that every measure space can be "completed."

Lemma 2.1.3 *Given a measure space $(E, \mathcal{B}, \mu)$, define $\overline{\mathcal{B}}^{\mu}$ to be the set of $\Gamma \subseteq E$ for which there exist C, $D \in \mathcal{B}$ satisfying $C \subseteq \Gamma \subseteq D$ and $\mu(D \setminus C) = 0$. Then $\overline{\mathcal{B}}^{\mu}$ is a σ-algebra over E and there is a unique extension $\bar{\mu}$ of μ to $\overline{\mathcal{B}}^{\mu}$ as a measure on $(E, \overline{\mathcal{B}}^{\mu})$. Furthermore, $(E, \overline{\mathcal{B}}^{\mu}, \bar{\mu})$ is a complete measure space, and if $(E, \mathcal{B}, \mu)$ is already complete, then $\overline{\mathcal{B}}^{\mu} = \mathcal{B}$.*

Proof. To see that $\overline{\mathcal{B}}^{\mu}$ is a σ-algebra, suppose that $\{\Gamma_n : n \geq 1\} \subseteq \overline{\mathcal{B}}^{\mu}$, and choose $\{C_n : n \geq 1\} \cup \{D_n : n \geq 1\} \subseteq \mathcal{B}$ accordingly. Then $C = \bigcup_{n=1}^{\infty} C_n$ and $D = \bigcup_{n=1}^{\infty} D_n$ are elements of $\mathcal{B}$, $C \subseteq \bigcup_{n=1}^{\infty} \Gamma_n \subseteq D$, and $\mu(D \setminus C) = 0$. Also, if $\Gamma \in \overline{\mathcal{B}}^{\mu}$ and C and D are associated elements of $\mathcal{B}$, then $D^{\complement}$, $C^{\complement} \in \mathcal{B}$, $D^{\complement} \subseteq \Gamma^{\complement} \subseteq C^{\complement}$, and $\mu(C^{\complement} \setminus D^{\complement}) = \mu(D \setminus C) = 0$.

Next, given $\Gamma \in \overline{\mathcal{B}}^{\mu}$, suppose that C, C', D, $D' \in \mathcal{B}$ satisfy $C \cup C' \subseteq \Gamma \subseteq D \cap D'$ and $\mu(D \setminus C) = 0 = \mu(D' \setminus C')$. Then $\mu(D' \setminus D) \leq \mu(D' \setminus C') = 0$ and so $\mu(D') \leq \mu(D) + \mu(D' \setminus D) = \mu(D)$. Similarly, $\mu(D) \leq \mu(D')$, which means that $\mu(D) = \mu(D')$, and, because $\mu(C) = \mu(D)$ and $\mu(C') = \mu(D')$, it follows that μ assigns the same measure to C, C', D and D'. Hence, we can unambiguously define $\bar{\mu}(\Gamma) = \mu(C) = \mu(D)$ when $\Gamma \in \overline{\mathcal{B}}^{\mu}$ and C, $D \in \mathcal{B}$ satisfy $C \subseteq \Gamma \subseteq D$ with $\mu(D \setminus C) = 0$. Furthermore, if $\{\Gamma_n : n \geq 1\}$ are mutually disjoint elements of $\overline{\mathcal{B}}^{\mu}$ and $\{C_n : n \geq 1\} \cup \{D_n : n \geq 1\} \subseteq \mathcal{B}$ are chosen accordingly, then the C_n's are mutually disjoint, and so

$$\bar{\mu}\left(\bigcup_{n=1}^{\infty}\Gamma_n\right) = \mu\left(\bigcup_{n=1}^{\infty}C_n\right) = \sum_{n=1}^{\infty}\mu(C_n) = \sum_{n=1}^{\infty}\bar{\mu}(\Gamma_n).$$

Hence, $\bar{\mu}$ is a measure on $(E, \overline{\mathcal{B}}^{\mu})$ whose restriction to $\mathcal{B}$ coincides with μ.

Finally, suppose that ν is any measure on $(E, \overline{\mathcal{B}}^{\mu})$ that extends μ. If Γ is a subset of E for which there exist Γ', $\Gamma'' \in \overline{\mathcal{B}}^{\mu}$ satisfying $\Gamma' \subseteq \Gamma \subseteq \Gamma''$ and $\nu(\Gamma'' \setminus \Gamma') = 0$, there exist C, $D \in \mathcal{B}$ satisfying $C \subseteq \Gamma'$ and $\Gamma'' \subseteq D$ such that $\nu(D \setminus \Gamma'') = 0 = \nu(\Gamma' \setminus C)$ and therefore

$$\mu(D \setminus C) = \nu(D \setminus C) = \nu(D \setminus \Gamma'') + \nu(\Gamma'' \setminus \Gamma') + \nu(\Gamma' \setminus C) = 0.$$

Hence, $\Gamma \in \overline{\mathcal{B}}^{\mu}$ and $\nu(\Gamma) = \mu(C) = \bar{\mu}(\Gamma)$. Thus, we now know that $\bar{\mu}$ is the only extension of μ as a measure on $(E, \overline{\mathcal{B}}^{\mu})$ and that $\overline{\mathcal{B}}^{\mu} = \mathcal{B}$ if $(E, \mathcal{B}, \mu)$ is complete. □

The measure space $(E, \overline{\mathcal{B}}^{\mu}, \bar{\mu})$ is called the **completion** of $(E, \mathcal{B}, \mu)$, and Lemma 2.1.3 says that every measure space has a unique completion. Elements of $\overline{\mathcal{B}}^{\mu}$ are said to be **μ-measurable**.

Given a topological space E, use $\mathfrak{G}(E)$ to denote the class of all open subsets of E and $\mathfrak{G}_\delta(E)$ the class of subsets that can be written as the countable intersection of open subsets. Analogously, $\mathfrak{F}(E)$ and $\mathfrak{F}_\sigma(E)$ will denote, respectively, the class of all closed subsets of E and the class of subsets that can be written as the countable union of closed subsets. Clearly $B \in \mathfrak{G}(E) \iff B^{\complement} \in \mathfrak{F}(E)$, $B \in \mathfrak{G}_\delta(E) \iff B^{\complement} \in \mathfrak{F}_\sigma(E)$, and $\mathfrak{G}_\delta(E) \cup \mathfrak{F}_\sigma(E) \subseteq \mathcal{B}_E$. Moreover, when the topology of E admits a metric ρ, it is easy to check that $\mathfrak{F}(E) \subseteq \mathfrak{G}_\delta(E)$ and therefore $\mathfrak{G}(E) \subseteq \mathfrak{F}_\sigma(E)$. Indeed, if $F \in \mathfrak{F}(E)$, then

$$\mathfrak{G}(E) \ni \{x : \rho(x, F) < \tfrac{1}{n}\} \searrow F \quad \text{as } n \to \infty.$$

Finally, a measure μ on $(E, \mathcal{B}_E)$ is called a **Borel measure** on E, and if μ is a Borel measure on E, a set $\Gamma \subseteq E$ is said to be **μ-regular** when, for each $\epsilon > 0$, there exist $F \in \mathfrak{F}(E)$ and $G \in \mathfrak{G}(E)$ such that $F \subseteq \Gamma \subseteq G$ and $\mu(G \setminus F) < \epsilon$. A Borel measure μ is said to be **regular** if every element of $\mathcal{B}_E$ is μ-regular.

Theorem 2.1.4 *Let E be a topological space and μ a Borel measure on E. If $\Gamma \subseteq E$ is μ-regular, then there exist $C \in \mathfrak{F}_\sigma(E)$ and $D \in \mathfrak{G}_\delta(E)$ for which $C \subseteq \Gamma \subseteq D$ and $\mu(D \setminus C) = 0$. In particular, $\Gamma \in \overline{\mathcal{B}_E}^{\mu}$ if Γ is μ-regular. Conversely, if μ is regular, then every element of $\overline{\mathcal{B}_E}^{\mu}$ is μ-regular. Moreover, if the topology on E admits a metric space and μ is a finite Borel measure on E, then μ is regular. (See Exercise 2.1.4 for a small extension.)*

Proof. To prove the first part, suppose that $\Gamma \subseteq E$ is μ-regular. Then, for each $n \geq 1$, there exist $F_n \in \mathfrak{F}(E)$ and $G_n \in \mathfrak{G}(E)$ such that $F_n \subseteq \Gamma \subseteq G_n$ and $\mu(G_n \setminus F_n) < \frac{1}{n}$. Thus, if $C = \bigcup_{n=1}^{\infty} F_n$ and $D = \bigcap_{n=1}^{\infty} G_n$, then $C \in \mathfrak{F}_\sigma(E)$, $D \in \mathfrak{G}_\delta(E)$, $C \subseteq \Gamma \subseteq D$, and, because $D \setminus C \subseteq G_n \setminus F_n$ for all

n, $\mu(D \setminus C) = 0$. Obviously, this means that $\Gamma \in \overline{\mathcal{B}_E}^{\mu}$. Conversely, if μ is regular and $\Gamma \in \overline{\mathcal{B}_E}^{\mu}$, then there exist Γ', $\Gamma'' \in \mathcal{B}_E$ for which $\Gamma' \subseteq \Gamma \subseteq \Gamma''$, $\mu(\Gamma''\setminus\Gamma') = 0$. By regularity, for each $\epsilon > 0$, there exist $F \in \mathfrak{F}(E)$, $G \in \mathfrak{G}(E)$ such that $F \subseteq \Gamma'$, $\Gamma'' \subseteq G$, and $\mu(G\setminus\Gamma'')\vee\mu(\Gamma'\setminus F) < \frac{\epsilon}{2}$. Hence, $F \subseteq \Gamma \subseteq G$ and

$$\mu(G \setminus F) = \mu(G \setminus \Gamma'') + \mu(\Gamma'' \setminus \Gamma') + \mu(\Gamma' \setminus F) < \epsilon,$$

and so Γ is μ-regular.

Now suppose the E admits a metric and that μ is finite, and let $\mathcal{R}$ be the collection of $B \in \mathcal{B}_E$ that are μ-regular. If we show that $\mathcal{R}$ is a σ-algebra that contains $\mathfrak{G}(E)$, then we will know that $\mathcal{R} = \mathcal{B}_E$ and therefore that μ is regular. Obviously $\mathcal{R}$ is closed under complementation. Next, suppose that $\{B_n : n \geq 1\} \subseteq \mathcal{R}$, and set $B = \bigcup_{n=1}^{\infty} B_n$. Given $\epsilon > 0$, for each n choose $F_n \in \mathfrak{F}(E)$ and $G_n \in \mathfrak{G}(E)$ such that $F_n \subseteq B_n \subseteq G_n$ and $\mu(G_n \setminus F_n) < 2^{-n-1}\epsilon$. Then $\mathfrak{G}(E) \ni G = \bigcup_{m=1}^{\infty} G_m \supseteq B$, $\mathfrak{F}_\sigma(E) \ni C = \bigcup_{m=1}^{\infty} F_m \subseteq B$, and

$$\mu(G \setminus C) \leq \mu\left(\bigcup_{m=1}^{\infty} (G_m \setminus F_m)\right) \leq \sum_{m=1}^{\infty} \mu(G_m \setminus F_m) < \frac{\epsilon}{2}.$$

Finally, because $\mu(C) < \infty$ and $C \setminus \bigcup_{m=1}^{n} F_m \searrow \emptyset$, (2.1.11) allows us to choose an $n \in \mathbb{Z}^+$ for which $\mu(C \setminus F) < \frac{\epsilon}{2}$ when $F = \bigcup_{m=1}^{n} F_m \in \mathfrak{F}(E)$. Hence, since $\mu(G \setminus F) = \mu(G \setminus C) + \mu(C \setminus F) < \epsilon$, we know that $B \in \mathcal{R}$ and therefore that $\mathcal{R}$ is a σ-algebra.

To complete the proof, it remains to show that $\mathfrak{G}(E) \subseteq \mathcal{R}$, and clearly this comes down to showing that for each open G and $\epsilon > 0$ there is a closed $F \subseteq G$ for which $\mu(G \setminus F) < \epsilon$. But, because E has a metric topology, we know that $\mathfrak{G}(E) \subseteq \mathfrak{F}_\sigma(E)$. Hence, if G is open, then there exists a non-decreasing sequence $\{F_n : n \geq 1\} \subseteq \mathfrak{F}(E)$ such that $F_n \nearrow G$, which, because $\mu(G) < \infty$, means that $\mu(G \setminus F_n) \searrow 0$. Thus, for any $\epsilon > 0$, there is an n for which $\mu(G \setminus F_n) < \epsilon$. □

2.1.3 Exercises for § 2.1

Exercise 2.1.1 The decomposition of the properties of a σ-algebra in terms of Π-systems and Λ-systems is not the traditional one. Instead, most of the early books on measure theory used algebras instead of Π-systems as the standard source of generating sets. An **algebra** over E is a collection $\mathcal{A} \subseteq \mathcal{P}(E)$ that contains E and is closed under finite unions and complementation. If one starts with an algebra $\mathcal{A}$, then the complementary notion is that of a **monotone class**: $\mathcal{M}$ is said to be a monotone class if $\Gamma \in \mathcal{M}$ whenever there exists $\{\Gamma_n : n \geq 1\} \subseteq \mathcal{M}$ such that $\Gamma_n \nearrow \Gamma$. Show that $\mathcal{B}$ is a σ-algebra over E if and only if it is both an algebra over E and a monotone class. In addition, show that if $\mathcal{A}$ is an algebra over E, then $\sigma(\mathcal{A})$ is the smallest monotone class containing $\mathcal{A}$.

Exercise 2.1.2 Given a measurable space $(E, \mathcal{B})$ and $\emptyset \neq E' \subseteq E$, show that $\mathcal{B}[E'] \equiv \{B \cap E' : B \in \mathcal{B}\}$ is a σ-algebra over E'. Further, show that if E is a topological space, then $\mathcal{B}_E[E'] = \mathcal{B}_{E'}$ when E' is given the topology that it inherits from E. Finally, if $\emptyset \neq E' \in \mathcal{B}$, show that $\mathcal{B}[E'] \subseteq \mathcal{B}$ and that the restriction to $\mathcal{B}[E']$ of any measure on $(E, \mathcal{B})$ is a measure on $\big(E', \mathcal{B}[E']\big)$. In particular, if E is a topological space and μ is a Borel measure on E, show that $\mu \restriction \mathcal{B}_{E'}$ is a Borel measure on E' and that it is regular if μ is regular.

Exercise 2.1.3 Given a map $\Phi : E \longrightarrow E'$, define $\Phi(\Gamma) = \{\Phi(x) : x \in \Gamma\}$ for $\Gamma \subseteq E$ and $\Phi^{-1}(\Gamma') = \{x \in E : \Phi(x) \in \Gamma'\}$ for $\Gamma' \subseteq E'$.

(i) Show that Φ and Φ^{-1} preserve unions in the sense that $\Phi\left(\bigcup_\alpha B_\alpha\right) = \bigcup_\alpha \Phi(B_\alpha)$ and $\Phi^{-1}\left(\bigcup_\alpha B'_\alpha\right) = \bigcup_\alpha \Phi^{-1}(B'_\alpha)$. In addition, show that Φ^{-1} preserves differences in the sense that $\Phi^{-1}(B' \setminus A') = \Phi^{-1}(B') \setminus \Phi^{-1}(A')$. On the other hand, show that Φ need not preserve differences, but that it will if it is one-to-one.

(ii) Suppose that $\mathcal{B}$ and $\mathcal{B}'$ are σ-algebras over, respectively, E and E' and that $\Phi : E \longrightarrow E'$. If $\mathcal{B}' = \sigma(\mathcal{C}')$ and $\Phi^{-1}(C') \in \mathcal{B}$ for every $C' \in \mathcal{C}'$, show that Φ is measurable. In particular, if E and E' are topological spaces and Φ is continuous, show that Φ is measurable as a map from $(E, \mathcal{B}_E)$ to $(E', \mathcal{B}_{E'})$. Similarly, if Φ is one-to-one and $\mathcal{B} = \sigma(\mathcal{C})$, show that $\Phi(B) \in \mathcal{B}'$ for all $B \in \mathcal{B}$ if $\Phi(C) \in \mathcal{B}'$ for all $C \in \mathcal{C} \cup \{E\}$.

(iii) Now suppose that μ is a measure on $(E, \mathcal{B})$ and that Φ is a measurable map from $(E, \mathcal{B})$ into $(E', \mathcal{B}')$. Define $\mu'(B') = \mu(\Phi^{-1}(B'))$ for $B' \in \mathcal{B}'$, and show that μ' is a measure on $(E', \mathcal{B}')$. This measure μ' is called the **pushforward** or **image** of μ under Φ and is denoted by either $\Phi_*\mu$ or $\mu \circ \Phi^{-1}$. Similarly, if $\Phi : E \longrightarrow E'$ is one-to-one and takes elements of $\mathcal{B}$ to elements of $\mathcal{B}'$ and if μ' is a measure on $(E', \mathcal{B}')$, show that $\Gamma \in \mathcal{B} \longmapsto \mu'\big(\Phi(\Gamma)\big) \in [0, \infty]$ is a measure on $(E, \mathcal{B})$. This measure is the **pullback** of μ' under Φ.

Exercise 2.1.4 Let E be a topological space and μ a Borel measure on E. Show that μ is regular if, for every $\Gamma \in \mathcal{B}_E$ and $\epsilon > 0$, there is an open $G \supseteq \Gamma$ for which $\mu(G \setminus \Gamma) < \epsilon$. In addition, if E is a metric space and there exists a non-decreasing sequence $\{G_n : n \geq 1\}$ of open sets such that $G_n \nearrow E$ and $\mu(G_n) < \infty$ for each $n \in \mathbb{Z}^+$, show that μ is regular.

Exercise 2.1.5 Let $(E, \mathcal{B}, \mu)$ be a finite measure space. Given $n \geq 2$ and $\{\Gamma_m : 1 \leq m \leq n\} \subseteq \mathcal{B}$, use (2.1.6) and induction to show that

$$\mu(\Gamma_1 \cup \cdots \cup \Gamma_n) = -\sum_F (-1)^{\operatorname{card}(F)} \mu(\Gamma_F),$$

where the summation is over non-empty subsets F of $\{1, \ldots, n\}$ and $\Gamma_F \equiv \bigcap_{i \in F} \Gamma_i$. Although this formula is seldom used except in the case $n = 2$, the following is an interesting application of the general result. Let E be the group of permutations of $\{1, \ldots, n\}$, $\mathcal{B} = \mathcal{P}(E)$, and $\mu\big(\{\pi\}\big) = \frac{1}{n!}$ for each $\pi \in E$. Denote by A the set of $\pi \in E$ such that $\pi(i) \neq i$ for any $1 \leq i \leq n$. Then one can interpret $\mu(A)$ as the probability that, when the numbers $1, \dots, n$ are randomly ordered, none of them is placed in the correct position. On the basis of this interpretation, one might suspect that $\mu(A)$ should tend to 0 as $n \to \infty$. However, by direct computation, one can see that this is not the case. Indeed, let Γ_i be the set of $\pi \in E$ for which $\pi(i) = i$. Then $A = (\Gamma_1 \cup \cdots \cup \Gamma_n)^{\complement}$, and therefore

$$\mu(A) = 1 - \mu(\Gamma_1 \cup \cdots \cup \Gamma_n) = 1 + \sum_F (-1)^{\text{card}(F)} \mu(\Gamma_F).$$

Show that $\mu(\Gamma_F) = \frac{(n-m)!}{n!}$ if $\text{card}(F) = m$, and conclude from this that $\mu(A) = \sum_0^n \frac{(-1)^m}{m!} \longrightarrow \frac{1}{e}$ as $n \to \infty$.

Exercise 2.1.6 Given a sequence $\{B_n : n \geq 1\}$ of sets, define their **limit inferior** to be the set

$$\varliminf_{n\to\infty} B_n = \bigcup_{m=1}^{\infty} \bigcap_{n=m}^{\infty} B_n,$$

or, equivalently, the set of $x \in E$ that are in all but finitely many of the B_n's. Also, define their **limit superior** to be the set

$$\varlimsup_{n\to\infty} B_n = \bigcap_{m=1}^{\infty} \bigcup_{n=m}^{\infty} B_n,$$

or, equivalently, the set of x that are in infinitely many of the B_n's. Show that $\varliminf_{n\to\infty} B_n \subseteq \varlimsup_{n\to\infty} B_n$, and say that the sequence $\{B_n : n \geq 1\}$ has a **limit** if equality holds, in which case $\varliminf_{n\to\infty} B_n = \varlimsup_{n\to\infty} B_n$ is said to be the limit $\lim_{n\to\infty} B_n$ of $\{B_n : n \geq 1\}$. Show that if $(E, \mathcal{B}, \mu)$ is a measure space and $\{B_n : n \geq 1\} \subseteq \mathcal{B}$, both $\varliminf_{n\to\infty} B_n \in \mathcal{B}$ and $\varlimsup_{n\to\infty} B_n$ are elements of $\mathcal{B}$. Further, show that

$$\mu\left(\varliminf_{n\to\infty} B_n\right) \leq \varliminf_{n\to\infty} \mu(B_n) \tag{2.1.12}$$

and that

$$\mu\left(\varlimsup_{n\to\infty} B_n\right) \geq \varlimsup_{n\to\infty} \mu(B_n) \quad \text{if } \mu\left(\bigcup_{n=1}^{\infty} B_n\right) < \infty. \tag{2.1.13}$$

Conclude that

$$\begin{aligned} &\lim_{n\to\infty} B_n \text{ exists } \& \ \mu\left(\bigcup_{n=1}^{\infty} B_n\right) < \infty \\ &\implies \mu\left(\lim_{n\to\infty} B_n\right) = \lim_{n\to\infty} \mu(B_n). \end{aligned} \tag{2.1.14}$$

Hint: Note that $\bigcap_{n=m}^{\infty} B_n \nearrow \underline{\lim}_{n\to\infty} B_n$ and $\bigcup_{n=m}^{\infty} B_n \searrow \overline{\lim}_{n\to\infty} B_n$ as $m \to \infty$.

Exercise 2.1.7 Let $(E, \mathcal{B}, \mu)$ be a measure space and $\{B_n : n \geq 1\} \subseteq \mathcal{B}$. Show that

$$\sum_{n=1}^{\infty} \mu(B_n) < \infty \implies \mu\left(\overline{\lim_{n\to\infty}} B_n\right) = 0.$$

This useful observation is usually attributed to E. Borel. More profound is the following converse statement, which is due to F. Cantelli. Assume that μ is a probability measure. Sets $\{B_n : n \geq 1\} \subseteq \mathcal{B}$ are said to be **independent** under μ or **μ-independent** if, for all $n \geq 1$ and choices of $C_m \in \{B_m, B_m^{\complement}\}$, $1 \leq m \leq n$, $\mu(C_1 \cap \cdots \cap C_n) = \mu(C_1) \cdots \mu(C_n)$. Cantelli's result says that if $\{B_n : n \geq 1\} \subseteq \mathcal{B}$ are μ-independent sets, then

$$\sum_{n=1}^{\infty} \mu(B_n) = \infty \implies \mu\left(\overline{\lim_{n\to\infty}} B_n\right) = 1.$$

Thus, for μ-independent sets, $\mu\left(\overline{\lim}_{n\to\infty} B_n\right)$ is either 0 or 1 according to whether $\sum_{n=1}^{\infty} \mu(B_n)$ is finite or infinite, a result that is referred to as the **Borel–Cantelli Lemma**. Give a proof of Cantelli's result. In doing so, the following outline might be helpful.

(**i**) Show that it suffices to prove that $\lim_{N\to\infty} \mu\left(\bigcap_{n=m}^{N} B_n^{\complement}\right) = 0$ for each $m \in \mathbb{Z}^+$.

(**ii**) Show that $1 - x \leq e^{-x}$ for all $x \geq 0$, and use this to check that

$$\mu\left(\bigcap_{n=m}^{N} B_n^{\complement}\right) \leq e^{-\sum_{n=m}^{N} \mu(B_n)} \quad \text{for } N \geq m.$$

Exercise 2.1.8 Given a pair of measures μ and ν on a measurable space $(E, \mathcal{B})$, one says that μ is **absolutely continuous** with respect to ν and writes $\mu \ll \nu$ if, for all $B \in \mathcal{B}$, $\nu(B) = 0 \implies \mu(B) = 0$. Assuming that μ is finite and that $\mu \ll \nu$, show that for each $\epsilon > 0$ there exists a $\delta > 0$ such that $\mu(B) < \epsilon$ whenever $\nu(B) < \delta$. Next, assume that E is a metric space, that both μ and ν are regular Borel measures on E, and that μ is finite. Show that $\mu \ll \nu$ if and only if for every $\epsilon > 0$ there exists a $\delta > 0$ for which $\nu(G) < \delta \implies \mu(G) < \epsilon$ whenever $G \in \mathfrak{G}(E)$.

Exercise 2.1.9 A pair of measures μ and ν on a measurable space $(E, \mathcal{B})$ are said to be **singular** to one another and one writes $\mu \perp \nu$ if there exists a $B \in \mathcal{B}$ such that $\mu(B) = 0 = \nu(B^{\complement})$. In words, μ and ν are singular to one another if they live on disjoint parts of E. Assuming that E is a metric space, that ν is a regular Borel measure on E, and that μ is finite, show that $\mu \perp \nu$ if and only if for every $\delta > 0$ there is a $G \in \mathfrak{G}(E)$ for which $\nu(G) < \delta$ and $\mu(G^{\complement}) = 0$.

2.2 A Construction of Measures

In this section I will first develop a procedure for constructing measures and will then apply that procedure to three important examples.

2.2.1 A Construction Procedure

Suppose that $\mathfrak{R}$ is a collection of compact subsets I of a metric space (E,ρ) and that V is a map from $\mathfrak{R}$ to $[0,\infty)$ that satisfy the following conditions:

(1) $\emptyset \in \mathfrak{R}$ and $I,\, I' \in \mathfrak{R} \implies I \cap I' \in \mathfrak{R}$.
(2) $V(\emptyset) = 0$ and $V(I) \le V(J)$ if $I,\, J \in \mathfrak{R}$ and $I \subseteq J$.
(3) For any $J \in \mathfrak{R}$, $n \in \mathbb{Z}^+$, and $\{I_1, \dots, I_n\} \subseteq \mathfrak{R}$, $V(J) \le \sum_{m=1}^n V(I_m)$ if $J \subseteq \bigcup_{m=1}^n I_m$ and $V(J) \ge \sum_{m=1}^n V(I_m)$ if the I_m's are non-overlapping (i.e., their interiors are mutually disjoint) and $J \supseteq \bigcup_{m=1}^n I_m$.
(4) For any $I \in \mathfrak{R}$ and $\epsilon > 0$, there exist $I',\, I'' \in \mathfrak{R}$ such that $I'' \subseteq \mathring{I}$, $I \subseteq \mathring{I}'$, and $V(I') \le V(I'') + \epsilon$.
(5) For any $G \in \mathfrak{G}(E)$, there is a sequence $\{I_n : n \ge 1\}$ of non-overlapping elements of $\mathfrak{R}$ such that $G = \bigcup_{n=1}^\infty I_n$.

An example to keep in mind is that for which $E = \mathbb{R}^N$, $\mathfrak{R}$ is the collection of all closed rectangles in $\mathbb{R}^N$ (one should consider $\emptyset$ to be a rectangle), and $V(I) = \operatorname{vol}(I)$.

The goal of this subsection is to prove that there is a unique Borel measure μ on E such that $\mu(I) = V(I)$ for all $I \in \mathfrak{R}$. Before getting started with the proof, recall the following elementary fact about double sums of non-negative numbers.

Lemma 2.2.1 *If* $\{a_{m,n} : (m,n) \in (\mathbb{Z}^+)^2\} \subseteq [0,\infty)$, *then*

$$\sum_{m=1}^\infty \sum_{n=1}^\infty a_{m,n} = \sum_{(m,n)\in(\mathbb{Z}^+)^2} a_{m,n} = \sum_{n=1}^\infty \sum_{m=1}^\infty a_{m,n}.$$

Proof. For each $M,\, N \in \mathbb{Z}^+$,

$$\left(\sum_{m=1}^\infty \sum_{n=1}^\infty a_{m,n}\right) \wedge \left(\sum_{n=1}^\infty \sum_{m=1}^\infty a_{m,n}\right) \ge \sum_{\{(m,n):\, m\le M \,\&\, n\le N\}} a_{m,n},$$

and therefore

$$\left(\sum_{m=1}^\infty \sum_{n=1}^\infty a_{m,n}\right) \wedge \left(\sum_{n=1}^\infty \sum_{m=1}^\infty a_{m,n}\right) \ge \sum_{(m,n)\in(\mathbb{Z}^+)^2} a_{m,n}.$$

Similarly,

$$\sum_{(m,n)\in(\mathbb{Z}^+)^2} a_{m,n} \geq \left(\sum_{m=1}^{\infty}\sum_{n=1}^{\infty} a_{m,n}\right) \vee \left(\sum_{n=1}^{\infty}\sum_{m=1}^{\infty} a_{m,n}\right),$$

and so all three must be equal. □

Now define $\tilde{\mu}(\Gamma)$ for $\Gamma \subseteq E$ to be the infimum of $\sum_{m=1}^{\infty} V(I_m)$ for all choices of $\{I_m : m \geq 1\} \subseteq \mathfrak{R}$ that cover Γ (i.e., $\Gamma \subseteq \bigcup_{m=1}^{\infty} I_m$). The strategy is to find a σ-algebra $\mathcal{L} \supseteq \mathcal{B}_E$ for which the restriction μ of $\tilde{\mu}$ to $\mathcal{L}$ is a measure. Thus, the first thing to check is that $\tilde{\mu}(I) = V(I)$ for all $I \in \mathfrak{R}$.

Lemma 2.2.2 *If $L \in \mathbb{Z}^+$ and $\Gamma = \bigcup_{\ell=1}^{L} J_\ell$ where the J_m's are non-overlapping elements of $\mathfrak{R}$, then $\tilde{\mu}(\Gamma) = \sum_{\ell=1}^{L} V(J_\ell)$. In particular, $\tilde{\mu}(I) = V(I)$ for all $I \in \mathfrak{R}$.*

Proof. Obviously $\tilde{\mu}(\Gamma) \leq \sum_{\ell=1}^{L} V(J_\ell)$. To prove the opposite inequality, let $\{I_m : m \geq 1\}$ be a cover of Γ by elements of $\mathfrak{R}$. Given an $\epsilon > 0$, choose I'_m for each $m \in \mathbb{Z}^+$ so that $I_m \subseteq \mathring{I}'_m$ and $V(I'_m) \leq V(I_m) + 2^{-m}\epsilon$. Because Γ is compact, there exists an $n \in \mathbb{Z}^+$ such that $\{I'_1, \dots, I'_n\}$ covers Γ.

Next, set $I_{m,\ell} = I'_m \cap J_\ell$ for $1 \leq m \leq n$ and $1 \leq \ell \leq L$. Then, for each ℓ, $J_\ell = \bigcup_{m=1}^{n} I_{m,\ell}$, and, for each m, the $I_{m,\ell}$'s are non-overlapping elements of $\mathfrak{R}$ with $\bigcup_{\ell=1}^{L} I_{m,\ell} \subseteq I'_m$. Hence, by (3),

$$\sum_{m=1}^{\infty} V(I_m) + \epsilon \geq \sum_{m=1}^{n} V(I'_m) \geq \sum_{m=1}^{n}\sum_{\ell=1}^{L} V(I_{m,\ell}) \geq \sum_{\ell=1}^{L} V(J_\ell).$$ □

In view of Lemma 2.2.2, I am justified in replacing $V(I)$ by $\tilde{\mu}(I)$ for $I \in \mathfrak{R}$.

The next result shows that half the equality in (2.1.4) is automatic, even before one restricts to Γ's from $\mathcal{L}$.

Lemma 2.2.3 *If $\Gamma \subseteq \Gamma'$, then $\tilde{\mu}(\Gamma) \leq \tilde{\mu}(\Gamma')$. In fact, if $\Gamma \subseteq \bigcup_{n=1}^{\infty} \Gamma_n$, then $\tilde{\mu}(\Gamma) \leq \sum_{n=1}^{\infty} \tilde{\mu}(\Gamma_n)$. In particular, if $\Gamma \subseteq \bigcup_{n=1}^{\infty} \Gamma_n$ and $\tilde{\mu}(\Gamma_n) = 0$ for each $n \geq 1$, then $\tilde{\mu}(\Gamma) = 0$.*

Proof. The first assertion follows immediately from the fact that every cover of Γ' is also a cover of Γ.

In order to prove the second assertion, let $\epsilon > 0$ be given, and choose for each $n \geq 1$ a cover $\{I_{m,n} : m \geq 1\} \subseteq \mathfrak{R}$ of Γ_n satisfying $\sum_{m=1}^{\infty} V(I_{m,n}) \leq \tilde{\mu}(\Gamma_n) + 2^{-n}\epsilon$. It is obvious that $\{I_{m,n} : (m,n) \in (\mathbb{Z}^+)^2\}$ is a countable cover of Γ. Hence, by Lemma 2.2.1,

$$\tilde{\mu}(\Gamma) \leq \sum_{(m,n)\in(\mathbb{Z}^+)^2} V(I_{m,n}) = \sum_{n=1}^{\infty}\sum_{m=1}^{\infty} V(I_{m,n}) \leq \sum_{n=1}^{\infty} \tilde{\mu}(\Gamma_n) + \epsilon.$$ □

As a consequence of Lemma 2.2.3, one has that

$$\tilde{\mu}(\Gamma) = \inf\{\tilde{\mu}(G) : \Gamma \subseteq G \in \mathfrak{G}(E)\}. \tag{2.2.1}$$

To see this, note that the left-hand side is certainly dominated by the right. Thus, it suffices to show that if $\{I_m : m \geq 1\}$ is a cover of Γ by elements of $\mathfrak{R}$ and $\epsilon > 0$, then there is a $\mathfrak{G}(E) \ni G \supseteq \Gamma$ such that $\tilde{\mu}(G) \leq \sum_{m=1}^{\infty} V(I_m) + \epsilon$. To this end, for each m choose $I'_m \in \mathfrak{R}$ such that $I_m \subseteq \mathring{I}'_m$ and $V(I'_m) \leq V(I_m) + 2^{-m}\epsilon$, and take $G = \bigcup_{m=1}^{\infty} \mathring{I}'_m$. Clearly $\Gamma \subseteq G \in \mathfrak{G}(E)$ and

$$\tilde{\mu}(G) \leq \sum_{m=1}^{\infty} V(I'_m) \leq \sum_{m=1}^{\infty} V(I_m) + \epsilon.$$

One important consequence of (2.2.1) is that it shows that for any $\Gamma \subseteq E$ there is a $\mathfrak{G}_\delta(E) \ni D \supseteq \Gamma$ for which $\tilde{\mu}(D) = \tilde{\mu}(\Gamma)$. Indeed, simply choose $\Gamma \subseteq G_n \in \mathfrak{G}(E)$ for which $\tilde{\mu}(G_n) \leq \tilde{\mu}(\Gamma) + \frac{1}{n}$, and take $D = \bigcap_{n=1}^{\infty} G_n$.

Another virtue of (2.2.1) is that it facilitates the proof of the second part of the following preliminary additivity result about $\tilde{\mu}$.

Lemma 2.2.4 *If G and G' are disjoint open subsets of E, then $\tilde{\mu}(G \cup G') = \tilde{\mu}(G) + \tilde{\mu}(G')$. Also, if K and K' are disjoint compact subsets of E, then $\tilde{\mu}(K \cup K') = \tilde{\mu}(K) + \tilde{\mu}(K')$.*

Proof. We begin by showing that if $\{I_m : m \geq 1\}$ is a sequence of non-overlapping elements of $\mathfrak{R}$, then

$$\tilde{\mu}\left(\bigcup_{m=1}^{\infty} I_m\right) = \sum_{m=1}^{\infty} V(I_m). \tag{2.2.2}$$

Because the left-hand side is dominated by the right, it suffices to show that the right-hand side is dominated by the left. However, by Lemmas 2.2.3 and 2.2.2, for each $n \in \mathbb{Z}^+$,

$$\tilde{\mu}\left(\bigcup_{m=1}^{\infty} I_m\right) \geq \tilde{\mu}\left(\bigcup_{m=1}^{n} I_m\right) = \sum_{m=1}^{n} V(I_m),$$

which completes the proof of (2.2.2).

Next, suppose that G and G' are disjoint open sets. By (5), there exist non-overlapping sequences $\{I_m : m \geq 1\}$ and $\{I'_m : m \geq 1\}$ of elements of $\mathfrak{R}$ such that $G = \bigcup_{m=1}^{\infty} I_m$ and $G' = \bigcup_{m=1}^{\infty} I'_{m'}$. Thus, if $I''_{2m-1} = I_m$ and $I''_{2m} = I'_m$ for $m \geq 1$, the I''_m's are non-overlapping elements of $\mathfrak{R}$ whose union is $G \cup G'$. Hence, by (2.2.2),

$$\tilde{\mu}(G \cup G') = \sum_{m=1}^{\infty} V(I''_m) = \sum_{m=1}^{\infty} V(I_m) + \sum_{m=1}^{\infty} V(I'_m) = \tilde{\mu}(G) + \tilde{\mu}(G').$$

To complete the proof, let compact sets K and K' be given. Because they are disjoint and compact, there exist disjoint open sets G and G' such that $K \subseteq G$ and $K' \subseteq G'$. Thus, for any open $H \supseteq K \cup K'$,

$$\tilde{\mu}(H) \geq \tilde{\mu}\big((H \cap G) \cup (H \cap G')\big) = \tilde{\mu}(H \cap G) + \tilde{\mu}(H \cap G') \geq \tilde{\mu}(K) + \tilde{\mu}(K'),$$

and therefore, by (2.2.1), $\tilde{\mu}(K \cup K') \geq \tilde{\mu}(K) + \tilde{\mu}(K')$. Because the opposite inequality always holds, there is nothing more to do. □

I am at last ready to describe the σ-algebra $\mathcal{L}$, although it will not be immediately obvious that it is a σ-algebra or that $\tilde{\mu}$ is countably additive on it. Be that as it may, take $\mathcal{L}$ to be the collection of $\Gamma \subseteq E$ with the property that, for each $\epsilon > 0$, there is an open $G \supseteq \Gamma$ for which $\tilde{\mu}(G \setminus \Gamma) < \epsilon$.

At first, one might be tempted to say that, in view of (2.2.1), every subset Γ is an element of $\mathcal{L}$. This is because one is inclined to think that $\tilde{\mu}(G) = \tilde{\mu}(G \setminus \Gamma) + \tilde{\mu}(\Gamma)$ when, in fact, $\tilde{\mu}(G) \leq \tilde{\mu}(G \setminus \Gamma) + \tilde{\mu}(\Gamma)$ is all that we know in general. Therein lies the subtlety of the definition! Nonetheless, it is clear that $\mathfrak{G}(E) \subseteq \mathcal{L}$. Furthermore, if $\tilde{\mu}(\Gamma) = 0$, then $\Gamma \in \mathcal{L}$, since, by (2.2.1), one can choose, for any $\epsilon > 0$, an open $G \supseteq \Gamma$ such that $\tilde{\mu}(G \setminus \Gamma) \leq \tilde{\mu}(G) < \epsilon$. Finally, if $\Gamma \in \mathcal{L}$, then there is a $D \in \mathfrak{G}_\delta(E)$ for which $\Gamma \subseteq D$ and $\tilde{\mu}(D \setminus \Gamma) = 0$. Indeed, simply choose $\{G_n : n \geq 1\} \subseteq \mathfrak{G}(E)$ for which $\Gamma \subseteq G_n$ and $\tilde{\mu}(G_n \setminus \Gamma) < \frac{1}{n}$, and take $D = \bigcap_{n=1}^{\infty} G_n$.

The next result shows that $\mathcal{L}$ is closed under countable unions.

Lemma 2.2.5 *If $\{\Gamma_n : n \geq 1\} \subseteq \mathcal{L}$, then $\Gamma = \bigcup_{n=1}^{\infty} \Gamma_n \in \mathcal{L}$, and, of course (cf. Lemma 2.2.3), $\tilde{\mu}(\Gamma) \leq \sum_{n=1}^{\infty} \tilde{\mu}(\Gamma_n)$.*

Proof. For each $n \geq 1$, choose $\mathfrak{G}(E) \ni G_n \supseteq \Gamma_n$ so that $\tilde{\mu}(G_n \setminus \Gamma_n) < 2^{-n}\epsilon$. Then $G \equiv \bigcup_1^{\infty} G_n$ is open, contains Γ, and (by Lemma 2.2.3) satisfies

$$\tilde{\mu}(G \setminus \Gamma) \leq \tilde{\mu}\left(\bigcup_{n=1}^{\infty} (G_n \setminus \Gamma_n)\right) \leq \sum_{n=1}^{\infty} \tilde{\mu}(G_n \setminus \Gamma_n) < \epsilon. \qquad \square$$

Knowing that $\mathcal{L}$ is closed under countable unions and that it contains $\mathfrak{G}(E)$, we will know that it is a σ-algebra and that $\mathcal{B}_E \subseteq \mathcal{L}$ as soon I show that $\mathcal{L}$ is closed under complementation.

Lemma 2.2.6 *If $K \subset\subset E$,[3] then $K \in \mathcal{L}$ and $\tilde{\mu}(K) < \infty$.*

Proof. The first step is to show that $\tilde{\mu}(K) < \infty$. For this purpose, choose a cover $\{I_m : m \geq 1\} \subseteq \mathfrak{R}$ of K, and then use (4) to choose $\{I'_m : m \geq 1\} \subseteq \mathfrak{R}$ so that $I_m \subseteq \mathring{I}'_m$ and $V(I'_m) \leq V(I_m) + 1$ for each m. Now apply the Heine–Borel property to find an $n \in \mathbb{Z}^+$ such that $K \subseteq \bigcup_{m=1}^{n} \mathring{I}'_m$. Then $\tilde{\mu}(K) \leq n + \sum_{m=1}^{n} V(I_m) < \infty$.

[3] I will often use the notation $K \subset\subset E$ to mean that K is a compact subset of E. When E is discrete, the notation means that K is a finite subset of E.

We will now show that $K \in \mathcal{L}$. To this end, let $\epsilon > 0$ be given, and choose an open set $G \supseteq K$ so that $\tilde{\mu}(G) \le \tilde{\mu}(K) + \epsilon$. Set $H = G \setminus K \in \mathfrak{G}(E)$, and choose a non-overlapping sequence $\{I_n : n \ge 1\} \subseteq \mathfrak{R}$ so that $H = \bigcup_1^\infty I_n$. Then, by (2.2.2), $\tilde{\mu}(H) = \sum_{m=1}^\infty V(I_m)$. In addition, for each $n \in \mathbb{Z}^+$, $K_n \equiv \bigcup_{m=1}^n I_m$ is compact and disjoint from K. Hence, by Lemmas 2.2.5 and 2.2.3,

$$\tilde{\mu}(K_n) + \tilde{\mu}(K) = \tilde{\mu}(K_n \cup K) \le \tilde{\mu}(G),$$

and so, because $\tilde{\mu}(K) < \infty$, $\sum_{m=1}^n V(I_m) = \tilde{\mu}(K_n) \le \epsilon$ for all n, and therefore $\tilde{\mu}(G \setminus K) = \tilde{\mu}(H) \le \sum_{m=1}^\infty \tilde{\mu}(I_m) \le \epsilon$. □

Lemma 2.2.7 *$\mathcal{L}$ is a σ-algebra over E that contains $\mathcal{B}_E$. Moreover, if $\Gamma \subseteq E$, then $\Gamma \in \mathcal{L}$ if and only if for every $\epsilon > 0$ there exist $F \in \mathfrak{F}(E)$ and $G \in \mathfrak{G}(E)$ such that $F \subseteq \Gamma \subseteq G$ and $\tilde{\mu}(G \setminus F) < \epsilon$. Alternatively, $\Gamma \in \mathcal{L}$ if there exist $C, D \in \mathcal{L}$ such that $C \subseteq \Gamma \subseteq D$ and $\tilde{\mu}(D \setminus C) = 0$, and if $\Gamma \in \mathcal{L}$, then there exist $C \in \mathfrak{F}_\sigma(E)$ and $D \in \mathfrak{G}_\delta(E)$ such that $C \subseteq \Gamma \subseteq D$ and $\tilde{\mu}(D \setminus C) = 0$.*

Proof. Because of Lemma 2.2.5, proving that $\mathcal{L}$ is a σ-algebra comes down to showing that it is closed under complementation. For this purpose, begin by observing that $\mathfrak{F}_\sigma(E) \subseteq \mathcal{L}$. To check this, choose $\{I_m : m \ge 1\} \subseteq \mathfrak{R}$ such that $E = \bigcup_{m=1}^\infty I_m$, and set $K_n = \bigcup_{m=1}^n I_m$. Then K_n is compact for each n. Given $F \in \mathfrak{F}(E)$ and $n \in \mathbb{Z}^+$, set $F_n = F \cap K_n$. Clearly F_n is compact and is therefore an element of $\mathcal{L}$. Hence, since $F = \bigcup_{n=1}^\infty F_n$ and $\mathcal{L}$ is closed under countable unions, we see first that $\mathfrak{F}(E) \subseteq \mathcal{L}$ and then that $\mathfrak{F}_\sigma(E) \subseteq \mathcal{L}$.

Next, suppose that $\Gamma \in \mathcal{L}$, and choose $D \in \mathfrak{G}_\delta(E)$ for which $\Gamma \subseteq D$ and $\tilde{\mu}(D \setminus \Gamma) = 0$. Then $C \equiv D^\complement \in \mathfrak{F}_\sigma(E)$, $C \subseteq \Gamma^\complement$, and $\tilde{\mu}(\Gamma^\complement \setminus C) = \tilde{\mu}(D \setminus \Gamma) = 0$. Hence, $\Gamma^\complement \setminus C \in \mathcal{L}$, and therefore so is $\Gamma^\complement = C \cup (\Gamma^\complement \setminus C)$, which means that $\mathcal{L}$ is closed under complementation and is therefore a σ-algebra over E.

Knowing that $\mathcal{L}$ contains $\mathfrak{G}(E)$ and is a σ-algebra, we know that $\mathcal{B}_E \subseteq \mathcal{L}$. In addition, if $\Gamma \in \mathcal{L}$, then for each $\epsilon > 0$ we can find an open $G \supseteq \Gamma$ and a closed F with $F^\complement \supseteq \Gamma^\complement$ for which $\tilde{\mu}(G \setminus \Gamma) \vee \tilde{\mu}(F^\complement \setminus \Gamma^\complement) < \frac{\epsilon}{2}$, which means that $F \subseteq \Gamma \subseteq G$ and $\tilde{\mu}(G \setminus F) < \epsilon$.

Finally, given the preceding, it is clear that if $\Gamma \in \mathcal{L}$ then there exist $C \in \mathfrak{F}_\sigma(E)$ and $D \in \mathfrak{G}_\delta(E)$ such that $C \subseteq \Gamma \subseteq D$ and $\tilde{\mu}(D \setminus C) = 0$. Conversely, if there exist such $C, D \in \mathcal{L}$, $C \subseteq \Gamma \subseteq D$, and $\tilde{\mu}(D \setminus C) = 0$, then $\tilde{\mu}(\Gamma \setminus C) \le \tilde{\mu}(D \setminus C) = 0$, and so $\Gamma = C \cup (\Gamma \setminus C) \in \mathcal{L}$. □

Theorem 2.2.8 *Refer to the preceding. Then there is a unique Borel measure μ on E for which $\mu(I) = V(I)$ for all $I \in \mathfrak{R}$. Moreover, μ is regular, $\mathcal{L} = \overline{\mathcal{B}_E}^\mu$, and (cf. Lemma 2.1.3) $\bar{\mu}$ is the restriction of $\tilde{\mu}$ to $\overline{\mathcal{B}_E}^\mu$.*

Proof. We will first show that $\tilde{\mu}$ is countably additive on $\mathcal{L}$. To this end, let $\{\Gamma_n : n \ge 1\} \subseteq \mathcal{L}$ be a sequence of mutually disjoint, relatively compact (i.e., their closures are compact) sets. By Lemma 2.2.7, for each $\epsilon > 0$ we can find a sequence $\{K_n : n \geq 1\}$ of compact sets such that $K_n \subseteq \Gamma_n$ and $\tilde{\mu}(\Gamma_n) \leq \tilde{\mu}(K_n) + 2^{-n}\epsilon$ for each n. Hence, by Lemma 2.2.5, for each $n \in \mathbb{Z}^+$,

$$\tilde{\mu}\left(\bigcup_{m=1}^{\infty} \Gamma_m\right) \geq \tilde{\mu}\left(\bigcup_{m=1}^{n} K_m\right) = \sum_{m=1}^{n} \tilde{\mu}(K_m),$$

and therefore

$$\tilde{\mu}\left(\bigcup_{m=1}^{\infty} \Gamma_m\right) \geq \sum_{m=1}^{\infty} \tilde{\mu}(K_m) \geq \sum_{m=1}^{\infty} \tilde{\mu}(\Gamma_m) - \epsilon,$$

which proves that

$$\tilde{\mu}\left(\bigcup_{m=1}^{\infty} \Gamma_m\right) \geq \sum_{m=1}^{\infty} \tilde{\mu}(\Gamma_m).$$

Since the opposite inequality is trivial, this proves the countable additivity of $\tilde{\mu}$ for relatively compact elements of $\mathcal{L}$.

To treat the general case, choose $\{I_m : m \geq 1\} \subseteq \mathfrak{R}$ for which that $E = \bigcup_{m=1}^{\infty} I_m$, and set $A_1 = I_1$ and $A_{n+1} = I_{n+1} \setminus \bigcup_{m=1}^{n} I_m$. Then the A_n's are mutually disjoint, relatively compact elements of $\mathcal{L}$. Hence, if $\Gamma_{m,n} = A_m \cap \Gamma_n$, then the $\Gamma_{m,n}$'s are also mutually disjoint and relatively compact. Furthermore, $\Gamma_n = \bigcup_{m=1}^{\infty} \Gamma_{m,n}$ for each n, and so, by the preceding and Lemma 2.2.1,

$$\tilde{\mu}\left(\bigcup_{n=1}^{\infty} \Gamma_n\right) = \sum_{(m,n)\in(\mathbb{Z}^+)^2} \tilde{\mu}(\Gamma_{m,n}) = \sum_{n=1}^{\infty} \tilde{\mu}(\Gamma_n).$$

Knowing that $\tilde{\mu}$ is countably additive on $\mathcal{L}$ and that $\mathcal{L} \supseteq \mathcal{B}_E$, we can take μ to be the restriction of $\tilde{\mu}$ to $\mathcal{B}_E$. Furthermore, the results in Lemma 2.2.7 show that this μ is regular, $\mathcal{L} = \overline{\mathcal{B}_E}^{\mu}$, and that $\tilde{\mu} \restriction \mathcal{L}$ equals $\bar{\mu}$.

To complete the proof, suppose that ν is a second Borel measure on E for which $\nu(I) = V(I)$ whenever $I \in \mathfrak{R}$. Given an open G, choose a non-overlapping $\{I_m : m \geq 1\} \subseteq \mathfrak{R}$ whose union is G, and apply (2.1.8) and (2.2.2) to conclude that

$$\nu(G) \leq \sum_{m=1}^{\infty} \nu(I_m) = \sum_{m=1}^{\infty} V(I_m) = \mu(G).$$

Next, given $\epsilon > 0$, choose for $m \in \mathbb{Z}^+$ an $I''_m \in \mathfrak{R}$ so that $I''_m \subseteq \mathring{I}_m$ and $V(I_m) \leq V(I''_m) + 2^{-m}\epsilon$. Then, because the I''_m's are disjoint,

$$\nu(G) \geq \nu\left(\bigcup_{m=1}^{\infty} I''_m\right) = \sum_{m=1}^{\infty} V(I''_m) \geq \sum_{m=1}^{\infty} V(I_m) - \epsilon \geq \mu(G) - \epsilon.$$

Hence, ν and μ are equal on $\mathfrak{G}(E)$. Finally, note that, by combining (4) and (5), we can produce a non-decreasing sequence of open sets $G_n \nearrow E$ such that $\mu(G_n) < \infty$. Hence, by Theorem 2.1.2, ν equals μ on $\mathcal{B}_{G_n}$ for each n, from which it follows easily that ν equals μ on $\mathcal{B}_E$. □

Corollary 2.2.9 *Suppose that $T : E \longrightarrow E$ is a transformation with the property that $T^{-1}(I) \in \mathfrak{R}$ and $V\big(T^{-1}(I)\big) = V(I)$ for all $I \in \mathfrak{R}$. Then $T^{-1}(\Gamma) \in \mathcal{B}_E$ and $\mu\big(T^{-1}(\Gamma)\big) = \mu(\Gamma)$ for all $\Gamma \in \mathcal{B}_E$. Equivalently, $T_*\mu = \mu$.*

Proof. By part (**i**) of Exercise 2.1.3, T^{-1} of a union of sets is the union of T^{-1} of each set over which the union is taken, and T^{-1} of a difference of sets is the difference of T^{-1} of each set. Next, let $\mathcal{B}$ be the set of $\Gamma \in \mathcal{B}_E$ with the property that $T^{-1}(\Gamma) \in \mathcal{B}_E$. By the preceding observation, $\mathcal{B}$ is a σ-algebra over E. In addition, $\mathfrak{R} \subseteq \mathcal{B}$. Thus, because for any open G there is a sequence $\{I_m : m \geq 1\} \subseteq \mathfrak{R}$ whose union is G, $\mathfrak{G}(E) \subseteq \mathcal{B}$. Since this means that $\mathcal{B}$ is a σ-algebra contained in $\mathcal{B}_E$ and containing $\mathfrak{G}(E)$, it follows that $\mathcal{B} = \mathcal{B}_E$.

Next, set $\nu(\Gamma) = \mu\big(T^{-1}(\Gamma)\big)$ for $\Gamma \in \mathcal{B}_E$. By part (**iii**) of Exercise 2.1.3, ν is a Borel measure on E. Moreover, by assumption, $\nu(I) = V(I)$ for $I \in \mathfrak{R}$. Hence, by the uniqueness statement in Theorem 2.2.8, $\nu = \mu$. □

2.2.2 Lebesgue Measure on $\mathbb{R}^N$

My first application of Theorem 2.2.8 will be to the construction of the father of all measures, Lebesgue measure on $\mathbb{R}^N$.

Endow $\mathbb{R}^N$ with the standard Euclidean metric, the one given by the Euclidean distance between points. Next, take $\mathfrak{R}$ to be the set of all rectangles I in $\mathbb{R}^N$ relative to the standard orthonormal coordinate system, include the empty rectangle in $\mathfrak{R}$, and define $V(I) = \text{vol}(I)$ if $I \neq \emptyset$ and $V(\emptyset) = 0$. In order to apply the results in § 2.2.1, I have to show that this choice of $\mathfrak{R}$ and V satisfies the hypotheses (1)–(5) listed at the beginning of that subsection. It is clear that they satisfy (1), (2), and (4). In addition, (3) follows from Lemma 1.1.1. To check (5), I will use the following lemma. In its statement and elsewhere, a **cube** will be a (multi-dimensional) rectangle all of whose edges have the same length. That is, a non-empty cube is a set Q of the form $x + [a, b]^N$ for some $x \in \mathbb{R}^N$ and $a \leq b$.

Lemma 2.2.10 *If G is an open set in $\mathbb{R}$, then G is the union of a countable number of mutually disjoint open intervals. More generally, if G is an open set in $\mathbb{R}^N$, then, for each $\delta > 0$, G admits a countable, non-overlapping, exact cover $\mathcal{C}$ by closed cubes Q with* $\text{diam}\,(Q) < \delta$.

Proof. If $G \subseteq \mathbb{R}$ is open and $x \in G$, let $\mathring{I}_x$ be the open connected component of G containing x. Then $\mathring{I}_x$ is an open interval and, for any $x, y \in G$, either $\mathring{I}_x \cap \mathring{I}_y = \emptyset$ or $\mathring{I}_x = \mathring{I}_y$. Hence, $\mathcal{C} \equiv \{\mathring{I}_x : x \in G \cap \mathbb{Q}\}$ ($\mathbb{Q}$ here denotes the set of rational numbers) is the required cover.

To handle the second assertion, set $Q_n = [0, 2^{-n}]^N$ and $\mathcal{K}_n = \{\frac{\mathbf{k}}{2^n} + Q_n : \mathbf{k} \in \mathbb{Z}^N\}$. For each n, $\mathcal{K}_n$ is a cover of $\mathbb{R}^N$. In addition, if $m \le n$, $Q \in \mathcal{K}_m$, and $Q' \in \mathcal{K}_n$, then either $Q' \subseteq Q$ or $\mathring{Q} \cap \mathring{Q}' = \emptyset$. Now let $G \in \mathfrak{G}(\mathbb{R}^N)$ and $\delta > 0$ be given, take n_0 to be the smallest $n \in \mathbb{Z}$ for which $2^{-n}\sqrt{N} < \delta$, and set $\mathcal{C}_{n_0} = \{Q \in \mathcal{K}_{n_0} : Q \subseteq G\}$. Next, define $\mathcal{C}_n$ inductively for $n \ge n_0$ so that

$$\mathcal{C}_{n+1} = \left\{ Q' \in \mathcal{K}_{n+1} : Q' \subseteq G \text{ and } \mathring{Q}' \cap \mathring{Q} = \emptyset \text{ for any } Q \in \bigcup_{m=n_0}^{n} \mathcal{C}_m \right\}.$$

Since if $m \le n$, $Q \in \mathcal{C}_m$, and $Q' \in \mathcal{C}_n$, either $Q = Q'$ or $\mathring{Q} \cap \mathring{Q}' = \emptyset$, the cubes in $\mathcal{C} \equiv \bigcup_{n=n_0}^{\infty} \mathcal{C}_n$ are non-overlapping, and certainly $\bigcup \mathcal{C} \subseteq G$. Finally, because G is open, for any $x \in G$ there exist $n \ge n_0$ and $Q' \in \mathcal{K}_n$ such that $x \in Q' \subseteq G$. If $Q' \notin \mathcal{C}_n$, then there exist an $n_0 \le m < n$ and a $Q \in \mathcal{C}_m$ for which $\mathring{Q} \cap \mathring{Q}' \ne \emptyset$. But this means that $Q' \subseteq Q$ and therefore that $x \in Q \subseteq \bigcup \mathcal{C}$. Thus $\mathcal{C}$ covers G. □

Knowing that $\mathfrak{R}$ and V satisfy hypotheses (1)–(5), we can apply Theorem 2.2.8 and thereby derive the following fundamental result.

Theorem 2.2.11 *There is one and only one Borel measure $\lambda_{\mathbb{R}^N}$ on $\mathbb{R}^N$ with the property that $\lambda_{\mathbb{R}^N}(Q) = \mathrm{vol}(Q)$ for all cubes Q in $\mathbb{R}^N$. Moreover, $\lambda_{\mathbb{R}^N}$ is regular.*

Proof. The existence of $\lambda_{\mathbb{R}^N}$ as well as its regularity are immediate consequences of Theorem 2.2.8. Furthermore, that theorem says that $\lambda_{\mathbb{R}^N}$ is the only Borel measure ν with the property that $\nu(I) = \mathrm{vol}(I)$ for all $I \in \mathfrak{R}$. Thus, to prove the uniqueness statement here, it suffices to check that $\nu(I) = \mathrm{vol}(I)$ for all rectangles I if it does for cubes. To this end, first note that if I is a rectangle, then there exists a sequence $\{I_n : n \ge 1\}$ of rectangles such that $I_n \nearrow \mathring{I}$ and $\mathrm{vol}(I_n) \nearrow \mathrm{vol}(I)$. Hence, by (2.1.10), $\nu(I) = \nu(\mathring{I})$. In particular, by (2.1.7), this means that [4] $\nu(\partial I) = \nu(I) - \nu(\mathring{I}) = 0$ for all rectangles I. Now apply Lemma 2.2.10 to write $\mathring{I} = \bigcup_{n=1}^{\infty} Q_n$, where the Q_n's are non-overlapping cubes, use the preceding to check that $\nu(Q_m \cap Q_n) = 0$ for $m \ne n$, and apply (2.1.9) to see that

$$\nu(I) = \nu(\mathring{I}) = \sum_{n=1}^{\infty} \mathrm{vol}(Q_n) = \lambda_{\mathbb{R}^N}(I) = \mathrm{vol}(I). \qquad \square$$

The Borel measure $\lambda_{\mathbb{R}^N}$ described in Theorem 2.2.11 is called **Lebesgue measure** on $\mathbb{R}^N$. In addition, elements of $\overline{\mathcal{B}_{\mathbb{R}^N}}^{\lambda_{\mathbb{R}^N}}$ are said to be **Lebesgue measurable** sets.

An important property of Lebesgue measure is that it is translation invariant. To be precise, for each $x \in \mathbb{R}^N$, define the **translation map** $T_x : \mathbb{R}^N \longrightarrow \mathbb{R}^N$ by $T_x(y) = x + y$. Obviously, T_x is one-to-one and onto.

[4] I use $\partial\Gamma$ to denote the boundary $\bar{\Gamma} \setminus \mathring{\Gamma}$ of a set Γ.

In fact, $T_x^{-1} = T_{-x}$. In addition, because $T_x = T_{-x}^{-1}$ and T_{-x} is continuous, T_x takes $\mathcal{B}_{\mathbb{R}^N}$ into itself. Finally, say that a Borel measure μ on $\mathbb{R}^N$ is **translation invariant** if $\mu\big(T_x(\Gamma)\big) = \nu(\Gamma)$ for all $x \in \mathbb{R}^N$ and $\Gamma \in \mathcal{B}_{\mathbb{R}^N}$. The following corollary provides an important characterization of Lebesgue measure in terms of translation invariance.

Corollary 2.2.12 *Lebesgue measure is the one and only translation invariant Borel measure on $\mathbb{R}^N$ that assigns the unit cube $[0,1]^N$ measure 1. Thus, if ν is a translation invariant Borel measure on $\mathbb{R}^N$ and $\alpha = \nu\big([0,1]^N\big) < \infty$, then $\nu = \alpha\lambda_{\mathbb{R}^N}$.*

Proof. That $\lambda_{\mathbb{R}^N}$ is translation invariant follows immediately from Corollary 2.2.9 and the fact that, for any rectangle I and $x \in \mathbb{R}^N$, $\mathrm{vol}\big(T_x(I)\big) = \mathrm{vol}(I)$.

To prove the uniqueness assertion, suppose that μ is a translation invariant Borel measure that gives measure 1 to $[0,1]^N$. We first show that $\mu(H) = 0$ for every H of the form $\{x \in \mathbb{R}^N : x_i = c\}$ for some $1 \le i \le N$ and $c \in \mathbb{R}$. Indeed, by translation invariance, it suffices to treat the case $c = 0$. In addition, by countable subadditivity and translation invariance, it is sufficient to show that $\mu(R) = 0$ when $R = \{x : x_i = 0 \text{ and } x_j \in [0,1) \text{ for } j \ne i\}$. But if $\mathbf{e}_i$ is the unit vector whose ith coordinate is 1 and whose other coordinates are 0, then, for any $n \ge 1$, the sets $\{\frac{m}{n}\mathbf{e}_i + R : 0 \le m \le n\}$ are mutually disjoint, have the same μ-measure as R, and are contained in $[0,1]^N$. Hence, $n\mu(R) \le 1$ for all $n \ge 1$, and so $\mu(R) = 0$.

Given the preceding, we know that $\mu(\partial I) = 0$ for all rectangles $I \subseteq \mathbb{R}^N$. Hence, if $(n_1, \dots, n_N) \in (\mathbb{Z}^+)^N$, then

$$\begin{aligned}\mu([0,1]^N) &= \mu\left(\bigcup\left\{\prod_{i=1}^N \left[\tfrac{k_i-1}{n_i}, \tfrac{k_i}{n_i}\right] : 1 \le k_i \le n_i \text{ for } 1 \le i \le N\right\}\right)\\ &= \left(\prod_{i=1}^N n_i\right)\mu\left(\prod_{i=1}^N \left[0, \tfrac{1}{n_i}\right]\right),\end{aligned}$$

and so $\mu\left(\prod_{i=1}^N \left[0, \frac{1}{n_i}\right]\right) = \prod_{i=1}^N \frac{1}{n_i}$. Starting from this, the same line of reasoning can be used to show that $\mu\left(\prod_{i=1}^N \left[0, \frac{m_i}{n_i}\right]\right) = \prod_{i=1}^N \frac{m_i}{n_i}$ for any pair $(m_1, \dots, m_N)$, $(n_1, \dots, n_N) \in (\mathbb{Z}^+)^N$. Hence, by translation invariance, for any rectangle whose sides have rational lengths, $\mu(I) = \mathrm{vol}(I)$. Finally, for any rectangle I, we can choose a non-increasing sequence $\{I_n : n \ge 1\}$ of rectangles with rational side lengths such that $I_n \searrow I$, and so $\mu(I) = \lim_{n\to\infty} \mathrm{vol}(I_n) = \mathrm{vol}(I)$. Now apply Theorem 2.2.8.

To prove the concluding assertion, first suppose that $\alpha = 0$. Then, because $\mathbb{R}^N$ can be covered by a countable number of translates of $[0,1]^N$, it follows that $\nu(\mathbb{R}^N) = 0$ and therefore that $\nu = \alpha\lambda_{\mathbb{R}^N}$. Next suppose that $\alpha > 0$. Then $\alpha^{-1}\nu$ is a translation invariant Borel measure on $\mathbb{R}^N$ and ν assigns the unit cube measure 1. Hence, by the earlier part, $\alpha^{-1}\nu = \lambda_{\mathbb{R}^N}$. □

Although the property of translation invariance was built into the construction of Lebesgue measure, it is not immediately obvious how Lebesgue measure responds to rotations of $\mathbb{R}^N$. One suspects that, as *the natural* measure on $\mathbb{R}^N$, Lebesgue measure should be invariant under the full group of Euclidean transformations (i.e., rotations as well as translations). However, because my description of Lebesgue's measure was based on rectangles and the rectangles were inextricably tied to a fixed set of coordinate axes, rotation invariance is not as obvious as translation invariance was.

The following corollary shows how Lebesgue measure transforms under an arbitrary linear transformation of $\mathbb{R}^N$, and rotation invariance will follow as an immediate corollary.

Given an $N \times N$, real matrix $A = \big((a_{ij})\big)_{1\le i,j\le N}$, define $T_A : \mathbb{R}^N \longrightarrow \mathbb{R}^N$ to be the action of A on x. That is, $(T_A x)_i = \sum_{j=1}^N a_{ij}x_j$ for $1 \le i \le N$. We can now prove the following.

Theorem 2.2.13 *For any $N \times N$, real matrix A and $\Gamma \in \overline{\mathcal{B}_{\mathbb{R}^N}}^{\lambda_{\mathbb{R}^N}}$, $T_A(\Gamma) \in \overline{\mathcal{B}_{\mathbb{R}^N}}^{\lambda_{\mathbb{R}^N}}$ and $\overline{\lambda_{\mathbb{R}^N}}(T_A\Gamma) = |\det(A)|\overline{\lambda_{\mathbb{R}^N}}(\Gamma)$. Moreover, if A is non-singular, then T_A takes $\mathcal{B}_{\mathbb{R}^N}$ into itself.*

Proof. I begin with the case in which A is non-singular. Then $T_{A^{-1}}$ is a continuous, one-to-one map from $\mathbb{R}^N$ onto itself, and $T_A = (T_{A^{-1}})^{-1}$. Hence, by **(iii)** of Exercise 2.1.3, T_A takes $\mathcal{B}_{\mathbb{R}^N}$ into itself. Next, define ν_A on $\mathcal{B}_{\mathbb{R}^N}$ by $\nu_A(\Gamma) = \lambda_{\mathbb{R}^N}\big(T_A(\Gamma)\big)$. Then, again, by part **(iii)** of Exercise 2.1.3, ν_A is a Borel measure on $\mathbb{R}^N$. Now set $\alpha(A) = \nu_A\big([0,1]^N\big)$. Because $T_A([0,1]^N)$ is compact, $\alpha(A) < \infty$. In addition, because

$$\nu_A\big(T_x(\Gamma)\big) = \lambda_{\mathbb{R}^N}\big(T_{Ax} + T_A(\Gamma)\big) = \lambda_{\mathbb{R}^N}\big(T_A(\Gamma)\big) = \nu_A(\Gamma),$$

ν_A is translation invariant. Thus Corollary (2.2.12) says that $\nu_A = \alpha(A)\lambda_{\mathbb{R}^N}$, and so all that we have to do is show that $\alpha(A) = |\det(A)|$. To this end, observe that there are cases in which $\alpha(A)$ can be computed by hand. The first of these is when A is diagonal with positive diagonal elements, in which case $T_A\big([0,1]\big) = \prod_{j=1}^N [0,a_{jj}]$ and therefore $\alpha(A) = \prod_{j=1}^N a_{jj} = \det(A)$. The second case is the one in which A is an orthogonal matrix. Then $T_A\big(\overline{B(0,1)}\big) = \overline{B(0,1)}$ and therefore, since $\lambda_{\mathbb{R}^N}\big(\overline{B(0,1)}\big) \in (0,\infty)$, $\alpha(A) = 1$. To go further, notice that, since $T_{AA'} = T_A \circ T_{A'}$, $\alpha(AA') = \alpha(A)\alpha(A')$. Hence, if A is symmetric and positive definite (i.e., all its eigenvalues are positive) and $\mathcal{O}$ is an orthogonal matrix for which[5] $D = \mathcal{O}^\top A\mathcal{O}$ is diagonal, then the diagonal entries of D are positive, $\det(A) = \det(D) = \alpha(D)$, and therefore $\alpha(A) = \alpha(\mathcal{O}^\top)\alpha(D)\alpha(\mathcal{O}) = \alpha(D) = \det(A)$. Finally, for any non-singular A, $AA^\top$ is a symmetric, positive definite matrix. Moreover, if $\mathcal{O} = A^{-1}(AA^\top)^{\frac{1}{2}}$, where $(AA^\top)^{\frac{1}{2}}$ denotes the positive definite, symmetric square root of $AA^\top$, then $\mathcal{O}$ satisfies $\mathcal{O}\mathcal{O}^\top = A^{-1}AA^\top(A^\top)^{-1} = I$ and is therefore orthogonal. Hence, since $A = (AA^\top)^{\frac{1}{2}}\mathcal{O}^\top$, we find that $\alpha(A) = \det\big((AA^\top)^{\frac{1}{2}}\big) = |\det(\mathrm{A})|$.

[5] Given a matrix A, I use $A^\top$ to denote the transpose matrix.

Finally, to show that T_A takes a $\Gamma \in \overline{\mathcal{B}_{\mathbb{R}^N}}^{\lambda_{\mathbb{R}^N}}$ into $\overline{\mathcal{B}_{\mathbb{R}^N}}^{\lambda_{\mathbb{R}^N}}$ and that $\overline{\lambda_{\mathbb{R}^N}}(T_A(\Gamma)) = |\det(A)|\overline{\lambda_{\mathbb{R}^N}}(\Gamma)$, choose $C, D \in \mathcal{B}_{\mathbb{R}^N}$ so that $C \subseteq \Gamma \subseteq D$ and $\lambda_{\mathbb{R}^N}(D \setminus C) = 0$. Then $T_A(C), T_A(D) \in \mathcal{B}_{\mathbb{R}^N}$, $T_A(C) \subseteq T_A(\Gamma) \subseteq T_A(D)$, $\lambda_{\mathbb{R}^N}\big(T_A(D)\setminus T_A(C)\big) = \lambda_{\mathbb{R}^N}\big(T_A(D\setminus C)\big) = 0$, and therefore $T_A(\Gamma) \in \overline{\mathcal{B}_{\mathbb{R}^N}}^{\lambda_{\mathbb{R}^N}}$ and $\overline{\lambda_{\mathbb{R}^N}}\big(T_A(\Gamma)\big) = \lambda_{\mathbb{R}^N}\big(T_A(C)\big) = |\det(A)|\lambda_{\mathbb{R}^N}(C) = |\det(A)|\overline{\lambda_{\mathbb{R}^N}}(\Gamma)$.

To treat the singular case, first observe that there is nothing to do when $N = 1$, since the singularity of A means that $T_A(\mathbb{R}) = \{0\}$ and $\lambda_{\mathbb{R}}(\{0\}) = 0$. Thus, assume that $N \geq 2$. Then, if A is singular, $T_A(\mathbb{R}^N)$ is contained in an $(N-1)$-dimensional subspace of $\mathbb{R}^N$. Therefore, what remains is to show that $\lambda_{\mathbb{R}^N}$ assigns measure 0 to an $(N-1)$-dimensional subspace H. This is clear if $H = \mathbb{R}^{N-1} \times \{0\}$, since in that case one can obviously cover H with a countable number of rectangles each of which has volume 0. To handle general H's, choose an orthogonal matrix $\mathcal{O}$ so that $H = T_{\mathcal{O}}\big(\mathbb{R}^{N-1} \times \{0\}\big)$, and use the preceding to conclude that $\lambda_{\mathbb{R}^N}(H) = \lambda_{\mathbb{R}^N}\big(\mathbb{R}^{N-1} \times \{0\}\big) = 0$.

□

Before concluding this preliminary discussion of Lebesgue measure, it may be appropriate to examine whether there are any sets that are not Lebesgue measurable. It turns out that the existence of such sets brings up some extremely delicate issues about the foundations of mathematics. Indeed, if one is willing to abandon the full axiom of choice, then R. Solovay has shown that there is a model of mathematics in which *every* subset of $\mathbb{R}^N$ is Lebesgue measurable. However, if one accepts the full axiom of choice, then the following argument, due to Vitali, shows that there are sets that are not Lebesgue measurable. The use of the axiom of choice comes in Lemma 2.2.15 below. It is not used in the proof of the next lemma, a result that is interesting in its own right. See Exercise 2.2.6 for a somewhat surprising application and Exercise 6.3.2 for another derivation of it.

Lemma 2.2.14 *If $\Gamma \in \overline{\mathcal{B}_{\mathbb{R}}}^{\lambda_{\mathbb{R}}}$ has positive Lebesgue measure, then the set $\Gamma - \Gamma \equiv \{y - x : x, y \in \Gamma\}$ contains an open interval $(-\delta, \delta)$ for some $\delta > 0$.*

Proof. Without loss of generality, we will assume that $\Gamma \in \mathcal{B}_{\mathbb{R}}$ and that $\lambda_{\mathbb{R}}(\Gamma) \in (0, \infty)$.

Choose an open set $G \supseteq \Gamma$ for which $\lambda_{\mathbb{R}}(G \setminus \Gamma) < \frac{1}{3}\lambda_{\mathbb{R}}(\Gamma)$, and let (cf. the first part of Lemma 2.2.10) $\mathcal{C}$ be a countable collection of mutually disjoint, non-empty, open intervals $\mathring{I}$ whose union is G. Then

$$\sum_{\mathring{I} \in \mathcal{C}} \lambda_{\mathbb{R}}(\mathring{I} \cap \Gamma) = \lambda_{\mathbb{R}}(\Gamma) \geq \tfrac{3}{4}\lambda_{\mathbb{R}}(G) = \tfrac{3}{4}\sum_{\mathring{I} \in \mathcal{C}} \lambda_{\mathbb{R}}(\mathring{I}).$$

Hence, there must be an $\mathring{I} = (a, b) \in \mathcal{C}$ for which $\lambda_{\mathbb{R}}(\mathring{I} \cap \Gamma) \geq \frac{3}{4}\lambda_{\mathbb{R}}(\mathring{I})$. Set $A = \mathring{I} \cap \Gamma$. If $d \in \mathbb{R}$ and $(d + A) \cap A = \emptyset$, then

$$2\lambda_{\mathbb{R}}(A) = \lambda_{\mathbb{R}}(d + A) + \lambda_{\mathbb{R}}(A) = \lambda_{\mathbb{R}}\big((d + A) \cup A\big) \leq \lambda_{\mathbb{R}}\big((d + \mathring{I}) \cup \mathring{I}\big).$$

At the same time, $(d+\mathring{I}) \cup \mathring{I} \subseteq (a, b+d)$ if $d \geq 0$ and $(d+\mathring{I}) \cup \mathring{I} \subseteq (a+d, b)$ if $d < 0$. Thus, in either case, $\lambda_{\mathbb{R}}\big((d+\mathring{I}) \cup \mathring{I}\big) \leq |d| + \lambda_{\mathbb{R}}(\mathring{I})$. Hence, if $(d+A) \cap A = \emptyset$, then $\frac{3}{2}\lambda_{\mathbb{R}}(\mathring{I}) \leq 2\lambda_{\mathbb{R}}(A) \leq |d| + \lambda_{\mathbb{R}}(\mathring{I})$, from which one sees that $|d| \geq \frac{1}{2}\lambda_{\mathbb{R}}(\mathring{I})$. In other words, if $|d| < \frac{1}{2}\lambda_{\mathbb{R}}(\mathring{I})$, then $(d+A) \cap A \neq \emptyset$. But this means that for every $d \in \big(-\frac{1}{2}\lambda_{\mathbb{R}}(\mathring{I}), \frac{1}{2}\lambda_{\mathbb{R}}(\mathring{I})\big)$ there exist $x, y \in A \subseteq \Gamma$ for which $d = y - x$. □

Lemma 2.2.15 *Let $\mathbb{Q}$ denote the set of rational real numbers. Assuming the axiom of choice, there is a subset A of $\mathbb{R}$ such that $(A - A) \cap \mathbb{Q} = \{0\}$ and yet $\mathbb{R} = \bigcup_{q \in \mathbb{Q}} T_q(A)$.*

Proof. Write $x \sim y$ if $y - x \in \mathbb{Q}$. Then "$\sim$" is an equivalence relation on $\mathbb{R}$, and, for each $x \in \mathbb{R}$, the equivalence class $[x]^{\sim}$ of x is $x + \mathbb{Q}$. Now, using the axiom of choice, choose A to be a set that contains precisely one element from each of the equivalence classes $[x]^{\sim}$, $x \in \mathbb{R}$. It is then clear that A has the required properties. □

Theorem 2.2.16 *Assuming the axiom of choice, every $\Gamma \in \overline{\mathcal{B}_{\mathbb{R}}}^{\lambda_{\mathbb{R}}}$ with positive Lebesgue measure contains a subset that is not Lebesgue measurable. (See part* **(iii)** *of Exercise 2.2.7 for another construction of non-measurable quantities.)*

Proof. Let A be the set constructed in Lemma 2.2.15, and suppose that $\Gamma \cap T_q(A)$ were Lebesgue measurable for each $q \in \mathbb{Q}$. Then we would have that $0 < \overline{\lambda_{\mathbb{R}}}(\Gamma) \leq \sum_{q \in \mathbb{Q}} \overline{\lambda_{\mathbb{R}}}\big(\Gamma \cap T_q(A)\big)$, and so there would exist a $q \in \mathbb{Q}$ such that $\overline{\lambda_{\mathbb{R}}}\big(\Gamma \cap T_q(A)\big) > 0$. But, by Lemma 2.2.14, we would then have that $(-\delta, \delta) \subseteq \{y - x : x, y \in T_q(A)\} \subseteq \{0\} \cup \mathbb{Q}^{\complement}$ for some $\delta > 0$, which cannot be. □

2.2.3 Distribution Functions and Measures

Given a finite Borel measure on $\mathbb{R}$, set $F_\mu(x) = \mu\big((-\infty, x]\big)$. Clearly F_μ is a non-negative, bounded, right-continuous, non-decreasing function that tends to 0 as $x \to -\infty$. The function F_μ is called the **distribution function** for μ. In this subsection I will show that every bounded, right-continuous, non-decreasing function F that tends to 0 at $-\infty$ is the distribution of a unique finite Borel measure μ_F on $\mathbb{R}$.

Let F be given. By Exercise 1.2.7, $F = F_{\mathrm{c}} + F_{\mathrm{d}}$, where F_{c} and F_{d} are bounded and non-decreasing, F_{c} is continuous, and F_{d} is a pure jump function. Further, it is easy to check that both F_{c} and F_{d} can be taken so that they tend to 0 at $-\infty$. Hence, to prove the existence of a μ for which $F = F_\mu$, it suffices to do so when F is either a continuous or a pure jump function and then take the sum of the measures corresponding to F_{c} and F_{d}.

When F is a pure jump function, there is nearly nothing to do. Simply define μ_F by

$$\mu_F(\Gamma) = \sum_{\{x\in\Gamma\cap D\}} \big(F(x) - F(x-)\big) \quad \text{for } \Gamma \in \mathcal{B}_{\mathbb{R}},$$

where D is the countable set consisting of the discontinuities of F. Without difficulty, one can check that μ_F is a finite Borel measure on $\mathbb{R}$ for which F is the distribution function.

Now assume that F is continuous, and take $\mathfrak{R}$ to be the set of all (including the empty interval) closed intervals I in $\mathbb{R}$, and define $V\big([a,b]\big) = F(b) - F(a)$ for $a \le b$. Checking that this choice of $\mathfrak{R}$ and V satisfies that hypotheses at the start of § 2.2.1 is easy. Both (1) and (2) are trivial, the same argument as we used to prove Lemma 1.1.1 when $N = 1$ shows that (3) holds, (4) follows from the continuity of F, and Lemma 2.2.10 proves (5). Thus, by Theorem 2.2.8, there is a Borel measure μ_F on $\mathbb{R}$ for which $\mu_F\big([a,b]\big) = F(b) - F(a)$ for all $a < b$. In particular,

$$F(x) = F(x) - \lim_{y\to-\infty} F(y) = \lim_{y\to-\infty} \mu_F\big([y,x]\big) = \mu_F\big((-\infty,x]\big),$$

and so F is the distribution function for μ_F.

Theorem 2.2.17 *Let F be a bounded, right-continuous, non-decreasing function on $\mathbb{R}$ that tends to 0 at $-\infty$. Then there is a unique Borel measure μ_F on $\mathbb{R}$ for which F is the distribution function. In particular, μ_F is finite and regular. (See Exercises* (3.1.2) *and* (8.2.1) *for other approaches.)*

Proof. The only assertions that have not been covered already are those of uniqueness and finiteness. However, the finiteness follows from $\mu_F(\mathbb{R}) = \lim_{x\nearrow\infty} F(x) < \infty$. To prove the uniqueness, suppose that ν is a second Borel measure on $\mathbb{R}$ satisfying $\nu\big((-\infty,x]\big) = F(x)$ for all $x \in \mathbb{R}$. Then, by the argument just given, ν is finite. In addition, for $a < b$,

$$\nu\big((a,b)\big) = \lim_{x\nearrow b}\big(F(x) - F(a)\big) = F(b-) - F(a) = \mu_F\big((a,b)\big).$$

Hence, $\nu(\mathring{I}) = \mu_F(\mathring{I})$ for all open intervals $\mathring{I}$, and so, by the first part of Lemma 2.2.10, $\nu(G) = \mu_F(G)$ for all $G \in \mathfrak{G}(\mathbb{R})$. By Theorem 2.1.2, this means that $\nu = \mu_F$ on $\mathcal{B}_{\mathbb{R}}$. □

2.2.4 Bernoulli Measure

Here is an application of the material in § 2.2.1 to a probabilistic model of coin tossing.

Set $\Omega = \{0,1\}^{\mathbb{Z}^+}$, the space of maps $\omega : \mathbb{Z}^+ \longrightarrow \{0,1\}$. In the model, Ω is thought of as the set of all possible outcomes of a countably infinite number of coin tosses: $\omega(i) = 1$ if the ith toss came up heads and $\omega(i) = 0$ if it came up tails. Similarly, given $\emptyset \neq F \subset\subset \mathbb{Z}^+$ (i.e., a non-empty, finite subset of $\mathbb{Z}^+$), think of $\{0,1\}^F$ as the set of possible outcomes of those tosses that occurred during F, and define $\Pi_F \Omega \longrightarrow \{0,1\}^F$ to be the projection map given by $\Pi_F \omega = \omega \upharpoonright F$. Then, for each F, $\mathcal{A}(F) \equiv \{\Pi_F^{-1}\Gamma : \Gamma \subseteq \{0,1\}^F\}$ is a σ-algebra over Ω, and, in the model, elements of $\mathcal{A}(F)$ are events (the probabilistic term for subsets) that depend only on the outcome of tosses corresponding to the times $i \in F$. Observe that if $F \subset F'$, then $\mathcal{A}(F) \subset \mathcal{A}(F')$. Indeed, if $\Gamma \subseteq \{0,1\}^F$ and $A = \Pi_F^{-1}\Gamma$, then $A = \Pi_{F'}^{-1}\Gamma'$ where $\Gamma' = \Gamma \times \{0,1\}^{F' \setminus F}$. Finally, set $\mathcal{A} = \bigcup_{F \subset\subset \mathbb{Z}^+} \mathcal{A}(F)$. Obviously $\Omega \in \mathcal{A}$ and $A \in \mathcal{A} \implies A^{\mathrm{C}} \in \mathcal{A}$. In addition, if $A = \Pi_F^{-1}\Gamma \in \mathcal{A}(F)$ and $A' = \Pi_{F'}^{-1}\Gamma' \in \mathcal{A}(F')$, then $A \cup A' = \Pi_{F \cup F'}^{-1}(\Gamma \cup \Gamma') \in \mathcal{A}(F \cup F')$, and so $\mathcal{A}$ is an algebra. However, it is not a σ-algebra since if $\eta \in \Omega$ and $F_n \nearrow \mathbb{Z}^+$, then $A_n = \Pi_{F_n}^{-1}(\Pi_{F_n}\eta) \in \mathcal{A}$ for all n, but $\bigcap_{n=1}^{\infty} A_n = \{\eta\} \notin \mathcal{A}$.

Now suppose that, on each toss, the coin comes up heads with probability $p \in (0,1)$ and tails with probability $q = 1 - p$. Further, assume that the outcomes of distinct tosses are independent of one another. That is, if $\eta \in \{0,1\}^F$, then the probability of the event

$$\Pi_F^{-1}(\{\eta\}) = \{\omega \in \Omega : \omega(i) = \eta(i) \text{ for } i \in F\}$$

is

$$p^{\sum_{i \in F} \eta(i)} q^{\sum_{i \in F}(1-\eta(i))}. \tag{2.2.3}$$

More generally, if $\beta_p^F : \mathcal{A}(F) \longrightarrow [0,1]$ is defined by the prescription

$$\beta_p^F\big(\Pi_F^{-1}\Gamma\big) = \sum_{\eta \in \Gamma} p^{\sum_{i \in F} \eta(i)} q^{\sum_{i \in F}(1-\eta(i))}, \tag{2.2.4}$$

where the sum over the empty set is taken to be 0, $\beta_p^F(A)$ should be the probability of the event $A \in \mathcal{A}(F)$.

To see that β_p^F is a probability measure on $\mathcal{A}(F)$, observe that it is the pullback under Π_F of the measure on $\{0,1\}^F$ which assigns each $\eta \in \{0,1\}^F$ the number in (2.2.3) and that

$$\sum_{\eta \in \{0,1\}^F} p^{\sum_{i \in \eta} \eta(i)} q^{\sum_{i \in \eta}(1-\eta_i)} = (p+q)^{\mathrm{card}(F)} = 1.$$

I now want to define β_p on the algebra $\mathcal{A}$ so that $\beta_p \upharpoonright \mathcal{A}(F) = \beta_p^F$ for all $F \subset\subset \mathbb{Z}^+$. However, to do so, I need to check that if $F \subset F'$ and $A \in \mathcal{A}(F)$, then $\beta_p^{F'}(A) = \beta_p^F(A)$. To do this it suffices to treat the case when $F' = F \cup \{j\}$ for some $j \notin F$. But in that case, if $A = \Pi_F^{-1}\Gamma$, then $A = \Pi_{F'}^{-1}\Gamma'$ where $\Gamma' = \big\{\eta' \in \{0,1\}^{F'} : \eta' \upharpoonright F \in \Gamma \;\&\; \eta'(j) \in \{0,1\}\big\}$, which means that

$$\beta_p^{F'}(A) = \sum_{\eta \in \Gamma} p^{\sum_{i \in \Gamma} \eta(i)} q^{\sum_{i \in \Gamma}(1-\eta(i))}(p+q) = \beta_p^F(A).$$

The challenge now is to show that β_p has a unique extension as a probability measure, which I will continue to denote by β_p, on $\sigma(\mathcal{A})$. Since $\mathcal{A}$ is a Π-system, Lemma 2.1.2 guarantees that there can be at most one such measure. To prove that there is at least one, I must first introduce a metric on Ω for which $\sigma(\mathcal{A})$ is the corresponding Borel σ-algebra and will then use Theorem 2.2.8 to complete the construction.

Define ρ on $\Omega \times \Omega$ by

$$\rho(\omega, \omega') = \sum_{i \in \mathbb{Z}^+} \frac{|\omega(i) - \omega'(i)|}{2^i}.$$

Because each term in the sum is symmetric and satisfies the triangle inequality, the same is true of ρ. In addition, $\rho(\omega, \omega') = 0 \iff \omega = \omega'$, and so ρ is a metric. Next observe that $\rho(\omega, \omega') < 2^{-n} \implies \omega(i) = \omega'(i)$ for $1 \le i \le n$ and that $\omega(i) = \omega'(i)$ for $1 \le i \le n$ implies $\rho(\omega, \omega') \le 2^{-n}$. In particular, $\rho(\omega_k, \omega) \longrightarrow 0$ if and only if $\omega_k(i) \longrightarrow \omega(i)$ for all $i \in \mathbb{Z}^+$. Thus, if $B_\Omega(\omega, r) = \{\omega' : \rho(\omega, \omega') < r\}$ is the ρ-ball of radius $r > 0$ centered at ω and $F_n = \{1, \dots, n\}$, then

$$2^{-n} < r \implies A_n(\omega) \equiv \{\omega' : \omega' \upharpoonright F_n = \omega \upharpoonright F_n\} \subseteq B_\Omega(\omega, r). \tag{2.2.5}$$

Finally, observe that every $A \in \mathcal{A}$ is both open and closed. Indeed, that A is closed is obvious. To see that it is open, set $F_n = \{1, \dots, n\}$ and suppose that $A \in \mathcal{A}$. Then there is an $n \ge 1$ such that $A = \{\omega : \omega \upharpoonright F_n \in \Gamma\}$ for some $\Gamma \subseteq \{0,1\}^{F_n}$. Thus, if $0 < r < 2^{-n}$ and $\omega \in A$, then $\omega' \in B_\Omega(\omega, r) \implies \omega' \upharpoonright F_n = \omega \upharpoonright F_n$ and therefore $\omega' \in A$. Combined with (2.2.5), this proves that $\mathcal{A}$ is a neighborhood basis for the ρ-topology on Ω. In fact, the sets $A_n(\omega)$ are a neighborhood basis for each $\omega \in \Omega$.

The following lemma will be used in the application of Theorem 2.2.8.

Lemma 2.2.18 *The metric space (Ω, ρ) is compact. In addition, if $G \in \mathfrak{G}(\Omega)$, then there exists a sequence $\{A_n : n \ge 1\} \subseteq \mathcal{A}$ of mutually disjoint sets such that $G = \bigcup_{n=1}^\infty A_n$.*

Proof. To prove compactness, let $\{\omega_m : m \ge 1\} \subseteq \Omega$ be given. Then there exists a strictly increasing sequence $\{m_{1,\ell} : \ell \ge 1\} \subseteq \mathbb{Z}^+$ such that $\omega_{m_{1,\ell}}(1) = \omega_{m_{1,1}}(1)$ for all $\ell \in \mathbb{Z}^+$. Knowing $\{m_{1,\ell} : \ell \ge 1\}$, choose a strictly increasing subsequence $\{m_{2,\ell} : \ell \ge 1\}$ of $\{m_{1,\ell} : \ell \ge 1\}$ for which $\omega_{m_{2,\ell}}(2) = \omega_{m_{2,1}}(2)$ for all $\ell \in \mathbb{Z}^+$. Proceeding by induction on $k \in \mathbb{Z}^+$, produce $\{m_{k,\ell} : (k, \ell) \in (\mathbb{Z}^+)^2\}$ such that $\{m_{k+1,\ell} : \ell \ge 1\}$ is a strictly increasing subsequence of $\{m_{k,\ell} : \ell \ge 1\}$ and $\omega_{m_{k,\ell}}(k) = \omega_{m_{k,1}}(k)$ for all $\ell \ge 1$. If $m_k = m_{k,k}$ and $\omega(i) = \omega_{m_i}(i)$, then $\{\omega_{m_i} : i \ge 1\}$ is a subsequence of $\{\omega_m : m \ge 1\}$ and $\omega_{m_j}(i) = \omega(i)$ for $j \ge i$.

Turning to the second assertion, there is nothing to do if $G = \emptyset$. Thus assume that $G \neq \emptyset$, and again set $F_n = \{1, \dots, n\}$. Take A_1 to be the largest element of $\mathcal{A}(F_1)$ contained G. Proceeding by induction, for $n \geq 2$ take A_n to be the largest element of $\mathcal{A}(F_n)$ contained in $G \setminus \bigcup_{m=1}^{n-1} A_m$. Obviously these A_m's are mutually disjoint elements of $\mathcal{A}$ contained in G. To see that their union is G, let $\omega \in G$, and choose $r > 0$ so that $B_\Omega(\omega, r) \subseteq G$ and $n \geq 1$ so that $2^{-n} < r$. By (2.2.5), $A_n(\omega) \subseteq B_\Omega(\Omega, r)$ and so $A_n(\omega)$ is an element of $\mathcal{A}(F_n)$ which is contained in G. Hence, either $A_n(\omega) \subseteq A_n$ or $n \geq 2$ and $A_n(\omega) \cap A_m \neq \emptyset$ for some $1 \leq m < n$. But if the latter is the case and $A_m = \Pi_{F_m}^{-1}\Gamma$, then there exists an $\omega' \in A_n(\omega)$ for which $\omega' \restriction F_m \in \Gamma$, which means that $\omega \restriction F_m \in \Gamma$ and therefore that $\omega \in A_m$. □

Theorem 2.2.19 *The Borel σ-algebra for (Ω, ρ) $\mathcal{B}_\Omega$ equals $\sigma(\mathcal{A})$, and there exists precisely one Borel probability measure β_p such that*

$$\beta_p\big(\Pi_F^{-1}\{\eta\}\big) = p^{\sum_{i \in F} \eta(i)} q^{\sum_{i \in F}(1-\eta(i))}$$

for all $\emptyset \neq F \subset\subset \mathbb{Z}^+$ and $\eta \in \{0,1\}^F$.

Proof. Since $\mathcal{A} \subseteq \mathfrak{G}(\Omega)$, it is clear that $\sigma(\mathcal{A}) \subseteq \mathcal{B}_\Omega$. At the same time, the second part of Lemma 2.2.18 says that $\mathfrak{G}(\Omega) \subseteq \sigma(\mathcal{A})$ and therefore that $\mathcal{B}_\Omega \subseteq \sigma(\mathcal{A})$.

To prove the uniqueness assertion, simply observe that if μ is a Borel probability measure with the stated property, then $\mu \restriction \mathcal{A} = \beta_p \restriction \mathcal{A}$ and therefore $\mu = \beta_p$. Having made these preparations, we can turn to the construction.

To prove the existence assertion, take $\mathfrak{R} = \mathcal{A}$, and define $V(A) = \beta_p(A)$ for $A \in \mathcal{A}$. Then $\mathfrak{R}$ and V satisfy the hypotheses in Theorem 2.2.8. Indeed, because each $A \in \mathcal{A}$ is closed and therefore, by Lemma 2.2.18, compact, the elements of $\mathfrak{R}$ are compact. Conditions (1), (2), and (3) are obvious from the facts that $\mathcal{A}$ is an algebra and that β_p is finitely additive on $\mathcal{A}$. As for (4), the fact that each $A \in \mathcal{A}$ is both open and closed means that there is nothing to check. Finally, Lemma 2.2.18 shows that (5) holds. □

Because Bernoulli (again Jacob) made seminal contributions to the study of coin tossing, the Borel probability measure β_p in Theorem 2.2.19 is called the **Bernoulli measure** with parameter p. Notice that $\beta_p(\{\omega\}) = 0$ for all $\omega \in \Omega$ since $\beta_p(\{\omega\}) \leq \beta_p(A_n(\omega)) \leq (p \vee q)^n$ for all $n \geq 1$.

Before closing this discussion of coin tossing, it should be pointed out that the independence on which (2.2.3) was based extends to β_p as a Borel measure. To describe what this means, suppose that $\emptyset \neq S \subsetneq \Omega$, and set $\mathcal{A}(S) = \bigcup_{F \subset\subset S} \mathcal{A}(F)$. Obviously, if S is finite, then this coincides with the earlier definition.

Theorem 2.2.20 *Let $\emptyset \neq S \subsetneq \mathbb{Z}^+$, and suppose that $B \in \sigma(\mathcal{A}(S))$ and $B' \in \sigma(\mathcal{A}(S^\complement))$. Then $\beta_p(B \cap B') = \beta_p(B)\beta_p(B')$.*

Proof. Suppose that $F \subset\subset S$ and $F' \subset\subset S^{\complement}$. Then, for any $\Gamma \subseteq \{0,1\}^F$ and $\Gamma' \subseteq \{0,1\}^{F'}$, $\omega \in \Pi_F^{-1}\Gamma \cap \Pi^{F'}\Gamma'$ if and only if $\omega \restriction F = \eta$ for some $\eta \in \Gamma$ and $\omega \restriction F' = \eta'$ for some $\eta' \in \Gamma'$. Thus, by (2.2.4),

$$\beta_p\big(\Pi_F^{-1}\Gamma \cap \Pi_{F'}^{-1}\Gamma'\big) = \sum_{(\eta,\eta')\in\Gamma\times\Gamma'} \beta_p(\{\eta\})\beta_p(\{\eta'\}) = \beta_p\big(\Pi_F^{-1}\Gamma\big)\beta_p\big(\Pi_{F'}^{-1}\Gamma'\big).$$

Hence the result holds for $B \in \mathcal{A}(S)$ and $B' \in \mathcal{A}(S^{\complement})$. Next set

$$\mathcal{B}_1 = \{B' \in \sigma(\mathcal{A}(S^{\complement})) : \beta_p(B \cap B') = \beta_p(B)\beta_p(B') \text{ for all } B \in \mathcal{A}(S)\}.$$

Then $\mathcal{B}_1$ is the set of elements of $\sigma(\mathcal{A}(S^{\complement}))$ on which the measures $\mu(B') = \beta_p(B \cap B')$ and $\nu(B') = \beta_p(B)\beta_p(B')$ are equal, and so $\mathcal{B}_1$ is a Λ-system. In addition, as we have just seen, $\mathcal{A}(S^{\complement}) \subseteq \mathcal{B}_1$, and so, since $\mathcal{A}(S^{\complement})$ is an algebra and therefore a Π-system, $\sigma(\mathcal{A}(S^{\complement})) \subseteq \mathcal{B}_1$. Thus $\beta_p(B \cap B') = \beta_p(B)\beta_p(B')$ for all B' for $B \in \mathcal{A}(S)$ and $B' \in \sigma(\mathcal{A}(S^{\complement}))$. Now let $B' \in \sigma(\mathcal{A}(S^{\complement}))$ be given, and set

$$\mathcal{B}_2 = \{B \in \sigma(\mathcal{A}(S)) : \beta_p(B \cap B') = \beta_p(B)\beta_p(B') \text{ for all } B' \in \mathcal{A}(S^{\complement})\}.$$

Just as before, $\mathcal{B}_2$ is a Λ-system. Further $\mathcal{A}(S)$ is a Π-system, and, by the result just proved, $\mathcal{A}(S) \subseteq \mathcal{B}_2$. Hence $\sigma(\mathcal{A}(S)) \subseteq \mathcal{B}_2$, and so the proof is complete. □

In the jargon of probability theory, Theorem 2.2.20 is saying that the σ-algebra $\sigma(\mathcal{A}(S))$ is *independent* under β_p of the σ-algebra $\sigma\big(\mathcal{A}(S^{\complement})\big)$.

2.2.5 Bernoulli and Lebesgue Measures

For obvious reasons, the case $p = \frac{1}{2}$ is thought of as the mathematical model of a coin tossing game in which the coin is fair (i.e., unbiased). Thus, one should hope that $\beta_{\frac{1}{2}}$ has special properties, and the purpose of this subsection is to prove one such property.

There is a natural continuous map $\Phi : \Omega \longrightarrow [0,1]$ given by

$$\Phi(\omega) = \sum_{n=1}^{\infty} 2^{-n}\omega(n). \tag{2.2.6}$$

By part (**ii**) of Exercise 2.1.3, Φ is measurable as a mapping from $(\Omega, \mathcal{B}_\Omega)$ to $\big([0,1], \mathcal{B}_{[0,1]}\big)$. What is less obvious is that (cf. part (**iii**) of Exercise 2.1.3) $\Phi_*\beta_{\frac{1}{2}} = \lambda_{[0,1]}$, where (cf. Exercise 2.1.2) $\lambda_{[0,1]}$ is the Borel measure on $[0,1]$ obtained by restricting $\lambda_{\mathbb{R}}$ to $\mathcal{B}_{[0,1]}$. To prove this, I will use the following lemma. In its statement and below, $\hat{\Omega}$ denotes the set of $\omega \in \Omega$ such that either $\omega = \mathbf{1}$, where $\mathbf{1}$ is the element Ω that is 1 at every $i \in \mathbb{Z}^+$, or $\omega(i) = 0$ for infinitely many $i \in \mathbb{Z}^+$. Observe that $\Omega \setminus \hat{\Omega}$ consists of $\omega \in \Omega \setminus \{\mathbf{1}\}$ for which there is an $m \in \mathbb{Z}^+$ such that $\omega(i) = 1$ for all $i \geq m$, and so $\Omega \setminus \hat{\Omega}$ is countable. Hence $\hat{\Omega} \in \mathcal{B}_\Omega$ and $\beta_p(\hat{\Omega}) = 1$ for all $p \in (0,1)$. As before, $F_n = \{1, \dots, n\}$.

Lemma 2.2.21 *For each $x \in [0,1]$ there is a unique $\omega_x \in \hat{\Omega}$ such that $\Phi(\omega_x) = x$; and $x \in [0,1)$ is of the form $k2^{-n}$ for some $0 \leq k < 2^n$ if and only if $\omega_x(i) = 0$ for $i > n$. Furthermore, if $\Psi : [0,1] \longrightarrow \hat{\Omega}$ is given by $\Psi(x) = \omega_x$, then $\Phi \circ \Psi$ is the identity map on $[0,1]$ and $\Psi \circ \Phi$ is the identity map on $\hat{\Omega}$. Finally, if $n \geq 1$, $\eta \in \{0,1\}^{F_n}$, and $k2^{-n} = \sum_{i=1}^n \eta(i)2^{-i}$, then*

$$\Phi\big(\{\omega \in \hat{\Omega} \setminus \{\mathbf{1}\} : \omega \restriction F_n = \eta\}\big) = \big[k2^{-n}, (k+1)2^{-n}\big).$$

Proof. Clearly $\mathbf{1}$ is the unique element of Ω that Φ takes to 1. To construct ω_x for $x \in [0,1)$, take $\omega_x(1) = 0$ if $x < \frac{1}{2}$ and $\omega_x(1) = 1$ otherwise. Next, having determined $\omega_x(m)$ for $1 \leq m \leq n$, take $\omega_x(n+1)$ equal to 0 or 1 depending on whether $x - \sum_{m=1}^n \omega_x(m)2^{-m} < 2^{-n-1}$ or $x - \sum_{m=1}^n \omega_x(m)2^{-m} \geq 2^{-n-1}$. Clearly $0 \leq x - \sum_{m=1}^n \omega_x(m) < 2^{-n}$ for all $n \geq 1$. Hence, $\Phi(\omega_x) = x$, and, if there existed an m such that $\omega_x(n) = 1$ for $n \geq m$, then we would have the contradiction

$$2^{-m} > x - \sum_{n=1}^{m} \omega_x(n) = \sum_{n=m+1}^{\infty} \omega_x(n) = 2^{-m}.$$

Further, if ω were another element of $\hat{\Omega} \setminus \{\mathbf{1}\}$ for which $\Phi(\omega) = x$, then there would be a smallest m at which ω differs from ω_x, and we need only treat the case when $\omega(m) = 1$ and $\omega_x(m) = 0$. But this would mean that

$$0 = \sum_{i=1}^{\infty} \omega(i) - \sum_{i=1}^{\infty} \omega_x(i)2^{-i} = 2^{-m} + \sum_{i>m} \big(\omega(i) - \omega_x(i)\big)2^{-i} \neq 0$$

since the absolute value of the last sum must be strictly less than 2^{-m}.

If $\omega(i) = 0$ for $i > n$, then $2^n\Phi(\omega)$ is an non-negative integer less than 2^n. Conversely, suppose that $x = k2^{-n}$ for some $1 \leq k < 2^{-n}$. Then $2^n \sum_{i=n+1}^{\infty} \omega_x(i)2^{-i}$ would have to be an integer, and so, since

$$\sum_{i=n+1}^{\infty} \omega_x(i)2^{-i} < 2^{-n},$$

this possible only if $\omega_x(i) = 0$ for $i > n$.

That $\Phi \circ \Psi(x) = x$ for all $x \in [0,1]$ is obvious. Now suppose that $\omega \in \hat{\Omega}$, and set $x = \Phi(\omega)$. Then, since ω_x is the unique $\omega' \in \hat{\Omega}$ for which $\Phi(\omega') = x$, it follows that $\omega = \omega_x$ and therefore that $\Psi \circ \Phi(\omega) = \omega$.

Finally, let $\eta \in \{0,1\}^{F_n}$ and $k2^{-n} = \sum_{i=1}^n \eta(i)2^{-i}$. For any $\omega \in \hat{\Omega} \setminus \{\mathbf{1}\}$ satisfying $\omega \restriction F_n = \eta$, $\Phi(\omega) - k2^{-n} < 2^{-n}$ and therefore

$$\Phi(\omega) \in [k2^{-n}, (k+1)2^{-n}).$$

Hence

$$\Phi\big(\{\omega \in \hat{\Omega} \setminus \{\mathbf{1}\} : \omega \restriction F_n = \eta\}\big) \subseteq \big[k2^{-n}, (k+1)2^{-n}\big).$$

Now suppose that $x \in [k2^{-n}, (k+1)2^{-n})$, and set $y = x - k2^{-n}$. Then $2^n y \in [0,1)$, and so $\omega_{2^n y} \in \hat{\Omega} \setminus \{\mathbf{1}\}$. If $\omega(i) = \eta(i)$ for $1 \le i \le n$ and $\omega(i) = \omega_{2^n y}(i-n)$ for $i > n$, then $\omega \in \hat{\Omega} \setminus \{\mathbf{1}\}$ and

$$\Phi(\omega) = k2^{-n} + \sum_{i=n+1}^{\infty} \omega_{2^n y}(i-n)2^{-i} = k2^{-n} + 2^{-n}\Phi(\omega_{2^n y}) = k2^{-n} + y = x,$$

which means that $\omega = \omega_x$ and therefore that

$$\Phi\big(\{\omega \in \hat{\Omega} \setminus \{\mathbf{1}\} : \omega \restriction F_n = \eta\}\big) \supseteq \big[k2^{-n}, (k+1)2^{-n}\big). \qquad \square$$

Theorem 2.2.22 $\Phi_* \beta_{\frac{1}{2}} = \lambda_{[0,1]}$. *In addition, for all* $A \in \mathcal{B}_\Omega$, $\Phi(A) \in \mathcal{B}_{[0,1]}$ *and* $\lambda\big(\Phi(A)\big) = \beta_{\frac{1}{2}}(A)$.

Proof. Since $\{[k2^{-n}, (k+1)2^{-n}) : n \ge 1 \;\&\; 0 \le k < 2^n\} \cup \{\mathbf{1}\}$ is a Π-system that generates $\mathcal{B}_{[0,1]}$ and $\lambda(\{1\}) = 0 = \beta_{\frac{1}{2}}(\{\mathbf{1}\})$, Theorem 2.1.2 says that we need only check that $\lambda_{[0,1]}([k2^{-n}, (k+1)2^{-n})) = \beta_{\frac{1}{2}}\big(\Phi^{-1}([k2^{-n}, (k+1)2^{-n}))\big)$ for $n \ge 1$ and $0 \le k < 2^n$. But $\lambda_{[0,1]}([k2^{-n}, (k+1)2^{-n})) = 2^{-n}$, and, by Lemma 2.2.21,

$$\Phi^{-1}\big([k2^{-n}, (k+1)2^{-n})\big) \cap (\hat{\Omega} \setminus \{\mathbf{1}\}) = \big\{\omega \in \hat{\Omega} \setminus \{\mathbf{1}\} : \omega \restriction F_n = \eta\big\},$$

where η is the element of $\{0,1\}^{F_n}$ for which $k2^{-n} = \sum_{i=1}^n \eta(i)2^{-i}$. Thus, since $\Omega \setminus (\hat{\Omega} \setminus \{\mathbf{1}\})$ is countable and therefore has $\beta_{\frac{1}{2}}$-measure 0,

$$\begin{aligned}\beta_{\frac{1}{2}}\big(\Phi^{-1}([k2^{-n}, (k+1)2^{-n})\big) &= \beta_{\frac{1}{2}}\Big(\big\{\omega \in \hat{\Omega} \setminus \{\mathbf{1}\} : \omega \restriction F_n = \eta\big\}\Big) \\ &= \beta_{\frac{1}{2}}\Big(\big\{\omega \in \Omega : \omega \restriction F_n = \eta\big\}\Big) = 2^{-n}.\end{aligned}$$

To prove that $\Phi(A) \in \mathcal{B}_{[0,1]}$ for $A \in \mathcal{B}_\Omega$, it suffices to do so when $A \subseteq \hat{\Omega}$, and so, since $\Phi = \Psi^{-1}$ on $\hat{\Omega}$, it suffices to show that Ψ is measurable. Further, because $\mathcal{B}_\Omega$ is generated by sets of the form $\{\omega : \omega \restriction F_n = \eta\}$ for some $n \ge 1$ and $\eta \in \{0,1\}^{F_n}$, and, by Lemma 2.2.21,

$$\Psi^{-1}\big(\{\omega \in \hat{\Omega} \setminus \{\mathbf{1}\} : \omega \restriction F_n = \eta\}\big) = \left[\sum_{i=1}^n \eta(i)2^{-i}, \sum_{i=1}^n \eta(i)2^{-i} + 2^{-n}\right) \in \mathcal{B}_{[0,1]},$$

we have shown that $\Phi(A) \in \mathcal{B}_{[0,1]}$ for all $A \in \mathcal{B}_{\Omega}$. Finally, if $A \in \mathcal{B}_{\Omega}$, then

$$\begin{aligned}\lambda_{[0,1]}\big(\Phi(A)\big) &= \beta_{\frac{1}{2}}\Big(\Phi^{-1}\big(\Phi(A)\big)\Big) \\ &= \beta_{\frac{1}{2}}(A) + \beta_{\frac{1}{2}}\big(\{\omega \notin \hat{\Omega} : \Phi(\omega) \in \Phi(A)\}\big) = \beta_{\frac{1}{2}}(A).\end{aligned}$$

□

The interest in Theorem 2.2.22 stems from the following considerations. For each $x \in [0,1]$, ω_x gives the coefficients in the dyadic expansion of x. Thus, if one is interested in the set Γ of those $x \in [0,1]$ whose dyadic coefficients possess a certain property that can be described in terms of an $A \in \mathcal{B}_{\Omega}$, then $\Gamma \in \mathcal{B}_{[0,1]}$, and one can use $\beta_{\frac{1}{2}}$ to calculate the Lebesgue measure of Γ. That is, as measured by $\lambda_{[0,1]}$, the behavior of the dyadic coefficients of $x \in [0,1]$ is the same as the outcomes of a coin tossing game played with a fair coin.

2.2.6 *Exercises for § 2.2*

Exercise 2.2.1 Suppose that G is an open subset of $\mathbb{R}^N$ and that $\Phi : G \longrightarrow \mathbb{R}^{N'}$ is **uniformly Lipschitz continuous** in the sense that there is an $L < \infty$ such that $|\Phi(y) - \Phi(x)| \le L|y - x|$ for all $x, y \in G$. Because Φ is continuous, it takes compact subsets of G to compact sets, and from this conclude that Φ takes elements of $\mathfrak{F}_\sigma(G)$ to elements of $\mathcal{B}_{\mathbb{R}^{N'}}$. Next, assuming that $N' \ge N$, show that if $\Gamma \in \overline{\mathcal{B}_G}^{\lambda_{\mathbb{R}^N}}$ has Lebesgue measure 0, then $\Phi(\Gamma)$ is an element of $\overline{\mathcal{B}_{\mathbb{R}^N}}^{\lambda_{\mathbb{R}^{N'}}}$ that has Lebesgue measure 0. Finally, combine these to show that $\Phi(\Gamma) \in \overline{\mathcal{B}_{\mathbb{R}^N}}^{\lambda_{\mathbb{R}^{N'}}}$ for every $\Gamma \in \overline{\mathcal{B}_G}^{\lambda_{\mathbb{R}^N}}$.

Exercise 2.2.2 Let $\mathcal{B}$ be a σ-algebra over E with the property that $\{x\} \in \mathcal{B}$ for all $x \in E$. A measure μ on $(E, \mathcal{B})$ is said to be **non-atomic** if $\mu(\{x\}) = 0$ for all $x \in E$. Show that if there is a non-trivial (i.e., not identically 0), non-atomic measure on $(E, \mathcal{B})$, then E must be uncountable. Next, apply this to show that the existence of $\lambda_{\mathbb{R}^N}$ and β_p implies that $\mathbb{R}^N$, Ω, and $\hat{\Omega}$ must all be uncountable.

Exercise 2.2.3 It is clear that any countable subset of $\mathbb{R}$ has Lebesgue measure zero. However, it is not so immediately clear that there are uncountable subsets of $\mathbb{R}$ whose Lebesgue measure is zero. The goal of this exercise is to show how to construct such a set. For this purpose, start with the set $C_0 = [0,1]$, and let C_1 be the set obtained by removing the open middle third of C_0 (i.e., $C_1 = C_0 \setminus \left(\frac{1}{3}, \frac{2}{3}\right) = \left[0, \frac{1}{3}\right] \cup \left[\frac{2}{3}, 1\right]$). Next, let C_2 be the set obtained from C_1 after removing the open middle third of each of the (two) intervals of which C_1 is the disjoint union. More generally, given C_n (which is the union of 2^n disjoint, closed intervals), let C_{n+1} be the set that one gets by removing from C_n the open middle third of each of the intervals of which

C_n is the disjoint union. Finally, set $C = \bigcap_{k=0}^{\infty} C_n$. The set C is called the **Cantor set**, and it turns out to be an extremely useful source of examples. In particular, show that C is an uncountable, closed subset of $[0,1]$ that has Lebesgue measure 0. See Exercise 8.3.5 for further information. Here are some steps that you might want to follow.

(i) Since each C_n is closed, C is also. Next, show that $\lambda_{\mathbb{R}}(C_n) = \left(\frac{2}{3}\right)^n$ and therefore that $\lambda_{\mathbb{R}}(C) = 0$.

(ii) To prove that C is uncountable, refer to the notation in § 2.2.4, and define $\Psi : \Omega \longrightarrow [0,1]$ by

$$\Psi(\omega) = \sum_{i=1}^{\infty} \frac{2\omega(i)}{3^i}.$$

Show that Ψ is one-to-one.

(iii) In view of Exercise 2.2.2, one will know that C must be uncountable if $\Psi(\hat{\Omega}) \subseteq C$. To this end, first show that $[0,1] \setminus C$ can be covered by open intervals of the form $\big((2k-1)3^{-n}, 2k3^{-n}\big)$, where $n \in \mathbb{Z}^+$ and $1 \le k \le \frac{3^n-1}{2}$. Next, show that $\Psi(\omega) > (2k-1)3^{-n} \implies \Psi(\omega) \ge 2k3^{-n}$ and therefore that $\Psi(\hat{\Omega}) \subseteq C$.

Exercise 2.2.4 Here is a rather easy application of Theorem 2.2.13. If $B_{\mathbb{R}^N}(c,r)$ is the open ball in $\mathbb{R}^N$ of radius r and center c, show that

$$\lambda_{\mathbb{R}^N}\big(B_{\mathbb{R}^N}(c,r)\big) = \lambda_{\mathbb{R}^N}\left(\overline{B_{\mathbb{R}^N}(c,r)}\right) = \Omega_N r^N, \quad \text{where } \Omega_N \equiv \lambda_{\mathbb{R}^N}\big(B_{\mathbb{R}^N}(0,1)\big)$$

is the (cf. **(iii)** in Exercise 5.1.3) the **volume of the unit ball**

Exercise 2.2.5 If $\mathbf{v}_1, \ldots,$ and $\mathbf{v}_N$ are vectors in $\mathbb{R}^N$, the **parallelepiped spanned by** $\{\mathbf{v}_1, \ldots, \mathbf{v}_N\}$ is the set

$$P(\mathbf{v}_1, \ldots, \mathbf{v}_N) \equiv \left\{\sum_1^N x_i \mathbf{v}_i : x \in [0,1]^N\right\}.$$

When $N \ge 2$, the classical prescription for computing the *volume* of a parallelepiped is to take the product of *the area of any one side* times the length of the corresponding *altitude*. In analytic terms, this means that the volume is 0 if the vectors $\mathbf{v}_1, \ldots, \mathbf{v}_N$ are linearly dependent and that otherwise the volume of $P(\mathbf{v}_1, \ldots, \mathbf{v}_N)$ can be computed by taking the product of the volume of $P(\mathbf{v}_1, \ldots, \mathbf{v}_{N-1})$, thought of as a subset of the hyperplane $H(\mathbf{v}_1, \ldots, \mathbf{v}_{N-1})$ spanned by $\mathbf{v}_1, \ldots, \mathbf{v}_{N-1}$, times the distance between the vector $\mathbf{v}_N$ and the hyperplane $H(\mathbf{v}_1, \ldots, \mathbf{v}_{N-1})$. Using Theorem 2.2.13, show that this prescription is correct when the *volume* of a set is interpreted as the Lebesgue measure of that set.

Hint: Take A to be the $N \times N$ matrix whose ith column is $\mathbf{v}_i$, and use Cramer's rule to compute $\det(A)$.

Exercise 2.2.6 Cauchy posed the problem of determining which functions $f : \mathbb{R} \longrightarrow \mathbb{R}$ are **additive** in the sense that $f(x+y) = f(x) + f(y)$ for all $x, y \in \mathbb{R}$. The goal of this exercise is to show that an additive function that is $\overline{\mathcal{B}_{\mathbb{R}}}^{\lambda_{\mathbb{R}}}$-measurable must be linear. That is, $f(x) = f(1)x$ for all $x \in \mathbb{R}$.

(i) Show that, for each $x \in \mathbb{R}$ and rational number q, $f(qx) = qf(x)$. In particular, conclude that any continuous, additive function is linear.

(ii) Show that if f is bounded on some non-empty open set, then f is continuous and therefore linear.

Hint: First show that f must be bounded on $[-R, R]$ for all $R > 0$, and then write $f(y) - f(x)$ as $n^{-1} f\big(n(y-x)\big)$.

(iii) Assume that f is a $\overline{\mathcal{B}_{\mathbb{R}}}^{\lambda_{\mathbb{R}}}$-measurable, additive function. Choose an $R > 0$ for which $\Gamma = \{x \in \mathbb{R} : |f(x)| \le R\}$ has strictly positive $\overline{\lambda_{\mathbb{R}}}$-measure, and use Lemma 2.2.14 and additivity to conclude that there is a $\delta > 0$ for which $|f(x)| \le 2R$ on $(-\delta, \delta)$. After combining this with **(ii)**, conclude that *every* $\overline{\mathcal{B}_{\mathbb{R}}}^{\lambda_{\mathbb{R}}}$*-measurable, additive function is linear.*

Exercise 2.2.7 In connection with Exercise 2.2.6, one should ask whether there are solutions to Cauchy's functional equation that are not linear. Because any such solution cannot be Lebesgue measurable, one should expect that its construction must require the axiom of choice. What follows is an outline of a construction.

(i) Let $\mathcal{A}$ denote the set of all subsets $A \subseteq \mathbb{R}$ that are linearly independent over the rational numbers $\mathbb{Q}$ in the sense that, for every finite subset $F \subseteq A$ and every choice of $\{\alpha_x : x \in F\} \subseteq \mathbb{Q}$, $\sum_{x \in F} \alpha_x x = 0 \implies \alpha_x = 0$ for all $x \in F$. Partially order $\mathcal{A}$ by inclusion, and show that every totally ordered subset of $\mathcal{A}$ admits an upper bound. That is, if $\mathcal{T} \subseteq \mathcal{A}$ and, for all $A, B \in \mathcal{T}$, either $A \subseteq B$ or $B \subseteq A$, then there exists an $M \in \mathcal{A}$ such that $A \subseteq M$ for all $A \in \mathcal{T}$. Now apply the Zorn's Lemma, which is one of the equivalent forms of the axiom of choice, to show that there exists an $M \in \mathcal{A}$ that is maximal in the sense that, for all $A \in \mathcal{A}$, $M \subseteq A \implies M = A$.

(ii) Referring to **(i)**, show that M is a Hamel basis for $\mathbb{R}$ over the rationals. That is, for all $y \in \mathbb{R} \setminus \{0\}$ there exist a unique finite $F(y) \subseteq M$ and a unique choice of $\{q_x(y) : x \in F(y)\} \subset \mathbb{Q} \setminus \{0\}$ for which $y = \sum_{x \in F(y)} q_x(y)x$.

(iii) Continuing **(i)** and **(ii)**, extend the definition of $q_x(y)$ so that $q_x(y) = 0$ if either $y = 0$ or $x \notin F(y)$. Then, for all $y \in \mathbb{R}$, $y = \sum_{x \in M} q_x(y)x$, where, for each y, all but a finite number of summands are 0. Next, let ψ be any $\mathbb{R}$-valued function on M, define $f : \mathbb{R} \longrightarrow \mathbb{R}$ so that $f(y) = \sum_{x \in M} \psi(x) q_x(y) x$, and show that f is always additive but that it is linear if and only if ψ is constant. In particular, for each $x \in M$, q_x is an additive, non-linear function. Conclude from this that, for each $x \in M$, $y \rightsquigarrow q_x(y)$ cannot be Lebesgue measurable and must be unbounded on each non-empty open interval.

Exercise 2.2.8 A right-continuous, non-decreasing function $F:\mathbb{R}\longrightarrow\mathbb{R}$ is said to be **absolutely continuous** if for every $\epsilon>0$ there exists a $\delta>0$ such that $\sum_{n=1}^{\infty}\bigl(F(b_n)-F(a_n)\bigr)<\epsilon$ whenever $\{(a_n,b_n):n\geq 1\}$ is a sequence of mutually disjoint open intervals satisfying $\sum_{n=1}^{\infty}(b_n-a_n)<\delta$. Show that an absolutely continuous F is uniformly continuous. Next, assume that F is bounded and tends to 0 at $-\infty$, and let (cf. Theorem 2.2.17) μ_F be the Borel measure on $\mathbb{R}$ for which F is the distribution function. Show that F is absolutely continuous as a function if and only if μ_F is (cf. Exercise 2.1.8) absolutely continuous with respect to $\lambda_{\mathbb{R}}$.

Exercise 2.2.9 Given a bounded, right-continuous, non-decreasing function F on $\mathbb{R}$, say that F is **singular** if for each $\delta>0$ there exists a sequence $\{(a_n,b_n):n\geq 1\}$ of mutually disjoint open intervals such that

$$\sum_{n=1}^{\infty}(b_n-a_n)<\delta \text{ and}$$

$$F(\infty)-F(-\infty)=\sum_{n=1}^{\infty}\bigl(F(b_n-)-F(a_n)\bigr).$$

Assuming that F tends to 0 at $-\infty$, show that F is singular if and only if the measure μ_F for which it is the distribution function is (cf. Exercise 2.1.9) singular to $\lambda_{\mathbb{R}}$.

Exercise 2.2.10 If $\{S_1,\dots,S_n\}$ are mutually disjoint subsets of $\mathbb{Z}^+$ for some $n\geq 2$, show that

$$\overline{\beta_p}(A_1\cap\cdots\cap A_n)=\overline{\beta_p}(A_1)\cdots\overline{\beta_p}(A_n)$$

for every choice of $\{A_1,\dots,A_n\}\subseteq\overline{\mathcal{B}_\Omega}^{\beta_p}$ with $A_m\in\mathcal{A}(S_m)$ for $1\leq m\leq n$.

Chapter 3
Lebesgue Integration

At the beginning of Chapter 2, I attempted to motivate the introduction of measures by indicating how important a role they play in Lebesgue's program for constructing integrals of functions. Now that we know that measures exist, in this chapter I will carry out his construction. Once I have introduced his integral, I will prove a few of the properties for which it is famous. In particular, I will show that it is remarkably continuous as a function of its integrand. Finally, in the last section, I will prove a beautiful theorem, again due to Lebesgue, which can be viewed as an extension of the Fundamental Theorem of Calculus to the setting of Lebesgue's theory of integration.

3.1 The Lebesgue Integral

In this section I will describe Lebesgue's theory of integration. Although his theory requires one to deal with a number of fussy points, I hope that my reader will, in the end, agree that it is such a natural theory that it could have been proposed by an intelligent child.

3.1.1 Some Miscellaneous Preliminaries

In order to handle certain measurability questions, it is best to introduce at this point a construction to which we will return in Chapter 4. Namely, given measurable spaces $(E_1, \mathcal{B}_1)$ and $(E_2, \mathcal{B}_2)$, define the **product** of $(E_1, \mathcal{B}_1)$ and $(E_2, \mathcal{B}_2)$ to be the measurable space $(E_1 \times E_2, \mathcal{B}_1 \times \mathcal{B}_2)$ where[1]

[1] Strictly speaking, the notation $\mathcal{B}_1 \times \mathcal{B}_2$ is incorrect and should be replaced by something even more hideous like $\sigma(\mathcal{B}_1 \times \mathcal{B}_2)$. Nonetheless, for the sake of aesthetics if nothing else, I have chosen to stick with $\mathcal{B}_1 \times \mathcal{B}_2$.

D. W. Stroock, *Essentials of Integration Theory for Analysis*, Graduate Texts in Mathematics 262, https://doi.org/10.1007/978-3-030-58478-8_3

$$\mathcal{B}_1 \times \mathcal{B}_2 \equiv \sigma\big(\{\Gamma_1 \times \Gamma_2 : \Gamma_1 \in \mathcal{B}_1 \text{ and } \Gamma_2 \in \mathcal{B}_2\}\big).$$

Also, if Φ_i is a measurable map on $(E_0, \mathcal{B}_0)$ into $(E_i, \mathcal{B}_i)$ for $i \in \{1,2\}$, then the **tensor product of Φ_1 times Φ_2** is the map $\Phi_1 \bigotimes \Phi_2 : E_0 \longrightarrow E_1 \times E_2$ given by $\Phi_1 \times \Phi_2(x) = \big(\Phi_1(x), \Phi_2(x)\big)$, $x \in E_0$.

Lemma 3.1.1 *Referring to the preceding, suppose that, for $i \in \{1,2\}$, Φ_i is a measurable map on $(E_0, \mathcal{B}_0)$ into $(E_i, \mathcal{B}_i)$. Then $\Phi_1 \bigotimes \Phi_2$ is a measurable map on $(E_0, \mathcal{B}_0)$ into $(E_1 \times E_2, \mathcal{B}_1 \times \mathcal{B}_2)$. Moreover, if E_1 and E_2 are second countable topological spaces,*[2] *then $\mathcal{B}_{E_1} \times \mathcal{B}_{E_2} = \mathcal{B}_{E_1 \times E_2}$.*

Proof. Because of Lemma 2.1.1, to prove the first assertion one need only note that if $\Gamma_i \in \mathcal{B}_i$, $i \in \{1,2\}$, then $\Phi_1 \bigotimes \Phi_2{}^{-1}(\Gamma_1 \times \Gamma_2) = \Phi_1^{-1}(\Gamma_1) \cap \Phi_2^{-1}(\Gamma_2) \in \mathcal{B}_0$. As for the second assertion, first observe that $G_1 \times G_2$ is open in $E_1 \times E_2$ for every pair of open sets G_1 in E_1 and G_2 in E_2. Hence, even without second countability, $\mathcal{B}_{E_1} \times \mathcal{B}_{E_2} \subseteq \mathcal{B}_{E_1 \times E_2}$. At the same time, with second countability, one can write every open G in $E_1 \times E_2$ as the countable union of sets of the form $G_1 \times G_2$ where G_i is open in E_i. Hence, in this case, the opposite inclusion also holds. □

In measure theory one is most interested in real-valued functions. However, for reasons of convenience, it is often handy to allow functions to take values in the **extended real line** $\overline{\mathbb{R}} \equiv [-\infty, \infty]$. Unfortunately, the introduction of $\overline{\mathbb{R}}$ introduces some annoying problems. These problems stem from the difficulty of extending addition to include ∞ and $-\infty$. In an attempt to avoid such problems in the future, I will spend a little time trying to lay them to rest now.

To begin with, it should be clear that the *natural* notion of convergence in $\overline{\mathbb{R}}$ is the one that says that $\{x_n : n \geq 1\} \subseteq \overline{\mathbb{R}}$ converges to $a \in \mathbb{R}$ if $\{x_{m+n} : n \geq 1\} \subseteq \mathbb{R}$ for some $m \geq 0$ and $x_{m+n} \longrightarrow a$ in $\mathbb{R}$ as $n \to \infty$ and that it converges to $\pm\infty$ if, for every $R > 0$, there is an $m_R \geq 0$ for which $\pm x_n \geq R$ whenever $n \geq m_R$. With this in mind, I introduce a metric on $\overline{\mathbb{R}}$ for which this *natural* notion is *the* notion of convergence and with which $\overline{\mathbb{R}}$ is compact. Namely, define

$$\overline{\rho}(\alpha, \beta) = \frac{2}{\pi}\big|\arctan(\beta) - \arctan(\alpha)\big|,$$

where $\arctan(\pm\infty) \equiv \pm\frac{\pi}{2}$. Clearly $(\overline{\mathbb{R}}, \overline{\rho})$ is a compact metric space: it is isometric to $[-1,1]$ under the map $t \in [-1,1] \longmapsto \tan\left(\frac{\pi t}{2}\right)$. Moreover, $\mathbb{R}$, with the usual Euclidean topology, is embedded in $\overline{\mathbb{R}}$ as a dense open set.

[2] A topological space is said to be second countable if it possesses a countable collection $\mathfrak{B}$ of open subsets that forms a basis in the sense that, for every open subset G and $x \in G$, there is a $B \in \mathfrak{B}$ with $x \in B \subseteq G$. For example, every separable metric space is second countable. Indeed, let $\{p_m : m \geq 1\}$ be a countable dense subset, and take $\mathfrak{B}$ to be the set of open balls centered at some p_m and having a radius n^{-1} for some $n \geq 1$.

In particular (cf. Exercise 2.1.2), $\mathcal{B}_{\mathbb{R}} = \mathcal{B}_{\overline{\mathbb{R}}}[\mathbb{R}]$. Having put a topology and measurable structure on $\overline{\mathbb{R}}$, we will next adopt the following extension to $\overline{\mathbb{R}}$ of multiplication: $(\pm\infty)\cdot 0 = 0 = 0\cdot(\pm\infty)$ and $(\pm\infty)\cdot\alpha = \alpha\cdot(\pm\infty) = \operatorname{sgn}(\alpha)\infty$ if $\alpha \in \overline{\mathbb{R}}\setminus\{0\}$. Although $(\alpha,\beta) \in \overline{\mathbb{R}}^2 \longmapsto \alpha\cdot\beta \in \overline{\mathbb{R}}$ is not continuous, one can easily check that it is a measurable map on $\bigl(\overline{\mathbb{R}}^2, \mathcal{B}_{\overline{\mathbb{R}}^2}\bigr)$ into $\bigl(\overline{\mathbb{R}}, \mathcal{B}_{\overline{\mathbb{R}}}\bigr)$. As mentioned earlier, the extension of addition presents a knottier problem. Indeed, because one does not know how to interpret $\pm\infty \mp \infty$ in general, we will simply avoid having to do so at all by restricting the domain of addition to the set $\widehat{\mathbb{R}^2}$ consisting of $\overline{\mathbb{R}}^2$ with the two points $(\pm\infty, \mp\infty)$ removed. Clearly $\widehat{\mathbb{R}^2}$ is an open subset of $\overline{\mathbb{R}}^2$, and so $\mathcal{B}_{\widehat{\mathbb{R}^2}} = \mathcal{B}_{\overline{R}^2}\bigl[\widehat{\mathbb{R}^2}\bigr]$. Moreover, there is no problem about defining the sum of $(\alpha,\beta) \in \widehat{\mathbb{R}^2} \longmapsto \alpha+\beta \in \overline{\mathbb{R}}$. Indeed, we will take $\alpha+\beta$ as usual when both α and β are in $\mathbb{R}$ and $(\pm\infty)+\alpha = \alpha+(\pm\infty) = \pm\infty$ if $\alpha \neq \mp\infty$. It is then easy to see that $(\alpha,\beta) \longmapsto \alpha+\beta$ is continuous on $\widehat{\mathbb{R}^2}$ into $\overline{\mathbb{R}}$, and therefore it is certainly a measurable map on $\bigl(\widehat{\mathbb{R}^2}, \mathcal{B}_{\widehat{\mathbb{R}^2}}\bigr)$ into $\bigl(\overline{\mathbb{R}}, \mathcal{B}_{\overline{\mathbb{R}}}\bigr)$. Finally, to complete this discussion of $\overline{\mathbb{R}}$, observe that the lattice operations "$\vee$" and "$\wedge$" both admit unique continuous extensions as maps from $\overline{\mathbb{R}}^2$ into $\overline{\mathbb{R}}$ and are therefore not a source of concern.

Having adopted these conventions, one sees that, for any pair of measurable functions f_1 and f_2 on a measurable space $(E,\mathcal{B})$ into $\bigl(\overline{\mathbb{R}}, \mathcal{B}_{\overline{\mathbb{R}}}\bigr)$, Lemma 3.1.1 guarantees the measurability of $x \in E \longmapsto \bigl(f_1(x), f_2(x)\bigr) \in \overline{\mathbb{R}}^2$ and therefore of the $\overline{\mathbb{R}}$-valued maps

$$\begin{aligned} x \in E &\longmapsto f_1\cdot f_2(x) \equiv f_1(x)\, f_2(x), \\ x \in E &\longmapsto (f_1+f_2)(x) \equiv f_1(x)+f_2(x) \in \overline{\mathbb{R}} \quad \text{if } \operatorname{Range}(f_1\times f_2) \subseteq \widehat{\mathbb{R}^2}, \\ x \in E &\longmapsto f_1 \vee f_2(x) \equiv f_1(x) \vee f_2(x), \\ \text{and } x \in E &\longmapsto f_1 \wedge f_2(x) \equiv f_1(x) \wedge f_2(x). \end{aligned}$$

Thus, of course, if f is measurable on $(E,\mathcal{B})$ into $\bigl(\overline{\mathbb{R}}, \mathcal{B}_{\overline{\mathbb{R}}}\bigr)$, then so are the functions $f^+ \equiv f \vee 0$, $f^- \equiv -(f\wedge 0)$, and $|f| = f^+ + f^-$. Finally, from now on, a measurable map on $(E,\mathcal{B})$ into $\bigl(\overline{\mathbb{R}}, \mathcal{B}_{\overline{\mathbb{R}}}\bigr)$ usually will be referred to simply as a **measurable function** on $(E,\mathcal{B})$.

From the measure-theoretic standpoint, the most elementary functions are those that are $\mathbb{R}$-valued and take on only a finite number of distinct values, and so it is reasonable that such a function is said to be **simple**. Note that the class of simple functions is closed under the lattice operations "$\vee$" and "$\wedge$", multiplication, and addition. Aside from constant functions, the simplest of the simple functions are those that take their values in $\{0,1\}$. Clearly there is a one-to-one correspondence between $\{0,1\}$-valued functions and subsets of E. Namely, with each $\Gamma \subseteq E$ one can associate the function $\mathbf{1}_\Gamma$ defined by

$$\mathbf{1}_\Gamma(x) = \begin{cases} 1 & \text{if} \quad x \in \Gamma \\ 0 & \text{if} \quad x \notin \Gamma. \end{cases}$$

The function $\mathbf{1}_\Gamma$ is called the **indicator** (or, in the older literature, **characteristic**) **function** of the set Γ.

The reason why simple functions play such a central role in measure theory is that their integrals are the easiest to describe. To be precise, let $(E, \mathcal{B}, \mu)$ be a measure space and f a non-negative (i.e., a $[0, \infty)$-valued) measurable function on $(E, \mathcal{B})$ that is simple. Then the **Lebesgue integral of f on E** is defined to be

$$\sum_{\alpha \in \text{Range}(f)} \alpha \mu(f = \alpha),$$

where $\mu(f = \alpha)$ is shorthand for $\mu(\{f = \alpha\})$ which, in turn, is shorthand for $\mu(\{x \in E : f(x) = \alpha\})$. There are various ways in which I will denote the Lebesgue integral of f, depending on how many details the particular situation demands. The various expressions that I will use are, in decreasing order of the information conveyed,

$$\int_E f(x)\, \mu(dx), \quad \int_E f\, d\mu, \quad \text{and} \quad \int f\, d\mu.$$

Further, for $\Gamma \in \mathcal{B}$, I will write

$$\int_\Gamma f(x)\, \mu(dx) \quad \text{or} \quad \int_\Gamma f\, d\mu$$

to denote the Lebesgue integral of $\mathbf{1}_\Gamma f$ on E. Observe that this latter notation is completely consistent, since one would get precisely the same number by restricting f to Γ and computing the Lebesgue integral of $f \restriction \Gamma$ relative to (cf. Exercise 2.1.2) $\big(\Gamma, \mathcal{B}[\Gamma], \mu \restriction \mathcal{B}[\Gamma]\big)$.

Although this definition is obviously the correct one, it is not immediately obvious that it results in an integration theory in which the integral is a linear function of the integrand. The following lemma addresses that question.

Lemma 3.1.2 *Let f and g be non-negative, simple, measurable functions on $(E, \mathcal{B}, \mu)$. Then, for any $\alpha, \beta \in [0, \infty)$, $\alpha f + \beta g$ is a non-negative, simple measurable function and*

$$\int (\alpha f + \beta g)\, d\mu = \alpha \int f\, d\mu + \beta \int g\, d\mu.$$

In particular, if $f \le g$, then $\int f\, d\mu \le \int g\, d\mu$. In fact, if $f \le g$ and $\int f\, d\mu < \infty$, then

$$\int (g - f)\, d\mu = \int g\, d\mu - \int f\, d\mu.$$

Proof. Clearly, it suffices to handle the case $\alpha = 1 = \beta$.

Let $\{a_1, \dots, a_m\}$, $\{b_1, \dots, b_n\}$, and $\{c_1, \dots, c_\ell\}$ be the distinct values taken by, respectively, f, g, and $f + g$. Next, set $A_i = \{f = a_i\}$, $B_j = \{g = b_j\}$, and $C_k = \{f + g = c_k\}$. Then

$$\mu(A_i) = \sum_{j=1}^{n} \mu(A_i \cap B_j),\ \mu(B_j) = \sum_{i=1}^{m} \mu(A_i \cap B_j),\ \mu(C_k) = \sum_{\substack{(i,j)\\ a_i+b_j=c_k}} \mu(A_i \cap B_j),$$

and therefore $\int (f+g)\,d\mu$ equals

$$\begin{aligned}\sum_{k=1}^{\ell} c_k\mu(C_k) &= \sum_{k=1}^{\ell} c_k \sum_{\substack{(i,j)\\ a_i+b_j=c_k}} \mu(A_i \cap B_j) = \sum_{(i,j)} (a_i + b_j)\mu(A_i \cap B_j) \\ &= \sum_{i=1}^{m} a_i\mu(A_i) + \sum_{j=1}^{n} b_j\mu(B_j) = \int f\,d\mu + \int g\,d\mu.\end{aligned}$$

When $f \le g$ and $\int g\,d\mu = \infty$, $\int f\,d\mu \le \int g\,d\mu$ trivially. When $\int f\,d\mu < \infty$, write $g = f + (g - f)$, and apply the preceding to see that

$$\int g\,d\mu = \int f\,d\mu + \int (g-f)\,d\mu \ge \int f\,d\mu$$

and that $\int (g-f)\,d\mu = \int g\,d\mu - \int f\,d\mu$ when $\int f\,d\mu < \infty$. □

In order to extend the definition of the Lebesgue integral to arbitrary non-negative measurable functions, we must use a limit procedure. The idea is to approximate such a function by ones that are simple. For example, if f is a non-negative measurable function on $(E, \mathcal{B}, \mu)$, then one might take the choice suggested in § 2.1.1:

$$\varphi_n = \sum_{k=0}^{4^n-1} 2^{-n}k\mathbf{1}_{\{2^n f\in[k,k+1)\}} + 2^n\mathbf{1}_{\{f\ge 2^n\}}$$

for $n \ge 1$. Then each φ_n is a non-negative, measurable, simple function, and $\varphi_n \nearrow f$ as $n \to \infty$. In fact, $\varphi_n \longrightarrow f$ uniformly in $(\overline{\mathbb{R}}, \overline{\rho})$. Thus it would seem reasonable to define the integral of f as

$$\lim_{n\to\infty} \int \varphi_n\,d\mu = \lim_{n\to\infty} \left[\sum_{k=1}^{4^n-1} \frac{k}{2^n}\mu\big(k \le 2^n f < k+1\big) + 2^n\mu\big(f \ge 2^n\big) \right]. \quad (3.1.1)$$

Note that, because $\varphi_n \le \varphi_{n+1}$, Lemma 3.1.2 guarantees that this limit exists. However, before adopting this definition, we must first check that it is not too dependent on the choice of the approximating sequence. In fact, at the moment, it takes a minute to check that this definition will coincide with the one we have already given for simple f's, and, as distinguished from Lemma 3.1.2, here we must use *countable*, as opposed to *finite*, additivity to do so.

Lemma 3.1.3 *Let $(E, \mathcal{B}, \mu)$ be a measure space, and suppose $\{\varphi_n : n \geq 1\}$ and ψ are non-negative, measurable, simple functions on $(E, \mathcal{B})$. If $\varphi_n \leq \varphi_{n+1}$ for all $n \geq 1$ and if $\psi \leq \lim_{n\to\infty} \varphi_n$, then $\int \psi\, d\mu \leq \lim_{n\to\infty} \int \varphi_n\, d\mu$. In particular, for any non-negative, measurable function f and any non-decreasing sequence $\{\psi_n : n \geq 1\}$ of non-negative, measurable, simple functions ψ_n that tend to f as $n \longrightarrow \infty$, $\lim_{n\to\infty} \int \psi_n\, d\mu$ is the same as the limit in* (3.1.1).

Proof. Note that the final statement is, as intimated, an easy consequence of what precedes it. Indeed, if the simple functions φ_n are those given above and $\{\psi_n : n \geq 1\}$ is any other sequence of simple functions satisfying $0 \leq \psi_n \nearrow f$, then, for every $m \geq 1$,

$$\varphi_m \leq \lim_{n\to\infty} \psi_n \quad \text{and} \quad \psi_m \leq \lim_{n\to\infty} \varphi_n.$$

Thus, by the first part,

$$\lim_{m\to\infty} \int \varphi_m\, d\mu \leq \lim_{n\to\infty} \int \psi_n\, d\mu \quad \text{and} \quad \lim_{m\to\infty} \int \psi_m\, d\mu \leq \lim_{n\to\infty} \int \varphi_n\, d\mu.$$

In order to prove the first part, I will begin by dispensing with the case in which $\mu(\psi > 0) = \infty$. Because ψ is simple and therefore there is an $\epsilon > 0$ for which $\psi > \epsilon$ whenever $\psi > 0$,

$$\int \psi\, d\mu = \infty \text{ and } \mu(\varphi_n > \epsilon) \nearrow \mu\left(\bigcup_{n=1}^{\infty} \{\varphi_n > \epsilon\}\right) \geq \mu(\psi > 0) = \infty,$$

which, because $\varphi_n \geq \epsilon \mathbf{1}_{\{\varphi_n > \epsilon\}}$, means that

$$\lim_{n\to\infty} \int \varphi_n\, d\mu \geq \lim_{n\to\infty} \epsilon\mu(\varphi_n > \epsilon) = \infty = \int \psi\, d\mu.$$

Now assume that $\mu(\psi > 0) < \infty$, and set $\hat{E} = \{0 < \psi < \infty\}$. Under the present conditions, $\mu(\hat{E}) < \infty$, $\int \psi\, d\mu = \int_{\hat{E}} \psi\, d\mu$, and $\int \varphi_n\, d\mu \geq \int_{\hat{E}} \varphi_n\, d\mu$ for all $n \geq 1$. Hence, without loss of generality, we will assume that $E = \hat{E}$. But then $\mu(E) < \infty$, and, because ψ is simple, there exist $\epsilon > 0$ and $M < \infty$ such that $\epsilon \leq \psi \leq M$. Now let $0 < \delta < \epsilon$ be given, and define $E_n = \{\varphi_n \geq \psi - \delta\}$. Then $E_n \nearrow E$ and so

$$\begin{aligned}
\lim_{n\to\infty} \int \varphi_n\, d\mu &\geq \lim_{n\to\infty} \int_{E_n} \varphi_n\, d\mu \geq \lim_{n\to\infty} \left[\int_{E_n} \psi\, d\mu - \delta\mu(E_n)\right] \\
&= \lim_{n\to\infty} \left[\int \psi\, d\mu - \int_{E_n{}^{\complement}} \psi\, d\mu - \delta\mu(E_n)\right] \\
&\geq \int \psi\, d\mu - M \lim_{n\to\infty} \mu(E_n{}^{\complement}) - \delta\mu(E) = \int \psi\, d\mu - \delta\mu(E),
\end{aligned}$$

since $\mu(E) < \infty$ and therefore, by (2.1.11), $\mu(E_n{}^{\complement}) \searrow 0$. Because this holds for arbitrarily small $\delta > 0$, we get our result upon letting $\delta \searrow 0$. □

The result in Lemma 3.1.3 allows us to complete the definition of the Lebesgue integral for non-negative, measurable functions. Namely, if f on $(E, \mathcal{B}, \mu)$ is a non-negative, measurable function, then we will define the **Lebesgue integral of f on E with respect to μ** to be the number in (3.1.1); and we will continue to use the same notation as we did for simple functions to denote its integral. Not only does Lemma 3.1.3 guarantee that this definition is consistent with my earlier one for simple f's, but it also makes clear that the value of $\int f\,d\mu$ does not depend on the particular way in which one chooses to approximate f by a non-decreasing sequence of non-negative, measurable, simple functions. Thus, for example, the following extension of Lemma 3.1.2 is trivial.

Lemma 3.1.4 *If f and g are non-negative, measurable functions on the measure space $(E, \mathcal{B}, \mu)$, then for every $\alpha, \beta \in [0, \infty]$,*

$$\int (\alpha f + \beta g)\,d\mu = \alpha \int f\,d\mu + \beta \int g\,d\mu.$$

In particular, if $f \le g$, then $\int f\,d\mu \le \int g\,d\mu$ and $\int (g-f)\,d\mu = \int g\,d\mu - \int f\,d\mu$ as long as $\int f\,d\mu < \infty$.

Obviously $\int f\,d\mu$ reflects the size of a non-negative measurable f. The result that follows makes this comment somewhat more quantitative.

Theorem 3.1.5 (Markov's inequality) *If f is a non-negative measurable function on $(E, \mathcal{B}, \mu)$, then*

$$\mu(f \ge \lambda) \le \frac{1}{\lambda} \int_{\{f \ge \lambda\}} f\,d\mu \le \frac{1}{\lambda} \int f\,d\mu \quad \text{for all } \lambda > 0. \tag{3.1.2}$$

In particular, $\int f\,d\mu = 0$ if and only if $\mu(f > 0) = 0$; and $\mu(f = \infty) = 0$ if $\int f\,d\mu < \infty$.

Proof. To prove (3.1.2), simply note that $\lambda \mathbf{1}_{\{f \ge \lambda\}} \le \mathbf{1}_{\{f \ge \lambda\}} f \le f$, and integrate. Hence, if $M = \int f\,d\mu < \infty$, then $\mu(f \ge \lambda) \le \frac{M}{\lambda}$ for all $\lambda > 0$, and therefore

$$\mu(f = \infty) \le \lim_{\lambda \to \infty} \mu(f \ge \lambda) = 0.$$

Finally, if $\int f\,d\mu = 0$, then, by (3.1.2), $\mu(f \ge n^{-1}) = 0$ for every $n \ge 1$ and therefore $\mu(f > 0) = \lim_{n\to\infty} \mu(f \ge n^{-1}) = 0$. Conversely, suppose that $\mu(f > 0) = 0$, and let $\{\varphi_n : n \ge 1\}$ be a sequence of non-negative simple functions that increases to f. Then $\mu(\varphi_n > 0) = 0$ for all $n \ge 1$, and so, since $\int \varphi_n\,d\mu \le \|\varphi_n\|_{\mathrm{u}} \mu(\varphi_n > 0)$, $\int \varphi_n\,d\mu = 0$ for all $n \ge 1$, which means that $\int f\,d\mu = \lim_{n\to\infty} \int \varphi_n\,d\mu = 0$. □

The final step in the definition of the Lebesgue integral is to extend the definition to cover measurable functions that can take both signs. To this end, let f be a measurable function on the measure space $(E, \mathcal{B}, \mu)$. Then both $\int f^+ \, d\mu$ and $\int f^- \, d\mu$ (recall that $f^\pm \equiv (\pm f) \vee 0$) are defined; and, if we want our integral to be linear, we can do nothing but take $\int f \, d\mu$ to be the difference between these two. However, before doing so, we must make sure that this difference is well-defined. With this consideration in mind, I now say that $\int f \, d\mu$ **exists** if $\int f^+ \, d\mu \wedge \int f^- \, d\mu < \infty$, in which case I define

$$\int_E f(x) \, \mu(dx) = \int f \, d\mu = \int f^+ \, d\mu - \int f^- \, d\mu$$

to be the **Lebesgue integral of** f **on** E. Observe that if $\int f \, d\mu$ exists, then so does $\int_\Gamma f \, d\mu \equiv \int \mathbf{1}_\Gamma f \, d\mu$ for every $\Gamma \in \mathcal{B}$, and in fact

$$\int_{\Gamma_1 \cup \Gamma_2} f \, d\mu = \int_{\Gamma_1} f \, d\mu + \int_{\Gamma_2} f \, d\mu$$

if Γ_1 and Γ_2 are disjoint elements of $\mathcal{B}$. Also, it is clear that, when $\int f \, d\mu$ exists,

$$\left| \int f \, d\mu \right| \le \int |f| \, d\mu.$$

In particular, $\int_\Gamma f \, d\mu = 0$ if $\mu(\Gamma) = 0$. Finally, when $\int f^+ \, d\mu \wedge \int f^- \, d\mu = \infty$, I will not even attempt to define $\int f \, d\mu$.

Once again, we need a consistency result before we will know for sure that our definition preserves linearity.

Lemma 3.1.6 *Let f and g be measurable functions on $(E, \mathcal{B}, \mu)$ for which $\int f \, d\mu$ and $\int g \, d\mu$ exist and $\left(\int f \, d\mu, \int g \, d\mu \right) \in \widehat{\mathbb{R}^2}$. Then $\mu\big((f,g) \notin \widehat{\mathbb{R}^2}\big) = 0$, $\int_{\{(f,g) \in \widehat{\mathbb{R}^2}\}} (f+g) \, d\mu$ exists, and*

$$\int_{\{(f,g) \in \widehat{\mathbb{R}^2}\}} (f+g) \, d\mu = \int f \, d\mu + \int g \, d\mu.$$

Proof. Set $\hat{E} = \{x \in E : \big(f(x), g(x)\big) \in \widehat{\mathbb{R}^2}\}$.

Note that, under the stated conditions, either

$$\int f^+ \, d\mu \vee \int g^+ \, d\mu < \infty \quad \text{or} \quad \int f^- \, d\mu \vee \int g^- \, d\mu < \infty.$$

For definiteness, assume that $\int f^- \, d\mu \vee \int g^- \, d\mu < \infty$. As a consequence,

$$\mu(\hat{E}^\complement) \le \mu(f^- \vee g^- = \infty) \le \mu(f^- = \infty) + \mu(g^- = \infty) = 0$$

and, because $(a+b)^- \le a^- + b^-$ for $(a,b) \in \widehat{\mathbb{R}^2}$, $\int_{\hat{E}} (f+g)^- \, d\mu < \infty$. Hence, all that remains is to prove the asserted equality, and, while doing so, we may and will assume that $E = \hat{E}$.

Set $E^+ = \{f+g \ge 0\}$ and $E^- = \{f+g<0\}$. Then

$$\int (f+g)\, d\mu \equiv \int (f+g)^+ \, d\mu - \int (f+g)^- \, d\mu = \int_{E^+} (f+g)\, d\mu + \int_{E^-} (f+g)\, d\mu.$$

By applying the last part of Lemma 3.1.4 to $(f^+ + g^+)\mathbf{1}_{E^+} - (f^- + g^-)\mathbf{1}_{E^+} = (f+g)\mathbf{1}_{E^+} \ge 0$, we see that

$$\int_{E^+} (f+g)\, d\mu = \int_{E^+} (f^+ + g^+)\, d\mu - \int_{E^+} (f^- + g^-)\, d\mu.$$

Thus, by the first part of Lemma 3.1.4,

$$\int_{E^+} (f+g)\, d\mu = \int_{E^+} f^+ \, d\mu + \int_{E^+} g^+ \, d\mu - \int_{E^+} f^- \, d\mu - \int_{E^+} g^- \, d\mu.$$

Similarly, because $(f^- + g^-)\mathbf{1}_{E^+} - (f^+ + g^+)\mathbf{1}_{E^-} = -(f+g)\mathbf{1}_{E^-} \ge 0$, $f^+ + g^+ \le f^- + g^-$ on E^-, and therefore $\int_{E^-} (f^+ + g^+)\, d\mu < \infty$ and

$$-\int_{E^-} (f+g)\, d\mu = \int_{E^-} (f^- + g^-)\, d\mu - \int_{E^-} (f^+ + g^+)\, d\mu,$$

which means that

$$\int_{E^-} (f+g)\, d\mu = \int_{E^-} f^+ \, d\mu + \int_{E^-} g^+ \, d\mu - \int_{E^-} f^- \, d\mu - \int_{E^-} g^- \, d\mu.$$

Finally, after adding these two and again applying the first part of Lemma (3.1.4), we arrive at

$$\int (f+g)\, d\mu = \int f^+ \, d\mu - \int f^- \, d\mu + \int g^+ \, d\mu - \int g^- \, d\mu = \int f \, d\mu + \int g \, d\mu.$$

□

3.1.2 The Space $L^1(\mu;\mathbb{R})$

Given a measurable function f on $(E,\mathcal{B},\mu)$, define $\|f\|_{L^1(\mu;\mathbb{R})} = \int |f| \, d\mu$, say that $f : E \longrightarrow \overline{\mathbb{R}}$ is **μ-integrable** if f is a measurable function on $(E,\mathcal{B})$ and $\|f\|_{L^1(\mu;\mathbb{R})} < \infty$, and use $L^1(\mu;\mathbb{R}) = L^1(E,\mathcal{B},\mu;\mathbb{R})$ to denote the set of all $\mathbb{R}$-valued, μ-integrable functions. Note that, from the integration-theoretic standpoint, there is no loss of generality to assume that an $f \in L^1(\mu;\mathbb{R})$ is $\mathbb{R}$-valued. Indeed, if f is a μ-integrable function, then $\mathbf{1}_{\{|f|<\infty\}} f \in L^1(\mu;\mathbb{R})$, $\|f - \mathbf{1}_{\{|f|<\infty\}} f\|_{L^1(\mu;\mathbb{R})} = 0$, and so integrals involving f and $\mathbf{1}_{\{|f|<\infty\}} f$ are indistinguishable. The main reason for insisting that f's in $L^1(\mu;\mathbb{R})$ be $\mathbb{R}$-valued is so that we have no problems taking linear combinations of them over $\mathbb{R}$. This simplifies the statement of results like the following.

Lemma 3.1.7 *For any measure space $(E,\mathcal{B},\mu)$, $L^1(\mu;\mathbb{R})$ is a vector space and*

$$\|\alpha f+\beta g\|_{L^1(\mu;\mathbb{R})}\le|\alpha|\,\|f\|_{L^1(\mu;\mathbb{R})}+|\beta|\,\|g\|_{L^1(\mu;\mathbb{R})} \tag{3.1.3}$$

whenever $\alpha,\beta\in\mathbb{R}$ and $f,g\in L^1(\mu;\mathbb{R})$.

Proof. Simply note that $|\alpha f+\beta g|\le|\alpha|\,|f|+|\beta|\,|g|$. □

Remark 3.1.1 As an application of the preceding inequality, we have that

$$\|f-h\|_{L^1(\mu;\mathbb{R})}\le\|f-g\|_{L^1(\mu;\mathbb{R})}+\|g-h\|_{L^1(\mu;\mathbb{R})}\quad\text{for } f,g,h\in L^1(\mu;\mathbb{R}).$$

To see this, take $\alpha=\beta=1$ and replace f and g by $f-g$ and $g-h$ in (3.1.3). Thus $\|f-g\|_{L^1(\mu;\mathbb{R})}$ looks like a good candidate to be chosen as a metric on $L^1(\mu;\mathbb{R})$. On the other hand, although, from the standpoint of integration theory, a measurable f for which $\|f\|_{L^1(\mu;\mathbb{R})}=0$ might as well be identically 0, there is, in general, no reason why f need be identically 0 as a function. This fact prevents $\|\cdot\|_{L^1(\mu;\mathbb{R})}$ from being a completely satisfactory measure of size. To overcome this problem, one adopts the time-honored subterfuge of *quotienting out by the offending subspace.* That is, denote by $\mathcal{N}(\mu)$ the set of $f\in L^1(\mu;\mathbb{R})$ for which (cf. Exercise 3.1.4 below) $\mu(f\neq 0)=0$, and, for f, $g\in L^1(\mu;\mathbb{R})$, write $f\overset{\mu}{\sim}g$ if $g-f\in\mathcal{N}(\mu)$. Since it is clear that $\mathcal{N}(\mu)$ is a linear subspace of $L^1(\mu;\mathbb{R})$, one sees that $\overset{\mu}{\sim}$ is an equivalence relation and that the quotient space $L^1(\mu;\mathbb{R})/\overset{\mu}{\sim}$ is again a vector space over $\mathbb{R}$. To be precise, for $f\in L^1(\mu;\mathbb{R})$, use $[f]^{\overset{\mu}{\sim}}$ to denote the $\overset{\mu}{\sim}$-equivalence class of f, and, for any f, $g\in L^1(\mu;\mathbb{R})$ and α, $\beta\in\mathbb{R}$, take

$$\alpha[f]^{\overset{\mu}{\sim}}+\beta[g]^{\overset{\mu}{\sim}}=[\alpha f+\beta g]^{\overset{\mu}{\sim}}.$$

Finally, since

$$f\overset{\mu}{\sim}g\implies|f|\overset{\mu}{\sim}|g|\implies\|f\|_{L^1(\mu;\mathbb{R})}=\|g\|_{L^1(\mu;\mathbb{R})},$$

we can define $\|[f]^{\overset{\mu}{\sim}}\|_{L^1(\mu;\mathbb{R})}=\|f\|_{L^1(\mu;\mathbb{R})}$ and thereby turn $L^1(\mu;\mathbb{R})/\overset{\mu}{\sim}$ into a bona fide metric space in which the distance between $[f]^{\overset{\mu}{\sim}}$ and $[g]^{\overset{\mu}{\sim}}$ is given by $\|f-g\|_{L^1(\mu;\mathbb{R})}$.

Having made this obeisance to rigor, I will now lapse into the usual, more casual practice of ignoring the niceties just raised. Thus, unless there is particular danger in doing so, I will not stress the distinction between f as a function and the equivalence class $[f]^{\overset{\mu}{\sim}}$ that f determines. For this reason, I will continue to write $L^1(\mu;\mathbb{R})$, even when I mean $L^1(\mu;\mathbb{R})/\overset{\mu}{\sim}$, and will continue to use f instead of $[f]_{\sim}^{\mu}$. In particular, $L^1(\mu;\mathbb{R})$ becomes in this way a vector space over $\mathbb{R}$ on which $\|f-g\|_{L^1(\mu;\mathbb{R})}$ is a metric. As we will see in the next section (cf. Lemma 3.2.9), this metric space is complete.

3.1.3 Exercises for § 3.1

Exercise 3.1.1 If $f:\mathbb{R}\longrightarrow\mathbb{R}$ is either right continuous or left continuous, show that f is $\mathcal{B}_{\mathbb{R}}$-measurable.

Exercise 3.1.2 Here is another construction of the measures μ_F in Theorem 2.2.17. Set $F(\infty)=\lim_{x\to\infty}F(x)$, and define $F^{-1}:\big[0,F(\infty)\big]\longrightarrow\mathbb{R}$ so that

$$F^{-1}(x)=\inf\{y\in\mathbb{R}:\ F(y)\geq x\}.$$

Check that F^{-1} is $\mathcal{B}_{[0,F(\infty)]}$-measurable, and set

$$\mu(\Gamma)=(F^{-1})_*\lambda_{\mathbb{R}}(\Gamma)=\lambda_{\mathbb{R}}\big(\{x\in[0,F(\infty)]:\ F^{-1}(x)\in\Gamma\}\big)\quad\text{for }\Gamma\in\mathcal{B}_{\mathbb{R}}.$$

Show that μ is a finite Borel measure on $\mathbb{R}$ whose distribution function is F. Hence, $\mu=\mu_F$.

Exercise 3.1.3 As we saw in Exercise 3.1.2, all finite Borel measures on $\mathbb{R}$ can be written as an image of $\lambda_{\mathbb{R}}$. This fact is a particular example of the much more general fact that, under mild technical conditions, nearly every measure can be written as the image of $\lambda_{\mathbb{R}}$. The purpose of this exercise is to construct a measurable $f:[0,1]\longrightarrow[0,1]^2$ such that $\lambda_{[0,1]^2}=f_*\lambda_{[0,1]}$, where $\lambda_{[0,1]^2}$ is the restriction of $\lambda_{\mathbb{R}^2}$ to $\mathcal{B}_{[0,1]^2}$. To construct f, first define π_0 and π_1 on (cf. § 2.2.4) Ω into itself so that $[\pi_0(\omega)](i)=\omega(2i)$ and $[\pi_1(\omega)](i)=\omega(2i-1)$ for $i\geq 1$. Next, define $f:[0,1]\longrightarrow[0,1]^2$ by (cf. (2.2.6))

$$f=\big(\Phi\circ\pi_0\circ\hat{\Phi}^{-1},\Phi\circ\pi_1\circ\hat{\Phi}^{-1}\big),$$

where (cf. § 2.2.5) $\hat{\Phi}=\Phi\restriction\hat{\Omega}$. Show that f is a measurable map from $([0,1];\mathcal{B}_{[0,1]})$ onto $([0,1]^2,\mathcal{B}_{[0,1]^2})$ and that $\lambda_{[0,1]^2}=f_*\lambda_{[0,1]}$.

Exercise 3.1.4 Let f be an $\overline{\mathbb{R}}$-valued function on the measurable space $(E,\mathcal{B})$. Show that f is measurable if and only if $\{f>a\}\in\mathcal{B}$ for every $a\in\mathbb{R}$ if and only if $\{f\geq a\}\in\mathcal{B}$ for every $a\in\mathbb{R}$. At the same time, check that "$>$" and "$<$" can be replaced by "$\leq$" and "$\geq$", respectively. In fact, show that one can restrict ones attention to a's from a dense subset of $\mathbb{R}$. Finally, if g is a second $\overline{\mathbb{R}}$-valued measurable function on $(E,\mathcal{B})$, show that each of the sets $\{f<g\}$, $\{f\leq g\}$, $\{f=g\}$, and $\{f\neq g\}$ is an element of $\mathcal{B}$.

Exercise 3.1.5 Let f be a non-negative, ν-integrable function on the measure space $(E,\mathcal{B},\nu)$, and define $\mu(\Gamma)=\int_\Gamma f\,d\nu$ for $\Gamma\in\mathcal{B}$. Show that μ is a finite measure on $(E, \mathcal{B})$ that is (cf. Exercise 2.1.8) absolutely continuous with respect to ν. In particular, by Exercise 2.1.8, this means that, for each $\epsilon > 0$, there is a $\delta > 0$ such that $\nu(\Gamma) < \delta \implies \int_\Gamma f\,d\nu < \epsilon$. In § 8.1.1 we will see that, under mild restrictions, every finite measure that is absolutely continuous with respect to ν can be represented in terms of integrals of a non-negative element of $L^1(\nu; \mathbb{R})$.

Exercise 3.1.6 Show that an integrable function is determined, up to a set of measure 0, by its integrals. To be more precise, let $(E, \mathcal{B}, \mu)$ be a measure space and f and g a pair of functions from $L^1(\mu; \mathbb{R})$. Show that $\mu(f > g) = 0$ if and only if $\int_\Gamma f\,d\mu \le \int_\Gamma g\,d\mu$ for every $\Gamma \in \mathcal{B}$, and conclude from this that $\mu(f \ne g) = 0$ if and only if $\int_\Gamma f\,d\mu = \int_\Gamma g\,d\mu$ for each $\Gamma \in \mathcal{B}$.

Exercise 3.1.7 Let β_p be the Bernoulli measure on $(\Omega, \mathcal{B}_\Omega)$ introduced in § 2.2.4.

(**i**) Show that

$$\beta_p\left(\sum_{i=1}^n \omega(i) = m\right) = \binom{n}{m} p^m q^{n-m} \quad \text{for } 0 \le m \le n,$$

and use this to derive

$$\int \exp\left(\lambda \sum_{i=1}^n (\omega(i) - p)\right) \beta_p(d\omega) = \left(pe^{\lambda q} + qe^{-\lambda p}\right)^n$$

for $n \ge 1$ and $\lambda \in \mathbb{R}$. Next, show that

$$pe^{\lambda q} + qe^{-\lambda p} \le e^{\frac{\lambda^2}{8}} \quad \text{for all } \lambda \in \mathbb{R},$$

and conclude that

$$\int \exp\left(\lambda \sum_{i=1}^n (\omega(i) - p)\right) \beta_p(d\omega) \le e^{\frac{n\lambda^2}{8}} \quad \text{for } \lambda \in \mathbb{R}.$$

Hint: Set $f(\lambda) = \log\left(pe^{\lambda q} + qe^{-\lambda p}\right)$, and show that $f(0) = f'(0) = 0$ and $f''(\lambda) \le \frac{1}{4}$.

(**ii**) Starting from the calculations in (**i**) and applying (3.1.2) to the function $e^{\lambda \sum_{i=1}^n (\omega(i) - p)}$, show first that

$$\beta_p\left(\sum_{i=1}^n (\omega(i) - p) \ge R\right) \le e^{-\lambda R + \frac{n\lambda^2}{8}} \quad \text{for } R \in [0, \infty) \text{ and } \lambda \in [0, \infty),$$

and then, after minimizing with respect to λ, that

$$\beta_p\left(\sum_{i=1}^{n}(\omega(i)-p)\geq R\right)\leq e^{-\frac{2R^2}{n}}.$$

Similarly, show that

$$\beta_p\left(\sum_{i=1}^{n}(\omega(i)-p)\leq -R\right)\leq e^{-\frac{2R^2}{n}},$$

and conclude that

$$\beta_p\left(\left|\frac{1}{n}\sum_{i=1}^{n}\omega(i)-p\right|\geq R\right)\leq 2e^{-n2R^2}\quad\text{for } R\in[0,\infty).$$

(iii) By combining the conclusion in **(ii)** with the first part of Exercise 2.1.7, show that

$$\beta_p\left(\exists m\ \forall n\geq m\quad \left|\frac{1}{n}\sum_{i=1}^{n}\omega(i)-p\right|\leq n^{-\frac{1}{4}}\right)=1.$$

This is an example of a general theorem known as the **Strong Law of Large Numbers**. Notice that when $p=\frac{1}{2}$, this result combined with Theorem 2.2.22 proves a famous theorem, due to E. Borel, which says that for $\lambda_{[0,1]}$-almost every $x\in[0,1]$ half (in the sense of averages) the coefficients in its dyadic expansion are 1 and therefore half are 0.

Exercise 3.1.8 Referring again to §§ 2.2.4 & 2.2.5, let Ω_p denote the set of $\omega\in\Omega$ for which $\frac{1}{n}\sum_{i=1}^{n}\omega(i)\longrightarrow p$, and use part **(iii)** of Exercise 3.1.7 to see that $\beta_p(\Omega_{p'})$ equals 1 if $p'=p$ and 0 otherwise. In particular, this means that if $p\neq p'$ then (cf. Exercise 2.1.9) $\beta_p\perp\beta_{p'}$. Next, let $\Phi:\Omega\longrightarrow[0,1]$ and $\hat{\Phi}:\hat{\Omega}\longrightarrow[0,1]$ be the maps defined in § 2.2.5, and set $\mu_p=\Phi_*\beta_p$. By Theorem 2.2.22, we know that $\mu_{\frac{1}{2}}=\lambda_{[0,1]}$. Show that if $B_p=[0,1)\cap\hat{\Phi}(\Omega_p)$, the set of $x\in[0,1)$ for which the coefficients in its dyadic expansion form an element of Ω_p, then $\mu_p(B_{p'})$ equals 0 or 1 according to whether $p=p'$ or $p\neq p'$. Thus, $\mu_p\perp\mu_{p'}$ if $p\neq p'$. In particular, if $p\neq\frac{1}{2}$ and $F_p(x)=\mu_p\big((0,x^+\wedge 1]\big)$, show that F_p is a non-decreasing function that is (cf. Exercise 2.2.9) singular even though it is continuous and strictly increasing on $[0,1]$. This is the example promised **(ii)** of Exercise 1.2.8.

3.2 Convergence of Integrals

One of the major advantages that Lebesgue's theory of integration has over Riemann's approach is that Lebesgue's integral is marvelously continuous with respect to convergence of integrands. In the present section I will explore some of these continuity properties, and I begin by showing that the class of measurable functions is closed under pointwise convergence.

Lemma 3.2.1 *Let $(E, \mathcal{B})$ be a measurable space, F a metric space, and $\{f_n : n \geq 1\}$ a sequence of measurable functions on $(E, \mathcal{B})$ into $(F, \mathcal{B}_F)$. If $f(x) = \lim_{n\to\infty} f_n(x)$ for each $x \in E$, then f is also measurable. Moreover, if F is separable and admits a complete metric ρ, then the set Δ of $x \in E$ for which $\lim_{n\to\infty} f_n(x)$ exists is an element of $\mathcal{B}$, and, for each $y_0 \in F$, the function f given by*

$$f(x) = \begin{cases} y_0 & \text{if } x \notin \Delta \\ \lim_{n\to\infty} f_n(x) & \text{if } x \in \Delta \end{cases}$$

is measurable on $(E, \mathcal{B})$. In particular, if $F = \overline{\mathbb{R}}$, then $\sup_{n\geq 1} f_n$, $\inf_{n\geq 1} f_n$, $\overline{\lim}_{n\to\infty} f_n$, and $\underline{\lim}_{n\to\infty} f_n$ are all measurable functions.

Proof. To prove the first assertion, observe (cf. Exercise 2.1.3) that the class of $\Gamma \subseteq F$ for which $\{f \in \Gamma\} \equiv \{x : f(x) \in \Gamma\} \in \mathcal{B}$ is a σ-algebra. Next, given an open set G, let D_k be the set of points that are at least a distance $\frac{1}{k}$ from $G^\complement$, and check that

$$\{x : f(x) \in G\} = \bigcup_{k=1}^{\infty} \bigcup_{m=1}^{\infty} \bigcap_{n\geq m} \{x : f_n(x) \in D_k\} \in \mathcal{B}$$

for all $G \in \mathfrak{G}(F)$. To prove the second assertion, assume that F is separable, and let ρ be a complete metric for F. Then, because, by Lemma 3.1.1, $x \in E \longmapsto \rho\big(f_n(x), f_m(x)\big) \in [0, \infty)$ is $\mathcal{B}$-measurable,

$$\Delta = \bigcap_{k=1}^{\infty} \bigcup_{m=1}^{\infty} \bigcap_{n\geq m} \{x : \rho\big(f_n(x), f_m(x)\big) < \tfrac{1}{k}\} \in \mathcal{B}.$$

Since, by Cauchy's criterion, Δ is the set of x for which $\lim_{n\to\infty} f_n(x)$ exists and, for each $\Gamma \in \mathcal{B}_F$,

$$\{x : f(x) \in \Gamma\} = \begin{cases} \{x \in \Delta : \lim_{n\to\infty} f(x) \in \Gamma\} & \text{if } y_0 \notin \Gamma \\ \Delta^\complement \cup \{x \in \Delta : \lim_{n\to\infty} f(x) \in \Gamma\} & \text{if } y_0 \in \Gamma, \end{cases}$$

it follows that f is measurable.

Turning to the case $F = \overline{\mathbb{R}}$, simply note that

$$f_1 \vee \cdots \vee f_n \nearrow \sup_{n\geq 1} f, \quad f_1 \wedge \cdots \wedge f_n \searrow \inf_{n\geq 1} f_n,$$

$$\sup_{n\geq m} f_n \searrow \overline{\lim_{n\to\infty}} f_n, \quad \text{and} \quad \inf_{n\geq m} f_n \nearrow \underline{\lim_{n\to\infty}} f_n.$$

□

3.2.1 The Big Three Convergence Results

We are now ready to prove the first of three fundamental continuity theorems about the Lebesgue integral. In some ways the first one is the least surprising in that it really only echoes the result obtained in Lemma 3.1.3 and is nothing more than the function version of the result in (2.1.10).

Theorem 3.2.2 (**The Monotone Convergence Theorem**) *Suppose that* $\{f_n : n \geq 1\}$ *is a sequence of non-negative, measurable functions on the measure space* $(E, \mathcal{B}, \mu)$ *and that* $f_n \nearrow f$ *(pointwise) as* $n \to \infty$*. Then* $\int f\,d\mu = \lim_{n\to\infty} \int f_n\,d\mu$*. In particular, if* $f \in L^1(\mu; \mathbb{R})$*, then* $\|f_n - f\|_{L^1(\mu;\mathbb{R})} \longrightarrow 0$.

Proof. Obviously $\int f\,d\mu \geq \lim_{n\to\infty} \int f_n\,d\mu$. To prove the opposite inequality, for each $m \geq 1$ choose a non-decreasing sequence $\{\varphi_{m,n} : n \geq 1\}$ of non-negative, measurable, simple functions for which $\varphi_{m,n} \nearrow f_m$ as $n \to \infty$. Next, define the non-negative, simple, measurable functions $\psi_n = \varphi_{1,n} \vee \cdots \vee \varphi_{n,n}$ for $n \geq 1$. One then has that

$$\psi_n \leq \psi_{n+1} \text{ and } \varphi_{m,n} \leq \psi_n \leq f_n \quad \text{for all } 1 \leq m \leq n;$$

and therefore

$$f_m \leq \lim_{n\to\infty} \psi_n \leq f \quad \text{for each } m \in \mathbb{Z}^+.$$

In particular, $\psi_n \nearrow f$, and therefore

$$\int f\,d\mu = \lim_{n\to\infty} \int \psi_n\,d\mu.$$

At the same time, $\psi_n \leq f_n$ for all $n \in \mathbb{Z}^+$, and therefore

$$\int f\,d\mu \leq \lim_{n\to\infty} \int f_n\,d\mu.$$

Finally, if $f \in L^1(\mu; \mathbb{R})$, then

$$\|f_n - f\|_{L^1(\mu;\mathbb{R})} = \int (f - f_n)\,d\mu = \int f\,d\mu - \int f_n d\,\mu \longrightarrow 0.$$

□

Being an inequality instead of an equality, the second continuity result is often more useful than the other two. It is the function version of (2.1.12) and (2.1.13).

Theorem 3.2.3 (**Fatou's lemma**) *Let* $\{f_n : n \geq 1\}$ *be a sequence of measurable functions on the measure space* $(E, \mathcal{B}, \mu)$*. If* $f_n \geq 0$ *for each* $n \geq 1$*, then*

$$\int \varliminf_{n\to\infty} f_n\,d\mu \leq \varliminf_{n\to\infty} \int f_n\,d\mu.$$

Also, if there exists a non-negative, μ-integrable function g such that $f_n \le g$ for each $n \ge 1$, then

$$\int \varlimsup_{n\to\infty} f_n\,d\mu \ge \varlimsup_{n\to\infty} \int f_n\,d\mu.$$

Proof. Assume that the f_n's are non-negative. To check the first assertion, set $h_m = \inf_{n\ge m} f_n$. Then $f_m \ge h_m \nearrow \varliminf_{n\to\infty} f_n$ and so, by the monotone convergence theorem,

$$\int \varliminf_{n\to\infty} f_n\,d\mu = \lim_{m\to\infty} \int h_m\,d\mu \le \varliminf_{m\to\infty} \int f_m\,d\mu.$$

Next, drop the non-negativity assumption, but impose $f_n \le g$ for some non-negative, μ-integrable g. Clearly, $\int \sup_{n\ge1} f_n^+\,d\mu < \infty$, and so, by Lemma 3.1.6,

$$\varliminf_{n\to\infty} \int (g - f_n)\,d\mu = \int g\,d\mu - \varlimsup_{n\to\infty} \int f_n\,d\mu$$

and

$$\int \varliminf_{n\to\infty} (g - f_n)\,d\mu = \int g\,d\mu - \int \varlimsup_{n\to\infty} f_n\,d\mu.$$

Thus, by the first part applied to the non-negative functions $g - f_n$,

$$\int g\,d\mu - \varlimsup_{n\to\infty} \int f_n\,d\mu \ge \int g\,d\mu - \int \varlimsup_{n\to\infty} f_n\,d\mu,$$

from which the desired result follows immediately. □

Before stating the third continuity result, I need to introduce a notion that is better suited to measure theory than ordinary pointwise equality. Namely, say that an x-dependent statement about quantities on the measure space $(E, \mathcal{B}, \mu)$ holds **μ-almost everywhere**, abbreviated by (a.e., μ), if the set Δ of x for which the statement fails is an element of $\overline{\mathcal{B}}^{\mu}$ that has $\overline{\mu}$-**measure** 0 (i.e., $\overline{\mu}(\Delta) = 0$). In particular, if $\{f_n : n \ge 1\}$ is a sequence of measurable functions on the measure space $(E, \mathcal{B}, \mu)$, I will say that $\{f_n : n \ge 1\}$ **converges μ-almost everywhere** and will write $\lim_{n\to\infty} f_n$ exists (a.e., μ), if

$$\mu\big(\{x \in E : \lim_{n\to\infty} f_n(x) \text{ does not exist}\}\big) = 0.$$

In keeping with this terminology, I will say that $\{f_n : n \ge 1\}$ **converges μ-almost everywhere** to a measurable f, and will write $f_n \longrightarrow f$ (a.e., μ), if

$$\mu\big(\{x \in E : f(x) \ne \lim_{n\to\infty} f_n(x)\}\big) = 0.$$

By Lemma 3.2.1, we know that if $\{f_n : n \ge 1\}$ converges μ-almost everywhere, then there is a measurable f to which $\{f_n : n \ge 1\}$ converges μ-almost everywhere. Similarly, if f and g are measurable functions, I write $f = g$ (a.e., μ), $f \leq g$ (a.e., μ), or $f \geq g$ (a.e., μ) if $\mu(f \neq g) = 0$, $\mu(f > g) = 0$, or $\mu(f < g) = 0$, respectively. Note that for $f, g \in L^1(\mu;\mathbb{R})$ $f = g$ (a.e., μ) is the same statement as $f \overset{\mu}{\sim} g$ discussed in Remark 3.1.1. In addition, check that if μ is a Borel measure on a topological space and $\mu(G) > 0$ for all non-empty $G \in \mathfrak{G}(E)$, then two continuous functions are equal (a.e.,μ) if and only if they are equal pointwise.

The following can be thought of as the function version of (2.1.14).

Theorem 3.2.4 (Lebesgue's dominated convergence theorem) *Let $\{f_n : n \geq 1\}$ be a sequence of measurable functions on $(E, \mathcal{B}, \mu)$, and suppose that f is a measurable function to which $\{f_n : n \geq 1\}$ converges μ-almost everywhere. If there is a μ-integrable function g such that $|f_n| \leq g$ (a.e., μ) for each $n \geq 1$, then f is integrable and $\lim_{n\to\infty} \int |f_n - f|\, d\mu = 0$. In particular, $\int f\, d\mu = \lim_{n\to\infty} \int f_n\, d\mu$.*

Proof. Let $\widehat{E}$ be the set of $x \in E$ for which $f(x) = \lim_{n\to\infty} f_n(x)$ and $\sup_{n\geq 1} |f_n(x)| \leq g(x)$. Then $\widehat{E}$ is measurable and $\mu(\widehat{E}^{\complement}) = 0$, and so integrals over $\widehat{E}$ are the same as those over E. Thus, without loss of generality, assume that all the assumptions hold for every $x \in E$. But then, $f = \lim_{n\to\infty} f_n$, $|f| \leq g$ and $|f - f_n| \leq 2g$. Hence, by the second part of Fatou's lemma,

$$\varlimsup_{n\to\infty} \left| \int f\, d\mu - \int f_n\, d\mu \right| \leq \varlimsup_{n\to\infty} \int |f - f_n|\, d\mu \leq \int \varlimsup_{n\to\infty} |f - f_n|\, d\mu = 0.$$

□

It is important to understand the role played by the *Lebesgue dominant* g. Namely, it acts as an *umbrella* to keep everything under control. To see that some such control is needed, consider Lebesgue measure $\lambda_{[0,1]}$ on $([0,1], \mathcal{B}_{[0,1]})$ and the functions $f_n = n\mathbf{1}_{(0,n^{-1})}$. Obviously, $f_n \longrightarrow 0$ everywhere on $[0,1]$, but $\int_{[0,1]} f_n\, d\lambda_{\mathbb{R}} = 1$ for all $n \in \mathbb{Z}^+$.

Unfortunately, in many circumstances it is difficult to find an appropriate Lebesgue dominant, and, for this reason, results like the following variation on Fatou's lemma are interesting and often helpful. See Exercises (3.2.6) and (3.2.9) for other variations.

Theorem 3.2.5 (Brezis–Lieb Version of Fatou's lemma) *Let $(E, \mathcal{B}, \mu)$ be a measure space, $\{f_n : n \geq 1\} \cup \{f\} \subseteq L^1(\mu;\mathbb{R})$, and assume that $f_n \longrightarrow f$ (a.e., μ). Then*

$$\begin{aligned} \lim_{n\to\infty} &\left| \|f_n\|_{L^1(\mu;\mathbb{R})} - \|f\|_{L^1(\mu;\mathbb{R})} - \|f_n - f\|_{L^1(\mu;\mathbb{R})} \right| \\ &= \lim_{n\to\infty} \int \Big| |f_n| - |f| - |f_n - f| \Big|\, d\mu = 0. \end{aligned} \tag{3.2.1}$$

In particular, if $\|f_n\|_{L^1(\mu;\mathbb{R})} \longrightarrow \|f\|_{L^1(\mu;\mathbb{R})}$, then $\|f - f_n\|_{L^1(\mu;\mathbb{R})} \longrightarrow 0$.

Proof. Since

$$\Big|\|f_n\|_{L^1(\mu;\mathbb{R})}-\|f\|_{L^1(\mu)}-\|f_n-f\|_{L^1(\mu;\mathbb{R})}\Big| \le \int \big||f_n|-|f|-|f_n-f|\big|\,d\mu, \quad n\ge 1,$$

we need check only the second equality in (3.2.1). But, because

$$\big||f_n|-|f|-|f_n-f|\big| \longrightarrow 0 \quad (\text{a.e.}, \mu)$$

and

$$\big||f_n|-|f|-|f_n-f|\big| \le \big||f_n|-|f_n-f|\big|+|f| \le 2|f|,$$

(3.2.1) follows from Lebesgue's dominated convergence theorem. □

3.2.2 Convergence in Measure

We now have a great deal of evidence that almost everywhere convergence of integrands often leads to convergence of the corresponding integrals. I next want to investigate what can be said about the opposite implication. To begin with, I point out that $\|f_n\|_{L^1(\mu;\mathbb{R})} \longrightarrow 0$ *does not* imply that $f_n \longrightarrow 0$ (a.e., μ). Indeed, define the functions $\{f_n : n \ge 1\}$ on $[0,1]$ so that, for $m \ge 0$ and $0 \le \ell < 2^m$,

$$f_{2^m+\ell} = \mathbf{1}_{[2^{-m}\ell, 2^{-m}(\ell+1)]}.$$

It is then clear that these f_n's are non-negative and measurable on $([0,1], \mathcal{B}_{[0,1]})$ and that $\overline{\lim}_{n\to\infty} f_n(x) = 1$ for every $x \in [0,1]$. On the other hand,

$$\int_{[0,1]} f_n\, d\lambda_{\mathbb{R}} = 2^{-m} \quad \text{if } 2^m \le n < 2^{m+1},$$

and therefore $\int_{[0,1]} f_n\, d\lambda_{\mathbb{R}} \longrightarrow 0$ as $n \to \infty$.

As the preceding discussion makes clear, it may be useful to consider other notions of convergence. Keeping in mind that we are looking for a type of convergence that can be tested using integrals, we should take a hint from Markov's inequality and say that the sequence $\{f_n : n \ge 1\}$ of measurable functions on the measure space $(E, \mathcal{B}, \mu)$ **converges in μ-measure** to the measurable function f if $\mu(|f_n - f| \ge \epsilon) \longrightarrow 0$ as $n \to \infty$ for every $\epsilon > 0$, in which case I will write $f_n \longrightarrow f$ **in μ-measure**. Note that, by Markov's inequality (3.1.2), if $\|f_n - f\|_{L^1(\mu;\mathbb{R})} \longrightarrow 0$ then $f_n \longrightarrow f$ in μ-measure. Hence, this sort of convergence can be easily tested with integrals, and, as such, must be quite different (cf. Exercise 3.2.8) from μ-almost everywhere convergence. In fact, it takes a moment to see in what sense the limit is even uniquely determined by convergence in μ-measure. For this reason, suppose that $\{f_n : n \ge 1\}$ converges to both f and g in μ-measure. Then, for $\epsilon > 0$,

$$\mu(|f-g|\geq\epsilon)\leq\mu\left(|f-f_n|\geq\tfrac{\epsilon}{2}\right)+\mu\left(|f_n-g|\geq\tfrac{\epsilon}{2}\right)\longrightarrow 0 \text{ as } n\to\infty.$$

Hence, $\mu(f\neq g)=\lim_{\epsilon\searrow 0}\mu\big(|f-g|\geq\epsilon\big)=0$, and so $f=g$ (a.e., μ). That is, convergence in μ-measure determines the limit function to precisely the same extent as does either μ-almost everywhere or $\|\cdot\|_{L^1(\mu;\mathbb{R})}$-convergence. In particular, from the standpoint of μ-integration theory, convergence in μ-measure has unique limits.

The following theorems are intended to further elucidate the notions of μ-almost everywhere convergence, convergence in μ-measure, and the relations between them.

Theorem 3.2.6 *Let $\{f_n : n\geq 1\}$ be a sequence of $\mathbb{R}$-valued measurable functions on the measure space $(E,\mathcal{B},\mu)$. Then there is an $\mathbb{R}$-valued, measurable function f for which*

$$\lim_{m\to\infty}\mu\left(\sup_{n\geq m}|f-f_n|\geq\epsilon\right)=0 \quad \textit{for all } \epsilon>0 \tag{3.2.2}$$

if and only if

$$\lim_{m\to\infty}\mu\left(\sup_{n\geq m}|f_n-f_m|\geq\epsilon\right)=0 \quad \textit{for all } \epsilon>0. \tag{3.2.3}$$

Moreover, (3.2.2) *implies that $f_n\longrightarrow f$ both* (a.e., μ) *and in μ-measure. Finally, when $\mu(E)<\infty$, $f_n\longrightarrow f$* (a.e., μ) *if and only if* (3.2.2) *holds. In particular, on a finite measure space, μ-almost everywhere convergence implies convergence in μ-measure.*

Proof. Set

$$\Delta=\left\{x\in E:\ \lim_{n\to\infty}f_n(x) \text{ does not exist in } \mathbb{R}\right\}.$$

For $m\geq 1$ and $\epsilon>0$, define $\Delta_m(\epsilon)=\big\{\sup_{n\geq m}|f_n-f_m|\geq\epsilon\big\}$. It is then easy to check (from Cauchy's convergence criterion for $\mathbb{R}$) that

$$\Delta=\bigcup_{\ell=1}^{\infty}\bigcap_{m=1}^{\infty}\Delta_m\left(\tfrac{1}{\ell}\right).$$

Since (3.2.3) implies that $\mu\big(\bigcap_{m=1}^{\infty}\Delta_m(\epsilon)\big)=0$ for every $\epsilon>0$, and, by the preceding,

$$\mu(\Delta)\leq\sum_{\ell=1}^{\infty}\mu\left(\bigcap_{m=1}^{\infty}\Delta_m\left(\tfrac{1}{\ell}\right)\right),$$

we see that (3.2.3) does indeed imply that $\{f_n : n\geq 1\}$ converges μ-almost everywhere. In addition, if (cf. the second part of Lemma 3.2.1) f is an $\mathbb{R}$-valued, measurable function to which $\{f_n : n\geq 1\}$ converges μ-almost everywhere, then

$$\sup_{n\geq m}|f_n - f| \leq \sup_{n\geq m}|f_n - f_m| + |f_m - f| \leq 2\sup_{n\geq m}|f_n - f_m| \quad \text{(a.e., }\mu);$$

and so (3.2.3) leads to the existence of an f for which (3.2.2) holds.

Next, suppose that (3.2.2) holds for some f. Then it is obvious that $f_n \longrightarrow f$ both (a.e., μ) and in μ-measure. In addition, (3.2.3) follows immediately from

$$\mu\left(\sup_{n\geq m}|f_n - f_m| \geq \epsilon\right) \leq \mu\left(\sup_{n\geq m}|f_n - f| \geq \tfrac{\epsilon}{2}\right) + \mu\left(\sup_{n\geq m}|f - f_m| \geq \tfrac{\epsilon}{2}\right).$$

Finally, suppose that $\mu(E) < \infty$ and that $f_n \longrightarrow f$ (a.e, μ). Then, by (2.1.11),

$$\lim_{m\to\infty}\mu\left(\sup_{n\geq m}|f_n - f| \geq \epsilon\right) = \mu\left(\bigcap_{m=1}^{\infty}\left\{\sup_{n\geq m}|f_n - f| \geq \epsilon\right\}\right) = 0$$

for every $\epsilon > 0$, and therefore (3.2.2) holds. In particular, this means that $f_n \longrightarrow f$ in μ-measure. □

Clearly, the first part of Theorem 3.2.6 provides a Cauchy criterion for μ-almost everywhere convergence. The following theorem gives a Cauchy criterion for convergence in μ-measure. In the process, it shows that, after passing to a subsequence, convergence in μ-measure leads to μ-almost everywhere convergence.

Theorem 3.2.7 *Again let $\{f_n : n \geq 1\}$ be a sequence of $\mathbb{R}$-valued, measurable functions on the measure space $(E, \mathcal{B}, \mu)$. Then there is an $\mathbb{R}$-valued, measurable function f to which $\{f_n : n \geq 1\}$ converges in μ-measure if and only if*

$$\lim_{m\to\infty}\sup_{n\geq m}\mu(|f_n - f_m| \geq \epsilon) = 0 \quad \textit{for all } \epsilon > 0. \tag{3.2.4}$$

Furthermore, if $f_n \longrightarrow f$ in μ-measure, then one can extract a subsequence $\{f_{n_j} : j \geq 1\}$ with the property that

$$\lim_{i\to\infty}\mu\left(\sup_{j\geq i}|f - f_{n_j}| \geq \epsilon\right) = 0 \quad \textit{for all } \epsilon > 0;$$

and therefore $f_{n_i} \longrightarrow f$ (a.e., μ) as well as in μ-measure.

Proof. To see that $f_n \longrightarrow f$ in μ-measure implies (3.2.4), simply note that

$$\mu(|f_n - f_m| \geq \epsilon) \leq \mu\left(|f - f_n| \geq \tfrac{\epsilon}{2}\right) + \mu\left(|f - f_m| \geq \tfrac{\epsilon}{2}\right).$$

Conversely, assume that (3.2.4) holds, and choose $1 \leq n_1 < \cdots < n_i < \cdots$ for which

$$\sup_{n\geq n_i} \mu\Big(|f_n - f_{n_i}| \geq 2^{-i-1}\Big) \leq 2^{-i-1}, \quad i \geq 1.$$

Then

$$\mu\left(\sup_{j\geq i}|f_{n_j} - f_{n_i}| > 2^{-i}\right) \leq \mu\left(\bigcup_{j\geq i}\{|f_{n_{j+1}} - f_{n_j}| \geq 2^{-j-1}\}\right)$$
$$\leq \sum_{j=i}^{\infty} \mu\big(|f_{n_{j+1}} - f_{n_j}| \geq 2^{-j-1}\big) \leq 2^{-i}.$$

From this it is clear that $\{f_{n_i} : i \geq 1\}$ satisfies (3.2.3) and therefore that there is an f for which (3.2.2) holds with $\{f_n : n \geq 1\}$ replaced by $\{f_{n_i} : i \geq 1\}$. Hence, $f_{n_i} \longrightarrow f$ both μ-almost everywhere and in μ-measure. In particular, when combined with (3.2.4), this means that

$$\mu(|f_m - f| > \epsilon) \leq \varlimsup_{i\to\infty} \mu\left(|f_m - f_{n_i}| \geq \tfrac{\epsilon}{2}\right) + \varlimsup_{i\to\infty} \mu\left(|f_{n_i} - f| \geq \tfrac{\epsilon}{2}\right)$$
$$\leq \sup_{n\geq m} \mu\left(|f_n - f_m| \geq \tfrac{\epsilon}{2}\right) \longrightarrow 0$$

as $m \to \infty$, and so $f_n \longrightarrow f$ in μ-measure.

Notice that the preceding argument proves the final statement as well. Namely, if $f_n \longrightarrow f$ in μ-measure, then (3.2.4) holds and therefore the argument just given shows that there exists a subsequence $\{f_{n_i} : i \geq 1\}$ that satisfies (3.2.3). But this means that $\{f_{n_i} : i \geq 1\}$ converges both (a.e., μ) and in μ-measure, and, as a subsequence of a sequence that is already converging in μ-measure to f, we conclude that f must be the function to which $\{f_{n_i} : i \geq 1\}$ is converging (a.e., μ). □

Because it is quite important to remember the relationships between the various sorts of convergence discussed in Theorems 3.2.6 and 3.2.7, I will summarize them as follows:

$$\|f_n - f\|_{L^1(\mu;\mathbb{R})} \longrightarrow 0 \Longrightarrow f_n \longrightarrow f \text{ in } \mu\text{-measure}$$
$$\Longrightarrow \lim_{i\to\infty} \mu\left(\sup_{j\geq i}|f_{n_j} - f| \geq \epsilon\right) = 0,\ \epsilon > 0, \text{ for some subsequence } \{f_{n_i}\}$$
$$\Longrightarrow f_{n_i} \longrightarrow f \text{ (a.e., } \mu)$$

and

$$\mu(E) < \infty \text{ and } f_n \longrightarrow f \text{ (a.e., } \mu) \Longrightarrow f_n \longrightarrow f \text{ in } \mu\text{-measure.}$$

Notice that, when $\mu(E) = \infty$, μ-almost everywhere convergence *does not* imply μ-convergence. For example, consider the functions $\mathbf{1}_{[n,\infty)}$ on $\mathbb{R}$ with Lebesgue measure.

I next show that, at least as far as Theorems 3.2.3 through 3.2.5 are concerned, convergence in μ-measure is just as good as μ-almost everywhere convergence.

Theorem 3.2.8 *Let f and $\{f_n : n \geq 1\}$ all be measurable, $\mathbb{R}$-valued functions on the measure space $(E, \mathcal{B}, \mu)$, and assume that $f_n \longrightarrow f$ in μ-measure.*

Fatou's lemma: *If $f_n \geq 0$ (a.e., μ) for each $n \geq 1$, then $f \geq 0$ (a.e., μ) and*

$$\int f \, d\mu \leq \varliminf_{n\to\infty} \int f_n \, d\mu.$$

Lebesgue's Dominated Convergence Theorem: *If there is an integrable g on $(E, \mathcal{B}, \mu)$ such that $|f_n| \leq g$ (a.e., μ) for each $n \geq 1$, then f is integrable, $\lim_{n\to\infty} \|f_n - f\|_{L^1(\mu;\mathbb{R})} = 0$, and so $\int f_n \, d\mu \longrightarrow \int f \, d\mu$ as $n \to \infty$.*

Brezis–Lieb Version of Fatou's lemma: *If $\sup_{n\geq 1} \|f_n\|_{L^1(\mu;\mathbb{R})} < \infty$, then f is integrable and*

$$\begin{aligned}\lim_{n\to\infty} &\Big| \|f_n\|_{L^1(\mu;\mathbb{R})} - \|f\|_{L^1(\mu;\mathbb{R})} - \|f_n - f\|_{L^1(\mu)} \Big| \\ &= \lim_{n\to\infty} \Big\| |f_n| - |f| - |f_n - f| \Big\|_{L^1(\mu;\mathbb{R})} = 0.\end{aligned}$$

In particular, $\|f_n - f\|_{L^1(\mu;\mathbb{R})} \longrightarrow 0$ if $\|f_n\|_{L^1(\mu;\mathbb{R})} \longrightarrow \|f\|_{L^1(\mu;\mathbb{R})} \in \mathbb{R}$.

Proof. Each of these results is obtained via the same trick from the corresponding result in § 3.2.1. Thus I will prove the preceding statement of Fatou's lemma and will leave the proofs of the other assertions to the reader.

Choose a subsequence $\{f_{n_m} : m \geq 1\}$ of $\{f_n : n \geq 1\}$ such that $\int f_{n_m} \, d\mu$ tends to $\varliminf_{n\to\infty} \int f_n \, d\mu$. Next, choose a subsequence $\{f_{n_{m_i}} : i \geq 1\}$ of $\{f_{n_m} : m \geq 1\}$ for which $f_{n_{m_i}} \longrightarrow f$ (a.e., μ). Because each of the $f_{n_{m_i}}$'s is non-negative (a.e., μ), it is now clear that $f \geq 0$ (a.e., μ). In addition, by restricting all integrals to the set $\hat{E}$ on which the $f_{n_{m_i}}$'s are non-negative and $f_{n_{m_i}} \longrightarrow f$, we can apply Theorem 3.2.3 to obtain

$$\int f \, d\mu = \int_{\hat{E}} f \, d\mu \leq \varliminf_{i\to\infty} \int_{\hat{E}} f_{n_{m_i}} \, d\mu = \lim_{m\to\infty} \int f_{n_m} \, d\mu = \varliminf_{n\to\infty} \int f_n \, d\mu.$$

□

3.2.3 Elementary Properties of $L^1(\mu; \mathbb{R})$

An important dividend of the considerations in § 3.2.2 is that they allow us to prove that $L^1(\mu; \mathbb{R})$ is a *complete metric space.*

Lemma 3.2.9 *Let $\{f_n : n \geq 1\} \subseteq L^1(\mu; \mathbb{R})$. If*

$$\lim_{m\to\infty} \sup_{n\geq m} \|f_n - f_m\|_{L^1(\mu;\mathbb{R})} = 0,$$

then there exists an $f \in L^1(\mu;\mathbb{R})$ *for which* $\|f_n - f\|_{L^1(\mu;\mathbb{R})} \longrightarrow 0$. *In other words,* $\big(L^1(\mu;\mathbb{R}), \|\cdot\|_{L^1(\mu;\mathbb{R})}\big)$ *is a complete metric space.*

Proof. By (3.1.2) we know that (3.2.4) holds. Hence, we can find a measurable f for which $f_n \longrightarrow f$ in μ-measure; and so, by Fatou's lemma,

$$\|f - f_m\|_{L^1(\mu;\mathbb{R})} \leq \varliminf_{n\to\infty} \|f_n - f_m\|_{L^1(\mu;\mathbb{R})} \leq \sup_{n\geq m} \|f_n - f_m\|_{L^1(\mu;\mathbb{R})} \longrightarrow 0$$

as $m \to \infty$. In particular, $f \in L^1(\mu;\mathbb{R})$. □

Having shown that $L^1(\mu;\mathbb{R})$ is complete, I turn next to the question of its separability. Before doing so, it will be convenient to have introduced the notion of σ-finiteness. A measure μ on a measurable space $(E,\mathcal{B})$ is said to be **σ-finite** if there is a sequence $\{E_n : n \geq 1\} \subseteq \mathcal{B}$ such that $E = \bigcup_{n=1}^{\infty} E_n$ and $\mu(E_n) < \infty$ for each $n \in \mathbb{Z}^+$. If μ is σ-finite, then the measure space $(E,\mathcal{B},\mu)$ is said to be a **σ-finite measure space**. Notice that if μ is σ-finite, one can always choose the E_n's to be either non-decreasing or mutually disjoint. In addition, observe that any measure constructed by the procedure in § 2.2 will be σ-finite. In fact, for such measures, the E_n's can be chosen so that they are either all open sets or all compact sets.

Theorem 3.2.10 *Let* $(E,\mathcal{B})$ *be a measurable space for which* $\mathcal{B}$ *is generated by the* Π*-system* $\mathcal{C}$, *and use* $\mathcal{S}$ *to denote the set of functions* $\sum_{m=1}^{n} \alpha_m \mathbf{1}_{\Gamma_m}$, *where* $n \in \mathbb{Z}^+$, $\{\alpha_m : 1 \leq m \leq n\} \subseteq \mathbb{Q}$, *and* $\{\Gamma_m : 1 \leq m \leq n\} \subseteq \mathcal{C} \cup \{E\}$. *If* μ *is a finite measure on* $(E,\mathcal{B})$, *then* $\mathcal{S}$ *is dense in* $L^1(\mu;\mathbb{R})$. *In particular, if* μ *is a* σ*-finite measure on* $(E,\mathcal{B})$ *and* $\mathcal{C}$ *is countable, then* $L^1(\mu;\mathbb{R})$ *is a separable metric space.*

Proof. Denote by $\overline{\mathcal{S}}$ the closure in $L^1(\mu;\mathbb{R})$ of $\mathcal{S}$. To prove the first assertion, what we have to show is that $\overline{\mathcal{S}} = L^1(\mu;\mathbb{R})$ when μ is finite.

Obviously, $\overline{\mathcal{S}}$ is a vector space over $\mathbb{R}$. In addition, if $f \in L^1(\mu;\mathbb{R})$ and both f^+ and f^- are elements of $\overline{\mathcal{S}}$, then $f \in \overline{\mathcal{S}}$, and so we need only check that every non-negative $f \in L^1(\mu;\mathbb{R})$ is in $\overline{\mathcal{S}}$. Further, since every such f is the limit in $L^1(\mu;\mathbb{R})$ of simple, $\mathbb{Q}$-valued elements of $L^1(\mu;\mathbb{R})$ and since $\overline{\mathcal{S}}$ is a vector space, it suffices to show that $\mathbf{1}_\Gamma \in \overline{\mathcal{S}}$ for every $\Gamma \in \mathcal{B}$. But it is easy to see that the class of $\Gamma \subseteq E$ for which $\mathbf{1}_\Gamma \in \overline{\mathcal{S}}$ is a Λ-system over E, and, by hypothesis, it contains the Π-system $\mathcal{C}$. Now apply Lemma 2.1.1.

To complete the proof, assume that μ is σ-finite, and choose a non-decreasing sequence $\{E_n : n \geq 1\}$ accordingly. Then, as an application of the first part to $\{C \cap E_n : C \in \mathcal{C}\}$, we know that, for each $n \in \mathbb{Z}^+$, $\mathcal{S}_n \equiv \{\mathbf{1}_{E_n}\varphi : \varphi \in \mathcal{S}\}$ is dense in the space $L^1\big(E_n, \mathcal{B}[E_n], \mu \restriction E_n; \mathbb{R}\big)$. Since $\mathcal{S}_n$ is countable for each n, so is $\mathcal{S}_\infty \equiv \bigcup_{n=1}^{\infty} \mathcal{S}_n$. Finally, given $f \in L^1(\mu;\mathbb{R})$, Lebesgue's dominated convergence theorem guarantees that

$\|f - f\mathbf{1}_{E_n}\|_{L^1(\mu;\mathbb{R})} \longrightarrow 0$ as $n \to \infty$. Hence, for any $\epsilon > 0$, we can choose a $\varphi \in \mathcal{S}_\infty$ for which $\|f - \varphi\|_{L^1(\mu;\mathbb{R})} < \epsilon$. □

As a more or less immediate corollary, we know that if μ is a σ-finite measure on a second countable topological space (e.g., a separable metric space), then $L^1(\mu;\mathbb{R})$ is separable. What follows is another important connection between Borel measures and the topology of the spaces on which they are defined.

Corollary 3.2.11 *Let (E,ρ) be a metric space and μ a σ-finite Borel measure on E for which one can choose the associated sequence $\{E_n : n \geq 1\}$ to be non-decreasing and consist of open sets. For each $n \in \mathbb{Z}^+$, take $\mathcal{U}_n$ to be the set of bounded, ρ-uniformly continuous, $\mathbb{R}$-valued functions on E that vanish identically off of E_n. Then $\mathcal{U} \equiv \bigcup_{n=1}^\infty \mathcal{U}_n$ is dense in $L^1(\mu;\mathbb{R})$.*

Proof. First assume that μ is finite and that $E_n = E$ for all $n \geq 1$. Then, by Theorem 3.2.10, the density of $\mathcal{U}$ in $L^1(\mu;\mathbb{R})$ comes down to showing that, for every $G \in \mathfrak{G}(E)$, $\mathbf{1}_G$ is the limit in $L^1(\mu;\mathbb{R})$ of bounded, ρ-uniformly continuous functions. To this end, set

$$\psi_{\ell,G}(x) = \left(\frac{\rho(x, E\setminus G)}{1+\rho(x,E\setminus G)}\right)^{\frac{1}{\ell}} \quad \text{for } \ell \in \mathbb{Z}^+,$$

and note that $\psi_{\ell,G}$ is ρ-uniformly continuous and that $0 \leq \psi_{\ell,G} \nearrow \mathbf{1}_G$. Hence, by the monotone convergence theorem, $\|\psi_{\ell,G} - \mathbf{1}_G\|_{L^1(\mu;\mathbb{R})} \longrightarrow 0$.

To handle the general case, let $f \in L^1(\mu;\mathbb{R})$. By the monotone convergence theorem,

$$\int_{E_n} |f|\,d\mu \nearrow \int |f|\,d\mu,$$

and so, for any $\epsilon > 0$, there is an $n \geq 1$ such that $\int_{E_n^\complement} |f|\,d\mu < \frac{\epsilon}{3}$. Further, by the preceding applied to the finite measure space $(E_n, \mathcal{B}_{E_n}, \mu \restriction E_n)$, there is a bounded, uniformly continuous φ on E_n for which $\int_{E_n} |f-\varphi|\,d\mu < \frac{\epsilon}{3}$. Finally, referring to the preceding paragraph, set $\eta_\ell = \psi_{\ell,E_n}$. Since $\eta_\ell \nearrow 1$ on E_n, Lebesgue's dominated convergence theorem says that

$$\lim_{\ell\to\infty} \int_{E_n} |\varphi - \eta_\ell\varphi|\,d\mu = 0,$$

and so we can choose $\ell \geq 1$ so that $\int_{E_n} |\varphi - \eta_\ell\varphi|\,d\mu < \frac{\epsilon}{3}$. Hence, if

$$\tilde{\varphi}(x) = \begin{cases} \eta_\ell(x)\varphi(x) & \text{for } x \in E_n \\ 0 & \text{for } x \in E \setminus E_n, \end{cases}$$

then $\tilde{\varphi} \in \mathcal{U}_n$ and $\int |f - \tilde{\varphi}|\,d\mu < \epsilon$. □

When applied to Lebesgue measure $\lambda_{\mathbb{R}^N}$ on $\mathbb{R}^N$, Corollary (3.2.11) says that for every $f \in L^1(\lambda_{\mathbb{R}^N};\mathbb{R})$ and $\epsilon > 0$ there is a uniformly continuous function φ such that φ vanishes off of a compact set and $\|f-\varphi\|_{L^1(\lambda_{\mathbb{R}^N};\mathbb{R})} < \epsilon$, and Theorem 6.3.9 shows that even more is true. This fact can be interpreted in either one of two ways: either measurable functions are not all that different from continuous ones or $\|\cdot\|_{L^1(\lambda_{\mathbb{R}^N};\mathbb{R})}$ provides a rather crude gage of size. Experience indicates that the latter interpretation may be the more accurate one.

3.2.4 Exercises for § 3.2

Exercise 3.2.1 Let $(E, \mathcal{B}, \mu)$ be a finite measure space, and show that $f_n \longrightarrow f$ in μ-measure if and only if $\int |f_n - f| \wedge 1\, d\mu \longrightarrow 0$.

Exercise 3.2.2 Suppose that $\varphi : \mathbb{R} \longrightarrow \mathbb{R}$ is a continuous function, and let $\{f_n : n \geq 1\}$ be a sequence of measurable functions on the measure space $(E, \mathcal{B}, \mu)$. It is clear that if f is a $\mathcal{B}$-measurable function and $f_n \longrightarrow f$ μ-almost everywhere, then $\varphi \circ f_n \longrightarrow \varphi \circ f$ μ-almost everywhere.

(i) If μ is finite, show that $f_n \longrightarrow f$ in μ-measure implies that $\varphi \circ f_n \longrightarrow \varphi \circ f$ in μ-measure.

(ii) Even if μ is infinite, show that the same implication holds as long as $\lim_{R\to\infty} \mu(|f| > R) = 0$.

(iii) Give an example in which $f_n \longrightarrow f$ both in μ-measure and μ-almost everywhere but $\{\varphi \circ f_n : n \geq 1\}$ fails to converge in μ-measure for some $\varphi \in C(\mathbb{R}; \mathbb{R})$.

Exercise 3.2.3 Let J be a compact rectangle in $\mathbb{R}^N$ and $f : J \longrightarrow \mathbb{R}$ a continuous function.

(i) Show that the Riemann integral $(\mathrm{R}) \int_J f(x)\, dx$ of f over J is equal to the Lebesgue integral $\int_J f(x)\, \lambda_{\mathbb{R}^N}(dx)$. Next, suppose that $f \in L^1(\lambda_{\mathbb{R}^N}; \mathbb{R})$ is continuous, and use the preceding to show that

$$\int f(x)\, \lambda_{\mathbb{R}^N}(dx) = \lim_{J \nearrow \mathbb{R}^N} (\mathrm{R}) \int_J f(x)\, dx,$$

where the limit means that, for any $\epsilon > 0$, there exists a rectangle J_ϵ such that

$$\left| \int f(x)\, \lambda_{\mathbb{R}^N}(dx) - (\mathrm{R}) \int_J f(x)\, dx \right| < \epsilon$$

whenever J is a rectangle containing J_ϵ. For this reason, even when f is not continuous, it is conventional to use $\int f(x)\, dx$ instead of $\int f\, d\lambda_{\mathbb{R}^N}$ to denote the Lebesgue integral of f when greater precision is not required.

(ii) Now assume that $N = 1$ and $J = [a, b]$. Given a right-continuous, non-decreasing function $\psi : J \longrightarrow \mathbb{R}$, let μ_ψ be the Borel measure on $\mathbb{R}$ for which $x \rightsquigarrow \psi((a \vee x) \wedge b) - \psi(a)$ is the distribution function. Show that for every $\varphi \in C(J;\mathbb{R})$,

$$(\mathrm{R})\int_J \varphi(x)\,d\psi(x) = \int \varphi\,d\mu_\psi.$$

Exercise 3.2.4 Let f be a non-negative, measurable function on $(E,\mathcal{B},\mu)$. Show that

$$f \in L^1(\mu;\mathbb{R}) \implies \lim_{R\to\infty} R\mu(f \ge R) = 0.$$

Next produce an example that shows that the preceding implication does not go in the opposite direction. Finally, show that

$$\sum_{n=0}^{\infty} \mu(f > n) < \infty \implies f \in L^1(\mu;\mathbb{R}).$$

See Exercise 5.1.2 for further information.

Exercise 3.2.5 Let $(E,\mathcal{B})$ be a measurable space and $-\infty \le a < b \le \infty$. Given a function $f:(a,b)\times E \longrightarrow \mathbb{R}$ with the properties that $f(\,\cdot\,,x)$ is once continuously differentiable on (a,b) for every $x \in E$ and $f(t,\,\cdot\,)$ is $\mathcal{B}$-measurable on $(E,\mathcal{B})$ for every $t \in (a,b)$, show that $\dot f(t,\,\cdot\,) = \partial_t f(t,\,\cdot\,)$ is measurable for each $t \in (a,b)$. Further, suppose that μ is a measure on $(E,\mathcal{B})$ and that there is a $g \in L^1(\mu;\mathbb{R})$ such that $|f(s,\,\cdot\,)| \le g(x)$ for some $s \in (a,b)$ and $|\dot f(t,x)| \le g(x)$ for all $(t,x) \in (a,b)\times E$. Show not only that $f(t,\,\cdot\,) \in L^1(\mu;\mathbb{R})$ for all $t \in (a,b)$, but also that $\int_E f(\,\cdot\,,x)\,\mu(dx) \in C^1((a,b);\mathbb{R})$ and

$$\frac{d}{dt}\int_E f(t,x)\,\mu(dx) = \int_E \dot f(t,x)\,\mu(dx).$$

Exercise 3.2.6 Let $(E,\mathcal{B},\mu)$ be a measure space and $\{f_n : n \ge 1\}$ a sequence of measurable functions on $(E,\mathcal{B})$. Suppose that $\{g_n : n \ge 1\} \subseteq L^1(\mu;\mathbb{R})$ and that $g_n \longrightarrow g \in L^1(\mu;\mathbb{R})$ in $L^1(\mu;\mathbb{R})$. The following variants of Fatou's lemma and Lebesgue's dominated convergence theorem are often useful.

(i) If $f_n \le g_n$ (a.e., μ) for each $n \ge 1$, show that

$$\varlimsup_{n\to\infty}\int f_n\,d\mu \le \int \varlimsup_{n\to\infty} f_n\,d\mu.$$

(ii) If $f_n \longrightarrow f$ either in μ-measure or μ-almost everywhere and if $|f_n| \le g_n$ (a.e., μ) for each $n \ge 1$, show that $\|f_n - f\|_{L^1(\mu;\mathbb{R})} \longrightarrow 0$ and therefore that $\lim_{n\to\infty}\int f_n\,d\mu = \int f\,d\mu$.

Exercise 3.2.7 Let (E,ρ) be a metric space and $\{E_n : n \ge 1\}$ a non-decreasing sequence of open subsets of E such that $E_n \nearrow E$. Let μ and ν be two measures on $(E,\mathcal{B}_E)$ with the properties that $\mu(E_n) = \nu(E_n) < \infty$

for every $n \geq 1$ and $\int \varphi \, d\mu = \int \varphi \, d\nu$ whenever φ is a bounded, ρ-uniformly continuous function that vanishes off one of the E_n's. Show that $\mu = \nu$ on $\mathcal{B}_E$.

Exercise 3.2.8 Although almost everywhere convergence does not follow from convergence in measure, it nearly does. Indeed, suppose $\{f_n : n \geq 1\}$ is a sequence of measurable, $\mathbb{R}$-valued functions on $(E, \mathcal{B}, \mu)$. Given an $\mathbb{R}$-valued, measurable function f, show that (3.2.2) holds, and therefore that $f_n \longrightarrow f$ both (a.e., μ) and in μ-measure if

$$\sum_1^\infty \mu\big(|f_n - f| \geq \epsilon\big) < \infty \quad \text{for every } \epsilon > 0.$$

In particular, if $\{f_n : n \geq 1\} \cup \{f\} \subseteq L^1(\mu;\mathbb{R})$ and $\sum_1^\infty \|f_n - f\|_{L^1(\mu;\mathbb{R})} < \infty$, conclude that $f_n \longrightarrow f$ (a.e., μ) and in $L^1(\mu;\mathbb{R})$

Exercise 3.2.9 Let $(E, \mathcal{B}, \mu)$ be a measure space. A family $\mathcal{K}$ of measurable functions f on $(E, \mathcal{B}, \mu)$ is said to be **uniformly μ-absolutely continuous** if, for each $\epsilon > 0$, there is a $\delta > 0$ such that $\int_\Gamma |f| \, d\mu \leq \epsilon$ for all $f \in \mathcal{K}$ whenever $\Gamma \in \mathcal{B}$ and $\mu(\Gamma) < \delta$, and it is said to be **uniformly μ-integrable** if for each $\epsilon > 0$ there is an $R < \infty$ such that $\int_{|f| \geq R} |f| \, d\mu \leq \epsilon$ for all $f \in \mathcal{K}$.

(i) Show that $\mathcal{K}$ is uniformly μ-integrable if it is uniformly μ-absolutely continuous and $\sup_{f \in \mathcal{K}} \|f\|_{L^1(\mu;\mathbb{R})} < \infty$. Conversely, suppose that $\mathcal{K}$ is uniformly μ integrable and show that it is then necessarily uniformly μ-absolutely continuous and, when $\mu(E) < \infty$, that $\sup_{f \in \mathcal{K}} \|f\|_{L^1(\mu;\mathbb{R})} < \infty$.

(ii) If $\sup_{f \in \mathcal{K}} \int |f|^{1+\delta} \, d\mu < \infty$ for some $\delta > 0$, show that $\mathcal{K}$ is uniformly μ-integrable.

(iii) Let $\{f_n : n \geq 1\} \subseteq L^1(\mu;\mathbb{R})$ be given. If $f_n \longrightarrow f$ in $L^1(\mu;\mathbb{R})$, show that $\{f_n : n \geq 1\} \cup \{f\}$ is uniformly μ-absolutely continuous and uniformly μ-integrable. Conversely, assuming that $\mu(E) < \infty$, show that $f_n \longrightarrow f$ in $L^1(\mu;\mathbb{R})$ if $f_n \longrightarrow f$ in μ-measure and $\{f_n : n \geq 1\}$ is uniformly μ-integrable.

(iv) Assume that $\mu(E) = \infty$. One says that a family $\mathcal{K}$ of measurable functions f on $(E, \mathcal{B}, \mu)$ is **tight** if, for each $\epsilon > 0$, there is a $\Gamma \in \mathcal{B}$ for which $\mu(\Gamma) < \infty$ and $\sup_{f \in \mathcal{K}} \int_{\Gamma \complement} |f| \, d\mu \leq \epsilon$. Assuming that $\mathcal{K}$ is tight, show that $\mathcal{K}$ is uniformly μ-integrable if and only if it is uniformly μ-absolutely continuous and $\sup_{f \in \mathcal{K}} \|f\|_{L^1(\mu;\mathbb{R})} < \infty$. Finally, suppose that $\{f_n : n \geq 1\} \subseteq L^1(\mu;\mathbb{R})$ is tight and that $f_n \longrightarrow f$ in μ-measure. Show that $\|f_n - f\|_{L^1(\mu;\mathbb{R})} \longrightarrow 0$ if and only if $\{f_n : n \geq 1\}$ is uniformly μ-integrable.

3.3 Lebesgue's Differentiation Theorem

Although it represents something of a departure from the spirit of this chapter, I return in this concluding section to Lebesgue measure on $\mathbb{R}$ and prove that every non-decreasing function on $\mathbb{R}$ has a derivative at $\lambda_{\mathbb{R}}$-almost every point. To be more precise, say that a function $F : \mathbb{R} \longrightarrow \mathbb{R}$ has a derivative $F'(x)$ at x if

$$F'(x) \equiv \lim_{|h|\searrow 0} \frac{F(x+h) - F(x)}{h} \quad \text{exists in } \mathbb{R}.$$

The main goal of this section is to show that $F'(x)$ exists at $\lambda_{\mathbb{R}}$-almost every $x \in \mathbb{R}$ if F is non-decreasing. Since the result is completely local and F is continuous at all but at most countably many points, it suffices to treat F's which are bounded and right continuous, and so I will be making those assumptions throughout. Further, because the result is unchanged when F is replaced by $F - F(-\infty)$, where $F(-\infty) = \lim_{x\searrow -\infty} F(x)$, I will also assume that F tends to 0 as x tends to $-\infty$. In particular, this means (cf. Theorem 2.2.17 or Exercise 3.1.2) that there is a unique Borel measure μ_F for which F is the distribution function.

3.3.1 The Sunrise Lemma

An $\mathbb{R}$-valued function g on a topological space E is said to be **upper semicontinuous** if $\{x \in E : g(x) < a\}$ is open for every $a \in \mathbb{R}$, which is equivalent to saying that $\overline{\lim}_{y\to x} g(y) \le g(x)$ for all $x \in E$. Similarly, g is **lower semicontinuous** if $\{x \in E : g(x) > a\} \in \mathfrak{G}(E)$ for all $a \in \mathbb{R}$, which is equivalent to $g(x) \le \underline{\lim}_{y\to x} g(y)$ for every $x \in E$. By Exercise 3.1.4, an upper or lower semicontinuous function is $\mathcal{B}_E$-measurable.

The following simple lemma, which is due to F. Riesz, is the key to everything that follows.

Theorem 3.3.1 (Sunrise Lemma[3]) *Let $g : \mathbb{R} \longrightarrow \mathbb{R}$ be a right-continuous, upper semicontinuous function with the property that $\lim_{x\to\pm\infty} g(x) = \mp\infty$, and set $G = \{x : \exists y > x\ g(y) > g(x)\}$. Then G is open, and each non-empty, open, connected component of G is a bounded interval (a, b) with $g(a) = g(b)$.*

Proof. The fact that G is open is an easy consequence of upper semicontinuity. Now suppose that (a, b) (possibly with $a = -\infty$ or $b = \infty$) is a non-empty, open, connected component of G, and let $c \in (a, b)$. Then, because $g(x) \longrightarrow -\infty$ as $x \to \infty$, $x = \sup\{y \ge c : g(y) \ge g(c)\} < \infty$. Moreover, by upper semicontinuity, $g(x) \ge g(c)$, and, by definition, $g(y) < g(c)$ for all $y > x$. Thus, $x \notin G$. In particular, this means that $b \le x < \infty$. In addition, it also means that $g(c) \le g(b)$. To see this, suppose that $g(c) > g(b)$. Then,

[3] The name derives from the following picture. The sun is rising infinitely far to the right in mountainous (one-dimensional) terrain, $g(x)$ is the elevation at x, and G is the region in shadow at the instant the sun comes over the horizon.

because $b \notin G$, and therefore $g(y) \le g(b)$ for all $y \ge b$, we would have that $x \in (c, b) \subseteq G$.

We now know that $b < \infty$ and that $g(c) \le g(b)$ for all $c \in (a, b)$. In particular, since $g(x) \longrightarrow \infty$ as $x \to -\infty$, this proves that $a > -\infty$. Finally, by right-continuity, $g(a) = \lim_{c \searrow a} g(c) \le g(b)$. On the other hand, because $a \notin G$, $g(b) \le g(a)$. □

To explain my initial application of the Sunrise Lemma, let F be a bounded, right-continuous, non-decreasing function that tends to 0 at $-\infty$, and define

$$\mathcal{L}_{\pm}F(x) = \sup_{h>0} \frac{F(x \pm h) - F(x)}{\pm h} \quad \text{and } \mathcal{L}F(x) = \mathcal{L}_{+}F(x) \vee \mathcal{L}_{-}F{-}(x) \text{ for } x \in \mathbb{R}.$$

Clearly, $\mathcal{L}F(x)$ is the Lipschitz constant for F at the point x.

Corollary 3.3.2 *Referring to the preceding, for each $R > 0$, the set $\{x \in \mathbb{R} : \mathcal{L}_{+}F(x) > R\}$ is open and*

$$\lambda_{\mathbb{R}}(\mathcal{L}_{+}F > R) \le \frac{F(\infty)}{R},$$

where $F(\infty) = \lim_{x\to\infty} F(x)$. Moreover, if F is continuous, then

$$\lambda_{\mathbb{R}}(\mathcal{L}_{+}F > R) = \frac{\mu_F(\mathcal{L}_{+}F > R)}{R} \le \frac{F(\infty)}{R},$$

where μ_F is the Borel measure on $\mathbb{R}$ whose distribution function is F.

Proof. Set $g_R(x) = F(x) - Rx$. Then g_R is right-continuous and upper semi-continuous, and $g_R(x)$ tends to $\mp\infty$ as $x \to \pm\infty$. Furthermore,

$$G_R \equiv \{x : \exists y > x : g_R(y) > g_R(x)\} = \{x : \mathcal{L}_{+}F(x) > R\}.$$

Thus $\{\mathcal{L}_{+}F > R\}$ is open, and its Lebesgue measure is that of G_R.

If $G_R = \emptyset$, there is nothing more to do. If $G_R \ne \emptyset$, apply the Sunrise Lemma to write it as the union of at most countably many bounded, mutually disjoint, non-empty open intervals (a_n, b_n) for which $g_R(b_n) = g_R(a_n)$, and therefore $b_n - a_n = \frac{F(b_n)-F(a_n)}{R}$. Then

$$\begin{aligned}\lambda_{\mathbb{R}}(G_R) &= \sum_n (b_n - a_n) = \frac{1}{R}\sum_n \big(F(b_n) - F(a_n)\big)\\ &= \frac{1}{R}\mu_F\left(\bigcup_n (a_n, b_n]\right) \le \frac{\mu_F(\mathbb{R})}{R} = \frac{F(\infty)}{R}.\end{aligned}$$

Moreover, if F is continuous, then $\mu_F\big((a_n, b_n]\big) = \mu_F\big((a_n, b_n)\big)$, and so

$$R\lambda_{\mathbb{R}}(G_R) = \sum_n \mu_F\big((a_n, b_n)\big) = \mu_F(G_R) \le F(\infty).$$

□

Corollary 3.3.3 *The function $\mathcal{L}F$ is $\mathcal{B}_{\mathbb{R}}$-measurable and, for each $R > 0$,*

$$\lambda_{\mathbb{R}}(\mathcal{L}F > R) \le \frac{2F(\infty)}{R}.$$

Moreover, if F is continuous, then

$$\lambda_{\mathbb{R}}(\mathcal{L}F > R) \le \frac{2\mu_F(\mathcal{L}F > R)}{R}.$$

Proof. Define $\breve{F}$ so that $\breve{F}(-x) = F(\infty) - F(x-)$, where $F(x-) \equiv \lim_{y \nearrow x} F(y)$. Then $\breve{F}$ is again a bounded, right-continuous, non-decreasing function that tends to 0 at $-\infty$, and so Corollary 3.3.2 applies and says that $\{\mathcal{L}_+\breve{F} > R\}$ is open,

$$\lambda_{\mathbb{R}}(\mathcal{L}_+\breve{F} > R) \le \frac{\breve{F}(\infty)}{R} = \frac{F(\infty)}{R},$$

and, when F is continuous,

$$\lambda_{\mathbb{R}}(\mathcal{L}_+\breve{F} > R) = \frac{\mu_{\breve{F}}(\mathcal{L}_+\breve{F} > R)}{R}.$$

In addition,

$$\mathcal{L}_+\breve{F}(-x) = \sup_{h<0} \frac{F(x+h) - F(x-)}{h}.$$

Hence $\mathcal{L}_-F(x) = \mathcal{L}_+\breve{F}(-x)$ for all but at most a countable number of $x \in \mathbb{R}$, and so $\mathcal{L}F$ is $\mathcal{B}_{\mathbb{R}}$-measurable and $\lambda_{\mathbb{R}}(\mathcal{L}_-F > R) = \lambda_{\mathbb{R}}(\mathcal{L}_+\breve{F} > R)$. Since $\breve{F}(\infty) = F(\infty)$ and $\lambda_{\mathbb{R}}(\mathcal{L}F > R) \le \lambda_{\mathbb{R}}(\mathcal{L}_+F > R) + \lambda_{\mathbb{R}}(\mathcal{L}_-F > R)$, this, in conjunction with Corollary 3.3.2, proves the first estimate. As for continuous F's, the preceding together with Corollary 3.3.2 implies that

$$\begin{aligned} R\lambda_{\mathbb{R}}(\mathcal{L}_-F > R) &= R\lambda_{\mathbb{R}}\big(\{x : \mathcal{L}_-F(-x) > R\}\big) = \mu_{\breve{F}}(\mathcal{L}_+\breve{F} > R) \\ &= \mu_{\breve{F}}\big(\{x : \mathcal{L}_-F(-x) > R\}\big) = \mu_F(\mathcal{L}_-F > R), \end{aligned}$$

since $\breve{F}(-x) = \mu_F\big([x, \infty)\big) = \mu_F\big(\{y : -y \in (-\infty, -x]\}\big)$, which means that the measure $\breve{\mu}_F$ determined by $\breve{\mu}_F(\Gamma) = \mu_F\big(\{x : -x \in \Gamma\}\big)$ has $\breve{F}$ as its distribution function and is therefore equal to $\mu_{\breve{F}}$. Hence, we now have that

$$\begin{aligned} \lambda_{\mathbb{R}}(\mathcal{L}F > R) &\le \lambda_{\mathbb{R}}(\mathcal{L}_+F > R) + \lambda_{\mathbb{R}}(\mathcal{L}_-F > R) \\ &= \frac{\mu_F(\mathcal{L}_+F > R) + \mu_F(\mathcal{L}_-F > R)}{R} \le \frac{2\mu_F(\mathcal{L}F > R)}{R}. \end{aligned}$$

□

3.3.2 The Absolutely Continuous Case

Recall that I am assuming throughout this discussion that F is bounded, right-continuous, non-decreasing, and tends to 0 at $-\infty$.

In this subsection I will treat F's that are (cf. Exercise 2.2.8) absolutely continuous. However, before I can do so, I need to show that every absolutely continuous F is the indefinite integral of a non-negative function f in the sense that

$$F(x) = \int_{(-\infty,x]} f\,d\lambda_{\mathbb{R}} \quad \text{for all } x \in \mathbb{R},$$

and I begin with the case in which F is uniformly Lipschitz continuous.

Lemma 3.3.4 *For any F there is a most one $f \in L^1(\lambda_{\mathbb{R}};\mathbb{R})$ of which it is the indefinite integral, and, if it exists, then $\mu_F(\Gamma) = \int_\Gamma f\,d\lambda_{\mathbb{R}}$ for all $\Gamma \in \mathcal{B}_{\mathbb{R}}$ and therefore $f \geq 0$ (a.e., $\lambda_{\mathbb{R}}$). Moreover, if $F(y) - F(x) \leq L(y-x)$ for some $L \in [0,\infty)$ and all $x < y$, then there is an $f \in L^1(\lambda_{\mathbb{R}};\mathbb{R})$ for which F is the indefinite integral, and f can be chosen to take values between 0 and L.*

Proof. Suppose that F is the indefinite integral of $f \in L^1(\lambda_{\mathbb{R}};\mathbb{R})$. Then F is continuous and so

$$\mu_F\big((a,b)\big) = F(b) - F(a) = \int_{(a,b)} f\,d\lambda_{\mathbb{R}}$$

for all open intervals (a,b). Furthermore, it is easy to check that the set of $\Gamma \in \mathcal{B}_{\mathbb{R}}$ for which $\mu_F(\Gamma) = \int_\Gamma f\,d\lambda_{\mathbb{R}}$ is a Λ-system. Thus, since the set of open intervals is a Π-system that generates $\mathcal{B}_{\mathbb{R}}$, it follows that this equality holds for all $\Gamma \in \mathcal{B}_{\mathbb{R}}$. As a consequence, Exercise 3.1.6 guarantees that, up to a set of $\lambda_{\mathbb{R}}$-measure 0, f is unique and non-negative.

Turning to the second part, for each $n \geq 0$, define $f_n : \mathbb{R} \longrightarrow [0,\infty)$ so that

$$f_n(x) = 2^n\big(F((k+1)2^{-n}) - F(k2^{-n})\big) \quad \text{when } k2^{-n} \leq x < (k+1)2^{-n}.$$

Obviously, $0 \leq f_n \leq L$ and $\|f_n\|_{L^1(\lambda_{\mathbb{R}};\mathbb{R})} = F(\infty)$. Hence, $f_n f_m \geq 0$ and $\int f_n f_m\,d\lambda_{\mathbb{R}} \leq LF(\infty)$ for all $m, n \in \mathbb{N}$. Moreover, if $m < n$, then

$$\begin{aligned}\int_{[k2^{-m},(k+1)2^{-m})} f_n f_m\,d\lambda_{\mathbb{R}} &= f_m(k2^{-m}) \sum_{j=k2^{n-m}}^{(k+1)2^{n-m}-1} \big(F((j+1)2^{-n}) - F(j2^{-n})\big)\\ &= 2^{-m} f_m(k2^{-n})^2 = \int_{[k2^{-m},(k+1)2^{-m})} f_m^2\,d\lambda_{\mathbb{R}},\end{aligned}$$

and therefore $\int f_n f_m\,d\lambda_{\mathbb{R}} = \int f_m^2\,d\lambda_{\mathbb{R}}$ for all $m \leq n$. In particular, this means that

$$\int (f_n - f_m)^2 \, d\lambda_{\mathbb{R}} = \int f_n^2 \, \lambda_{\mathbb{R}} - \int f_m^2 \, d\lambda_{\mathbb{R}}, \tag{$*$}$$

and so the sequence of integrals $\int f_n^2 \, d\lambda_{\mathbb{R}}$ is non-decreasing, bounded above by $LF(\infty)$, and therefore convergent as $n \to \infty$. Hence, by $(*)$ and (3.1.2), we now know that, for any $\epsilon > 0$,

$$\sup_{n \ge m} \lambda_{\mathbb{R}}\big(|f_n - f_m| \ge \epsilon\big) \le \epsilon^{-2} \sup_{n \ge m} \int (f_n - f_m)^2 \, d\lambda_{\mathbb{R}} \longrightarrow 0 \quad \text{as } m \to \infty,$$

and therefore, by Theorem 3.2.7, that there exists a $\mathcal{B}_{\mathbb{R}}$ measurable f to which $\{f_n : n \ge 1\}$ converges in $\lambda_{\mathbb{R}}$-measure. Furthermore, because $0 \le f_n \le L$, we may assume that $0 \le f \le L$. Also, Lebesgue's dominated convergence theorem for convergence in measure (cf. Theorem 3.2.8) implies that, for all $m \in \mathbb{Z}^+$ and $k < \ell$,

$$F(\ell 2^{-m}) - F(k 2^{-m}) = \int_{(k2^{-m}, \ell 2^{-m}]} f_{n \vee m} \, d\lambda_{\mathbb{R}} \longrightarrow \int_{(k2^{-m}, \ell 2^{-m}]} f \, d\lambda_{\mathbb{R}}$$

as $n \to \infty$, and therefore that

$$F(y) - F(x) = \int_{(x,y]} f \, d\lambda_{\mathbb{R}},$$

first for $x < y$ that are dyadic numbers (i.e., of the form $k2^{-m}$) and then, by continuity, for all $x < y$. Thus, after letting $x \to -\infty$, we see that F is the indefinite integral of f. □

Lemma 3.3.5 *Given $L \in [0, \infty)$, define $F_L : \mathbb{R} \longrightarrow \mathbb{R}$ by*

$$F_L(x) = \mu_F\big((-\infty, x] \cap \{\mathcal{L}F \le L\}\big) \quad \textit{for } x \in \mathbb{R}.$$

Then F_L is a bounded, non-decreasing function that tends to 0 at $-\infty$ and satisfies $F_L(y) - F_L(x) \le L(y - x)$ for all $x < y$.

Proof. Clearly, it suffices to prove the final inequality.

For any $x < y$, $F_L(y) - F_L(x) = \mu_F\big((x, y] \cap \{\mathcal{L}F \le L\}\big)$. Thus, if $(x, y] \cap \{\mathcal{L}F \le L\} = \emptyset$, then $F_L(y) - F_L(x) = 0$; and if $c \in (x, y] \cap \{\mathcal{L}F \le L\}$, then

$$\begin{aligned} 0 \le F_L(y) - F_L(x) \le F(y) - F(x) &= \big(F(y) - F(c)\big) + \big(F(c) - F(x)\big) \\ &\le L(y - c) + L(c - x) = L(y - x). \end{aligned}$$

□

Theorem 3.3.6 *The following properties are equivalent.*

(a) *F is absolutely continuous.*

(b) *$\mu_F(\mathcal{L}F = \infty) = 0$.*

(**c**) *There exists a sequence* $\{F_n : n \geq 1\}$ *of bounded, non-decreasing, uniformly Lipschitz continuous functions, each of which tends to* 0 *at* $-\infty$*, such that* $F_{n+1} - F_n$ *is non-decreasing for each* $n \in \mathbb{Z}^+$ *and* $F_n(\infty) \nearrow F(\infty)$.

(**d**) *There is a non-negative* $f \in L^1(\lambda_{\mathbb{R}};\mathbb{R})$ *for which* F *is the indefinite integral.*

Moreover, the f *in* (**d**) *is the only element in* $L^1(\lambda_{\mathbb{R}};\mathbb{R})$ *of which* F *is its indefinite integral. (The implication* (**a**) $\implies$ (**d**) *is a special case of Theorem 8.1.2.)*

Proof. If F is absolutely continuous, then (cf. Exercise 2.2.8) $\mu_F \ll \lambda_{\mathbb{R}}$, and therefore, because $\lambda_{\mathbb{R}}(\mathcal{L}F = \infty) = 0$, $\mu_F(\mathcal{L}F = \infty) = 0$. Hence, (**a**) $\implies$ (**b**). Next, assume that $\mu_F(\mathcal{L}F = \infty) = 0$, and set

$$F_n(x) = \mu_F\big((-\infty, x] \cap \{\mathcal{L}F \leq n\}\big) \quad \text{for } x \in \mathbb{R}.$$

Then it is clear that, for each $n \in \mathbb{Z}^+$, F_n is a bounded, non-decreasing function that tends to 0 at $-\infty$. In addition, by Lemma (3.3.5),

$$0 \leq F_n(y) - F_n(x) \leq n(y - x) \quad \text{for all } x < y,$$

and

$$F_{n+1}(x) - F_n(x) = \mu_F\big((-\infty, x] \cap \{n < \mathcal{L}F \leq (n+1)\}\big)$$

is non-negative and non-decreasing. Finally,

$$F_n(\infty) = \mu_F(\mathcal{L}F \leq n) \nearrow \mu_F(\mathcal{L}F < \infty) = \mu_F(\mathbb{R}) = F(\infty).$$

Thus (**b**) $\implies$ (**c**).

To prove that (**c**) $\implies$ (**d**), for each $n \in \mathbb{Z}^+$ use Lemma (3.3.4) to find a non-negative $f_n \in L^1(\lambda_{\mathbb{R}};\mathbb{R})$ for which F_n is the indefinite integral. Then, because $F_{n+1}(x) - F_n(x) = \int_{(-\infty,x]}(f_{n+1} - f_n)\, d\lambda_{\mathbb{R}}$, we know that $f_{n+1} - f_n$ is the unique element of $L^1(\lambda_{\mathbb{R}};\mathbb{R})$ for which $F_{n+1} - F_n$ is the indefinite integral and, as such, is non-negative (a.e., $\lambda_{\mathbb{R}}$). Hence, $f_{n+1} \geq f_n$ (a.e., $\lambda_{\mathbb{R}}$). In other words, we can now assume that the f_n's are non-negative and $f_{n+1} \geq f_n$ everywhere. Finally, set $f = \lim_{n\to\infty} f_n$. Because $F_n \longrightarrow F$ pointwise (in fact, uniformly), it follows from the monotone convergence theorem that

$$F(x) = \lim_{n\to\infty} F_n(x) = \lim_{n\to\infty} \int_{(-\infty,x]} f_n\, d\lambda_{\mathbb{R}} = \int_{(-\infty,x]} f\, d\lambda_{\mathbb{R}} \quad \text{for all } x \in \mathbb{R}.$$

That (**d**) $\implies$ (**a**) is (cf. Exercise 3.1.5) trivial, and the concluding uniqueness statement is covered by Lemma 3.3.4. □

Now that we know that all absolutely continuous F's are indefinite integrals, the $\lambda_{\mathbb{R}}$-almost everywhere existence of F' for such F's becomes a matter of extending the Fundamental Theorem of Calculus to measurable functions. To this end, for $f \in L^1(\lambda_{\mathbb{R}};\mathbb{R})$, define the **Hardy–Littlewood Maximal**

Function Mf of f by

$$Mf(x) = \sup_{I \ni x} \fint_I |f| \, d\lambda_{\mathbb{R}}, \quad x \in \mathbb{R},$$

where the supremum is over closed intervals $I \ni x$ with $\mathring{I} \neq \emptyset$ and

$$\fint_I f \, d\lambda_{\mathbb{R}} \equiv \frac{1}{\mathrm{vol}(I)} \int_I f \, d\lambda_{\mathbb{R}}$$

is the average value of f on I. Obviously, if F is the indefinite integral of $|f|$, then $Mf = \mathcal{L}F$, and so Corollary (3.3.3) says that

$$\lambda_{\mathbb{R}}(Mf > R) \leq \frac{2\|f\|_{L^1(\lambda_{\mathbb{R}};\mathbb{R})}}{R} \quad \text{for all } R > 0. \tag{3.3.1}$$

(See Exercise 6.2.5 for an important consequence.)

The inequality (3.3.1) is the famous **Hardy–Littlewood inequality** that often plays a crucial role in the analysis of almost everywhere convergence results. For us, its importance is demonstrated in the following statement of the Fundamental Theorem of Calculus.

Theorem 3.3.7 *For each* $f \in L^1(\lambda_{\mathbb{R}}; \mathbb{R})$,

$$\lim_{I \searrow \{x\}} \fint_I |f - f(x)| \, d\lambda_{\mathbb{R}} = 0 \quad \textit{for } \lambda_{\mathbb{R}}\textit{-almost every } x \in \mathbb{R}, \tag{3.3.2}$$

where the limit is taken over intervals $I \ni x$ *as* $\mathrm{vol}(I) \searrow 0$. *In particular, if* F *is the indefinite integral of* f, *then* $F'(x)$ *exists and is equal to* $f(x)$ *for* $\lambda_{\mathbb{R}}$*-almost every* $x \in \mathbb{R}$.

Proof. There is nothing to do when $f \in C(\mathbb{R}; \mathbb{R}) \cap L^1(\lambda_{\mathbb{R}}; \mathbb{R})$. Furthermore, by Corollary (3.2.11), we know that $C(\mathbb{R}; \mathbb{R}) \cap L^1(\lambda_{\mathbb{R}}; \mathbb{R})$ is dense in $L^1(\lambda_{\mathbb{R}}; \mathbb{R})$. Hence, we will be done once we show that the set of f's for which (3.3.2) holds is closed under convergence in $L^1(\lambda_{\mathbb{R}}; \mathbb{R})$. To prove this, suppose that $\{f_n : n \geq 1\} \subseteq L^1(\lambda_{\mathbb{R}}; \mathbb{R})$ is a sequence of functions for which (3.3.2) holds and that $f_n \longrightarrow f$ in $L^1(\lambda_{\mathbb{R}}; \mathbb{R})$. Then, for any $\epsilon > 0$ and $n \in \mathbb{Z}^+$,

$$\begin{aligned}
&\lambda_{\mathbb{R}}\left(\left\{x : \varlimsup_{I \searrow \{x\}} \fint_I |f - f(x)|\, d\lambda_{\mathbb{R}} \geq 3\epsilon\right\}\right) \\
&\quad \leq \lambda_{\mathbb{R}}\left(\left\{x : \varlimsup_{I \searrow \{x\}} \fint_I |f - f_n|\, d\lambda_{\mathbb{R}} \geq \epsilon\right\}\right) \\
&\qquad + \lambda_{\mathbb{R}}\left(\left\{x : \varlimsup_{I \searrow \{x\}} \fint_I |f_n - f_n(x)|\, d\lambda_{\mathbb{R}} \geq \epsilon\right\}\right) + \lambda_{\mathbb{R}}\big(|f_n - f| \geq \epsilon\big).
\end{aligned}$$

By (3.3.1) and (3.1.2) applied to $f - f_n$, the first and third terms on the right are dominated by, respectively, $2\epsilon^{-1}\|f - f_n\|_{L^1(\lambda_{\mathbb{R}};\mathbb{R})}$ and $\epsilon^{-1}\|f - f_n\|_{L^1(\lambda_{\mathbb{R}};\mathbb{R})}$, and, by assumption, the second term on the right vanishes. Hence,

$$\lambda_{\mathbb{R}}\left(\left\{x : \varlimsup_{I \searrow \{x\}} \fint_I |f - f(x)|\, d\lambda_{\mathbb{R}} \geq 3\epsilon\right\}\right) \leq \frac{3\|f - f_n\|_{L^1(\lambda_{\mathbb{R}};\mathbb{R})}}{\epsilon}$$

for every $n \in \mathbb{Z}^+$, and so the desired conclusion follows after we let $n \to \infty$.

□

Before closing this subsection, there are several comments that should be made. First, one should recognize that the conclusions drawn in Theorem (3.3.7) remain true for any Lebesgue measurable f that is integrable on each compact subset of $\mathbb{R}$. Indeed, all the assertions there are completely local and therefore follow by replacing f with $f\mathbf{1}_{(-R,R)}$, restricting ones attention to $x \in (-R, R)$, and then letting $R \nearrow \infty$.

Second, one should notice that (3.3.1) would be a trivial consequence of Markov's inequality if we had the estimate $\|Mf\|_{L^1(\lambda_{\mathbb{R}};\mathbb{R})} \leq C\|f\|_{L^1(\lambda_{\mathbb{R}};\mathbb{R})}$ for some $C < \infty$. Thus, one is tempted to ask whether such an estimate is true. That the answer is a resounding *no* can be most easily seen from the observation that, if $\|f\|_{L^1(\lambda_{\mathbb{R}};\mathbb{R})} \neq 0$, then $\alpha \equiv \int_{(-r,r)} |f(t)|\, \lambda_{\mathbb{R}}(dt) > 0$ for some $r > 0$ and therefore $Mf(x) \geq \frac{\alpha}{|x|+r}$ for all $x \in \mathbb{R}$. Therefore, *if* $f \in L^1(\lambda_{\mathbb{R}}; \mathbb{R})$ *does not vanish almost everywhere, then* Mf *is not integrable.* (To see that the situation is even worse and that, in general, Mf need not be integrable over bounded sets, see Exercise 3.3.1 below.) Thus, in a very real sense, (3.3.1) is about as well as one can do. Because this sort of situation arises quite often, inequalities of the form in (3.3.1) have been given a special name: they are called *weak-type inequalities* to distinguish them from inequalities of the form $\|Mf\|_{L^1(\lambda_{\mathbb{R}};\mathbb{R})} \leq C\|f\|_{L^1(\lambda_{\mathbb{R}};\mathbb{R})}$, which are called *strong-type inequalities.* See Exercise 6.2.5 below for related information.

Finally, it should be clear that, except for the derivation of (3.3.1), the arguments given in the proof of Theorem 3.3.7 would work equally well for functions on $\mathbb{R}^N$. Thus, we would know that, for each Lebesgue integrable f on $\mathbb{R}^N$,

$$\lim_{B\searrow\{x\}} \fint_B \big|f(y)-f(x)\big|\,\lambda_{\mathbb{R}^N}(dy)=0 \quad \text{for } \lambda_{\mathbb{R}^N}\text{-almost every } x\in\mathbb{R}^N \tag{3.3.3}$$

if we knew that, for some $C<\infty$,

$$\lambda_{\mathbb{R}^N}(Mf>\epsilon)\le\frac{C}{\epsilon}\|f\|_{L^1(\lambda_{\mathbb{R}^N};\mathbb{R})}, \tag{3.3.4}$$

where Mf is the Hardy–Littlewood maximal function

$$Mf(x)\equiv\sup_{B\ni x}\fint_B \big|f(y)\big|\,\lambda_{\mathbb{R}^N}(dy)$$

and B denotes a generic open ball in $\mathbb{R}^N$. It turns out that (3.3.4), and therefore (3.3.3), are both true. However, because there is no multidimensional analog of the Sunrise Lemma, the proof[4] of (3.3.4) for $N\ge 2$ is somewhat more involved than the one that I have given of (3.3.1).

3.3.3 The General Case

In this subsection I will complete the program of differentiating non-decreasing functions F. As a consequence of Theorems 3.3.6 and 3.3.7, we already know that F' exists $\lambda_{\mathbb{R}}$-almost everywhere when F is absolutely continuous. To handle general F's, I will begin by showing that F can be written as the sum of an absolutely continuous and a (cf. Exercise 2.2.9) singular function.

Theorem 3.3.8 *If F is a bounded, right-continuous, non-decreasing function that tends to 0 at $-\infty$, then there exist a unique absolutely continuous, non-decreasing function F_{a} and a unique singular, right-continuous, non-decreasing function F_{s}, both of which tend to 0 at $-\infty$, for which $F=F_{\mathrm{a}}+F_{\mathrm{s}}$. In fact,*

$$F_{\mathrm{a}}(x)=\mu_F\big((-\infty,x]\cap\{\mathcal{L}F<\infty\}\big)\quad\textit{and}\quad F_{\mathrm{s}}(x)=\mu_F\big((-\infty,x]\cap\{\mathcal{L}F=\infty\}\big).$$

Thus, F itself is singular if and only if $\mu_F(\mathcal{L}F<\infty)=0$.

Proof. Set $B=\{\mathcal{L}F<\infty\}$, and define $\mu_{\mathrm{a}}(\Gamma)=\mu_F(\Gamma\cap B)$ and $\mu_{\mathrm{s}}(\Gamma)=\mu_F(\Gamma\cap B^{\complement})$ for $\Gamma\in\mathcal{B}_{\mathbb{R}}$. Because $\lambda_{\mathbb{R}}(\mathcal{L}F=\infty)=0$, μ_{s} is singular to $\lambda_{\mathbb{R}}$, and therefore, by Exercise 2.2.9, its distribution function F_{s} is singular. On the other hand, if F_{a} is the distribution function for μ_{a}, then $\mathcal{L}F_{\mathrm{a}}\le\mathcal{L}F$ and therefore $\mu_{\mathrm{a}}(\mathcal{L}F_{\mathrm{a}}=\infty)\le\mu_F(\infty=\mathcal{L}F<\infty)=0$. Hence, by Theorem 3.3.6,

[4] See, for example, E.M. Stein's *Singular Integrals and Differentiability Properties of Functions*, published by Princeton Univ. Press in 1970.

the distribution function F_a of μ_a is absolutely continuous, and so, since $F = F_a + F_s$, we have proved the existence assertion.

Turning to the question of uniqueness, suppose that $F = F_1 + F_2$ is any other decomposition of the prescribed sort. Then $\mu_{F_1} \ll \lambda_{\mathbb{R}}$ and $\mu_{F_2} \perp \lambda_{\mathbb{R}}$. Now choose $A \in \mathcal{B}$ such that $\lambda_{\mathbb{R}}(A^{\complement}) = 0$ and $\mu_s(A) = 0 = \mu_{F_2}(A)$. Then, because $\mu_a + \mu_s = \mu_F = \mu_{F_1} + \mu_{F_2}$, we have that

$$\mu_a(\Gamma) = \mu_a(\Gamma \cap A) = \mu_F(\Gamma \cap A) = \mu_{F_1}(\Gamma \cap A) = \mu_{F_1}(\Gamma)$$

for all $\Gamma \in \mathcal{B}_{\mathbb{R}}$. That is, $\mu_a = \mu_{F_1}$, and therefore $F_a = F_1$ and $F_s = F_2$.

Finally, because the decomposition is unique and $F_a \equiv 0$ if and only if $\mu_F(\mathcal{L}F < \infty) = 0$, the last assertion is clear. □

The result in Theorem 3.3.8 is called the **Lebesgue decomposition** of F into its absolutely continuous and singular parts. See Theorem 8.1.2 for an abstract generalization of this result.

In view of Theorem 3.3.8 and the comments preceding it, what remains is to prove that F' exists $\lambda_{\mathbb{R}}$-almost everywhere when F is singular, and for this purpose I will use the following lemma. In its statement, all F's are assumed to be bounded, right-continuous, non-decreasing functions that tend to 0 at $-\infty$.

Lemma 3.3.9 *Let F be given, and suppose that there is a sequence $\{F_n : n \geq 1\}$ with the properties that $F_{n+1} - F_n$ is non-decreasing for each $n \in \mathbb{Z}^+$ and that $F_n(\infty) \nearrow F(\infty)$ as $n \to \infty$. Further, assume that $B \in \mathcal{B}_{\mathbb{R}}$ has the property that, for each $n \in \mathbb{Z}^+$, $F_n'(x)$ exists for every $x \in B$, and set $\tilde{B} = \{x \in B : F'(x) \text{ exists}\}$. Then $\lambda_{\mathbb{R}}(B \setminus \tilde{B}) = 0$ and $\int_{\tilde{B}} |F' - F_n'|\, d\lambda_{\mathbb{R}} \longrightarrow 0$ as $n \to \infty$.*

Proof. Obviously, $F'(x) \geq 0$ for any x at which it exists. Further, if F' exists on a set $B \in \mathcal{B}_{\mathbb{R}}$, then $\int_B F'(x)\, dx \leq F(\infty)$. To see this, define f_n from F as in the proof of Lemma 3.3.4. Clearly $f_n \geq 0$ and $\int f_n\, d\lambda_{\mathbb{R}} \leq F(\infty)$. Thus, since $f_n \longrightarrow F'$ on B, Fatou's lemma guarantees that $\int_B F'\, d\lambda_{\mathbb{R}} \leq F(\infty)$.

Turning to the proof of the lemma, let $m < n$ be given, apply the preceding remark to $F_n - F_m$, and conclude that

$$\int_B |F_n' - F_m'|\, d\lambda_{\mathbb{R}} = \int_B |(F_n - F_m)'|\, d\lambda_{\mathbb{R}} \leq F_n(\infty) - F_m(\infty).$$

Hence $\{\mathbf{1}_B F_n' : n \geq 1\}$ is Cauchy convergent in $L^1(\lambda_{\mathbb{R}}; \mathbb{R})$, and therefore there exists an $f \in L^1(\lambda_{\mathbb{R}}; \mathbb{R})$ to which it converges, and, for each $n \in \mathbb{Z}^+$, $\int_B |f - F_n'|\, d\lambda_{\mathbb{R}} \leq F(\infty) - F_n(\infty)$. Now, reasoning as we did in the proof of Theorem 3.3.7, we find that, for each $\epsilon > 0$ and $n \in \mathbb{Z}^+$,

$$\lambda_{\mathbb{R}}\left(\left\{x \in B : \overline{\lim_{|h|\searrow 0}} \left|\frac{F(x+h)-F(x)}{h} - f(x)\right| \geq 3\epsilon\right\}\right)$$
$$\leq \lambda_{\mathbb{R}}\big(\mathcal{L}(F-F_n) \geq \epsilon\big) + \lambda_{\mathbb{R}}\big(\{x \in B : |F_n'(x) - f(x)| \geq \epsilon\}\big)$$
$$\leq \frac{3(F(\infty) - F_n(\infty))}{\epsilon},$$

and clearly this shows that F' exists and is equal to f $\lambda_{\mathbb{R}}$-almost everywhere on B. □

Theorem 3.3.10 *If F is a bounded, right-continuous, singular, non-decreasing function, then $F'(x)$ exists and is equal to 0 for $\lambda_{\mathbb{R}}$-almost every $x \in \mathbb{R}$.*

Proof. Without loss in generality, assume that F tends to 0 at $-\infty$.

By Exercise 2.2.8, μ_F is a finite Borel measure on $\mathbb{R}$ that is singular to $\lambda_{\mathbb{R}}$, and therefore there exists a $B \in \mathcal{B}_{\mathbb{R}}$ for which $\mu_F(B) = 0 = \lambda_{\mathbb{R}}(B^{\complement})$. Now choose a non-decreasing sequence $\{K_n : n \geq 1\}$ of compact subsets of $B^{\complement}$ such that $\mu_F(B^{\complement} \setminus K_n) \leq \frac{1}{n}$ for each $n \in \mathbb{Z}^+$, and set

$$F_n(x) = \mu_F\big((-\infty, x] \cap K_n\big) \quad \text{for } x \in \mathbb{R}.$$

If $x \in B$, then $|x - K_n| > 0$ and therefore $F_n'(x) = 0$. Hence, for all n, F_n' exists and is equal to 0 on B. Thus, because $F_n(\infty) \nearrow F(\infty)$ and $F_{n+1} - F_n$ is non-decreasing for each $n \in \mathbb{Z}^+$, Lemma 3.3.9 says that F' exists and is equal to 0 $\lambda_{\mathbb{R}}$-almost everywhere on B. □

I close by summarizing these results in the following statement.

Theorem 3.3.11 (Lebesgue's Differentiation Theorem) [5] *Let F be a bounded, right-continuous, non-decreasing function on $\mathbb{R}$ that tends to 0 at $-\infty$. Then $F'(x)$ exists and is non-negative for $\lambda_{\mathbb{R}}$-almost every $x \in \mathbb{R}$. Moreover, $\int F'\,d\lambda_{\mathbb{R}} \leq F(\infty)$, $\int F'\,d\lambda_{\mathbb{R}} = 0$ if and only if F is singular, and $\int F'\,d\lambda_{\mathbb{R}} = F(\infty)$ if and only if F is absolutely continuous.*

3.3.4 Exercises for § 3.3

Exercise 3.3.1 Define $f : \mathbb{R} \longrightarrow [0,\infty)$ so that $f(x) = \big(x(\log x)^2\big)^{-1}$ if $x \in (0, e^{-1})$ and $f(x) = 0$ if $x \notin (0, e^{-1})$. Using part (**ii**) of Exercise 3.2.3 and the Fundamental Theorem of Calculus , check that $f \in L^1(\lambda_{\mathbb{R}};\mathbb{R})$ and that

$$\int_{(0,x)} f(t)\,\lambda_{\mathbb{R}}(dt) = \frac{-1}{\log x}, \quad \text{for } x \in (0, e^{-1}).$$

[5] F. Riesz and B. Nage give a less informative but beautiful derivation of the almost sure existence of F' in their book *Functional Analysis*, now available in paperback from Dover Press. Like the one here, theirs is an application of the Sunrise Lemma.

In particular, conclude that $\int_{(0,r)} Mf(x)\,dx = \infty$ for every $r > 0$, even though $f \in L^1(\lambda_{\mathbb{R}};\mathbb{R})$. See Exercise 6.2.5 for more information.

Exercise 3.3.2 Given $f \in L^1(\lambda_{\mathbb{R}};\mathbb{R})$, define the **Lebesgue set** $\mathrm{Leb}(f)$ of f to be the set of those $x \in \mathbb{R}$ for which the limit in (3.3.2) is 0. Clearly, (3.3.2) is the statement that $\mathrm{Leb}(f)^{\complement}$ has Lebesgue measure 0, and clearly $\mathrm{Leb}(f)$ is the set on which f is *well behaved* in the sense that its averages $⨍_I f\,d\lambda_{\mathbb{R}}$ converge to $f(x)$ as $I \searrow \{x\}$ for $x \in \mathrm{Leb}(f)$. The purpose of this exercise is to show that other averaging procedures converge to f on $\mathrm{Leb}(f)$. To be precise, let ρ be a bounded continuous function on $\mathbb{R}$ having one bounded, continuous derivative ρ'. Further, assume that $\rho \in L^1(\lambda_{\mathbb{R}};\mathbb{R})$, $\int \rho\,d\lambda_{\mathbb{R}} = 1$, $\int |t\rho'(t)|\,\lambda_{\mathbb{R}}(dt) < \infty$, and $\lim_{|t|\to\infty} t\rho(t) = 0$. Next, for each $\epsilon > 0$, set $\rho_\epsilon(t) = \epsilon^{-1}\rho(\epsilon^{-1}t)$, and define

$$f_\epsilon(x) = \int \rho_\epsilon(x-t)f(t)\,\lambda_{\mathbb{R}}(dt), \quad x \in \mathbb{R} \text{ and } f \in L^1(\mathbb{R};\mathbb{R}).$$

The goal here to show that

$$f_\epsilon(x) \longrightarrow f(x) \quad \text{as } \epsilon \searrow 0 \text{ for each } x \in \mathrm{Leb}(f).$$

See § 6.3.3 for related results.

(**i**) Show that, for any $f \in L^1(\lambda_{\mathbb{R}};\mathbb{R})$ and $x \in \mathrm{Leb}(f)$,

$$\lim_{\delta\searrow 0}\frac{1}{\delta}\int_{[x,x+\delta)} f\,d\lambda_{\mathbb{R}} = f(x) = \lim_{\delta\searrow 0}\frac{1}{\delta}\int_{(x-\delta,x]} f\,d\lambda_{\mathbb{R}}.$$

(**ii**) Assuming that f is continuous and vanishes off of a compact set, first show that

$$f_\epsilon(x) = \int_{[0,\infty)} \rho(-t)f(x+\epsilon t)\,\lambda_{\mathbb{R}}(dt) + \int_{[0,\infty)} \rho(t)f(x-\epsilon t)\,\lambda_{\mathbb{R}}(dt),$$

and then (using Exercise 3.2.3 and Theorem 1.2.1) verify the following equalities:

$$\int_{[0,\infty)} \rho(-t)f(x+\epsilon t)\,\lambda_{\mathbb{R}}(dt) = \int_{(0,\infty)} \rho'(-t)\left(\frac{1}{\epsilon}\int_{[x,x+\epsilon t)} f(s)\,\lambda_{\mathbb{R}}(ds)\right)\lambda_{\mathbb{R}}(dt)$$

and

$$\int_{[0,\infty)} \rho(t)f(x-\epsilon t)\,\lambda_{\mathbb{R}}(dt) = -\int_{(0,\infty)} \rho'(t)\left(\frac{1}{\epsilon}\int_{(x-\epsilon t,x]} f(s)\,\lambda_{\mathbb{R}}(ds)\right)\lambda_{\mathbb{R}}(dt).$$

Next (using Corollary 3.2.11) argue that these continue to hold for every $f \in L^1(\lambda_{\mathbb{R}};\mathbb{R})$.

(**iii**) Combining parts (**i**) and (**ii**), conclude that

$$\lim_{\epsilon\searrow 0} f_\epsilon(x) = -f(x)\int t\rho'(t)\,\lambda_{\mathbb{R}}(dt) \quad \text{for } f\in L^1(\lambda_{\mathbb{R}};\mathbb{R}) \text{ and } x\in \mathrm{Leb}(f),$$

and, as another application of Exercise 3.2.3 and Theorem 1.2.1, observe that

$$-\int t\rho'(t)\,\lambda_{\mathbb{R}}(dt) = \int \rho\, d\lambda_{\mathbb{R}} = 1.$$

Exercise 3.3.3 Recall the Bernoulli measures β_p introduced in § 2.2.3 and the associated distribution functions F_p introduced in Exercise 3.1.8. As was pointed out there, for each $p\neq \frac{1}{2}$, F_p is a continuous, singular function that is strictly increasing on $[0,1]$, and obviously $F_p(0)=0$ and $F_p(1)=1$. The purpose of this exercise is to show that, since $F_p'=0$ (a.e.,$\lambda_{[0,1]}$), such functions have got to be pretty inscrutable. Namely, by Exercise 1.2.8, we know that for any right-continuous, non-decreasing function $F:[0,1]\longrightarrow[0,1]$ that is 0 at 0 and 1 at 1, $\mathrm{Arc}\big(F;[0,1]\big)$ lies between $\sqrt{2}$ and 2. Moreover, it was shown there that the lower bound is achieved when F is linear and that the upper is achieved when F is pure jump. Here, it will be seen that the upper bound is also achieved by any continuous, non-decreasing function that, like F_p for $p\neq\frac{1}{2}$, is singular. To be precise, show that if $F:[0,1]\longrightarrow[0,1]$ is a non-decreasing, continuous function satisfying $F(0)=0$ and $F(1)=1$, then $\mathrm{Arc}\big(F;[0,1]\big)=2$ if and only if F is singular. The following are steps that you might want to follow.

(**i**) Set

$$L_n=\sum_{k=0}^{2^n-1}\sqrt{4^{-n}+\big(F((k+1)2^{-n})-F(k2^{-n})\big)^2},$$

and apply part (**iii**) of Exercise 1.2.8 to show that $\mathrm{Arc}\big(F;[0,1]\big)=\lim_{n\to\infty}L_n$.

(**ii**) Define $f_n:[0,1]\longrightarrow[0,\infty)$ so that

$$f_n(x)=2^n\big(F((k+1)2^{-n})-F(k2^{-n})\big)\quad\text{when } x\in[k2^{-n},(k+1)2^{-n}),$$

and show that

$$2-L_n=2^{-n}\sum_{k=0}^{2^n-1}\Big[\big(1+f_n(k2^{-n})\big)-\sqrt{1+f_n(k2^{-n})^2}\Big].$$

(**iii**) Show that $(1+a)-\sqrt{1+a^2}$ lies between $\frac{a}{1+a}$ and $\frac{2a}{1+a}$ for any $a\geq 0$, and use this together with (**ii**) to show that

$$\frac{2-L_n}{2}\leq\int_{[0,1]}\frac{f_n(x)}{1+f_n(x)}\,\lambda_{\mathbb{R}}(dx)\leq 2-L_n.$$

(**iv**) Starting from (**iii**), show that

$$\frac{2 - \mathrm{Arc}\big(F;[0,1]\big)}{2} \le \int_{[0,1]} \frac{F'(x)}{1+F'(x)}\,\lambda_{\mathbb{R}}(dx) \le 2 - \mathrm{Arc}\big(F;[0,1\big),$$

and conclude from this that $\mathrm{Arc}\big(F;[0,1]\big) = 2$ if and only if F is singular.

The reader who finds these computations amusing might want to consult my article "Doing Analysis by Tossing a Coin," which appeared in vol. 22 #2 of *The Mathematical Intelligencer*, published by Springer-Verlag in 2000.

$$\frac{\operatorname{Arc}(F;[0,1])}{2} \; ... \; \int ... \frac{F'(x)}{\sqrt{1+F'(x)^2}} \; ... \; \operatorname{Arc}(F;[0,1])$$

and conclude from this that $\operatorname{Arc}(F;[0,1]) = 4$ ... if F is singular.

The reader who finds these computations amusing ... article "Nonsmooth Analysis ..." ... of F ...

Chapter 4
Products of Measures

Just before Lemma 3.1.1, I introduced the product $(E_1 \times E_2, \mathcal{B}_1 \times \mathcal{B}_2)$ of two measurable spaces $(E_1, \mathcal{B}_1)$ and $(E_2, \mathcal{B}_2)$. In the present chapter I will show that if μ_i, $i \in \{1,2\}$, is a measure on $(E_i, \mathcal{B}_i)$, then, under reasonable conditions, there is a unique measure $\nu = \mu_1 \times \mu_2$ on $(E_1 \times E_2, \mathcal{B}_1 \times \mathcal{B}_2)$ with the property that $\nu(\Gamma_1 \times \Gamma_2) = \mu_1(\Gamma_1)\,\mu_2(\Gamma_2)$ for all $\Gamma_i \in \mathcal{B}_i$. In addition, I will derive several important properties relating integrals with respect to $\mu_1 \times \mu_2$ to iterated integrals with respect to μ_1 and μ_2. Finally, in § 4.2, I will apply these properties to derive the isodiametric inequality.

4.1 Fubini's Theorem

The key to the construction of $\mu_1 \times \mu_2$ is found in the following function analog of Λ-systems (cf. Lemma 2.1.1).

Lemma 4.1.1 *Let $(E, \mathcal{B})$ be a measurable space and $\mathcal{L}$ a collection of functions $f : E \longrightarrow [0, \infty)$ with the properties that*

(a) $\alpha f + \beta g \in \mathcal{L}$ *if* $f, g \in \mathcal{L}$ *and* $\alpha, \beta \in [0, \infty)$.
(b) *If* $f, g \in \mathcal{L}$ *and* $f \le g$, *then* $g - f \in \mathcal{L}$.
(c) *If* $\{f_n : n \ge 1\} \subseteq \mathcal{L}$ *and* $f_n \nearrow f$, *then* $f \in \mathcal{L}$.

If $\mathcal{C}$ is a Π-system that generates $\mathcal{B}$ and $\mathbf{1}_\Gamma \in \mathcal{L}$ for all $\Gamma \in \mathcal{C}$, then $\mathcal{L}$ contains all non-negative, $\mathcal{B}$-measurable functions on E.

Proof. First note that $\{\Gamma \subseteq E : \mathbf{1}_\Gamma \in \mathcal{L}\}$ is a Λ-system that contains $\mathcal{C}$. Hence, by Lemma 2.1.1, $\mathbf{1}_\Gamma \in \mathcal{L}$ for every $\Gamma \in \mathcal{B}$. Combined with **(a)** above, this means that $\mathcal{L}$ contains every non-negative, measurable, simple function on $(E, \mathcal{B})$. Finally, use **(c)** to complete the proof. □

I will call collection of functions that satisfy the conditions **(a)**–**(c)** an $\mathcal{L}$-system. The power of Lemma 4.1.1 to handle questions involving products is already apparent in the following.

D. W. Stroock, *Essentials of Integration Theory for Analysis*, Graduate Texts in Mathematics 262, https://doi.org/10.1007/978-3-030-58478-8_4

Lemma 4.1.2 *Let $(E_1,\mathcal{B}_1)$ and $(E_2,\mathcal{B}_2)$ be measurable spaces, and suppose that f is an $\overline{\mathbb{R}}$-valued measurable function on $(E_1\times E_2,\mathcal{B}_1\times\mathcal{B}_2)$. Then for each $x_1\in E_1$ and $x_2\in E_2$, $f(x_1,\cdot\,)$ and $f(\,\cdot\,,x_2)$ are measurable functions on $(E_2,\mathcal{B}_2)$ and $(E_1,\mathcal{B}_1)$, respectively. Next, suppose that μ_i, $i\in\{1,2\}$, is a finite measure on $(E_i,\mathcal{B}_i)$. Then for every non-negative, measurable function f on $(E_1\times E_2,\mathcal{B}_1\times\mathcal{B}_2)$, the functions*

$$\int_{E_2} f(\,\cdot\,,x_2)\,\mu_2(dx_2) \quad \text{and} \quad \int_{E_1} f(x_1,\cdot\,)\,\mu_1(dx_1)$$

are measurable on $(E_1,\mathcal{B}_1)$ and $(E_2,\mathcal{B}_2)$, respectively.

Proof. Clearly, by the monotone convergence theorem, it is enough to check all these assertions when f is bounded.

Let $\mathcal{L}$ be the collection of all bounded functions on $E_1\times E_2$ that have all the asserted properties. It is easy to check the $\mathcal{L}$ is an $\mathcal{L}$-system, and it is clear that $\mathbf{1}_{\Gamma_1\times\Gamma_2}\in\mathcal{L}$ for all $\Gamma_i\in\mathcal{B}_i$. Hence, by Lemma 4.1.1 with $\mathcal{C}=\{\Gamma_1\times\Gamma_2:\Gamma_i\in\mathcal{B}_i \text{ for } i\in\{1,2\}\}$, we are done. □

Lemma 4.1.3 *Given a pair $(E_1,\mathcal{B}_1,\mu_1)$ and $(E_2,\mathcal{B}_2,\mu_2)$ of finite measure spaces, there exists a unique measure ν on $(E_1\times E_2,\mathcal{B}_1\times\mathcal{B}_2)$ for which*

$$\nu(\Gamma_1\times\Gamma_2)=\mu_1(\Gamma_1)\,\mu_2(\Gamma_2) \quad \text{for all} \quad \Gamma_i\in\mathcal{B}_i.$$

Moreover, for every non-negative, $\mathcal{B}_1\times\mathcal{B}_2$-measurable function f on $E_1\times E_2$,

$$\begin{aligned}\int_{E_1\times E_2} & f(x_1,x_2)\,\nu(dx_1\times dx_2)\\ &=\int_{E_2}\left(\int_{E_1} f(x_1,x_2)\,\mu_1(dx_1)\right)\mu_2(dx_2)\\ &=\int_{E_1}\left(\int_{E_2} f(x_1,x_2)\,\mu_2(dx_2)\right)\mu_1(dx_1).\end{aligned}\tag{4.1.1}$$

Proof. The uniqueness of ν is guaranteed by Theorem (2.1.2). To prove the existence of ν, define

$$\nu_{1,2}(\Gamma)=\int_{E_2}\left(\int_{E_1}\mathbf{1}_\Gamma(x_1,x_2)\,\mu_1(dx_1)\right)\mu_2(dx_2)$$

and

$$\nu_{2,1}(\Gamma)=\int_{E_1}\left(\int_{E_2}\mathbf{1}_\Gamma(x_1,x_2)\,\mu_2(dx_2)\right)\mu_1(dx_1)$$

for $\Gamma\in\mathcal{B}_1\times\mathcal{B}_2$. Using the monotone convergence theorem, one sees that both $\nu_{1,2}$ and $\nu_{2,1}$ are finite measures on $(E_1\times E_2,\mathcal{B}_1\times\mathcal{B}_2)$. Moreover, by the same sort of argument as was used to prove Lemma 4.1.2, for every non-negative measurable function f on $(E_1\times E_2,\mathcal{B}_1\times\mathcal{B}_2)$,

$$\int f\, d\nu_{1,2} = \int_{E_1} \left(\int_{E_2} f(x_1, x_2)\, \mu_1(dx_1) \right) \mu_2(dx_2)$$

and

$$\int f\, d\nu_{2,1} = \int_{E_2} \left(\int_{E_1} f(x_1, x_2)\, \mu_2(dx_2) \right) \mu_1(dx_1).$$

Finally, since $\nu_{1,2}(\Gamma_1 \times \Gamma_2) = \mu(\Gamma_1)\,\mu(\Gamma_2) = \nu_{2,1}(\Gamma_1 \times \Gamma_2)$ for all $\Gamma_i \in \mathcal{B}_i$, we see that both $\nu_{1,2}$ and $\nu_{2,1}$ fulfill the requirements placed on ν. Hence, not only does ν exist, but it is also equal to both $\nu_{1,2}$ and $\nu_{2,1}$; and so the preceding equalities lead to (4.1.1). □

Recall that a measure space $(E, \mathcal{B}, \mu)$ is said to be σ-finite if E is the countable union of $\mathcal{B}$-measurable sets having finite μ-measure.

Theorem 4.1.4 (Tonelli's Theorem) *Let $(E_1, \mathcal{B}_1, \mu_1)$ and $(E_2, \mathcal{B}_2, \mu_2)$ be a pair of σ-finite measure spaces. Then there is a unique measure ν on $(E_1 \times E_2, \mathcal{B}_1 \times \mathcal{B}_2)$ such that $\nu(\Gamma_1 \times \Gamma_2) = \mu_1(\Gamma_1)\,\mu_2(\Gamma_2)$ for all $\Gamma_i \in \mathcal{B}_i$. In addition, for every non-negative measurable function f on $(E_1 \times E_2, \mathcal{B}_1 \times \mathcal{B}_2)$, $\int f(\,\cdot\,, x_2)\,\mu_2(dx_2)$ and $\int f(x_1, \,\cdot\,)\,\mu_1(dx_1)$ are measurable on $(E_1, \mathcal{B}_1)$ and $(E_2, \mathcal{B}_2)$, respectively, and* (4.1.1) *continues to hold.*

Proof. Choose $\{E_{i,n} : n \geq 1\} \subseteq \mathcal{B}_i$ for $i \in \{1, 2\}$ such that $\mu_i(E_{i,n}) < \infty$ for each $n \geq 1$ and $E_i = \bigcup_{n=1}^{\infty} E_{i,n}$. Without loss of generality, assume that $E_{i,m} \cap E_{i,n} = \emptyset$ for $m \neq n$. For each $n \in \mathbb{Z}^+$, define $\mu_{i,n}(\Gamma_i) = \mu_i(\Gamma_i \cap E_{i,n})$, $\Gamma_i \in \mathcal{B}_i$; and, for $(m, n) \in {\mathbb{Z}^+}^2$, let $\nu_{(m,n)}$ on $(E_1 \times E_2, \mathcal{B}_1 \times \mathcal{B}_2)$ be the measure constructed from $\mu_{1,m}$ and $\mu_{2,n}$ as in Lemma 4.1.3.

Clearly, by Lemma 4.1.2, for any non-negative measurable function f on $(E_1 \times E_2, \mathcal{B}_1 \times \mathcal{B}_2)$,

$$\int_{E_2} f(\,\cdot\,, x_2)\,\mu_2(dx_2) = \sum_{n=1}^{\infty} \int_{E_{2,n}} f(\,\cdot\,, x_2)\,\mu_{2,n}(dx_2)$$

is measurable on $(E_1, \mathcal{B}_1)$; and, similarly, $\int_{E_1} f(x_1, \,\cdot\,)\,\mu_1(dx_1)$ is measurable on $(E_2, \mathcal{B}_2)$. Finally, the map $\Gamma \in \mathcal{B}_1 \times \mathcal{B}_2 \longmapsto \sum_{m,n=1}^{\infty} \nu_{(m,n)}(\Gamma)$ defines a measure ν_0 on $(E_1 \times E_2, \mathcal{B}_1 \times \mathcal{B}_2)$, and it is easy to check that ν_0 has all the required properties. At the same time, if ν is any other measure on $(E_1 \times E_2, \mathcal{B}_1 \times \mathcal{B}_2)$ for which $\nu(\Gamma_1 \times \Gamma_2) = \mu_1(\Gamma_1)\,\mu_2(\Gamma_2)$, $\Gamma_i \in \mathcal{B}_i$, then, by the uniqueness assertion in Lemma 4.1.3, for each $(m, n) \in (\mathbb{Z}^+)^2$, ν coincides with $\nu_{(m,n)}$ on $\mathcal{B}_1 \times \mathcal{B}_2\big[E_{1,m} \times E_{2,n}\big]$ and is therefore equal to ν_0 on $\mathcal{B}_1 \times \mathcal{B}_2$. □

The measure ν constructed in Theorem 4.1.4 is called the **product of** μ_1 **and** μ_2 and is denoted by $\mu_1 \times \mu_2$.

Theorem 4.1.5 (Fubini's Theorem) *Let $(E_1, \mathcal{B}_1, \mu_1)$ and $(E_2, \mathcal{B}_2, \mu_2)$ be σ-finite measure spaces and f a measurable function on $(E_1 \times E_2, \mathcal{B}_1 \times \mathcal{B}_2)$. Then f is $\mu_1 \times \mu_2$-integrable if and only if*

$$\int_{E_1}\left(\int_{E_2}|f(x_1,x_2)|\,\mu_2(dx_2)\right)\,\mu_1(dx_1)<\infty$$

if and only if

$$\int_{E_2}\left(\int_{E_1}|f(x_1,x_2)|\,\mu_1(dx_1)\right)\,\mu_2(dx_2)<\infty.$$

Next, set

$$\Lambda_1=\left\{x_1\in E_1:\int_{E_2}|f(x_1,x_2)|\,\mu_2(dx_2)<\infty\right\}$$

and

$$\Lambda_2=\left\{x_2\in E_2:\int_{E_1}|f(x_1,x_2)|\,\mu_1(dx_1)<\infty\right\};$$

and define f_i on E_i, $i\in\{1,2\}$, by

$$f_1(x_1)=\begin{cases}\int_{E_2}f(x_1,x_2)\,\mu_2(dx_2) & \text{if } x_1\in\Lambda_1\\ 0 & \text{otherwise}\end{cases}$$

and

$$f_2(x_2)=\begin{cases}\int_{E_1}f(x_1,x_2)\,\mu_1(dx_1) & \text{if } x_2\in\Lambda_2\\ 0 & \text{otherwise.}\end{cases}$$

Then f_i is an $\mathbb{R}$-valued, measurable function on $(E_i,\mathcal{B}_i)$. Finally, if f is $\mu_1\times\mu_2$-integrable, then $\mu_i(\Lambda_i{}^{\complement})=0$, $f_i\in L^1(\mu_i;\mathbb{R})$, and

$$\int_{E_i}f_i(x_i)\,\mu_i(dx_i)=\int_{E_1\times E_2}f(x_1,x_2)\,(\mu_1\times\mu_2)(dx_1\times dx_2)\quad \textit{for } i\in\{1,2\}.$$

Proof. The first assertion is an immediate consequence of Theorem 4.1.4. Moreover, because $\Lambda_i\in\mathcal{B}_i$, it is easy (cf. Lemma 4.1.2) to check that f_i is an $\mathbb{R}$-valued, measurable function on $(E_i,\mathcal{B}_i)$. Finally, if f is $\mu_1\times\mu_2$-integrable, then, by the first assertion, $\mu_i(\Lambda_i^{\complement})=0$ and $f_i\in L^1(\mu_i;\mathbb{R})$. Hence, by Theorem 4.1.4 applied to f^+ and f^-, we see that

$$
\begin{aligned}
&\int_{E_1 \times E_2} f(x_1, x_2)\,(\mu_1 \times \mu_2)(dx_1 \times dx_2) \\
&\quad = \int_{\Lambda_1 \times E_2} f^+(x_1, x_2)\,(\mu_1 \times \mu_2)(dx_1 \times dx_2) \\
&\qquad\qquad - \int_{\Lambda_1 \times E_2} f^-(x_1, x_2)\,(\mu_1 \times \mu_2)(dx_1 \times dx_2) \\
&\quad = \int_{\Lambda_1} \left(\int_{E_2} f^+(x_1, x_2)\,\mu_1(dx_2) \right) \mu_1(dx_1) \\
&\qquad\qquad - \int_{\Lambda_1} \left(\int_{E_2} f^-(x_1, x_2)\,\mu_2(dx_2) \right) \mu_1(dx_1) \\
&\quad = \int_{\Lambda_1} f_1(x_1)\,\mu_1(dx_1) = \int_{E_1} f_1(x_1)\,\mu_1(dx_1),
\end{aligned}
$$

and the same line of reasoning applies to f_2. □

One may well wonder why I have separated the statement in Tonelli's Theorem from the statement in Fubini's Theorem. The reason is that Tonelli's Theorem requires no a priori information about integrability. Thus, for example, Tonelli's Theorem allowed me to show that f is $\mu_1 \times \mu_2$-integrable if and only if

$$
\begin{aligned}
&\left(\int_{E_2} \left(\int_{E_1} |f(x_1, x_2)|\,\mu_1(dx_1) \right) \mu_2(dx_2) \right) \\
&\qquad \wedge \left(\int_{E_1} \left(\int_{E_2} |f(x_1, x_2)|\,\mu_2(dx_2) \right) \mu_1(dx_1) \right) < \infty.
\end{aligned}
$$

4.1.1 Exercises for § 4.1

Exercise 4.1.1 Let $(E, \mathcal{B}, \mu)$ be a σ-finite measure space. Given a non-negative measurable function f on $(E, \mathcal{B})$, define

$$\Gamma(f) = \big\{(x, t) \in E \times [0, \infty) : t \le f(x)\big\}$$

and

$$\widehat{\Gamma}(f) = \big\{(x, t) \in E \times [0, \infty) : t < f(x)\big\}.$$

Show that both $\Gamma(f)$ and $\widehat{\Gamma}(f)$ are elements of $\mathcal{B} \times \mathcal{B}_{[0,\infty)}$ and, in addition, that

$$\mu \times \lambda_{\mathbb{R}}\big(\widehat{\Gamma}(f)\big) = \int_E f\,d\mu = \mu \times \lambda_{\mathbb{R}}\big(\Gamma(f)\big). \tag{$*$}$$

Clearly $(*)$ can be interpreted as the statement that *the integral of a non-negative function is the area under its graph.*

Hint: In proving measurability, consider the function $(x,t) \in E \times [0,\infty) \longmapsto f(x) - t \in (-\infty,\infty]$, and get $(*)$ as an application of Tonelli's Theorem.

Exercise 4.1.2 Let $(E_1, \mathcal{B}_1, \mu_1)$ and $(E_2, \mathcal{B}_2, \mu_2)$ be σ-finite measure spaces and assume that, for $i \in \{1,2\}$, $\mathcal{B}_i = \sigma(E_i; \mathcal{C}_i)$, where $\mathcal{C}_i$ is a Π-system containing a sequence $\{E_{i,n} : n \geq 1\}$ for which $E_i = \bigcup_{n=1}^{\infty} E_{i,n}$ and $\mu_i(E_{i,n}) < \infty$, $n \geq 1$. Show that if ν is a measure on $(E_1 \times E_2, \mathcal{B}_1 \times \mathcal{B}_2)$ with the property that $\nu(\Gamma_1 \times \Gamma_2) = \mu_1(\Gamma_1)\,\mu_2(\Gamma_2)$ for all $\Gamma_i \in \mathcal{C}_i$, then $\nu = \mu_1 \times \mu_2$. Use this fact to show that, for any $M, N \in \mathbb{Z}^+$, $\lambda_{\mathbb{R}^{M+N}} = \lambda_{\mathbb{R}^M} \times \lambda_{\mathbb{R}^N}$ on $\mathcal{B}_{\mathbb{R}^{M+N}} = \mathcal{B}_{\mathbb{R}^M} \times \mathcal{B}_{\mathbb{R}^N}$.

Exercise 4.1.3 Let $(E_1, \mathcal{B}_1, \mu_1)$ and $(E_2, \mathcal{B}_2, \mu_2)$ be σ-finite measure spaces. Given $\Gamma \in \mathcal{B}_1 \times \mathcal{B}_2$, define

$$\Gamma_{(1)}(x_2) \equiv \big\{x_1 \in E_1 : (x_1, x_2) \in \Gamma\big\} \quad \text{for} \quad x_2 \in E_2$$

and

$$\Gamma_{(2)}(x_1) \equiv \big\{x_2 \in E_2 : (x_1, x_2) \in \Gamma\big\} \quad \text{for} \quad x_1 \in E_1.$$

If $i \neq j$, check both that $\Gamma_{(i)}(x_j) \in \mathcal{B}_i$ for each $x_j \in E_j$ and that $x_j \in E_j \longmapsto \mu_i\big(\Gamma_{(i)}(x_j)\big) \in [0,\infty]$ is measurable on $(E_j, \mathcal{B}_j)$. Finally, show that $\mu_1 \times \mu_2(\Gamma) = 0$ if and only if $\mu_i\big(\Gamma_{(i)}(x_j)\big) = 0$ for μ_j-almost every $x_j \in E_j$, and conclude that $\mu_1\big(\Gamma_{(1)}(x_2)\big) = 0$ for μ_2-almost every $x_2 \in E_2$ if and only if $\mu_2\big(\Gamma_{(2)}(x_1)\big) = 0$ for μ_1-almost every $x_1 \in E_1$. In other words, $\Gamma \in \mathcal{B}_1 \times \mathcal{B}_2$ has $\mu_1 \times \mu_2$-measure 0 if and only if μ_1-almost every *vertical slice* (μ_2-almost every *horizontal slice*) has μ_2-measure (μ_1-measure) 0. In particular, μ_2-almost every horizontal slice has μ_1-measure 0 if and only if μ_1-almost every vertical slice has μ_2-measure 0.

Exercise 4.1.4 The condition that the measure spaces of which one is taking a product be σ-finite is essential if one wants to carry out the program in this section. To see this, let $E_1 = E_2 = (0,1)$ and $\mathcal{B}_1 = \mathcal{B}_2 = \mathcal{B}_{(0,1)}$. Take μ_1 to be the **counting measure** (i.e., $\mu(\Gamma) = \text{card}(\Gamma)$) and μ_2 to be Lebesgue measure $\lambda_{(0,1)}$ on $(E_{(0,1)}, \mathcal{B}_{(0,1)})$. Show that there is a set $\Gamma \in \mathcal{B}_1 \times \mathcal{B}_2$ such that

$$\int_{E_2} \mathbf{1}_\Gamma(x_1, x_2)\,\mu_2(dx_2) = 0 \quad \text{for every } x_1 \in E_1$$

but

$$\int_{E_1} \mathbf{1}_\Gamma(x_1, x_2)\,\mu_1(dx_1) = 1 \quad \text{for every } x_2 \in E_2.$$

Notice that what fails is not so much the existence statement as the uniqueness one in Lemma 4.1.3.

4.2 Steiner Symmetrization

In order to provide an example that displays the power of Fubini's Theorem, I will prove in this section an elementary but important inequality about Lebesgue measure. I will then use that inequality to give another description of Lebesgue measure.

4.2.1 The Isodiametric inequality

In this subsection I will show that, for any bounded $\Gamma \in \mathcal{B}_{\mathbb{R}^N}$,

$$\lambda_{\mathbb{R}^N}(\Gamma) \leq \Omega_N \mathrm{rad}(\Gamma)^N, \tag{4.2.1}$$

where Ω_N denotes the volume (cf. **(iii)** in Exercise 5.1.3) of the unit ball $B(0,1)$ in $\mathbb{R}^N$ and

$$\mathrm{rad}(\Gamma) \equiv \sup\left\{\tfrac{|y-x|}{2} : x,\, y \in \Gamma\right\}$$

is the **radius** (i.e., half the **diameter**) of Γ. Notice (cf. Exercise 2.2.4) that (4.2.1) says that, among all the measurable subsets of $\mathbb{R}^N$ with a given diameter, the ball of that diameter has the largest volume. It is for this reason that (4.2.1) is called the **isodiametric inequality**.

At first glance one might be inclined to think that there is nothing to (4.2.1). Indeed, one might carelessly suppose that every Γ is a subset of a closed ball of radius $\mathrm{rad}(\Gamma)$ and therefore that (4.2.1) is trivial. This is true when $N = 1$. However, after a moment's thought, one realizes that, for $N > 1$, although Γ is always contained in a closed ball whose radius is equal to the diameter of Γ, it is not necessarily contained in one with the same radius as Γ. (For example, consider an equilateral triangle in $\mathbb{R}^2$.) Thus, the inequality $\lambda_{\mathbb{R}^N}(\Gamma) \leq \Omega_N\big(2\mathrm{rad}(\Gamma)\big)^N$ is trivial, but the inequality in (4.2.1) is not! Nonetheless, there are many Γ's for which (4.2.1) is easy. In particular, if Γ is *symmetric* in the sense that $\Gamma = -\Gamma \equiv \{-x : x \in \Gamma\}$, then it is clear that

$$x \in \Gamma \implies 2|x| = |x + x| \leq 2\mathrm{rad}(\Gamma) \quad \text{and therefore} \quad \Gamma \subseteq \overline{B\big(0, \mathrm{rad}(\Gamma)\big)}.$$

Hence (4.2.1) is trivial when Γ is symmetric, and what we have to do is devise a procedure for reducing the general case to the symmetric one.

The method with which we will perform this reduction is based on a famous construction known as the **Steiner symmetrization procedure**. To describe Steiner's procedure, it is necessary to introduce a little notation. Given $\mathbf{e}$ from the unit $(N-1)$-**sphere** $\mathbb{S}^{N-1} \equiv \{x \in \mathbb{R}^N : |x| = 1\}$, let $\mathbf{L}(\mathbf{e})$ denote the line $\{t\mathbf{e} : t \in \mathbb{R}\}$, $\mathbf{P}(\mathbf{e})$ the $(N-1)$-dimensional subspace $\{\xi \in \mathbb{R}^N : \xi \perp \mathbf{e}\}$, and define

$$\mathcal{S}(\Gamma;\mathbf{e}) \equiv \left\{\xi + t\mathbf{e} : \xi \in \mathbf{P}(\mathbf{e}) \text{ and } |t| < \tfrac{1}{2}\ell(\Gamma;\mathbf{e},\xi)\right\},$$

where

$$\ell(\Gamma;\mathbf{e},\xi) \equiv \lambda_{\mathbb{R}}\big(\{t \in \mathbb{R} : \xi + t\mathbf{e} \in \Gamma\}\big)$$

is the length of the intersection of the line $\xi + \mathbf{L}(\mathbf{e})$ with Γ. Notice that, in the creation of $\mathcal{S}(\Gamma;\mathbf{e})$ from Γ, what I have done is take the intersection of Γ with $\xi + \mathbf{L}(\mathbf{e})$, squashed it to remove all gaps, and then slid the resulting interval along $\xi + \mathbf{L}(\mathbf{e})$ until its center point lay at ξ. In particular, $\mathcal{S}(\Gamma;\mathbf{e})$ is the *symmetrization* of Γ with respect to the subspace $\mathbf{P}(\mathbf{e})$ in the sense that, for each $\xi \in \mathbf{P}(\mathbf{e})$ and $t \in \mathbb{R}$,

$$\xi + t\mathbf{e} \in \mathcal{S}(\Gamma;\mathbf{e}) \iff \xi - t\mathbf{e} \in \mathcal{S}(\Gamma;\mathbf{e}). \tag{4.2.2}$$

What is only slightly less obvious is that $\mathcal{S}(\Gamma;\mathbf{e})$ possesses the properties proved in the next lemma.

Lemma 4.2.1 *Let Γ be a bounded element of $\mathcal{B}_{\mathbb{R}^N}$. Then, for each $\mathbf{e} \in \mathbb{S}^{N-1}$, $\mathcal{S}(\Gamma;\mathbf{e})$ is also a bounded element of $\mathcal{B}_{\mathbb{R}^N}$, $\operatorname{rad}\big(\mathcal{S}(\Gamma;\mathbf{e})\big) \le \operatorname{rad}(\Gamma)$, and $\lambda_{\mathbb{R}^N}\big(\mathcal{S}(\Gamma;\mathbf{e})\big) = \lambda_{\mathbb{R}^N}(\Gamma)$. Finally, if $\mathbf{R} : \mathbb{R}^N \longrightarrow \mathbb{R}^N$ is an orthogonal transformation (i.e., in the notation of Theorem 2.2.13, $\mathbf{R} = T_{\mathcal{O}}$ for some orthogonal matrix $\mathcal{O}$) for which $\mathbf{L}(\mathbf{e})$ and Γ are invariant (i.e., $\mathbf{R}\big(\mathbf{L}(\mathbf{e})\big) = \mathbf{L}(\mathbf{e})$ and $\mathbf{R}(\Gamma) = \Gamma$), then $\mathbf{R}\mathcal{S}(\Gamma;\mathbf{e}) = \mathcal{S}(\Gamma;\mathbf{e})$.*

Proof. There is nothing to do when $N = 1$, and, because none of the quantities under consideration depends on the particular choice of coordinate axes, we will assume both that $N > 1$ and that $\mathbf{e} = \mathbf{e}_N \equiv (0,\dots,0,1)$.

By Lemma 4.1.2,

$$\xi \in \mathbb{R}^{N-1} \longmapsto f(\xi) \equiv \frac{1}{2}\int_{\mathbb{R}} \mathbf{1}_\Gamma\big((\xi,t)\big)\,dt = \tfrac{1}{2}\ell\big(\Gamma;\mathbf{e},(\xi,0)\big) \in [0,\infty)$$

is $\mathcal{B}_{\mathbb{R}^{N-1}}$-measurable. In addition, by Exercise 4.1.1, because $\mathcal{S}(\Gamma;\mathbf{e}_N)$ is equal to

$$\big\{(\xi,t) \in \mathbb{R}^{N-1} \times [0,\infty) : t < f(\xi)\big\} \cup \big\{(\xi,t) \in \mathbb{R}^{N-1} \times (-\infty,0] : -t < f(\xi)\big\},$$

we know both that $\mathcal{S}(\Gamma;\mathbf{e}_N)$ is an element of $\mathcal{B}_{\mathbb{R}^N}$ and that

$$\lambda_{\mathbb{R}^N}\big(\mathcal{S}(\Gamma;\mathbf{e}_N)\big) = 2\int_{\mathbb{R}^{N-1}} f(\xi)\,\lambda_{\mathbb{R}^{N-1}}(d\xi)$$
$$= \int_{\mathbb{R}^{N-1}}\left(\int_{\mathbb{R}} \mathbf{1}_\Gamma\big((\xi,t)\big)\,\lambda_{\mathbb{R}}(dt)\right)\lambda_{\mathbb{R}^{N-1}}(d\xi) = \lambda_{\mathbb{R}^N}(\Gamma),$$

where, in the final step, Tonelli's Theorem was applied.

I turn next to the proof that $\mathrm{rad}\big(\mathcal{S}(\Gamma;\mathbf{e}_N)\big) \le \mathrm{rad}(\Gamma)$. Because $\mathrm{rad}(\Gamma) = \mathrm{rad}(\overline{\Gamma})$ and $\mathrm{rad}\big(S(\Gamma,\mathbf{e})\big) \le \mathrm{rad}\big(S(\overline{\Gamma},\mathbf{e})\big)$, we may and will assume that Γ is compact. Now suppose that $x,\,y \in \mathcal{S}(\Gamma;\mathbf{e}_N)$ are given, and choose $\xi,\,\eta \in \mathbb{R}^{N-1}$ and $s,\,t \in \mathbb{R}$ for which $x = (\xi,s)$ and $y = (\eta,t)$. Next, set

$$M^\pm(x) = \pm\sup\{\tau : (\xi,\pm\tau)\in\Gamma\} \quad\text{and}\quad M^\pm(y) = \pm\sup\{\tau : (\eta,\pm\tau)\in\Gamma\},$$

and note that, because Γ is compact, all four of the points $X^\pm \equiv \big(\xi, M^\pm(x)\big)$ and $Y^\pm = \big(\eta, M^\pm(y)\big)$ are elements of Γ. Moreover, $2|s| \le M^+(x) - M^-(x)$ and $2|t| \le M^+(y) - M^-(y)$, and therefore

$$\begin{aligned}&\big(M^+(y) - M^-(x)\big) \vee \big(M^+(x) - M^-(y)\big)\\ &\quad\ge \frac{M^+(y) - M^-(x)}{2} + \frac{M^+(x) - M^-(\mathbf{y})}{2}\\ &\quad= \frac{M^+(y) - M^-(y)}{2} + \frac{M^+(x) - M^-(x)}{2} \ge |t| + |s|.\end{aligned}$$

In particular, this means that

$$\begin{aligned}|y-x|^2 &= |\eta-\xi|^2 + |t-s|^2 \le |\eta-\xi|^2 + \big(|s|+|t|\big)^2\\ &\le |\eta-\xi|^2 + \Big(\big(M^+(y) - M^-(x)\big) \vee \big(M^+(x) - M^-(y)\big)\Big)^2\\ &= \big(|Y^+ - X^-| \vee |X^+ - Y^-|\big)^2 \le 4\,\mathrm{rad}(\Gamma)^2.\end{aligned}$$

Finally, let $\mathbf{R}$ be an orthogonal transformation. It is then an easy matter to check that $\mathbf{P}(\mathbf{R}\mathbf{e}) = \mathbf{R}\big(\mathbf{P}(\mathbf{e})\big)$ and that $\ell(\mathbf{R}\Gamma;\mathbf{R}\mathbf{e},\mathbf{R}\xi) = \ell(\Gamma;\mathbf{e},\xi)$ for all $\xi \in \mathbf{P}(\mathbf{e})$. Hence, $\mathcal{S}(\mathbf{R}\Gamma,\mathbf{R}\mathbf{e}) = \mathbf{R}\mathcal{S}(\Gamma,\mathbf{e})$. In particular, if $\Gamma = \mathbf{R}\Gamma$ and $\mathbf{L}(\mathbf{e}) = \mathbf{R}\big(\mathbf{L}(\mathbf{e})\big)$, then $\mathbf{R}\mathbf{e} = \pm\mathbf{e}$, and so the preceding (together with (4.2.2)) leads to $\mathbf{R}\mathcal{S}(\Gamma,\mathbf{e}) = \mathcal{S}(\Gamma,\mathbf{e})$. □

Theorem 4.2.2 *The inequality in* (4.2.1) *holds for every bounded* $\Gamma \in \mathcal{B}_{\mathbb{R}^N}$.

Proof. Clearly it suffices to treat Γ's which are compact. Thus, let a compact Γ be given, choose an orthonormal basis $\{\mathbf{e}_1,\dots,\mathbf{e}_N\}$ for $\mathbb{R}^N$, set $\Gamma_0 = \Gamma$, and define $\Gamma_n = \mathcal{S}(\Gamma_{n-1};\mathbf{e}_n)$ for $1 \le n \le N$. By repeated application of (4.2.2) and Lemma 4.2.1, we know that $\lambda_{\mathbb{R}^N}(\Gamma_n) = \lambda_{\mathbb{R}^N}(\Gamma)$, $\mathrm{rad}(\Gamma_n) \le \mathrm{rad}(\Gamma)$, and that $\mathbf{R}_m\Gamma_n = \Gamma_n$, $1 \le m \le n \le N$, where $\mathbf{R}_m$ is the orthogonal transformation given by $\mathbf{R}_m x = x - 2(x,\mathbf{e}_m)_{\mathbb{R}^N}\mathbf{e}_m$ for each $x \in \mathbb{R}^N$. In particular, this means that $\mathbf{R}_m\Gamma_N = \Gamma_N$ for all $1 \le m \le N$ and therefore that $-\Gamma_N = \Gamma_N$. Hence, by the discussion preceding the introduction of Steiner's procedure,

$$\lambda_{\mathbb{R}^N}(\Gamma) = \lambda_{\mathbb{R}^N}(\Gamma_N) \le \Omega_N \mathrm{rad}(\Gamma_N)^N \le \Omega_N \mathrm{rad}(\Gamma)^N.$$

□

4.2.2 Hausdorff's Description of Lebesgue's Measure

I will now use (4.2.1) to give a description, due to F. Hausdorff, of Lebesgue measure on $\mathbb{R}^N$. As distinguished from the one given in § 2.2.2, this one is completely coordinate free.

For $\mathcal{C} \subseteq \mathcal{P}(\mathbb{R}^N)$, set $\|\mathcal{C}\| = \sup\{\mathrm{diam}(C) : C \in \mathcal{C}\}$, and, given $\delta > 0$, define

$$\mathbf{H}^N(\Gamma) = \lim_{\delta \searrow 0} \mathbf{H}^{N,\delta}(\Gamma), \text{ where } \mathbf{H}^{N,\delta}(\Gamma) \text{ is}$$

$$\inf\left\{\sum_{C\in\mathcal{C}} \Omega_N \,\mathrm{rad}(C)^N : \mathcal{C} \text{ a countable cover of } \Gamma \text{ with } \|\mathcal{C}\| \le \delta\right\}. \tag{4.2.3}$$

I emphasize that I have placed *no restriction* on the sets C making up the covers $\mathcal{C}$. On the other hand, it should be clear (cf. Exercise 4.2.2) that $\mathbf{H}^N(\Gamma)$ would have been unchanged if I had restricted myself to covers either by closed sets or by open sets. Also, it is obvious that $\mathbf{H}^{N,\delta}(\Gamma)$ is a non-increasing function of $\delta > 0$ and therefore that there is no question that the indicated limit exists.

Directly from its definition, one sees that $\mathbf{H}^{N,\delta}$ is monotone and countably subadditive in the sense that

$$\mathbf{H}^{N,\delta_1}(\Gamma_1) \le \mathbf{H}^{N,\delta_2}(\Gamma_2) \quad \text{whenever } \delta_2 \le \delta_1 \text{ and } \Gamma_1 \subseteq \Gamma_2$$

and

$$\mathbf{H}^{N,\delta}\left(\bigcup_{n=1}^{\infty} \Gamma_n\right) \le \sum_{n=1}^{\infty} \mathbf{H}^{N,\delta}(\Gamma_n) \quad \text{for all } \{\Gamma_n : n \ge 1\} \subseteq \mathcal{P}(\mathbb{R}^N).$$

Indeed, the first of these is completely trivial, and the second follows by choosing, for a given $\epsilon > 0$, $\{\mathcal{C}_n : n \ge 1\}$ for which

$$\|\mathcal{C}_n\| \le \delta,\ \Gamma_n \subseteq \bigcup \mathcal{C}_m, \quad \text{and} \sum_{C\in\mathcal{C}_n} \Omega_N\big(\mathrm{rad}(C)\big)^N \le \mathbf{H}^{N,\delta}(\Gamma_n) + 2^{-n}\epsilon$$

and noting that

$$\mathbf{H}^{N,\delta}\left(\bigcup_{n=1}^{\infty}\Gamma_n\right) \le \sum_{n=1}^{\infty}\sum_{C\in\mathcal{C}_n}\Omega_N\big(\mathrm{rad}(C)\big)^N \le \sum_{n=1}^{\infty}\mathbf{H}^{N,\delta}(\Gamma_n)+\epsilon.$$

Moreover, because

$$\lambda_{\mathbb{R}^N}(\Gamma) \le \sum_{C\in\mathcal{C}}\lambda_{\mathbb{R}^N}(C) \quad \text{for any countable cover } \mathcal{C}\subseteq\mathcal{B}_{\mathbb{R}^N} \text{ of } \Gamma,$$

the inequality $\lambda_{\mathbb{R}^N}(\Gamma) \le \mathbf{H}^{N,\delta}(\Gamma) \le \mathbf{H}^N(\Gamma)$ is an essentially trivial consequence of (4.2.1). In order to prove the opposite inequality, I will use the following lemma.

Lemma 4.2.3 *For any $\delta > 0$ and open set G in $\mathbb{R}^N$ with $\lambda_{\mathbb{R}^N}(G) < \infty$, there exists a sequence $\{B_n : n \ge 1\}$ of mutually disjoint closed balls contained in G with the properties that* $\mathrm{rad}(B_n) < \frac{\delta}{2}$ *and*

$$\lambda_{\mathbb{R}^N}\left(G\setminus\bigcup_{n=1}^{\infty}B_n\right)=0.$$

Proof. If $G=\emptyset$, there is nothing to do. Thus, assume that $G\neq\emptyset$, and set $G_0=G$. Using Lemma 2.2.10, choose a countable, exact cover $\mathcal{C}_0$ of G_0 by non-overlapping cubes Q with diameter strictly less than δ. Next, given $Q\in\mathcal{C}_0$, choose $\mathbf{a}=(a_1,\dots,a_N)\in\mathbb{R}^N$ and $\rho\in[0,\delta)$ for which

$$Q=\prod_{i=1}^{N}\big[a_i-\rho,a_i+\rho\big],$$

and set $B_Q=\overline{B\left(\mathbf{a},\frac{\rho}{2}\right)}$. Clearly, the B_Q's are mutually disjoint closed balls. At the same time, there is a dimensional constant $\alpha_N\in(0,1)$ for which $\lambda_{\mathbb{R}^N}(B_Q)\ge\alpha_N\mathrm{vol}(Q)$. Therefore there exists a finite subset $\{B_{0,1},\dots,B_{0,n_0}\}$ $\subseteq\{B_Q : Q\in\mathcal{C}_0\}$ for which

$$\lambda_{\mathbb{R}^N}\left(G_0\setminus\bigcup_{m=1}^{n_0}B_{0,m}\right)\le\beta_N\lambda_{\mathbb{R}^N}(G_0), \quad \text{where } \beta_N\equiv 1-\tfrac{\alpha_N}{2}\in(0,1).$$

Now set $G_1=G_0\setminus\bigcup_1^{n_0}B_{0,m}$. Noting that G_1 is again non-empty and open, we can repeat the preceding argument to find a finite collection of mutually disjoint closed balls $B_{1,m}\subset G_1$, $1\le m\le n_1$, of radius less than $\frac{\delta}{2}$ for which

$$\lambda_{\mathbb{R}^N}\left(G_1\setminus\bigcup_{m=1}^{n_1}B_{1,m}\right)\le\beta_N\lambda_{\mathbb{R}^N}(G_1).$$

More generally, we can use induction on $\ell\in\mathbb{Z}^+$ to construct open sets $G_\ell\subseteq G_{\ell-1}$ and finite collections $B_{\ell,1},\dots,B_{\ell,n_\ell}$ of mutually disjoint closed balls

$B \subset G_\ell$ with $\mathrm{rad}(B) < \frac{\delta}{2}$ so that

$$\lambda_{\mathbb{R}^N}(G_{\ell+1}) \le \beta_N \lambda_{\mathbb{R}^N}(G_\ell) \quad \text{where } G_{\ell+1} = G_\ell \setminus \bigcup_{m=1}^{n_\ell} B_{\ell,m}.$$

Clearly the collection

$$\{B_{\ell,m} : \ell \in \mathbb{N} \text{ and } 1 \le m \le n_\ell\}$$

has the required properties. □

Theorem 4.2.4 *For every* $\Gamma \in \mathcal{B}_{\mathbb{R}^N}$ *and* $\delta > 0$, $\mathbf{H}^N(\Gamma) = \lambda_{\mathbb{R}^N}(\Gamma) = \mathbf{H}^{N,\delta}(\Gamma)$.

Proof. As has been already pointed out, the inequality $\lambda_{\mathbb{R}^N}(\Gamma) \le \mathbf{H}^{N,\delta}(\Gamma) \le \mathbf{H}^N(\Gamma)$ is an immediate consequence of the definition combined with (4.2.1). To get the opposite inequality, it suffices to show that $\lambda_{\mathbb{R}^N}(\Gamma) \ge \mathbf{H}^{N,\delta}(\Gamma)$ for every $\delta > 0$ and $\Gamma \in \mathcal{B}_{\mathbb{R}^N}$. To this end, begin by observing that $\mathbf{H}^{N,\delta}$ is countably subadditive, and therefore, since $\lambda_{\mathbb{R}^N}$ is countably additive, it suffices to check $\mathbf{H}^{N,\delta}(\Gamma) \le \lambda_{\mathbb{R}^N}(\Gamma)$ for bounded sets $\Gamma \in \mathcal{B}_{\mathbb{R}^N}$. Thus, suppose that $\Gamma \in \mathcal{B}_{\mathbb{R}^N}$ is bounded, and let G be any open superset of Γ with $\lambda_{\mathbb{R}^N}(G) < \infty$. By Lemma 4.2.3, there exists a sequence $\{B_n : n \ge 1\}$ of mutually disjoint closed balls contained in G for which $\mathrm{diam}(B_n) \le \delta$ and

$$\lambda_{\mathbb{R}^N}(G \setminus A) = 0 \quad \text{where } A = \bigcup_{n=1}^{\infty} B_n.$$

Hence, because

$$\mathbf{H}^{N,\delta}(\Gamma) \le \mathbf{H}^{N,\delta}(G) \le \mathbf{H}^{N,\delta}(A) + \mathbf{H}^{N,\delta}(G \setminus A) = \mathbf{H}^{N,\delta}(A),$$

we see that

$$\mathbf{H}^{N,\delta}(\Gamma) \le \sum_{n=1}^{\infty} \Omega_N \mathrm{rad}(B_n)^N = \sum_{n=1}^{\infty} \lambda_{\mathbb{R}^N}(B_n) = \lambda_{\mathbb{R}^N}(A) = \lambda_{\mathbb{R}^N}(G)$$

for every open $G \supseteq \Gamma$, and, after taking the infimum over such G's, we arrive at the desired conclusion. □

Knowing Theorem 4.2.4, the reader may be wondering why I did not simply define

$$\mathbf{H}^N(\Gamma) = \inf\left\{\sum_{C \in \mathcal{C}} \Omega_N \mathrm{rad}(C)^N : \mathcal{C} \text{ a countable cover of } \Gamma\right\}.$$

Indeed, as the theorem shows, this definition is the same as the more complicated one that I gave and would have obviated the need for introducing

$\mathbf{H}^{N,\delta}$ and taking the limit as $\delta \searrow 0$. The reason for introducing the more complicated definition will not become clear until § 8.3.3, where $\mathbf{H}^N$ will be seen as a member of a continuously parameterized family of measures on $\mathbb{R}^N$, and $\mathbf{H}^N$ is the only member for which the more complicated definition is not needed.

4.2.3 Exercises for § 4.2

Exercise 4.2.1 Assume that $N \geq 2$. Using the definition of $\mathbf{H}^N$ in (4.2.3), give a direct (i.e., one that does not use Theorem 4.2.4) proof that $\mathbf{H}^N(P) = 0$ for every hyperplane P (i.e., the translate of an $(N-1)$-dimensional subspace) in $\mathbb{R}^N$.

Exercise 4.2.2 Show that, for each $\delta > 0$, $\mathbf{H}^{N,\delta}(\Gamma)$ is unchanged when one restricts to covers by closed sets, and from this conclude that $\mathbf{H}^N(\Gamma)$ is unchanged if one restricts to covers by open sets or by closed sets. Less obvious than the preceding is the fact that, at least for $\Gamma \in \mathcal{B}_{\mathbb{R}^N}$, one can restrict to covers by closed balls or by open balls. To be precise, let $\mathbf{H}_{\mathrm{b}}^{N,\delta}(\Gamma)$ be the quantity in (4.2.3) obtained by restricting to covers by closed balls. Given $\Gamma \in \mathcal{B}_{\mathbb{R}^N}$, use Lemma 4.2.3 and argue as in the proof of Theorem 4.2.4 to show that, for each $\delta > 0$, $\mathbf{H}_{\mathrm{b}}^{N,\delta}(\Gamma) = \mathbf{H}^{N,\delta}(\Gamma)$. Finally, conclude from this that the same equality holds if one restricts to covers by open balls.

Chapter 5
Changes of Variable

I have now developed the basic theory of Lebesgue integration, but I have provided nearly no tools with which to compute the integrals that have been shown to exist. The purpose of the present chapter is to introduce a technique that often makes evaluation, or at least estimation, possible. The technique is that of changing variables.

5.1 Riemann vs. Lebesgue, Distributions, and Polar Coordinates

The content of this section is applications of the following general principle.

Let $(E, \mathcal{B}, \mu)$ be a measure space, $(E', \mathcal{B}')$ a measurable space, $\Phi : E \longrightarrow E'$ a measurable map, and recall (cf. part **(iii)** of Exercise 2.1.3) that the pushforward of μ under Φ is the measure $\Phi_*\mu$ on $(E', \mathcal{B}')$ given by $\Phi_*\mu(\Gamma') = \mu\big(\Phi^{-1}(\Gamma')\big)$ for $\Gamma' \in \mathcal{B}'$. The following lemma is an essentially immediate consequence of this definition.

Lemma 5.1.1 *Refer to the preceding. Then, for every non-negative measurable function φ on $(E', \mathcal{B}')$,*

$$\int_{E'} \varphi \, d(\Phi_*\mu) = \int_E \varphi \circ \Phi \, d\mu. \tag{5.1.1}$$

Moreover, $\varphi \in L^1(\Phi_*\mu; \mathbb{R})$ if and only if $\varphi \circ \Phi \in L^1(\mu; \mathbb{R})$, and (5.1.1) *holds for all $\varphi \in L^1(\Phi_*\mu; \mathbb{R})$.*

Proof. Clearly it suffices to prove the first assertion. To this end, note that, by definition, (5.1.1) holds when φ is the indicator of a set $\Gamma' \in \mathcal{B}'$. Hence, it also holds when φ is a non-negative measurable simple function on $(E', \mathcal{B}')$, and so, by the monotone convergence theorem, it must hold for all non-negative measurable functions on $(E', \mathcal{B}')$. □

D. W. Stroock, *Essentials of Integration Theory for Analysis*, Graduate Texts in Mathematics 262, https://doi.org/10.1007/978-3-030-58478-8_5

5.1.1 Riemann vs. Lebesgue

My first significant example of a change of variables will relate integrals over an arbitrary measure space to integrals over the real line. Namely, given a measurable $\overline{\mathbb{R}}$-valued function f on a measure space $(E, \mathcal{B}, \mu)$, define the **distribution of** f under μ to be the measure $\mu_f \equiv f_*\mu$ on $(\overline{\mathbb{R}}, \mathcal{B}_{\overline{\mathbb{R}}})$.[1] We then have that, for any non-negative measurable φ on $(\overline{\mathbb{R}}, \mathcal{B}_{\overline{\mathbb{R}}})$,

$$\int_E \varphi \circ f(x)\, \mu(dx) = \int_{\overline{\mathbb{R}}} \varphi(t)\, \mu_f(dt). \tag{5.1.2}$$

The reason why it is sometimes useful to make this change of variables is that the integral on the right-hand side of (5.1.2) can often be evaluated as the limit of Riemann integrals, to which all the powerful tools of calculus apply. In order to see how the right-hand side of (5.1.2) leads to Riemann integrals, I will prove a general fact about the relationship between Lebesgue and Riemann integrals on the line. From a theoretical standpoint, the most interesting feature of this result is that it shows that a complete description of the class of Riemann integrable functions in terms of continuity properties defies a totally Riemannian solution and requires the Lebesgue notion of *almost everywhere*.

Theorem 5.1.2 *Let ν be a finite measure on $\big((a,b], \mathcal{B}_{(a,b]}\big)$, where $-\infty < a < b < \infty$, and set $\psi(t) = \nu((a,t])$ for $t \in [a,b]$ ($\psi(a) = \nu(\emptyset) = 0$). Then, ψ is right-continuous on $[a,b)$, non-decreasing on $[a,b]$, and, for each $t \in (a,b]$, $\psi(t) - \psi(t-) = \nu(\{t\})$. Furthermore, if φ is a bounded function on $[a,b]$, then φ is Riemann integrable on $[a,b]$ with respect to ψ if and only if φ is continuous* (a.e., ν) *on $(a,b]$, in which case, φ is measurable on $\big((a,b], \overline{\mathcal{B}_{(a,b]}}^{\nu}\big)$ and*

$$\int_{(a,b]} \varphi\, d\overline{\nu} = (\mathrm{R}) \int_{[a,b]} \varphi(t)\, d\psi(t). \tag{5.1.3}$$

Proof. It will be convenient to think of ν as being defined on $\big([a,b], \mathcal{B}_{[a,b]}\big)$ by $\nu(\Gamma) \equiv \nu(\Gamma \cap (a,b])$ for $\Gamma \in \mathcal{B}_{[a,b]}$. Thus, we will do so.

Obviously ψ is right-continuous and non-increasing on $[a,b]$, and therefore (cf. Exercise 3.1.1) it is also $\mathcal{B}_{[a,b]}$-measurable there. Moreover, for each $t \in (a,b]$, $\psi(t) - \psi(t-) = \lim_{s \nearrow t} \nu\big((s,t]\big) = \nu(\{t\})$.

Now assume that φ is Riemann integrable on $[a,b]$ with respect to ψ. To see that φ is continuous (a.e., ν) on $(a,b]$, choose (cf. Lemma 1.2.6), for each $n \geq 1$, a finite, non-overlapping, exact cover $\mathcal{C}_n$ of $[a,b]$ by closed intervals $I = [a_I, b_I]$ such that $\|\mathcal{C}_n\| < \frac{1}{n}$ and, for each $I \in \mathcal{C}_n$, ψ is continuous at a_I. If $\Delta = \bigcup_{n=1}^{\infty} \{a_I : I \in \mathcal{C}_n\}$, then $\nu(\Delta) = 0$. Given $m \geq 1$, let $\mathcal{C}_{m,n}$ be the set of those $I \in \mathcal{C}_n$ for which $\sup_I \varphi - \inf_I \varphi \geq \frac{1}{m}$. It is then easy to check that

[1] In probability theory, distributions take on particular significance. In fact, from the point of view of a probabilistic purist, it is the distribution of a function, as opposed to the function itself, that is its distinguishing feature.

$$\{t \in (a,b] \setminus \Delta : \varphi \text{ is not continuous at } t\} \subseteq \bigcup_{m=1}^{\infty} \bigcap_{n=1}^{\infty} \bigcup_{I \in \mathcal{C}_{m,n}} I.$$

But, by Exercises 2.1.6 and 1.2.4, for each $m \geq 1$,

$$\nu\Biggl(\bigcap_{n=1}^{\infty} \bigcup_{I \in \mathcal{C}_{m,n}} I\Biggr) \leq \lim_{n\to\infty} \sum_{I \in \mathcal{C}_{m,n}} \Delta_I \psi = 0,$$

and therefore $\nu\bigl(\bigcup_{m=1}^{\infty} \bigcap_{n=1}^{\infty} \bigcup_{I\in\mathcal{C}_{m,n}} I\bigr) = 0$. Hence, we have now shown that φ is continuous (a.e., ν) on $(a,b]$.

Conversely, suppose that φ is continuous (a.e., ν) on $(a,b]$. Given a sequence $\{\mathcal{C}_n : n \geq 1\}$ of finite, non-overlapping, exact covers of $[a,b]$ by compact intervals I with $\|\mathcal{C}_n\| \longrightarrow 0$, for each $n \geq 1$, define $\overline{\varphi}_n(t) = \sup_I \varphi$ and $\underline{\varphi}_n(t) = \inf_I \varphi$ for $t \in (a_I, b_I]$ and $I \in \mathcal{C}_n$. Clearly, both $\overline{\varphi}_n$ and $\underline{\varphi}_n$ are measurable on $\bigl((a,b], \mathcal{B}_{(a,b]}\bigr)$. Moreover,

$$\inf_{(a,b]} \varphi \leq \underline{\varphi}_n \leq \varphi \leq \overline{\varphi}_n \leq \sup_{(a,b]} \varphi$$

for all $n \geq 1$. Finally, since φ is continuous (a.e., ν),

$$\varphi = \lim_{n\to\infty} \underline{\varphi}_n = \lim_{n\to\infty} \overline{\varphi}_n \quad \text{(a.e., } \nu);$$

and so, not only is φ equal to $\varliminf_{n\to\infty} \underline{\varphi}_n$ (a.e., ν) and therefore measurable on $((a,b], \overline{\mathcal{B}_{(a,b]}}^{\nu})$, but also

$$\lim_{n\to\infty} \int_{(a,b]} \underline{\varphi}_n \, d\nu = \int_{(a,b]} \varphi \, d\overline{\nu} = \lim_{n\to\infty} \int_{(a,b]} \overline{\varphi}_n \, d\nu.$$

In particular, we conclude that

$$\sum_{I \in \mathcal{C}_n} \Bigl(\sup_I \varphi - \inf_I \varphi\Bigr) \Delta_I \psi = \int_{(a,b]} \bigl(\overline{\varphi}_n - \underline{\varphi}_n\bigr)\, d\nu \longrightarrow 0$$

as $n \to \infty$. From this it is clear both that φ is Riemann integrable on $[a,b]$ with respect to ψ and that (5.1.3) holds. □

I am now ready to prove the main result to which this subsection is devoted.

Theorem 5.1.3 *Let $(E, \mathcal{B}, \mu)$ be a measure space and $f : E \longrightarrow [0,\infty]$ a $\mathcal{B}$-measurable function. Then $t \in (0,\infty) \longmapsto \mu(f > t) \in [0,\infty]$ is a right-continuous, non-increasing function. In particular, it is measurable on $\bigl((0,\infty), \mathcal{B}_{(0,\infty)}\bigr)$ and has at most a countable number of discontinuities. Next, assume that $\varphi \in C\bigl([0,\infty);\mathbb{R}\bigr) \cap C^1\bigl((0,\infty);\mathbb{R}\bigr)$ is a non-decreasing function satisfying $\varphi(0) = 0 < \varphi(t)$, $t > 0$, and set $\varphi(\infty) = \lim_{t\to\infty} \varphi(t)$. Then*

$$\int_E \varphi \circ f(x)\,\mu(dx) = \int_{(0,\infty)} \varphi'(t)\mu(f > t)\,\lambda_{\mathbb{R}}(dt). \tag{5.1.4}$$

Hence, either $\mu(f > \delta) = \infty$ *for some* $\delta \in (0,\infty)$, *in which case both sides of* (5.1.4) *are infinite, or, for each* $0 < \delta < r < \infty$, *the map* $t \in [\delta, r] \longmapsto \varphi'(t)\mu(f > t)$ *is Riemann integrable and*

$$\int_E \varphi \circ f(x)\,\mu(dx) = \lim_{\substack{\delta \searrow 0 \\ r \nearrow \infty}} (\mathrm{R}) \int_{[\delta,r]} \varphi'(t)\mu(f > t)\,dt.$$

Proof. It is clear that $t \in (0,\infty) \longmapsto \mu(f > t)$ is right-continuous, non-increasing, and therefore $\mathcal{B}_{(0,\infty)}$-measurable.

We turn next to the proof of (5.1.4). Since, by Theorems 5.1.2 and 1.2.1,

$$\lim_{\alpha \searrow 0} \int_{(\alpha,t]} \varphi'(s)\,\lambda_{\mathbb{R}}(ds) = \lim_{\alpha \searrow 0} (\mathrm{R}) \int_\alpha^t \varphi'(s)\,ds = \varphi(t),$$

Theorem 4.1.4 says that

$$\int_E \varphi \circ f\,d\mu = \int_E \left(\int_{(0,f(x)]} \varphi'(t)\,dt \right) \mu(dx) = \int_{(0,\infty)} \varphi'(t)\mu(f > t)\,dt.$$

To prove the last part of the theorem, first observe that both sides of the asserted equation are infinite if $\mu(f > \delta) = \infty$ for some $\delta > 0$. When $\mu(f > \delta) < \infty$ for every $\delta > 0$, simply note that

$$\int_{(0,\infty)} \varphi'(t)\mu(f > t)\,\lambda_{\mathbb{R}}(dt) = \lim_{\substack{\delta \searrow 0 \\ r \nearrow \infty}} \int_{(\delta,r]} \varphi'(t)\mu(f > t)\,\lambda_{\mathbb{R}}(dt),$$

and apply Theorem 5.1.2. □

5.1.2 Polar Coordinates

In this subsection I will examine the change of variables that is referred to casually as "switching to polar coordinates."

Warning: From now on, at least whenever the meaning is clear from the context, I will use the notation "dx" instead of the more cumbersome "$\lambda_{\mathbb{R}^N}(dx)$" when doing Lebesgue integration with respect to Lebesgue measure on $\mathbb{R}^N$.

Let $\mathbb{S}^{N-1}$ denote the unit $(N-1)$-sphere $\{x \in \mathbb{R}^N : |x| = 1\}$ in $\mathbb{R}^N$, and define $\Phi : \mathbb{R}^N \setminus \{0\} \longrightarrow \mathbb{S}^{N-1}$ by $\Phi(x) = \frac{x}{|x|}$. Clearly Φ is continuous. Next, define the **surface measure** $\lambda_{\mathbb{S}^{N-1}}$ on $\mathbb{S}^{N-1}$ to be the image under Φ of $N\lambda_{\mathbb{R}^N}$ restricted to $\mathcal{B}_{B_{\mathbb{R}^N}(0,1)\setminus\{0\}}$. Because $\Phi(rx) = \Phi(x)$ for all $r > 0$ and $x \in \mathbb{R}^N \setminus \{0\}$, one can use Theorem 2.2.13 or Exercise 2.2.4 to check that

$$\int_{B_{\mathbb{R}^N}(0,r)\setminus\{0\}} f \circ \Phi(x)\, dx = r^N \int_{B_{\mathbb{R}^N}(0,1)\setminus\{0\}} f \circ \Phi(x)\, dx$$

and therefore that

$$\int_{B_{\mathbb{R}^N}(0,r)\setminus\{0\}} f \circ \Phi(x)\, dx = \frac{r^N}{N} \int_{\mathbb{S}^{N-1}} f(\omega)\, \lambda_{\mathbb{S}^{N-1}}(d\omega) \tag{$*$}$$

for every non-negative measurable f on $(\mathbb{S}^{N-1}, \mathcal{B}_{\mathbb{S}^{N-1}})$. In particular, if (cf. **(iii)** in Exercise 5.1.3) $\omega_{N-1} \equiv \lambda_{\mathbb{S}^{N-1}}(\mathbb{S}^{N-1})$ is the **surface area** of $\mathbb{S}^{N-1}$, we have that $\Omega_N = \frac{\omega_{N-1}}{N}$, where Ω_N is the volume (i.e., Lebesgue measure) of the unit ball in $\mathbb{R}^N$.

Next, define $\Psi : (0,\infty) \times \mathbb{S}^{N-1} \longrightarrow \mathbb{R}^N \setminus \{0\}$ by $\Psi(r,\omega) = r\omega$. Note that Ψ is a homeomorphism from $(0,\infty) \times \mathbb{S}^{N-1}$ onto $\mathbb{R}^N \setminus \{0\}$ and that its inverse is given by $\Psi^{-1}(x) = (|x|, \Phi(x))$. That is, the components of $\Psi^{-1}(x)$ are the **polar coordinates** of the point $x \in \mathbb{R}^N \setminus \{0\}$. Finally, define the measure R_N on $\big((0,\infty), \mathcal{B}_{(0,\infty)}\big)$ by $R_N(\Gamma) = \int_\Gamma r^{N-1}\, dr$.

The importance here of these considerations is contained in the following result.

Theorem 5.1.4 *Referring to the preceding, one has that*

$$\lambda_{\mathbb{R}^N} = \Psi_*\big(R_N \times \lambda_{\mathbb{S}^{N-1}}\big) \text{ on } \mathcal{B}_{\mathbb{R}^N\setminus\{0\}}.$$

In particular, if f is a non-negative, measurable function on $(\mathbb{R}^N, \mathcal{B}_{\mathbb{R}^N})$, then

$$\begin{aligned} \int_{\mathbb{R}^N} f(x)\, dx &= \int_{(0,\infty)} r^{N-1} \left(\int_{\mathbb{S}^{N-1}} f(r\omega)\, \lambda_{\mathbb{S}^{N-1}}(d\omega) \right) dr \\ &= \int_{\mathbb{S}^{N-1}} \left(\int_{(0,\infty)} f(r\omega) r^{N-1}\, dr \right) \lambda_{\mathbb{S}^{N-1}}(d\omega). \end{aligned} \tag{5.1.5}$$

Proof. By Corollary 3.2.11 and Theorem 4.1.5, all that we have to do is check that the first equality in (5.1.5) holds for every

$$f \in C_c(\mathbb{R}^N;\mathbb{R}) \equiv \big\{ f \in C(\mathbb{R}^N;\mathbb{R}) : f \equiv 0 \text{ off of some compact set} \big\}.$$

To this end, let $f \in C_c(\mathbb{R}^N;\mathbb{R})$ be given, and set $F(r) = \int_{B_{\mathbb{R}^N}(0,r)} f(x)\, dx$ for $r > 0$. Then, by $(*)$, for all $r, h > 0$, $F(r+h) - F(r)$ equals

$$\begin{aligned}
&\int_{B_{\mathbb{R}^N}(0,r+h)\setminus B_{\mathbb{R}^N}(0,r)} f(x)\,dx \\
&= \int_{B_{\mathbb{R}^N}(0,r+h)\setminus B_{\mathbb{R}^N}(0,r)} f\circ\Psi(r,\Phi(x))\,dx \\
&\quad + \int_{B_{\mathbb{R}^N}(0,r+h)\setminus B_{\mathbb{R}^N}(0,r)} \big(f(x) - f\circ\Psi(r,\Phi(x))\big)\,dx \\
&= \frac{(r+h)^N - r^N}{N}\int_{\mathbb{S}^{N-1}} f\circ\Psi(r,\omega)\,\lambda_{\mathbb{S}^{N-1}}(d\omega) + o(h),
\end{aligned}$$

where "$o(h)$" denotes a function that tends to 0 faster than h. Hence, F is continuously differentiable on $(0,\infty)$ and its derivative at $r \in (0,\infty)$ is given by $r^{N-1}\int_{\mathbb{S}^{N-1}} f\circ\Psi(r,\omega)\,\lambda_{\mathbb{S}^{N-1}}(d\omega)$. Since $F(r) \longrightarrow 0$ as $r \searrow 0$, the desired result now follows from Theorem 5.1.2 and the Fundamental Theorem of Calculus. □

5.1.3 Exercises for § 5.1

Exercise 5.1.1 Suppose that $F : \mathbb{R} \longrightarrow \mathbb{R}$ is a bounded, continuous, non-decreasing function that tends to 0 at $-\infty$, and let μ_F be the Borel measure on $\mathbb{R}$ for which F is the distribution function. Show that for $-\infty < a < b < \infty$

$$(\mathrm{R})\int_{[a,b]} \varphi\circ F(s)\,dF(s) = (\mathrm{R})\int_{[F(a),F(b)]} \varphi(t)\,dt \quad \text{if } \varphi \in C\big([F(a),F(b)];\mathbb{R}\big),$$

which is the classical *change of variables formula*. Next, apply Theorem 5.1.2 to conclude from this that

$$\int_{\mathbb{R}} \varphi\circ F\,d\mu_F = \int_{[0,F(\infty)]} \varphi(t)\,dt$$

for all $\mathcal{B}_{\mathbb{R}}$-measurable $\varphi : \mathbb{R} \longrightarrow [0,\infty)$.

Exercise 5.1.2 A particularly important case of Theorem 5.1.3 is the one in which $\varphi(t) = t^p$ for some $p \in (0,\infty)$, in which case (5.1.5) yields

$$\int_E |f(x)|^p\,\mu(dx) = p\int_{(0,\infty)} t^{p-1}\mu(|f| > t)\,dt. \tag{5.1.6}$$

Use (5.1.6) to show that $|f|^p$ is μ-integrable if and only if

$$\sum_{n=1}^{\infty}\frac{1}{n^{p+1}}\mu\left(|f| > \tfrac{1}{n}\right) + \sum_{n=1}^{\infty} n^{p-1}\mu(|f| > n) < \infty.$$

Compare this result to the one obtained in the last part of Exercise 3.2.4.

Exercise 5.1.3 Perform the calculations outlined in the following.

(**i**) Justify Gauss's trick (cf. Exercise 5.2.1)

$$\left(\int_{\mathbb{R}} e^{-\frac{x^2}{2}}\, dx\right)^2 = \int_{\mathbb{R}^2} e^{-\frac{|x|^2}{2}}\, dx = 2\pi \int_{(0,\infty)} re^{-\frac{r^2}{2}}\, dr = 2\pi$$

and conclude that for any $N \in \mathbb{Z}^+$ and symmetric $N \times N$ matrix A that is strictly positive definite (i.e., all the eigenvalues of A are strictly positive),

$$\int_{\mathbb{R}^N} \exp\left[-\frac{1}{2}(x, A^{-1}x)_{\mathbb{R}^N}\right] dx = (2\pi)^{\frac{N}{2}} \left(\det(A)\right)^{\frac{1}{2}}.$$

(**ii**) Define $\Gamma(s) = \int_{(0,\infty)} t^{s-1}e^{-t}\, dt$ for $s \in (0,\infty)$. Show that, for any $s \in (0,\infty)$, $\Gamma(s+1) = s\Gamma(s)$. Also, show that $\Gamma\left(\frac{1}{2}\right) = \pi^{\frac{1}{2}}$. The function $\Gamma(\,\cdot\,)$ is called Euler's **gamma function**. Notice that it provides an extension of the factorial function in the sense that $\Gamma(n+1) = n!$ for integers $n \geq 0$.

(**iii**) Show that

$$\omega_{N-1} \equiv \lambda_{\mathbb{S}^{N-1}}(\mathbb{S}^{N-1}) = \frac{2\pi^{\frac{N}{2}}}{\Gamma\left(\frac{N}{2}\right)},$$

and conclude that the volume Ω_N of the N-dimensional unit ball is given by

$$\Omega_N = \frac{\pi^{\frac{N}{2}}}{\Gamma\left(\frac{N}{2}+1\right)}.$$

Hint: Use Theorems 4.1.4 and 5.1.4 to compute $\left(\int_{\mathbb{R}} e^{-\frac{x^2}{2}}\, dx\right)^N = \int_{\mathbb{R}^N} e^{-\frac{|x|^2}{2}}\, dx$.

(**iv**) Given $\alpha, \beta \in (0,\infty)$, show that

$$\int_{(0,\infty)} t^{-\frac{1}{2}} \exp\left[-\alpha^2 t - \frac{\beta^2}{t}\right] dt = \frac{\pi^{\frac{1}{2}}e^{-2\alpha\beta}}{\alpha}.$$

Finally, use the preceding to show that

$$\int_{(0,\infty)} t^{-\frac{3}{2}} \exp\left[-\alpha^2 t - \frac{\beta^2}{t}\right] dt = \frac{\pi^{\frac{1}{2}}e^{-2\alpha\beta}}{\beta}.$$

Hint: Define $F(s)$ for $s \in \mathbb{R}$ to be the unique $t \in (0,\infty)$ satisfying $s = \alpha t^{\frac{1}{2}} - \beta t^{-\frac{1}{2}}$, and use Exercise 5.1.1 to show that

$$\int_{(0,\infty)} t^{-\frac{1}{2}} \exp\left[-\alpha^2 t - \frac{\beta^2}{t}\right] dt = \frac{e^{-2\alpha\beta}}{\alpha} \int_{\mathbb{R}} e^{-s^2}\, ds.$$

Exercise 5.1.4 This exercise deals with a few of the elementary properties of $\lambda_{\mathbb{S}^{N-1}}$.

(**i**) Show that if Γ is a non-empty, connected, open subset of $\mathbb{S}^{N-1}$, then $\lambda_{\mathbb{S}^{N-1}}(\Gamma) > 0$. Next, show that $\lambda_{\mathbb{S}^{N-1}}$ is **orthogonally invariant**. That is, show that if $\mathcal{O}$ is an $N \times N$ orthogonal matrix and $T_{\mathcal{O}}$ is the associated transformation on $\mathbb{R}^N$ (cf. the paragraph preceding Theorem 2.2.13), then $(T_{\mathcal{O}})_* \lambda_{\mathbb{S}^{N-1}} = \lambda_{\mathbb{S}^{N-1}}$. Finally, use this fact to show that

$$\int_{\mathbb{S}^{N-1}} (\xi, \omega)_{\mathbb{R}^N} \, \lambda_{\mathbb{S}^{N-1}}(d\omega) = 0$$

and

$$\int_{\mathbb{S}^{N-1}} (\xi, \omega)_{\mathbb{R}^N} (\eta, \omega)_{\mathbb{R}^N} \, \lambda_{\mathbb{S}^{N-1}}(d\omega) = \Omega_N \, (\xi, \eta)_{\mathbb{R}^N}$$

for any ξ, $\eta \in \mathbb{R}^N$.

Hint: In proving these, for given $\xi \in \mathbb{R}^N \setminus \{0\}$, consider the orthogonal transformation that sends ξ to $-\xi$ but acts as the identity on the orthogonal complement of ξ.

(**ii**) Define $\Psi : [0, 2\pi] \longrightarrow \mathbb{S}^1$ by $\Psi(\theta) = \binom{\cos\theta}{\sin\theta}$, and set $\mu = \Psi_* \lambda_{[0,2\pi]}$. Given any rotation invariant finite measure ν on $(\mathbb{S}^1, \mathcal{B}_{\mathbb{S}^1})$, show that $\nu = \frac{\nu(\mathbb{S}^1)}{2\pi}\mu$. In particular, conclude that $\lambda_{\mathbb{S}^1} = \mu$. (Cf. Exercise 5.2.1 below.)

Hint: Define

$$\mathcal{O}_\theta = \begin{pmatrix} \cos\theta & \sin\theta \\ -\sin\theta & \cos\theta \end{pmatrix} \quad \text{for} \quad \theta \in [0, 2\pi],$$

and note that

$$\int_{\mathbb{S}^1} f \, d\nu = \frac{1}{2\pi} \int_{[0,2\pi]} \left(\int_{\mathbb{S}^1} f \circ T_{\mathcal{O}_\theta}(\omega) \, \nu(d\omega) \right) d\theta.$$

5.2 Jacobi's Transformation and Surface Measure

I begin this section with a derivation of Jacobi's famous generalization to non-linear maps of the result in Theorem 2.2.13. I will then apply Jacobi's result to obtain surface measure by *differentiating Lebesgue measure across a smooth hypersurface.* Throughout, I will be assuming that $N \geq 2$.

5.2.1 Jacobi's Transformation Formula

Given an open set $G \subseteq \mathbb{R}^N$ and a continuously differentiable map

$$x \in G \longmapsto \Phi(x) = \begin{pmatrix} \Phi_1(x) \\ \vdots \\ \Phi_N(x) \end{pmatrix} \in \mathbb{R}^N,$$

define the **Jacobian matrix** $\frac{\partial \Phi}{\partial x}(x)$ **of** Φ **at** x to be the $N \times N$ matrix whose jth column is the vector

$$\partial_{x_j}\Phi(x) \equiv \begin{pmatrix} \frac{\partial \Phi_1}{\partial x_j} \\ \vdots \\ \frac{\partial \Phi_N}{\partial x_j} \end{pmatrix}.$$

In addition, call $J\Phi(x) \equiv \left|\det\left(\frac{\partial \Phi}{\partial x}(x)\right)\right|$ the **Jacobian of** Φ **at** x.

Lemma 5.2.1 *If G is an open set in $\mathbb{R}^N$ and Φ an element of $C^1(G;\mathbb{R}^N)$ whose Jacobian never vanishes on G, then Φ maps open (or $\mathcal{B}_G$-measurable) subsets of G into open (or $\mathcal{B}_{\mathbb{R}^N}$-measurable) sets in $\mathbb{R}^N$. In addition, if $\Gamma \in \mathcal{B}_G$ with $\lambda_{\mathbb{R}^N}(\Gamma) = 0$, then $\lambda_{\mathbb{R}^N}\big(\Phi(\Gamma)\big) = 0$; and if $\Gamma \in \mathcal{B}_{\Phi(G)}$ with $\lambda_{\mathbb{R}^N}(\Gamma) = 0$, then $\lambda_{\mathbb{R}^N}\big(\Phi^{-1}(\Gamma)\big) = 0$. In particular,*

$$\Gamma \in \overline{\mathcal{B}_G}^{\lambda_{\mathbb{R}^N}} \quad \textit{if and only if} \quad \Phi(\Gamma) \in \overline{\mathcal{B}_{\Phi(G)}}^{\lambda_{\mathbb{R}^N}}.$$

Proof. By the Inverse Function Theorem,[2] for each $x \in G$ there is an open neighborhood $U \subseteq G$ of x such that $\Phi \restriction U$ is invertible and its inverse has first derivatives that are bounded and continuous. Hence, G can be written as the union of a countable number of open sets on each of which Φ admits an inverse having bounded continuous first-order derivatives; and so, without loss of generality, we may and will assume that $\Phi \in C^1_{\mathrm{b}}(G;\mathbb{R}^N)$ and that it admits an inverse $\Phi^{-1} \in C^1_{\mathrm{b}}\big(\Phi(G);\mathbb{R}^N\big)$. But, in that case, both Φ and Φ^{-1} take open sets to open sets and therefore $\mathcal{B}_{\mathbb{R}^N}$-measurable sets to $\mathcal{B}_{\mathbb{R}^N}$-measurable sets. Finally, to see that Φ takes sets of $\lambda_{\mathbb{R}^N}$-measure 0 to sets of $\lambda_{\mathbb{R}^N}$-measure 0, note that Φ is uniformly Lipschitz continuous and apply (5.1.5). The argument for Φ^{-1} is precisely the same. □

A continuously differentiable map Φ on an open set $U \subseteq \mathbb{R}^N$ into $\mathbb{R}^N$ is called a **diffeomorphism** if it is **injective** (i.e., one-to-one) and $J\Phi$ never vanishes. If Φ is a diffeomorphism on the open set U and if $W = \Phi(U)$, then I will say that Φ is **diffeomorphic from** U **onto** W.

[2] See, for example, W. Rudin's *Principles of Mathematical Analysis*, McGraw Hill (1976).

In what follows, and elsewhere, for any given set $\Gamma \subseteq \mathbb{R}^N$ and $\delta > 0$, I will use

$$\Gamma^{(\delta)} \equiv \{x \in \mathbb{R}^N : |x - \Gamma| < \delta\} \tag{5.2.1}$$

to denote the **open δ-hull of Γ**.

Theorem 5.2.2 (Jacobi's Transformation Formula) *Let G be an open set in $\mathbb{R}^N$ and Φ an element of $C^2(G;\mathbb{R}^N)$.*[3] *If the Jacobian $J\Phi$ of Φ never vanishes, then, for every measurable function f on $\big(\Phi(G), \overline{\mathcal{B}_{\Phi(G)}}^{\lambda_{\mathbb{R}^N}}\big)$, $f \circ \Phi$ is measurable on $\big(G, \overline{\mathcal{B}_G}^{\lambda_{\mathbb{R}^N}}\big)$ and*

$$\int_{\Phi(G)} f(y)\, dy \le \int_G f \circ \Phi(x)\, J\Phi(x)\, dx \tag{5.2.2}$$

whenever f is non-negative. Moreover, if Φ is a diffeomorphism on G, then (5.2.2) can be replaced by

$$\int_{\Phi(G)} f(y)\, dy = \int_G f \circ \Phi(x)\, J\Phi(x)\, dx. \tag{5.2.3}$$

Proof. First note that (5.2.3) is a consequence of (5.2.2) when Φ is one-to-one. Indeed, if Φ is one-to-one, then the Inverse Function Theorem guarantees that $\Phi^{-1} \in C^2(\Phi(G);\mathbb{R}^N)$. In addition, by the chain rule,

$$\frac{\partial \Phi^{-1}}{\partial y}(y) = \left(\frac{\partial \Phi}{\partial x}(\Phi^{-1}(y))\right)^{-1} \quad \text{for} \quad y \in \Phi(G).$$

Hence one can apply (5.2.2) to Φ^{-1} and thereby obtain

$$\int_G f \circ \Phi(x)\, J\Phi(x)\, dx \le \int_{\Phi(G)} f(y)\, (J\Phi) \circ \Phi^{-1}(y)\, J\Phi^{-1}(y)\, dy = \int_{\Phi(G)} f(y)\, dy,$$

which, in conjunction with (5.2.2), yields (5.2.3).

Next observe that it suffices to prove (5.2.2) under the assumptions that G is bounded, Φ on G has bounded first and second order derivatives, and $J\Phi$ is uniformly positive on G. In fact, if this is not already the case, then one can choose a non-decreasing sequence of bounded open sets G_n such that $\Phi \restriction G_n$ has these properties for each $n \ge 1$ and $G_n \nearrow G$. Clearly, the result for Φ on G follows from the result for $\Phi \restriction G_n$ on G_n for every $n \ge 1$. Thus, from now on, we will assume that G is bounded, the first and second derivatives of Φ are bounded, and $J\Phi$ is uniformly positive on G.

Set $Q = Q(c;r) = \prod_1^N [c_i - r, c_i + r] \subseteq G$, the cube of side length $2r$ centered at $c = (c_1, \dots, c_N) \in \mathbb{R}^N$. Then, by Taylor's Theorem, there is an $L \in [0,\infty)$ (depending only on the bound on the second derivatives of Φ)

[3] By being a little more careful, one can get the same conclusion for $\Phi \in C^1(G, \mathbb{R}^N)$.

such that[4] (cf. § 2.2.2 for the notation here)

$$\Phi\big(Q(c;r)\big) \subseteq T_{\Phi(c)}\Big(T_{\frac{\partial\Phi}{\partial x}(c)}\big(Q(0;r)\big)^{(Lr^2)}\Big).$$

At the same time, there is an $M < \infty$ (depending only on L, the lower bound on $J\Phi$, and the upper bounds on the first derivatives of Φ) such that

$$\big(T_{\frac{\partial\Phi}{\partial x}(c)}Q(0;r)\big)^{(Lr^2)} \subseteq T_{\frac{\partial\Phi}{\partial \mathbf{x}}(c)}Q\big(0,r+Mr^2\big).$$

Hence, by Theorem 2.2.13,

$$\lambda_{\mathbb{R}^N}\big(\Phi(Q)\big) \le J\Phi(c)\,\lambda_{\mathbb{R}^N}\big(Q(0,r+Mr^2)\big) = (1+Mr)^N J\Phi(c)\lambda_{\mathbb{R}^N}(Q).$$

Now define $\mu(\Gamma) = \int_\Gamma J\Phi(x)\,dx$ for $\Gamma \in \overline{\mathcal{B}_G}^{\lambda_{\mathbb{R}^N}}$, set $\nu = \Phi_*\mu$, note that ν is finite, and apply Theorem 2.1.4 to see that it is regular. Given an open set $H \subseteq \Phi(G)$, use Lemma 2.2.10 to choose, for each $m \in \mathbb{Z}^+$, a countable, non-overlapping, exact cover $\mathcal{C}_m$ of $\Phi^{-1}(H)$ by cubes Q with $\operatorname{diam}(Q) < \frac{1}{m}$. Then, by the preceding paragraph,

$$\lambda_{\mathbb{R}^N}(H) \le \sum_{Q\in\mathcal{C}_m} \lambda_{\mathbb{R}^N}\big(\Phi(Q)\big) \le \left(1+\frac{M}{m}\right)^N \sum_{Q\in\mathcal{C}_m} J\Phi(c_Q)\lambda_{\mathbb{R}^N}(Q),$$

where c_Q denotes the center of the cube Q. After letting $m \to \infty$ in the preceding, we conclude that $\lambda_{\mathbb{R}^N}(H) \le \nu(H)$ for open $H \subseteq \Phi(G)$; and so, because $\lambda_{\mathbb{R}^N}$ and ν are regular, it follows that $\lambda_{\mathbb{R}^N}(\Gamma) \le \nu(\Gamma)$ for all $\Gamma \in \overline{\mathcal{B}_{\Phi(G)}}^{\lambda_{\mathbb{R}^N}}$.

Starting from the preceding, working first with simple functions, and then passing to monotone limits, one concludes that (5.2.2) holds for all non-negative, measurable functions f on $\big(\Phi(G), \overline{\mathcal{B}_{\Phi(G)}}^{\lambda_{\mathbb{R}^N}}\big)$. □

As an essentially immediate consequence of Theorem 5.2.2, we have the following.

Corollary 5.2.3 *Let G be an open set in $\mathbb{R}^N$ and $\Phi \in C^2\big(G;\mathbb{R}^N\big)$ a diffeomorphism, and set*

$$\mu_\Phi(\Gamma) = \int_\Gamma J\Phi(x)\,dx \quad \text{for} \quad \Gamma \in \overline{\mathcal{B}_G}^{\lambda_{\mathbb{R}^N}}.$$

Then $\Phi_*\mu_\Phi$ coincides with the restriction $\lambda_{\Phi(G)}$ of $\lambda_{\mathbb{R}^N}$ to $\overline{\mathcal{B}_{\Phi(G)}}^{\lambda_{\mathbb{R}^N}}$. In particular,

$$f \in L^1\big(\lambda_{\Phi(G)};\mathbb{R}\big) \iff f\circ\Phi \in L^1\big(\mu_\Phi;\mathbb{R}\big),$$

[4] It is at this point that the argument has to be modified when Φ is only once continuously differentiable. Namely, the estimate that follows must be replaced by one involving the modulus of continuity of Φ's first derivatives.

in which case (5.2.3) *holds.*

As a mnemonic device, it is useful to represent the conclusion of Corollary (5.2.3) as the change of variables statement

$$f(y)\,dy = f \circ \Phi(x)\, J\Phi(x)\, dx \quad \text{when} \quad y = \Phi(x).$$

5.2.2 *Surface Measure*

Aside from its power to facilitate the calculation of various integrals, because it says how Lebesgue measure transforms under changes of coordinates, Theorem 5.2.2 plays an essential role in applications of measure theory to differential geometry. In particular, it allows one to construct an intrinsic, Lebesgue-like measure on a Riemannian manifold, and in this subsection I discuss a special case of this construction, the one for hypersurfaces in $\mathbb{R}^N$. However, because it is not intrinsic, my construction will not be one that would please a differential geometer. Indeed, in order to highlight the analytic meaning of what I am doing, I have chosen to take a concertedly extrinsic approach. Specifically, I will construct surface measure by *differentiating Lebesgue measure.*

To get started, say that $M \subset \mathbb{R}^N$ is a **hypersurface** if, for each $p \in M$, there exist an $r > 0$ and a three times continuously differentiable[5] $F : B_{\mathbb{R}^N}(p, r) \longrightarrow \mathbb{R}$ with the properties that

$$\begin{gathered} B_{\mathbb{R}^N}(p,r) \cap M = \big\{y \in B_{\mathbb{R}^N}(p,r) : F(y) = 0\big\} \\ \text{and } |\nabla F(y)| \neq 0 \text{ for any } y \in B_{\mathbb{R}^N}(p,r), \end{gathered} \tag{5.2.4}$$

where

$$\nabla F(y) \equiv \big(\partial_{y_1} F(y), \dots, \partial_{y_N} F(y)\big) \in \mathbb{R}^N$$

is the **gradient** of F at y. Given $p \in M$, the **tangent space** $\mathbf{T}_p(M)$ to M at p is the set of $\mathbf{v} \in \mathbb{R}^N$ for which there exist an $\epsilon > 0$ and a twice continuously differentiable curve $\gamma : (-\epsilon, \epsilon) \longrightarrow M$ for which $\gamma(0) = p$ and $\dot\gamma(0) = \mathbf{v}$. (If γ is a differentiable curve, I use $\dot\gamma(t)$ to denote its velocity $\frac{d\gamma}{dt}$ at t.)

Canonical Example: The unit sphere $\mathbb{S}^{N-1}$ is a hypersurface in $\mathbb{R}^N$. In fact, at every point $p \in \mathbb{S}^{N-1}$ one can use the function $F(y) = |y|^2 - 1$ and can identify $\mathbf{T}_p(\mathbb{S}^{N-1})$ with the subspace of $\mathbf{v} \in \mathbb{R}^N$ for which $\mathbf{v} - p$ is orthogonal to p.

Lemma 5.2.4 *Every hypersurface M can be written as the countable union of compact sets and is therefore $\mathcal{B}_{\mathbb{R}^N}$-measurable. In addition, for each $p \in M$, $\mathbf{T}_p(M)$ is an $(N-1)$-dimensional subspace of $\mathbb{R}^N$. In fact, if $r > 0$ and*

[5] Because we will be dealing here with balls in different dimensional Euclidean spaces, I will use the notation $B_{\mathbb{R}^N}(a, r)$ to emphasize that the ball is in $\mathbb{R}^N$.

$F \in C^3\big(B_{\mathbb{R}^N}(p,r);\mathbb{R}\big)$ satisfies (5.2.4), then, for every $q \in B_{\mathbb{R}^N}(p,r) \cap M$, $\mathbf{T}_q(M)$ coincides with the space of the vectors $\mathbf{v} \in \mathbb{R}^N$ that are orthogonal to $\nabla F(q)$. Finally, if, for $\Gamma \subseteq M$ and $\rho > 0$,

$$\Gamma(\rho) \equiv \big\{y \in \mathbb{R}^N : \exists p \in \Gamma \ (y-p) \perp \mathbf{T}_p(M) \text{ and } |y-p| < \rho\big\}, \qquad (5.2.5)$$

then $\Gamma(\rho) \in \overline{\mathcal{B}_{\mathbb{R}^N}}^{\lambda_{\mathbb{R}^N}}$ whenever $\Gamma \in \mathcal{B}_M$.

Proof. To see that M is the countable union of compact sets, choose, for each $p \in M$, an $r(p) > 0$ and a function F_p for which (5.2.4) holds. Next, select a countable subset $\{p_n : n \geq 1\}$ from M so that $M \subseteq \bigcup_1^\infty B_{\mathbb{R}^N}\big(p_n, \frac{r_n}{2}\big)$ with $r_n = r(p_n)$. Clearly, for each $n \in \mathbb{Z}^+$, $K_n \equiv \Big\{y \in \overline{B_{\mathbb{R}^N}\big(p_n, \frac{r_n}{2}\big)} : F_{p_n}(y) = 0\Big\}$ is compact; and $M = \bigcup_1^\infty K_n$.

Now let $p \in M$ be given, choose associated r and F for which (5.2.4) holds, and let $q \in B_{\mathbb{R}^N}(p,r) \cap M$. To see that $\nabla F(q) \perp \mathbf{T}_q(M)$, let $\mathbf{v} \in \mathbf{T}_q(M)$ be given and choose $\epsilon > 0$ and γ accordingly. Then, because $\gamma : (-\epsilon,\epsilon) \longrightarrow M$,

$$\big(\nabla F(q), \mathbf{v}\big)_{\mathbb{R}^N} = \frac{d}{dt} F \circ \gamma(t)\big|_{t=0} = 0.$$

Conversely, if $\mathbf{v} \in \mathbb{R}^N$ satisfying $\big(\mathbf{v}, \nabla F(q)\big)_{\mathbb{R}^N} = 0$ is given, set

$$V(x) = \mathbf{v} - \frac{\big(\mathbf{v}, \nabla F(x)\big)_{\mathbb{R}^N}}{|\nabla F(x)|^2} \nabla F(x)$$

for $x \in B_{\mathbb{R}^N}(p,r)$. By the basic existence theory for ordinary differential equations,[6] we can then find an $\epsilon > 0$ and a twice continuously differentiable curve $\gamma : (-\epsilon,\epsilon) \longrightarrow B_{\mathbb{R}^N}(p,r)$ such that $\gamma(0) = q$ and $\dot{\gamma}(t) = V\big(\gamma(t)\big)$ for all $t \in (-\epsilon,\epsilon)$. Clearly $\dot{\gamma}(0) = \mathbf{v}$, and it is an easy matter to check that

$$\frac{d}{dt} F \circ \gamma(t) = \Big(\nabla F\big(\gamma(t)\big), \dot{\gamma}(t)\Big)_{\mathbb{R}^N} = 0 \quad \text{for} \quad t \in (-\epsilon,\epsilon).$$

Hence, $\gamma : (-\epsilon,\epsilon) \longrightarrow M$ and so $\mathbf{v} \in \mathbf{T}_q(M)$.

To prove the final assertion, note that a covering argument (just like the one given at the beginning of this proof) allows us to reduce to the case in which there exist a $p \in M$, an $r > 0$, and an $F \in C^3_{\mathrm{b}}\big(B_{\mathbb{R}^N}(p,r);\mathbb{R}\big)$ such that $M = \big\{x \in B_{\mathbb{R}^N}(p,r) : F(x) = 0\big\}$ and $|\nabla F|$ is uniformly positive. But in that case,

$$\Gamma(\rho) = \left\{x + \xi \frac{\nabla F(x)}{|\nabla F(x)|} : x \in \Gamma \text{ and } |\xi| < \rho\right\},$$

and so the desired measurability follows as an application of Exercise 2.2.1 to the Lipschitz function

---

[6] See Chapter 1 of E. Coddington and N. Levinson's *Theory of Ordinary Differential Equations*, McGraw Hill (1955).

$$(x,\xi) \in B_{\mathbb{R}^N}(p,r) \times \mathbb{R} \longmapsto x + \xi \frac{\nabla F}{|\nabla F|}(x).$$

□

I am at last in a position to say where I am going. Namely, I want to show that there is a unique measure λ_M on $\big(M, \mathcal{B}_{\mathbb{R}^N}[M]\big)$ with the property that (cf. (5.2.5))

$$\lambda_M(\Gamma) = \lim_{\rho \searrow 0} \frac{1}{2\rho} \lambda_{\mathbb{R}^N}\big(\Gamma(\rho)\big) \quad \text{for bounded } \Gamma \in \mathcal{B}_{\mathbb{R}^N} \text{ with } \overline{\Gamma} \subseteq M. \quad (5.2.6)$$

There are several aspects of this definition that should be noticed. In the first place, it is important to recognize that the set $\Gamma(\rho)$, which is sometimes called a **tubular neighborhood** of Γ, can be very different from the ρ-hull $\Gamma^{(\rho)}$. Obviously, $\Gamma(\rho) \subseteq \Gamma^{(\rho)}$. However, it can be much smaller. For example, suppose M is the hypersurface $\{(x,0) : x \in \mathbb{R}\}$ in $\mathbb{R}^2$ and let $\Gamma = \{(n^{-1},0) : n \in \mathbb{Z}^+\}$. If $\rho < n^{-1}$, then $\Gamma^{(\rho)}$ contains the disjoint balls $B_{\mathbb{R}^2}(m^{-1},\rho)$ for $1 \le m \le n$ and therefore

$$\frac{\lambda_{\mathbb{R}^2}\big(\Gamma^{(\rho)}\big)}{2\rho} \ge \frac{\Omega_2 n^2}{2(n+1)^2} \quad \text{if } (n+1)^{-1} \le \rho \le n^{-1}.$$

Hence, $\underline{\lim}_{\rho \searrow 0} \frac{\lambda_{\mathbb{R}^2}(\gamma^{(\rho)})}{2\rho} \ge \frac{\Omega_2}{2}$. On the other hand, $\Gamma(\rho)$ is the union over $n \in \mathbb{Z}^+$ of the line segments $\{n^{-1}\} \times (-\rho,\rho)$, and therefore $\frac{\lambda_{\mathbb{R}^2}(\Gamma(\rho))}{2\rho} = 0$ for all $\rho > 0$. Secondly, aside from the obvious question about whether the limit exists at all, there is a serious question about the additivity of the resulting map $\Gamma \longmapsto \lambda_M(\Gamma)$. Indeed, just because Γ_1 and Γ_2 are disjoint subsets of M, in general it need will not be true that $\Gamma_1(\rho)$ and $\Gamma_2(\rho)$ will be disjoint. For example, when $M = \mathbb{S}^{N-1}$ and $\rho > 1$, $\Gamma_1(\rho)$ and $\Gamma_2(\rho)$ will intersect as soon as both are non-empty. On the other hand, at least in this example, everything will be all right when $\rho \le 1$; and, in fact, we already know from § 5.1.2 that (5.2.6) defines a measure when $M = \mathbb{S}^{N-1}$. To see this latter fact, observe that, when $\rho \in (0,1)$ and $M = \mathbb{S}^{N-1}$,

$$\Gamma(\rho) = \left\{ y : 1-\rho < |y| < 1+\rho \text{ and } \frac{y}{|y|} \in \Gamma \right\},$$

and apply (5.1.5) to see that

$$\frac{\lambda_{\mathbb{R}^N}\big(\Gamma(\rho)\big)}{2\rho} = \lambda_{\mathbb{S}^{N-1}}(\Gamma) \frac{(1+\rho)^N - (1-\rho)^N}{2N\rho},$$

where the measure $\lambda_{\mathbb{S}^{N-1}}$ is the one described in § 5.1.2. Hence, after letting $\rho \searrow 0$, one sees not only that the required limit exists but that it also gives the measure $\lambda_{\mathbb{S}^{N-1}}$. In other words, the program works in the case $M = \mathbb{S}^{N-1}$, and, perhaps less important, the notation used here is consistent with the notation used earlier.

In order to handle the problems raised in the preceding paragraph for general hypersurfaces, I am going to use Theorem 5.2.2 to reduce, at least locally, to the essentially trivial case $M = \mathbb{R}^{N-1} \times \{0\}$. In this case, it is clear that we can identify M with $\mathbb{R}^{N-1}$ and $\Gamma(\rho)$ with $\Gamma \times (-\rho, \rho)$. Hence, even before passing to a limit, we see that in this case

$$\frac{1}{2\rho}\lambda_{\mathbb{R}^N}\big(\Gamma(\rho)\big) = \lambda_{\mathbb{R}^{N-1}}(\Gamma).$$

In the lemmas that follow, we will develop the requisite machinery with which to make the reduction.

Lemma 5.2.5 *For each $p \in M$ there are an open neighborhood U of the origin 0 in $\mathbb{R}^{N-1}$ and a three times continuously differentiable injection (i.e., one-to-one) $\Psi : U \longrightarrow M$ with the properties that $p = \Psi(0)$ and, for each $u \in U$, the set $\{\partial_{u_1}\Psi(u), \dots, \partial_{u_{N-1}}\Psi(u)\}$ forms a basis in $\mathbf{T}_{\Psi(u)}(M)$.*

Proof. Choose r and F for which (5.2.4) holds. After renumbering the coordinates if necessary, we may and will assume that $\partial_{x_N} F(p) \neq 0$. Now consider the map

$$y \in B_{\mathbb{R}^N}(p,r) \longmapsto \Phi(y) \equiv \begin{pmatrix} y_1 - p_1 \\ \vdots \\ y_{N-1} - p_{N-1} \\ F(y) \end{pmatrix} \in \mathbb{R}^N.$$

Clearly, Φ is three times continuously differentiable. In addition,

$$\frac{\partial \Phi}{\partial x}(p) = \begin{pmatrix} \mathbf{I}_{\mathbb{R}^{N-1}} & 0 \\ \mathbf{v} & \partial_{x_N} F(p) \end{pmatrix},$$

where $\mathbf{v} = \big(\partial_{x_1}F(p), \dots, \partial_{x_{N-1}}F(p)\big)$. In particular, $J\Phi(p) \neq 0$, and so the Inverse Function Theorem guarantees the existence of a $\rho \in (0, r]$ such that $\Phi \restriction B_{\mathbb{R}^N}(p,\rho)$ is diffeomorphic and Φ^{-1} has three continuous derivatives on the open set $W \equiv \Phi\big(B_{\mathbb{R}^N}(p,\rho)\big)$. Thus, if $U \equiv \{u \in \mathbb{R}^{N-1} : (u,0) \in W\}$, then U is an open neighborhood of the origin in $\mathbb{R}^{N-1}$, and

$$u \in U \longmapsto \Psi(u) \equiv \Phi^{-1}(u,0) \in M$$

is one-to-one and has three continuous derivatives. Finally, it is obvious that $\Psi(0) = p$, and, because Ψ takes its values in M

$$\partial_{u_j}\Psi(u) = \frac{d}{dt}\Psi(u + t\mathbf{e}_j)\big|_{t=0} \in \mathbf{T}_{\Psi(u)}(M) \quad \text{for each } 1 \le j \le N-1.$$

At the same time, as the first $(N-1)$ columns in the non-degenerate Jacobian matrix of Φ^{-1} at $(u,0)$, the vectors $\partial_{u_j}\Psi(u)$ must be linearly independent and therefore form a basis in $\mathbf{T}_{\Psi(u)}(M)$. □

Given a non-empty, connected, open set U in $\mathbb{R}^{N-1}$ and a three times continuously differentiable injection $\Psi : U \longrightarrow M$ with the property that

$$\{\partial_{u_1}\Psi(u), \dots, \partial_{u_{N-1}}\Psi(u)\} \text{ forms a basis in } \mathbf{T}_{\Psi(u)}(M)$$

for every $u \in U$, the pair (Ψ, U) is said to be a **coordinate chart** for M.

Lemma 5.2.6 *Suppose that (Ψ, U) is a coordinate chart for M, and define*

$$J\Psi = \left[\det\Big(\Big(\big(\partial_{u_i}\Psi, \partial_{u_j}\Psi\big)_{\mathbb{R}^N} \Big)\Big)_{1\le i,j\le N-1}\right]^{\frac{1}{2}}. \tag{5.2.7}$$

Then $J\Psi$ never vanishes, and there exists a unique twice continuously differentiable $\mathbf{n} : U \longrightarrow \mathbb{S}^{N-1}$ with the properties that $\mathbf{n}(u) \perp \mathbf{T}_{\Psi(u)}(M)$ and[7]

$$\det\Big(\partial_{u_1}\Psi(u), \dots, \partial_{u_{N-1}}\Psi(u), \mathbf{n}(u)^\top\Big) = J\Psi(u)$$

for every $u \in U$. Finally, define

$$\tilde{\Psi}(u, \xi) = \Psi(u) + \xi\mathbf{n}(u)^\top \quad \textit{for } u \in U.$$

Then

$$J\tilde{\Psi}(u, 0) = J\Psi(u) \quad \textit{for } u \in U \textit{ and } \xi \in \mathbb{R}, \tag{5.2.8}$$

and there exists an open set $\tilde{U}$ in $\mathbb{R}^N$ such that

$$U = \{u \in \mathbb{R}^{N-1} : (u,0) \in \tilde{U}\},$$

$\Psi(U) = \tilde{\Psi}(\tilde{U}) \cap M$, and $\tilde{\Psi} \restriction \tilde{U}$ is a diffeomorphism. In particular, if x and y are distinct elements of $\Psi(U)$, then $\{x\}(\rho) \cap \tilde{\Psi}(\tilde{U})$ is disjoint from $\{y\}(\rho) \cap \tilde{\Psi}(\tilde{U})$ for all $\rho > 0$.

Proof. Given a $u \in U$ and an $\mathbf{n} \in \mathbb{S}^{N-1}$ that is orthogonal to $\mathbf{T}_{\Psi(u)}(M)$, $\{\partial_{u_1}\Psi(u), \dots, \partial_{u_{N-1}}\Psi(u), \mathbf{n}^\top\}$ is a basis for $\mathbb{R}^N$, and therefore

$$\det\Big(\partial_{u_1}\Psi(u), \dots, \partial_{u_{N-1}}\Psi(u), \mathbf{n}^\top\Big) \ne 0.$$

Hence, for each $u \in U$ there is precisely one $\mathbf{n}(u) \in \mathbb{S}^{N-1} \cap \mathbf{T}_{\Psi(u)}(M)^\perp$ with the property that

$$\det\Big(\partial_{u_1}\Psi(u), \dots, \partial_{u_{N-1}}\Psi(u), \mathbf{n}(u)^\top\Big) > 0.$$

[7] Thinking of a vector $\mathbf{v}$ as a rectangular matrix, I use $\mathbf{v}^\top$ to denote the column (row) vector corresponding to a row (column) vector $\mathbf{v}$.

To see that $u \in U \longmapsto \mathbf{n}(u) \in \mathbb{S}^{N-1}$ is twice continuously differentiable, set $p = \Psi(u)$ and choose r and F for p so that (5.2.4) holds. Then $\mathbf{n}(u) = \pm\frac{\nabla F(p)}{|\nabla F(p)|}$, and so, by continuity, we know that, with the same sign throughout,

$$\mathbf{n}(w) = \pm\frac{\nabla F(\Psi(w))}{\big|\nabla F(\Psi(w))\big|}$$

for every w in a neighborhood of u.

Turning to the function $\tilde{\Psi}$, note that

$$\begin{aligned} J\tilde{\Psi}(u,0)^2 &= \Big(\det\Big(\partial_{u_1}\Psi(u),\dots,\partial_{u_{N-1}}\Psi(u),\mathbf{n}(u)^\top\Big)\Big)^2 \\ &= \det\begin{pmatrix} \partial_{u_1}\Psi(u)^\top \\ \vdots \\ \partial_{u_{N-1}}\Psi(u)^\top \\ \mathbf{n}(u) \end{pmatrix}\Big(\partial_{u_1}\Psi(u),\dots,\partial_{u_{N-1}}\Psi(u),\mathbf{n}(u)^\top\Big) \\ &= \det\begin{pmatrix} \Big(\Big(\big(\partial_{u_i}\Psi(u),\partial_{u_j}\Psi(u)\big)_{\mathbb{R}^N}\Big)\Big) & \mathbf{0} \\ \mathbf{0} & 1 \end{pmatrix} = J\Psi(u)^2. \end{aligned}$$

Hence, (5.2.8) is proved. In particular, by the Inverse Function Theorem, this means that, for each $u \in U$, there is a neighborhood of $(u,0)$ in $\mathbb{R}^N$ on which $\tilde{\Psi}$ is a diffeomorphism. In fact, given $u \in U$, choose r and F as in (5.2.4) for $p = \Psi(u)$, and take $\rho > 0$ such that

$$B_{\mathbb{R}^{N-1}}(u,2\rho) \subseteq U \quad \text{and} \quad \Psi\big(B_{\mathbb{R}^{N-1}}(u,2\rho)\big) \subseteq B_{\mathbb{R}^N}\left(p,\tfrac{r}{2}\right).$$

Then, because $\mathbf{n}(w) = \pm\frac{\nabla F(\Psi(w))}{|\nabla F(\Psi(w))|}$ with the same sign for all $w \in B_{\mathbb{R}^{N-1}}(u,\rho)$,

$$\begin{aligned} F\big(\tilde{\Psi}(w,\xi)\big) &= \pm\,\xi\big|\nabla F(\Psi(w))\big| + E(w,\xi) \\ &\qquad \text{for } (w,\xi) \in B_{\mathbb{R}^{N-1}}(u,\rho) \times \left(-\tfrac{r}{2},\tfrac{r}{2}\right), \end{aligned}$$

where $|E(w,t)| \le C\xi^2$ for some $C \in (0,\infty)$. Hence, by readjusting the choice of $\rho > 0$, we can guarantee that $\tilde{\Psi} \restriction B_{\mathbb{R}^{N-1}}(u,\rho) \times (-\rho,\rho)$ is diffeomorphic and also satisfies

$$\tilde{\Psi}\Big(B_{\mathbb{R}^{N-1}}(u,\rho) \times (-\rho,\rho)\Big) \cap M = \Psi\big(B_{\mathbb{R}^{N-1}}(u,\rho)\big).$$

In order to prove the final assertion, we must find an open $\tilde{U}$ in $\mathbb{R}^N$ for which $\Psi(U) = \tilde{\Psi}(\tilde{U}) \cap M$ and $\tilde{\Psi} \restriction \tilde{U}$ is a diffeomorphism. To this end, for each $u \in U$, use the preceding to choose $\rho(u) > 0$ such that: $B_{\mathbb{R}^{N-1}}\big(u,\rho(u)\big) \subseteq U$ and $\tilde{\Psi} \restriction B_{\mathbb{R}^{N-1}}\big(u,\rho(u)\big) \times \big(-\rho(u),\rho(u)\big)$ is both a diffeomorphism and satisfies

$$\tilde{\Psi}\Big(B_{\mathbb{R}^{N-1}}(u,\rho(u))\times(-\rho(u),\rho(u))\Big)\cap M=\Psi\Big(B_{\mathbb{R}^{N-1}}(u,\rho(u))\Big).$$

Next, choose a countable set $\{u_n : n\ge 1\}\subseteq U$ so that

$$U=\bigcup_1^\infty B_{\mathbb{R}^{N-1}}\big(u_n,\tfrac{\rho_n}{3}\big),\ \text{where}\ \rho_n=\rho(u_n);$$

and set

$$U_n=\bigcup_1^n B_{\mathbb{R}^{N-1}}\big(u_m,\tfrac{\rho_m}{3}\big)\quad\text{and}\quad R_n=\rho_1\wedge\cdots\wedge\rho_n.$$

To construct the open set $\tilde{U}$, proceed by induction as follows. Set $\epsilon_1=\frac{\rho_1}{3}$ and $\tilde{K}_1=\overline{U_1}\times[-\epsilon_1,\epsilon_1]$. Given $\tilde{K}_n$ and ϵ_n, define ϵ_{n+1} by

$$2\epsilon_{n+1}=R_{n+1}\wedge\epsilon_n\wedge\Bigg(\inf\Big\{\big|\Psi(u)-\tilde{\Psi}(w,t)\big| : u\in\overline{U_{n+1}}\setminus U_n,\ (w,t)\in\tilde{K}_n,\ \text{and}\ |u-w|\ge\tfrac{R_{n+1}}{3}\Big\}\Bigg),$$

and take

$$\tilde{K}_{n+1}=\tilde{K}_n\cup\Big((\overline{U_{n+1}}\setminus U_n)\times\big[-\epsilon_{n+1},\epsilon_{n+1}\big]\Big).$$

Clearly, for each $n\in\mathbb{Z}^+$, $\tilde{K}_n$ is compact, $\tilde{K}_n\subseteq\tilde{K}_{n+1}$, $U_n\times(-\epsilon_n,\epsilon_n)\subseteq\tilde{K}_n$, and $J\tilde{\Psi}$ never vanishes on $\tilde{K}_n$. Thus, if we can show that $\epsilon_n>0$ and that $\tilde{\Psi}\restriction\tilde{K}_n$ is one-to-one for each $n\in\mathbb{Z}^+$, then we can take $\tilde{U}=\bigcup_{n=1}^\infty U_n\times(-\epsilon_n,\epsilon_n)$. With this in mind, first observe that there is nothing to do when $n=1$. Furthermore, if $\epsilon_n>0$ and $\tilde{\Psi}\restriction\tilde{K}_n$ is one-to-one, then $\epsilon_{n+1}=0$ is possible only if there exist a $u\in B_{\mathbb{R}^{N-1}}(u_{n+1},\rho_{n+1})$ and a $(w,\xi)\in B_{\mathbb{R}^{N-1}}(u_m,\rho_m)\times(-\rho_m,\rho_m)$ for some $1\le m\le n$ for which $\Psi(u)=\tilde{\Psi}(w,\xi)$ and $|u-w|\ge\frac{R_{n+1}}{3}$. But, because

$$(w,\xi)\in B_{\mathbb{R}^{N-1}}(u_m,\rho_m)\times(-\rho_m,\rho_m)\ \text{and}\ \tilde{\Psi}(w,\xi)=\Psi(u)\in M\implies\xi=0,$$

this would mean that $\Psi(w)=\Psi(u)$ and therefore, since Ψ is one-to-one, it would lead to the contradiction that $0=|w-u|\ge\frac{R_{n+1}}{3}$. Hence, $\epsilon_{n+1}>0$. Finally, to see that $\tilde{\Psi}\restriction\tilde{K}_{n+1}$ is one-to-one, we need only check that

$$(u,\xi)\in(\overline{U_{n+1}}\setminus U_n)\times[-\epsilon_{n+1},\epsilon_{n+1}]\ \text{and}\ (w,\eta)\in\tilde{K}_n\implies\tilde{\Psi}(u,\xi)\ne\tilde{\Psi}(w,\eta).$$

But, if $|u-w|<\frac{R_{n+1}}{3}$, then both (u,ξ) and (w,η) are in $B_{\mathbb{R}^{N-1}}(u_{n+1},\rho_{n+1})\times(-\rho_{n+1},\rho_{n+1})$ and $\tilde{\Psi}$ is one to one there. On the other hand, if $|u-w|\ge\frac{R_{n+1}}{3}$, then

$$\big|\tilde{\Psi}(u,\xi)-\tilde{\Psi}(w,\eta)\big|\ge\big|\Psi(u)-\tilde{\Psi}(w,\eta)\big|-|\xi|\ge 2\epsilon_{n+1}-\epsilon_{n+1}=\epsilon_{n+1}>0.$$

□

Lemma 5.2.7 *If (Ψ, U) is a coordinate chart for M and Γ a bounded element of $\mathcal{B}_{\mathbb{R}^N}$ with $\overline{\Gamma} \subseteq \Psi(U)$, then (cf. (5.2.7))*

$$\lim_{\rho \searrow 0} \frac{1}{2\rho} \overline{\lambda_{\mathbb{R}^N}}\big(\Gamma(\rho)\big) = \int_{\Psi^{-1}(\Gamma)} J\Psi(u)\, \lambda_{\mathbb{R}^{N-1}}(du). \tag{5.2.9}$$

Proof. Choose $\tilde{U}$ and $\tilde{\Psi}$ as in Lemma 5.2.6. Since $\overline{\Gamma}$ is a compact subset of the open set $G = \tilde{\Psi}(\tilde{U})$, $\epsilon = \operatorname{dist}\big(\Gamma, G^{\complement}\big) > 0$. Hence, for $\rho \in (0, \epsilon)$,

$$\Gamma(\rho) = \tilde{\Psi}\Big(\Psi^{-1}(\Gamma) \times (-\rho, \rho)\Big) \in \mathcal{B}_{\mathbb{R}^N},$$

which, by (5.2.3), (5.2.8), and Tonelli's Theorem, means that $\frac{\lambda_{\mathbb{R}^N}(\Gamma(\rho))}{2\rho}$ equals

$$\int_{\Psi^{-1}(\Gamma)} \left(\frac{1}{2\rho} \int_{(-\rho,\rho)} J\tilde{\Psi}(u, \xi)\, d\xi \right) \lambda_{\mathbb{R}^{N-1}}(du) \longrightarrow \int_{\Psi^{-1}(\Gamma)} J\Psi(u) \lambda_{\mathbb{R}^{N-1}}(du),$$

as $\rho \to 0$. □

Theorem 5.2.8 *Let M be a hypersurface in $\mathbb{R}^N$. Then there exists a unique measure λ_M on $\big(M, \mathcal{B}_M\big)$ for which (5.2.6) holds. In fact, $\lambda_M(K) < \infty$ for every compact subset of M, and so λ_M is σ-finite. Finally, if (Ψ, U) is a coordinate chart for M and f is a non-negative, $\mathcal{B}_M$-measurable function, then (cf. (5.2.7))*

$$\int_{\Psi(U)} f(x)\, \lambda_M(dx) = \int_U f \circ \Psi(u) J\Psi(u)\, \lambda_{\mathbb{R}^{N-1}}(du). \tag{5.2.10}$$

Proof. For each $p \in M$, use Lemma 5.2.5 to produce an $r(p) > 0$ and a coordinate chart $\big(\Psi_p, U_p\big)$ for M for which $B_{\mathbb{R}^N}\big(p, 3r(p)\big)$ is contained in (cf. Lemma 5.2.6) $\tilde{\Psi}(\tilde{U}_p)$. Next, select a countable set $\{p_n : n \geq 1\} \subseteq M$ such that

$$M \subseteq \bigcup_{n=1}^{\infty} B_{\mathbb{R}^N}(p_n, r_n), \quad \text{where } r_n \equiv r(p_n),$$

set $M_1 = B_{\mathbb{R}^N}(p_1, r_1) \cap M$, and

$$M_n = \big(B_{\mathbb{R}^N}(p_n, r_n) \cap M\big) \setminus \bigcup_{m=1}^{n-1} M_m \quad \text{for } n \geq 2.$$

Finally, for each $n \in \mathbb{Z}^+$, define the finite measure μ_n on $\big(M, \mathcal{B}_M\big)$ by

$$\mu_n(\Gamma) = \int_{\Psi_n^{-1}(\Gamma\cap M_n)} J\Psi_n(u)\,\lambda_{\mathbb{R}^{N-1}}(du), \quad \text{where } \Psi_n \equiv \Psi_{p_n},$$

and set

$$\lambda_M = \sum_1^\infty \mu_n. \tag{$*$}$$

Given a compact $K \subseteq M$, choose an $n \in \mathbb{Z}^+$ for which

$$K \subseteq \bigcup_1^n B_{\mathbb{R}^N}\big(p_m, \tfrac{r_m}{2}\big),$$

and set $r = r_1 \wedge \cdots \wedge r_n$ and $\epsilon = \frac{r}{3}$. It is then an easy matter to check that, for any pair of distinct elements x and y from K, either $|x-y| \geq r$, in which case it is obvious that $\{x\}(\epsilon) \cap \{y\}(\epsilon) = \emptyset$, or $|x-y| < r$, in which case both $\{x\}(\epsilon)$ and $\{y\}(\epsilon)$ lie in $\tilde{\Psi}_m(\tilde{U}_{p_m})$ for some $1 \leq m \leq n$ and, therefore, the last part of Lemma 5.2.6 applies and says that $\{x\}(\epsilon) \cap \{y\}(\epsilon) = \emptyset$. Thus, if $\Gamma \in \mathcal{B}_M$ is a subset of K and $\Gamma_m = \Gamma \cap M_m$, then, for each $\rho \in (0,\epsilon)$, $\{\Gamma_1(\rho), \dots, \Gamma_n(\rho)\}$ is an exact cover of $\Gamma(\rho)$ by mutually disjoint measurable sets, and so, for each $0 < \rho < \epsilon$,

$$\lambda_{\mathbb{R}^N}\big(\Gamma(\rho)\big) = \sum_1^n \lambda_{\mathbb{R}^N}\big(\Gamma_m(\rho)\big).$$

At the same time, by Lemma 5.2.7,

$$\frac{1}{2\rho}\lambda_{\mathbb{R}^N}\big(\Gamma_m(\rho)\big) \longrightarrow \mu_m(\Gamma_m) \quad \text{for each } 1 \leq m \leq n.$$

In particular, we have now proved that the measure λ_M defined in ($*$) satisfies (5.2.6) and that it is finite on compacts. Moreover, since (cf. Lemma 5.2.4) M is a countable union of compacts, it is clear that there can be only one measure satisfying (5.2.6).

Finally, if (Ψ, U) is a coordinate chart for M and $\Gamma \in \mathcal{B}_M$ with $\overline{\Gamma} \subset\subset \Psi(U)$, then (5.2.10) with $f = \mathbf{1}_\Gamma$ is an immediate consequence of (5.2.9). Hence, (5.2.10) follows in general by taking linear combinations and monotone limits. □

The measure λ_M produced in Theorem 5.2.8 is called the **surface measure** on M.

5.2.3 Exercises for § 5.2

Exercise 5.2.1 In part (**ii**) in Exercise 5.1.4, I tacitly accepted the equality of π, the volume Ω_2 of the unit ball $B_{\mathbb{R}^2}(0,1)$ in $\mathbb{R}^2$, with π, the half-period of the sine and cosine functions. We are now in a position to justify this identification. To this end, define $\Phi : G \equiv (0,1) \times (0,2\pi) \longrightarrow \mathbb{R}^2$ (the π

here is the half-period of sin and cos) by $\Phi(r,\theta) = (r\cos\theta, r\sin\theta)^\top$. Note that $\Phi(G) \subseteq B_{\mathbb{R}^2}(0,1)$ and $B_{\mathbb{R}^2}(\mathbf{0},1) \setminus G \subseteq \{(x_1,x_2) : x_2 = 0\}$ and therefore that $\lambda_{\mathbb{R}^2}(\Phi(G)) = \Omega_2$. Now use Jacobi's Transformation Formula to compute $\lambda_{\mathbb{R}^2}(\Phi(G))$.

Exercise 5.2.2 Let M be a hypersurface in $\mathbb{R}^N$. Show that, for each $p \in M$, the tangent space $\mathbf{T}_p(M)$ coincides with the set of $\mathbf{v} \in \mathbb{R}^N$ such that

$$\varlimsup_{\xi \to 0} \frac{\operatorname{dist}(p + \xi\mathbf{v}, M)}{\xi^2} < \infty.$$

Hint: Given $\mathbf{v} \in \mathbf{T}_p(M)$, choose a twice continuously differentiable associated curve γ, and consider $\xi \rightsquigarrow |p + \xi\mathbf{v} - \gamma(\xi)|$.

Exercise 5.2.3 In this exercise I introduce a function that is intimately related to Euler's gamma function Γ introduced in part (**ii**) of Exercise 5.1.3.

(**i**) For $(\alpha,\beta) \in (0,\infty)^2$, define

$$B(\alpha,\beta) = \int_{(0,1)} u^{\alpha-1}(1-u)^{\beta-1}\,du.$$

Show that

$$B(\alpha,\beta) = \frac{\Gamma(\alpha)\Gamma(\beta)}{\Gamma(\alpha+\beta)}.$$

(See part (**iv**) of Exercise 6.3.3 for another derivation.) The function B is called the **Euler's beta function**. Clearly it provides an extension of the binomial coefficients in the sense that

$$\frac{1}{(m+n+1)B(m+1,n+1)} = \binom{m+n}{m}$$

for all non-negative integers m and n.

Hint: Think of $\Gamma(\alpha)\,\Gamma(\beta)$ as an integral in (s,t) over $(0,\infty)^2$, and consider the map

$$(u,v) \in (0,\infty) \times (0,1) \longmapsto \begin{pmatrix} uv \\ u(1-v) \end{pmatrix} \in (0,\infty)^2.$$

(**ii**) For $\lambda > \frac{N}{2}$ show that

$$\int_{\mathbb{R}^N} \frac{1}{(1+|x|^2)^\lambda}\,dx = \frac{\omega_{N-1}}{2} B\left(\frac{N}{2}, \lambda - \frac{N}{2}\right) = \frac{\pi^{\frac{N}{2}}\Gamma(\lambda - \frac{N}{2})}{\Gamma(\lambda)},$$

where (cf. part (**ii**) of Exercise 5.1.3) ω_{N-1} is the surface area of $\mathbb{S}^{N-1}$. In particular, conclude that

$$\int_{\mathbb{R}^N} \frac{1}{(1+|x|^2)^{\frac{N+1}{2}}}\,dx = \frac{\omega_N}{2}.$$

Hint: Use polar coordinates and then try the change of variable $\Phi(r) = \frac{r^2}{1+r^2}$.

(**iii**) For $\lambda \in (0,\infty)$, show that

$$\int_{(-1,1)} \left(1-\xi^2\right)^{\lambda-1} d\xi = \frac{\pi^{\frac{1}{2}}\Gamma(\lambda)}{\Gamma\left(\lambda+\frac{1}{2}\right)};$$

and conclude that, for any $N \in \mathbb{Z}^+$,

$$\int_{(-1,1)} \left(1-\xi^2\right)^{\frac{N}{2}-1} d\xi = \frac{\omega_N}{\omega_{N-1}}.$$

Exercise 5.2.4 In this exercise, you are to compute the derivative Γ' of the gamma function at positive integers. Clearly,

$$\Gamma'(s+1) = \int_{[0,\infty)} e^{-t}t^s \log t \, dt \quad \text{for } s > 0.$$

(**i**) Using integration by parts, show that

$$\frac{\Gamma'(n+1)}{n!} = \frac{\Gamma'(n)}{(n-1)!} + \frac{1}{n} \quad \text{for } n \in \mathbb{Z}^+,$$

and conclude that

$$\Gamma'(n+1) = n!\big(\Gamma'(1) + S_n\big) \text{ where } S_n = \sum_{m=1}^{n} \frac{1}{m}.$$

(**ii**) Using the equation $\Gamma(n+1+s) = \frac{\Gamma(n)\Gamma(1+s)}{B(n,1+s)}$, show that

$$\frac{\Gamma'(n+1)}{n!} = \Gamma'(1) - n\int_{[0,1]} (1-u)^{n-1} \log u \, du,$$

and conclude that

$$-n\int_{[0,1]} (1-u)^{n-1} \log u \, du = S_n.$$

(**iii**) From (**ii**), show that it

$$-\int_{[0,n]} (1-t)^{n-1} \log t \, dt = S_n - \log n,$$

and use Lebesgue's dominated convergence theorem to show that

$$\lim_{n\to\infty} \int_{[0,n]} (1-t)^{n-1} \log t \, dt = \Gamma'(1).$$

Combining this with the preceding, conclude that

$$\Gamma'(n+1) = n!\big(-\gamma + S_n\big),$$

where (cf. part (**iii**) of Exercise 1.2.2) $\gamma = \lim_{n\to\infty}(S_n - \log n)$ is Euler's constant.

Exercise 5.2.5 For $N \in \mathbb{Z}^+$, define $\Xi : (-1,1) \times \mathbb{S}^{N-1} \longrightarrow \mathbb{S}^N$ by

$$\Xi(\rho,\omega) = \begin{pmatrix} (1-\rho^2)^{\frac{1}{2}}\omega \\ \rho \end{pmatrix},$$

and set

$$\mu_N(\Gamma) = \int_\Gamma (1-\rho^2)^{\frac{N}{2}-1}\,(\lambda_{(-1,1)} \times \lambda_{\mathbb{S}^{N-1}})(d\rho \times d\omega)$$

for $\Gamma \in \mathcal{B}_{(-1,1)} \times \mathcal{B}_{\mathbb{S}^{N-1}}$. The goal of this exercise is to show that $\lambda_{\mathbb{S}^N} = \Xi_*\mu_N$.

(**i**) Show that for $F \in C_c(\mathbb{R}^{N+1};\mathbb{R})$,

$$\int_{(0,\infty)} r^N \left(\iint\limits_{(-1,1)\times\mathbb{S}^{N-1}} F\big(r\,\Xi(\rho,\omega)\big)\mu_N(d\rho\times d\omega)\right) dr$$
$$= \int\limits_{\mathbb{R}^N\times(-1,1)} |x|(1-y^2)^{\frac{N}{2}-1} F\big((1-y^2)^{\frac{1}{2}}x, |x|y\big)\,\lambda_{\mathbb{R}^{N+1}}(dx\times dy),$$

where elements of $\mathbb{R}^{N+1}$ are written as $(x,y) \in \mathbb{R}^N \times \mathbb{R}$.

(**ii**) Define $\Phi : (\mathbb{R}^N\setminus\{\mathbf{0}\}) \times (-1,1) \longrightarrow \mathbb{R}^{N+1}\setminus\{(\mathbf{0},0)\}$ by

$$\Phi(x,y) = \begin{pmatrix} (1-y^2)^{\frac{1}{2}}x \\ |x|y \end{pmatrix},$$

and show that Φ is a diffeomorphism and that $J\Phi(x,y) = |x|(1-y^2)^{\frac{N}{2}-1}$. By combining this with (**i**), conclude that

$$\int_{(0,\infty)} r^N \left(\iint\limits_{(-1,1)\times\mathbb{S}^{N-1}} F\big(r\,\Xi(\rho,\omega)\big)\mu_N(d\rho\times d\omega)\right) dr = \int_{\mathbb{R}^{N+1}\setminus\{0\}} F\,d\lambda_{\mathbb{R}^{N+1}}.$$

(**iii**) By applying (**ii**) to functions F of the form $F(z) = \varphi(|z|)\psi\big(\frac{z}{|z|}\big)$, where $\varphi \in C_c\big((0,\infty);\mathbb{R}\big)$ and $\psi \in C(\mathbb{S}^N;\mathbb{R})$, complete the proof that $\lambda_{\mathbb{S}^N} = \Xi_*\mu_N$.

(**iv**) Let $f \in C_b\big((-1,1);\mathbb{R}\big)$ and $\theta \in \mathbb{S}^N$, and show that

$$\int_{\mathbb{S}^N} f\big((\theta,\omega)_{\mathbb{R}^{N+1}}\big)\,\lambda_{\mathbb{S}^N}(d\omega) = \omega_{N-1}\int_{(-1,1)} (1-\rho^2)^{\frac{N}{2}-1} f(\rho)\,d\rho.$$

Hint: Reduce to the case in which θ is the $(N+1)$st unit coordinate vector.

Exercise 5.2.6 Show that if M is a hypersurface, then $\lambda_M(\{x\}) = 0$ for all $x \in M$ and that $\lambda_M(\Gamma) > 0$ for all non-empty, open $\Gamma \subseteq M$.

Exercise 5.2.7 Given $r > 0$, set $\mathbb{S}^{N-1}(r) = \big\{x \in \mathbb{R}^N : |x| = r\big\}$, and observe that $\mathbb{S}^{N-1}(r)$ is a hypersurface. Next, define $\Phi_r : \mathbb{S}^{N-1} \longrightarrow \mathbb{S}^{N-1}(r)$ by $\Phi_r(\omega) = r\omega$; and show that $\lambda_{\mathbb{S}^{N-1}(r)} = r^{N-1}(\Phi_r)_*\lambda_{\mathbb{S}^{N-1}}$.

Exercise 5.2.8 The purpose of this exercise is to examine some properties of the measure $\lambda_{\mathbb{S}^{N-1}}$ for large N.

(i) To begin, use Exercise 5.2.5 to see that, for $N \geq 2$ and bounded, $\mathcal{B}_{\mathbb{R}}$-measurable $f : \mathbb{R} \longrightarrow \mathbb{R}$,

$$\fint_{\mathbb{S}^{N-1}(\sqrt{N})} f(\omega_1)\,\lambda_{\mathbb{S}^{N-1}(\sqrt{N})}(d\omega) = \frac{\omega_{N-2}}{\omega_{N-1}\sqrt{N}}\int_{(-\sqrt{N},\sqrt{N})} f(\rho)\big(1-\tfrac{\rho^2}{N}\big)^{\frac{N-3}{2}}\,d\rho,$$

where $\fint$ is used to denote averaging: the ratio of the integral over a set to the measure of that set.

(ii) Starting from **(i)** and using the computation in **(i)** of Exercise 5.1.3, show that

$$\lim_{N\to\infty}\frac{\sqrt{N}\,\omega_{N-1}}{\omega_{N-2}} = \sqrt{2\pi}.$$

(iii) By combining **(i)** and **(ii)**, show that

$$\lim_{N\to\infty}\fint_{\mathbb{S}^{N-1}(\sqrt{N})} f(\omega_1)\,\lambda_{\mathbb{S}^{N-1}(\sqrt{N})}(d\omega) = \frac{1}{\sqrt{2\pi}}\int_{\mathbb{R}} f(x)e^{-\frac{x^2}{2}}\,dx$$

for all bounded, $\mathcal{B}_{\mathbb{R}}$-measurable f's.

Exercise 5.2.9 Let a non-empty, open subset U of $\mathbb{R}^{N-1}$ and $f \in C^3(U;\mathbb{R})$ be given, and take

$$M = \big\{(u, f(u)) : u \in U\big\} \quad\text{and}\quad \Psi(u) = \begin{pmatrix} u \\ f(u)\end{pmatrix}, \quad u \in U.$$

That is, M is the *graph* of f.

(i) Check that M is a hypersurface and that (Ψ, U) is a coordinate chart for M that is **global** in the sense that $M = \Psi(U)$.

(ii) Show that

$$\left(\left((\partial_{u_i}\Psi, \partial_{u_j}\Psi)_{\mathbb{R}^N}\right)\right)_{1\le i,j\le N-1} = \mathbf{I}_{\mathbb{R}^{N-1}} + (\nabla f)^\top \nabla f,$$

and conclude that $J\Psi = \sqrt{1+|\nabla f|^2}$.

Hint: Given a non-zero row vector $\mathbf{v}$, set $A = \mathbf{I}+\mathbf{v}^\top\mathbf{v}$ and $\mathbf{e} = \frac{\mathbf{v}^\top}{|\mathbf{v}|}$. Note that $A\mathbf{e} = \big(1+|\mathbf{v}|^2\big)\mathbf{e}$ and that $A\mathbf{w} = \mathbf{w}$ for $\mathbf{w} \perp \mathbf{e}$. Thus, A has one eigenvalue equal to $1+|\mathbf{v}|^2$ and all its other eigenvalues equal 1.

(**iii**) From the preceding, arrive at

$$\int_M \varphi\, d\lambda_M = \int_U \varphi\big(u, f(u)\big)\sqrt{1+|\nabla f(u)|^2}\, du, \quad \varphi \in C_c(M;\mathbb{R}),$$

a formula that should be familiar from elementary calculus.

Exercise 5.2.10 Let G be a non-empty, open set in $\mathbb{R}^N$, $F \in C^3\big(G;\mathbb{R}\big)$, and assume that $M \equiv \{x \in G : F(x) = 0\}$ is a non-empty, connected set on which $\partial_{x_N} F$ never vanishes.

(**i**) Show that there is a connected, open neighborhood $H \subseteq G$ of M on which $\partial_{x_N} F$ never vanishes. Next, define $\Phi : H \longrightarrow \mathbb{R}^N$ by

$$\Phi(x) = \begin{pmatrix} x_1 \\ \vdots \\ x_{N-1} \\ F(x) \end{pmatrix}.$$

Show that Φ is a diffeomorphism, set $f(u) = \Phi^{-1}(u,0)_N$ for

$$u \in U \equiv \{u \in \mathbb{R}^{N-1} : (u,\xi) \in M \text{ for some } \xi \in \mathbb{R}\},$$

and conclude that $f \in C^3(U;\mathbb{R})$ and that $F\big(u, f(u)\big) = 0$. In particular, M is the graph of f.

(**ii**) Conclude that, for any non-negative, $\mathcal{B}_M$-measurable φ,

$$\int_M \varphi\, d\lambda_M = \int_U \frac{\varphi|\nabla F|}{|\partial_{x_N} F|}\big(u, f(u)\big)\, du.$$

Exercise 5.2.11 Again let G be a non-empty, open set in $\mathbb{R}^N$ and $F \in C^3(G;\mathbb{R})$, but this time assume that $\partial_{x_N} F$ vanishes nowhere on G, not just on $\{F = 0\}$. Set $\Xi = \text{Range}(F)$ and, for $\xi \in \Xi$, $M_\xi = \{x \in G : F(x) = \xi\}$ and $U_\xi = \{u \in \mathbb{R}^{N-1} : (u,\xi) \in M_\xi\}$.

(**i**) Define Φ on G as in part (**i**) of Exercise 5.2.10, and show that Φ is a diffeomorphism. As an application of Exercise 5.2.10, show that, for each $\xi \in \Xi$, M_ξ is a hypersurface and that, for each $\varphi \in C_c(G;\mathbb{R})$,

$$\int_{M_\xi} \varphi \, d\lambda_{M_\xi} = \int_{\mathbb{R}^{N-1}} \mathbf{1}_{\Phi(G)}(u,\xi) \frac{\varphi|\nabla F|}{|\partial_{x_N} F|} \circ \Phi^{-1}(u,\xi)\, du \quad \text{for } \xi \in \Xi.$$

In particular, conclude that

$$\xi \in \Xi \longmapsto \int_{M_\xi} \varphi \, d\lambda_{M_\xi} \in \mathbb{R}$$

is bounded and $\mathcal{B}_{\mathbb{R}}$-measurable.

(**ii**) Using Theorems 4.1.5 and 5.2.2, show that, for $\varphi \in C_c(G;\mathbb{R})$,

$$\begin{aligned} \int_G \varphi(x)\, dx &= \int_{\Phi(G)} \frac{\varphi}{|\partial_{x_N} F|} \circ \Phi^{-1}(y)\, dy \\ &= \int_\Xi \left(\int_{\mathbb{R}^{N-1}} \mathbf{1}_{\Phi(G)}(u,\xi) \frac{\varphi}{|\partial_{x_N} F|} \circ \Phi^{-1}(u,\xi)\, du \right) d\xi. \end{aligned}$$

After combining this with part (**i**), arrive at the following (somewhat primitive) version of the **co-area formula**:

$$\int_G \varphi(x)\, dx = \int_\Xi \left(\int_{M_\xi} \frac{\varphi}{|\nabla F|} \, d\lambda_{M_\xi} \right) d\xi, \quad \varphi \in C_c(G;\mathbb{R}). \tag{5.2.11}$$

(**iii**) Take $G = \{x \in \mathbb{R}^N : x_N \neq 0\}$, $F = |x|$ for $x \in G$, and show that (5.1.5) can be easily derived from (5.2.11).

5.3 The Divergence Theorem

Again let $N \geq 2$. Perhaps the single most striking application of the construction made in the second part of § 5.2.2 is to multidimensional integration by parts formulas, and this section is devoted to the derivation of one of the most useful of these, the one known as the divergence theorem.

5.3.1 Flows Generated by Vector Fields

Let $V : \mathbb{R}^N \longrightarrow \mathbb{R}^N$, and think of V as a **vector field** that at each point prescribes the velocity of a particle passing through that point. To describe mathematically the trajectory of such a particle, consider the ordinary differential equation

$$\dot{\Phi}(t,x) = V\big(\Phi(t,x)\big) \quad \text{with } \Phi(0,x) = x. \tag{5.3.1}$$

Assuming that V is uniformly Lipschitz continuous, one knows that, for each $x \in \mathbb{R}^N$, there is precisely one solution to (5.3.1) and that that solution exists for all time, both in the future, $t \in [0,\infty)$, and the past, $t \in (-\infty, 0]$.

As a consequence of uniqueness, one also knows that Φ satisfies the **flow property**

$$\Phi(s+t,x) = \Phi\big(t,\Phi(s,x)\big) \quad \text{for } (s,t,x) \in \mathbb{R}\times\mathbb{R}\times\mathbb{R}^N. \tag{5.3.2}$$

In particular, $\Phi\big(t,\Phi(-t,x)\big) = x = \Phi\big(-t,\Phi(t,x)\big)$, and so $\Phi(t,\,\cdot\,)$ is one-to-one and onto. Furthermore, if, in addition, V is twice continuously differentiable, then so is $\Phi(t,\,\cdot\,)$, and, by the chain rule,

$$\frac{\partial\Phi(-t,\,\cdot\,)}{\partial x}\circ\Phi(t,x)\frac{\partial\Phi(t,x)}{\partial x} = \mathbf{I}.$$

Hence $\Phi(t,\,\cdot\,)$ is a diffeomorphism,

$$\frac{1}{J\Phi(t,x)} = J\Phi\big(-t,\Phi(t,x)\big),$$

and so, by Theorem 5.2.2,

$$\int f\circ\Phi(t,x)\,dx = \int f(y)J\Phi(-t,y)\,dy$$

for any non-negative, $\mathcal{B}_{\mathbb{R}^N}$-measurable f. Finally, starting from (5.3.1), one has

$$\frac{d}{dt}\frac{\partial\Phi(t,x)}{\partial x} = \frac{\partial V}{\partial y}\big(\Phi(t,x)\big)\frac{\partial\Phi(t,x)}{\partial x}, \tag{5.3.3}$$

and from this one can derive

$$J\Phi(t,x) = \exp\left(\int_0^t \operatorname{div}V\big(\Phi(\tau,x)\big)\,d\tau\right), \tag{5.3.4}$$

where $\operatorname{div}(V) \equiv \sum_{i=1}^N \partial_{y_i}V_i$ is the **divergence** of V. To see how to pass from (5.3.3) to (5.3.4), one can use Cramer's rule to verify that, for any $N\times N$ matrix $A = \big((a_{ij})\big)$, $\partial_{a_{ij}}\det(A) = A^{(ij)}$, where $A^{(ij)}$ is the (i,j)th cofactor of A. Hence, from (5.3.3) and the fact that $\sum_{j=1}^N a_{kj}A^{(ij)} = \delta_{k,i}\det(A)$,

$$\begin{aligned}\frac{d}{dt}\det\left(\frac{\Phi(t,x)}{\partial x}\right) &= \sum_{i,j,k=1}^N \left(\frac{\partial V}{\partial y}\big(\Phi(t,x)\big)\right)_{ik}\left(\frac{\partial\Phi(t,x)}{\partial x}\right)_{kj}\left(\frac{\partial\Phi(t,x)}{\partial x}\right)^{(ij)}\\ &= \operatorname{div}(V)\big(\Phi(t,x)\big)\det\left(\frac{\Phi(t,x)}{\partial x}\right),\end{aligned}$$

from which (5.3.4) is obvious.

After combining (5.3.4) with the preceding, we now know that

$$\int f \circ \Phi(t,x)\,dx = \int f(x) \exp\left(- \int_0^t \operatorname{div}(V)\big(\Phi(-\tau,x)\,d\tau\big) \right) dx \tag{5.3.5}$$

for all non-negative, $\mathcal{B}_{\mathbb{R}^N}$-measurable f on $\mathbb{R}^N$.

5.3.2 Mass Transport

I now want to apply (5.3.5) to measure the rate at which the flow generated by V moves mass in and out of an open set G. To be precise, if one interprets $\int \mathbf{1}_G(x)\,dx - \int \mathbf{1}_G \circ \Phi(t,x)\,dx$ as the net loss or gain due to the flow at time t, then (5.3.5) says that $\int_G \operatorname{div}\big(V(x)\big)\,dx$ is the rate of loss or gain at time $t = 0$. On the other hand, there is another way in which to think about this computation. Namely,

$$\begin{aligned} &\int \mathbf{1}_G(x)\,dx - \int \mathbf{1}_G \circ \Phi(t,x)\,dx \\ &\quad = \int_G \mathbf{1}_{G^\complement}\big(\Phi(t,x)\big)\,dx - \int_{G^\complement} \mathbf{1}_G\big(\Phi(t,x)\big)\,dx, \end{aligned} \tag{$*$}$$

which indicates that one should be able to do the same calculation by observing how much mass has moved in each direction across the boundary of G during the time interval $[0,t]$. To carry out this approach, I will assume that G is a non-empty, bounded open set that is a **smooth region** in the sense that for each $p \in \partial G$ there exist an open neighborhood $W \ni p$ and an $F \in C^3(W;\mathbb{R})$ such that $|\nabla F| > 0$ and $G \cap W = \{x \in W : F(x) < 0\}$. In particular, ∂G is a compact hypersurface and, for $x \in W \cap \partial G$, $\frac{\nabla F}{|\nabla F|}(x)$ is the *outward pointing unit normal* to ∂G at x.

Let (U, Ψ) be a coordinate chart for some subset of ∂G, and, referring to Lemma 5.2.6, take $\mathbf{n}(u) = \frac{\nabla F \circ \Psi(u)}{|\nabla F \circ \Psi(u)|}$ to be the outward point normal, and define $\tilde{\Psi}$ on $\tilde{U} \times (-\rho, \rho)$ accordingly, Given $(u,\xi) \in \tilde{U}$, $x = \tilde{\Psi}(u,\xi) \in G$ if and only if $\xi < 0$. Since $|\Phi(t,x) - x| \le |t|\|V\|_{\mathrm{u}}$, $\Phi(t,x) \in \tilde{\Psi}(\tilde{U})$ if $|t| < T = \frac{\rho}{\|V\|_{\mathrm{u}}}$. We can therefore define the the map

$$t \in [0,T] \longmapsto \big(u(t), \xi(t)\big) = (\tilde{\Psi})^{-1}\big(\Phi(t,x)\big) \in \tilde{U},$$

in which case $\Phi(t,x) \notin G$ if and only if $\xi(t) \ge 0$.

Observe that, because $|\Psi(u) - x| \le t\|V\|_{\mathrm{u}}$ if $\Phi(t,x) \notin G$, $\Phi(t,x) - x = tV\big(\Psi(u)\big) + E_0(t,x)$, where $|E_0(t,x)| \le C_0 t^2$ for some $C_0 < \infty$, and using this and the fact that $x = \Psi(u) + \xi \mathbf{n}\big(\Psi(u)\big)$, one sees that

$$\xi = \Big(\mathbf{n}\big(\Psi(u)\big), \Phi(t,x) - \Psi(u)\Big)_{\mathbb{R}^N} - t\Big(\mathbf{n}\big(\Psi(u)\big), V\big(\Psi(u)\big)\Big)_{\mathbb{R}^N} - E_0(t,u).$$

Next write $\Big(\mathbf{n}(\Psi(u)), \Phi(t,x)-\Psi(u)\Big)_{\mathbb{R}^N}$ as the sum

$$\Big(\mathbf{n}(\Psi(u)), \Psi\big(u(t)\big)-\Psi(u)\Big)_{\mathbb{R}^N} + \Big(\mathbf{n}(\Psi(u)), \Psi\big(u(t)\big)-\Psi(u)\Big)_{\mathbb{R}^N}$$
$$+ \Big(\mathbf{n}\big(\Psi(u(t))\big), \Phi(t,x)-\Psi\big(u(t)\big) + \Big(\mathbf{n}\big(\Psi(u)\big)-\mathbf{n}\big(\Psi(u(t))\big), \Phi(t,x)-\Psi\big(u(t)\big)\Big)_{\mathbb{R}^N}.$$

Since the middle term equals $\xi(t)$, $\Phi(t,x)\notin G$ if and only if this term is non-negative. Further, it is clear that if $\Phi(t,x)\notin G$, then the absolute value of the last term is dominated by a constant times t^2. Finally, the first term vanishes at $t=0$, and its derivative at $t=0$ equals

$$\Big(\mathbf{n}\big(\Psi(u)\big), v\Big)_{\mathbb{R}^N} = 0$$

since $v=\frac{d}{dt}\Psi\big(u(t)\big)\big|_{t=0}\in T_{\Psi(u)}\partial G$. Hence, this term is also of order t^2, which means that

$$x\in G\,\&\,\Phi(t,x)\notin G \iff 0\ge\xi\ge -t\Big(\mathbf{n}\big(\Psi(u)\big), V\big(\Psi(u)\big)\Big)_{\mathbb{R}^N} - E(t,u),$$

where $|E(t,u)|\le Ct^2$ for some $C<\infty$.

Now let $\Gamma(t)=\big\{(u,\xi)\in\tilde U : \xi<0\;\&\;\xi(t)\ge 0\big\}$. Then, since

$$(u,\xi)\in\Gamma(t) \iff u\in U\;\&\;0\ge\xi\ge -t\Big(\mathbf{n}\big(\Psi(u)\big), V\big(\Psi(u)\big)\Big)_{\mathbb{R}^N} - E(t,u),$$

$$\frac{1}{t}\int_{G\cap\tilde\Psi(\Gamma(t))} \mathbf{1}_{G^{\complement}}\big(\Phi(t,x)\big)\,\lambda_{\mathbb{R}^N}(dx) = \frac{1}{t}\int\left(\int \mathbf{1}_{\Gamma(t)}(u,\xi) J\tilde\Psi(u,\xi)\,d\xi\right)du$$
$$= \int_U \Big(\mathbf{n}\big(\Psi(u)\big), V\big(\Psi(u)\big)\Big)^{+}_{\mathbb{R}^N} J\Psi(u)\,du + \mathcal{O}(t).$$

Hence, after using an elementary covering argument, we see that

$$\lim_{t\searrow 0}\frac{1}{t}\int_G \mathbf{1}_{G^{\complement}}(x)\,dx = \int_{\partial G}\big(\mathbf{n}(x), V(x)\big)^{+}_{\mathbb{R}^N}\,\lambda_{\partial G}(dx).$$

By essentially the same argument, one can show that

$$\lim_{t\searrow 0}\frac{1}{t}\int_{G^{\complement}} \mathbf{1}_{G}(x)\,dx = \int_{\partial G}\big(\mathbf{n}(x), V(x)\big)^{-}_{\mathbb{R}^N}\,\lambda_{\partial G}(dx).$$

Thus, by combining these calculations with the one that results from differentiating (5.3.5) at $t=0$, we arrive at the following statement.

Theorem 5.3.1 (Divergence Theorem) *Let G be a bounded, smooth region in $\mathbb{R}^N$ and $V:\mathbb{R}^N\longrightarrow\mathbb{R}^N$ a twice continuously differentiable vector field with uniformly bounded first derivative. Then*

$$\int_G \mathrm{div}(V)(x)\,dx = \int_{\partial G} \big(\mathbf{n}(x), V(x)\big)_{\mathbb{R}^N}\,\lambda_{\partial G}(dx),$$

where $\mathbf{n}(x)$ *is the outward pointing unit normal to* ∂G *at* $x \in \partial G$.

There are so many applications of Theorem 5.3.1 that it is hard to choose among them. However, here is one that is particularly useful. In its statement, L_V is the directional derivative operator $\sum_{i=1}^N V_i \partial_{x_i}$ determined by V, and $L_V^\top$ is the corresponding formal adjoint operator given by

$$L_V^\top f = -\sum_{i=1}^N \partial_{x_i}(fV_i) = -L_V f - f\mathrm{div}(V).$$

Corollary 5.3.2 *Referring to the preceding, one has*

$$\int_G fL_V g\,d\lambda_{\mathbb{R}^N} = \int_G gL_V^\top f\,d\lambda_{\mathbb{R}^N} + \int_{\partial G} fg\big(\mathbf{n}, V\big)_{\mathbb{R}^N}\,d\lambda_{\partial G}$$

for all $f,\ g \in C_b^2(\mathbb{R}^N;\mathbb{R})$.

Proof. Simply observe that $\mathrm{div}(fgV) = gL_V f - fL_V^\top g$, and apply Theorem 5.3.1 to the vector field fgV. □

5.3.3 Exercises for § 5.3

Exercise 5.3.1 Let G and V be as in Theorem 5.3.1. Choose $\rho > 0$, $p : (\partial G)^{(\rho)} \longrightarrow \partial G$, $\mathbf{n} : (\partial G)^{(\rho)} \longrightarrow \mathbb{S}^{N-1}$, and $\xi : (\partial G)^{(\rho)} \longrightarrow (-\rho, \rho)$ as we did at the beginning of § 5.3.2, and assume that $\big(V(p), \mathbf{n}(p)\big)_{\mathbb{R}^N} = 0$ for all $p \in \partial G$.

(i) Show that $\nabla\xi = \mathbf{n}$ on $(\partial G)^{(\rho)}$.

Hint: From $x = p(x) + \xi(x)\mathbf{n}(x)$, show that $\mathbf{e}_i = \partial_{x_i} p + (\partial_{x_i}\xi)\mathbf{n} + \xi\partial_{x_i}\mathbf{n}$, where $(\mathbf{e}_i)_j = \delta_{i,j}$. Show that $\partial_{x_i} p(x) \in \mathbf{T}_{p(x)}(M)$ and that $\big(\partial_{x_i}\mathbf{n}, \mathbf{n}\big)_{\mathbb{R}^N} = \frac{1}{2}\partial_{x_i}|\mathbf{n}(x)|^2 = 0$, and conclude that $\mathbf{n}(x)_i = \partial_{x_i}\xi$.

(ii) Show that there is a $C < \infty$ such that $\big|\big(V(x), \mathbf{n}(p(x))\big)_{\mathbb{R}^N}\big| \le C|\xi(x)|$ for all $x \in (\partial G)^{(\rho)}$.

(iii) Let Φ be the function determined by (5.3.1). Show that there is a $T > 0$ such that $\Phi(t,p) \in \partial G^{(\rho)}$ for all $p \in \partial G$ and $|t| \le T$. Next, set $u(t,p) = \xi \circ \Phi(t,p)$ for $|t| \le T$, and show that there exists a $C < \infty$ such that $|\dot{u}(t,p)| \le C|u(t,p)|$ and therefore that $\Phi(t,p) \in \partial G$ for all $p \in \partial G$ and $|t| \le T$.

Hint: Use induction on $n \ge 0$ to show that $|u(t,p)| \le \frac{\rho(Ct)^n}{n!}$ for $|t| \le T$.

(iv) Show that $\Phi(t,x) \in \partial G$ for all $t \in \mathbb{R}$ if $\Phi(s,x) \in \partial G$ for some $s \in \mathbb{R}$, and use this to conclude that, depending on whether $x \in G$ or $x \notin \bar{G}$, $\Phi(t,x) \in G$ or $\Phi(t,x) \notin \bar{G}$ for all $t \in \mathbb{R}$.

Hint: Use (**iii**) and the flow property in (5.3.2).

(**v**) Under the additional assumption that $\mathrm{div}(V) = 0$ on G, show that $(\Phi(t,\cdot\,)\restriction G)_*\lambda_G = \lambda_G$ for all $t \in \mathbb{R}$, where $\lambda_G = \lambda_{\mathbb{R}^N} \restriction \mathcal{B}_G$.

Exercise 5.3.2 Use $\Delta = \sum_{i=1}^N \partial_{x_i}^2$ to denote the Euclidean **Laplacian** on $\mathbb{R}^N$. Given a pair of functions $u, v \in C^2_{\mathrm{b}}(\mathbb{R}^N;\mathbb{R}^N)$ and a bounded smooth region G in $\mathbb{R}^N$, prove **Green's formula**:

$$\int_G (u\Delta v - v\Delta u)\, d\lambda_{\mathbb{R}^N} = \int_{\partial G} \Big(u\big(\mathbf{n}, \nabla v\big)_{\mathbb{R}^N} - v\big(\mathbf{n}, \nabla u\big)_{\mathbb{R}^N}\Big)\, d\lambda_{\partial G},$$

where $\mathbf{n}$ denotes the outward pointing unit normal. In particular,

$$\int_G \Delta u\, d\lambda_{\mathbb{R}^N} = \int_{\partial G} \big(\mathbf{n}, \nabla u\big)_{\mathbb{R}^N}\, d\lambda_{\partial G}.$$

Hint: Note that $u\Delta v - v\Delta u = \mathrm{div}\big(u\nabla v - v\nabla u\big)$.

Exercise 5.3.3 Take $N = 2$, and assume that ∂G is a closed, simple curve in the sense that there is a $\gamma \in C^2\big([0,1];\mathbb{R}^2\big)$ with the properties that

$$\begin{gathered} t \in [0,1) \longmapsto \gamma(t) \in \partial G \text{ is an injective surjection,} \\ \gamma(0) = \gamma(1), \quad \dot\gamma(0) = \dot\gamma(1), \quad \ddot\gamma(0) = \ddot\gamma(1), \quad \text{and } |\dot\gamma(t)| > 0 \text{ for } t \in [0,1]. \end{gathered}$$

(**i**) Show that

$$\int_{\partial G} \varphi\, \lambda_{\partial G} = \int_{[0,1]} \varphi \circ \gamma(t)\, |\dot\gamma(t)|\, dt$$

for all bounded measurable φ on ∂G.

(**ii**) Let $\mathbf{n}(t)$ denote the outer unit normal to G at $\gamma(t)$, check that

$$\mathbf{n}(t) = \pm|\dot\gamma(t)|^{-1}\big(\dot\gamma_2(t), -\dot\gamma_1(t)\big),$$

with the same sign for all $t \in [0,1)$, and assume that γ has been parameterized so that the plus sign is the correct one. Next, suppose that $u, v \in C^2_{\mathrm{b}}\big(\mathbb{R}^2;\mathbb{R}\big)$, and set $f = u + iv$. If $\partial_{\bar z} \equiv \frac{1}{2}(\partial_x + i\partial_y)$, show that

$$2i \int_G \partial_{\bar z} f\, d\lambda_{\mathbb{R}^2} = \int_0^1 f\big(\gamma(t)\big)\, dz(t),$$

where $z(t) \equiv \gamma_1(t) + i\gamma_2(t)$ and $dz(t) = \dot z(t)dt$. When f is analytic in G, the Cauchy–Riemann equations imply that $\partial_{\bar z} f = 0$ there, and the preceding becomes the renowned **Cauchy Integral Theorem**. See Exercise 5.3.5 for a continuation of this exercise.

Hint: Check that $2\partial_{\bar z} f = \mathrm{div}(V) + i\mathrm{div}(W)$, where

$$V = \begin{pmatrix} u \\ -v \end{pmatrix} \quad \text{and} \quad W = \begin{pmatrix} v \\ u \end{pmatrix}.$$

Now apply the Divergence Theorem, and check that $(V, \mathbf{n})_{\mathbb{R}^2} + i(W, \mathbf{n})_{\mathbb{R}^2} = -if\dot{z}$.

Exercise 5.3.4 Suppose that $u \in C^2_{\mathrm{b}}(\mathbb{R}^N; \mathbb{R})$ and that $\Delta u = 0$ on $B_{\mathbb{R}^N}(x, R)$. The goal of this exercise is to prove the **mean-value property**

$$u(x) = \frac{1}{\omega_{N-1}} \int_{\mathbb{S}^{N-1}} u(x + R\omega)\, \lambda_{\mathbb{S}^{N-1}}(d\omega), \tag{5.3.6}$$

and the first step is to show that it suffices to handle the case in which x is the origin. Second, observe that there is hardly anything to do when $N = 1$, since in that case there exist a, $b \in \mathbb{R}$ for which $u(x) = ax + b$ for $|x| \le R$, and so $u(0) = \frac{u(-R)+u(R)}{2}$. Thus, assume that $N \ge 2$, and define

$$g_N(r) = \begin{cases} \log \frac{1}{r} & \text{if } N = 2 \\ (N-2)^{-1} r^{2-N} & \text{if } N \ge 3 \end{cases} \quad \text{for } r > 0,$$

and, for $\epsilon > 0$, set

$$v_\epsilon(x) = g_N\big(\sqrt{\epsilon^2 + |x|^2}\big) - g_N\big(\sqrt{\epsilon^2 + R^2}\big).$$

(**i**) Given $0 < r < R$, set $G_r = B_{\mathbb{R}^N}(0, R) \setminus \overline{B_{\mathbb{R}^N}(0, r)}$, and show that

$$\begin{aligned} &\int_{\partial G_r} \Big(\big(\mathbf{n}, \nabla v_\epsilon\big)_{\mathbb{R}^N} u - \big(\mathbf{n}, \nabla u\big)_{\mathbb{R}^N} v_\epsilon \Big)\, d\lambda_{\partial G_r} \\ &\quad = \left(\frac{r}{\sqrt{\epsilon^2 + r^2}} \right)^N \int_{\mathbb{S}^{N-1}} u(r\omega)\, \lambda_{\mathbb{S}^{N-1}}(d\omega) \\ &\quad\quad - \left(\frac{R}{\sqrt{\epsilon^2 + R^2}} \right)^N \int_{\mathbb{S}^{N-1}} u(R\omega)\, \lambda_{\mathbb{S}^{N-1}}(d\omega) \\ &\quad\quad + r^{N-1} A_\epsilon(r, R) \int_{\mathbb{S}^{N-1}} \big(\omega, \nabla u(r\omega)\big)_{\mathbb{R}^N}\, \lambda_{\mathbb{S}^{N-1}}(d\omega), \end{aligned}$$

where $A_\epsilon(r, R) \equiv g_N\big(\sqrt{\epsilon^2 + R^2}\big) - g_N\big(\sqrt{\epsilon^2 + r^2}\big)$.

(**ii**) Using the fact that $\Delta w(x) = \varphi''(|x|) + (N-1)\frac{\varphi'(|x|)}{|x|}$ if $w(x) = \varphi(|x|)$, show that

$$\Delta v_\epsilon(x) = -\frac{N\epsilon^2}{(\epsilon^2 + |x|^2)^{1+\frac{N}{2}}},$$

and combine this with (cf. Exercise 5.3.2) Green's formula and (**i**) to conclude that

$$\begin{aligned} N\epsilon^2 \int_{G_r} \frac{u(x)}{(\epsilon^2+|x|^2)^{1+\frac{N}{2}}}\,\lambda_{\mathbb{R}^N}(dx) &= \left(\frac{r}{\sqrt{\epsilon^2+r^2}}\right)^N \int_{\mathbb{S}^{N-1}} u(r\omega)\,\lambda_{\mathbb{S}^{N-1}}(d\omega) \\ &\quad - \left(\frac{R}{\sqrt{\epsilon^2+R^2}}\right)^N \int_{\mathbb{S}^{N-1}} u(R\omega)\,\lambda_{\mathbb{S}^{N-1}}(d\omega) \\ &\quad + r^{N-1}A_\epsilon(r,R) \int_{\mathbb{S}^{N-1}} \big(\omega, \nabla u(r\omega)\big)_{\mathbb{R}^N}\,\lambda_{\mathbb{S}^{N-1}}(d\omega). \end{aligned}$$

Now let $\epsilon \searrow 0$ and then $r \searrow 0$ to arrive at (5.3.6).

Exercise 5.3.5 Refer to the setting in Exercise 5.3.3, especially part **(ii)**, and use $f(z)$ to denote $f(x,y)$ when $z = x+iy$. The goal of this exercise is to show that

$$2i\int_G \frac{\partial_{\bar z} f(z)}{z-\zeta}\,d\lambda_{\mathbb{R}^2} = \int_0^1 \frac{f\big(z(t)\big)}{z(t)-\zeta}\,dz(t) - 2\pi i f(\zeta) \quad \text{for } \zeta \in G.$$

In particular, when f is analytic in G, this proves the **Cauchy Integral Formula.**

(i) First reduce to the case $0 \in G$ and $\zeta = 0$. Second, show that $\partial_{\bar z}\frac{f(z)}{z} = \frac{\partial_{\bar z} f(z)}{z}$ for $z \neq 0$.

(ii) Define

$$\eta(z) = \begin{cases} \frac{1}{z} & \text{if } |z| > 1 \\ \frac{1}{z}\big(1-16(1-|z|)^4\big)^4 & \text{if } \frac{1}{2} < |z| \le 1 \\ 0 & \text{if } |z| \le \frac{1}{2}, \end{cases}$$

and check that $\eta \in C^2_{\mathrm{b}}(\mathbb{R}^2;\mathbb{C})$.

(iii) Given $r > 0$ with $\overline{B(0,r)} \subset\subset G$, set $f_r(z) = \eta(r^{-1}z)f(z)$, and apply parts **(i)** here and **(ii)** of Exercise 5.3.3 to f_r to see that

$$2i\int_{G\setminus\overline{B(0,r)}} \frac{\partial_{\bar z} f(z)}{z}\,d\lambda_{\mathbb{R}^2} = \int_{\partial G} \frac{f\big(z(t)\big)}{z(t)}\,dz(t) - 2\pi i \int_0^1 f\big(re^{i2\pi t}\big)\,dt.$$

Finally, let $r \searrow 0$.

Chapter 6
Basic Inequalities and Lebesgue Spaces

I have already introduced (cf. §§ 3.1.2 and 3.2.3) the vector space $L^1(\mu;\mathbb{R})$ with the norm[1] $\|\cdot\|_{L^1(\mu;\mathbb{R})}$ and shown it to be a **Banach space**: that is, a normed vector space that is complete with respect to the metric determined by its norm. Although, from the measure-theoretic point of view, $L^1(\mu;\mathbb{R})$ is an natural space with which to deal, from a geometric standpoint, it is flawed. To understand its flaw, consider the two point space $E = \{1,2\}$ and the measure μ that assigns measure 1 to both points. Then $L^1(\mu;\mathbb{R})$ is easily identified with $\mathbb{R}^2$, and the length that $\|\cdot\|_{L^1(\mu;\mathbb{R})}$ assigns $x = (x_1, x_2) \in \mathbb{R}^2$ is $|x_1| + |x_2|$. Hence, the unit ball in this space is the equilateral diamond whose center is the origin and whose vertices lie on the coordinate axes, and, as such, its boundary has nasty corners. For this reason, it is reasonable to ask whether there are measure-theoretically natural Banach spaces that have better geometric properties.

In the first part of this chapter I will develop a few inequalities that will allow me in the second part to introduce the sort of Banach spaces alluded to in the preceding. Once I have done so, I will conclude with a cursory presentation of results about the boundedness properties of linear maps between these spaces.

6.1 Jensen, Minkowski, and Hölder

In this section I will derive some inequalities that generalize the inequalities, like the triangle inequality, which are familiar in the Euclidean context.

Since all the inequalities here are consequences of convexity considerations, I will begin by reviewing a few elementary facts about convex sets and concave

[1] Given a vector space V, a norm $\|\cdot\|$ on V is a non-negative map with the properties that $\|v\| = 0$ if and only if $v = 0$, $\|\alpha v\| = |\alpha|\|v\|$ for all $\alpha \in \mathbb{R}$ and $v \in V$, and $\|v+w\| \le \|v\|+\|w\|$ for all v, $w \in V$. The metric on V determined by the norm $\|\cdot\|$ is the one for which $\|w - v\|$ gives the distance between v and w.

D. W. Stroock, *Essentials of Integration Theory for Analysis*, Graduate Texts in Mathematics 262, https://doi.org/10.1007/978-3-030-58478-8_6

functions on them. Let V be a real or complex vector space. A subset $C \subseteq V$ is said to be **convex** if $(1-\alpha)x + \alpha y \in C$ whenever $x, y \in C$ and $\alpha \in [0,1]$. Given a convex set $C \subseteq V$, $g : C \longrightarrow \mathbb{R}$ is said to be a **concave function** on C if

$$g\big((1-\alpha)x + \alpha y\big) \geq (1-\alpha)g(x) + \alpha g(y) \quad \text{for all} \quad x, y \in C \text{ and } \alpha \in [0,1].$$

Note that g is concave on C if and only if $\big\{(x,t) \in C \times \mathbb{R} : t \leq g(x)\big\}$ is a convex subset of $V \oplus \mathbb{R}$. In addition, one can use induction on $n \geq 2$ to see that

$$\sum_1^n \alpha_k y_k \in C \quad \text{and} \quad g\left(\sum_1^n \alpha_k y_k\right) \geq \sum_1^n \alpha_k g(y_k)$$

for all $n \geq 2$, $\{y_1, \ldots, y_n\} \subseteq C$ and $\{\alpha_1, \ldots, \alpha_n\} \subseteq [0,1]$ with $\sum_1^n \alpha_k = 1$. Namely, if $n = 2$ or $\alpha_n \in \{0,1\}$, then there is nothing to do. On the other hand, if $n \geq 3$ and $\alpha_n \in (0,1)$, set $x = (1-\alpha_n)^{-1} \sum_{k=1}^{n-1} \alpha_k y_k$, and, assuming the result for $n-1$, conclude that

$$\begin{aligned} g\left(\sum_1^n \alpha_k y_k\right) &= g\big((1-\alpha_n)x + \alpha_n y_n\big) \\ &\geq (1-\alpha_n) g\left(\sum_{k=1}^{n-1} \alpha_k (1-\alpha_n)^{-1} y_k\right) + \alpha_n g(y_n) \geq \sum_{k=1}^n \alpha_k g(y_k). \end{aligned}$$

The essence of the relationship between these notions and measure theory is contained in the following.

Theorem 6.1.1 (Jensen's inequality) *Let C be a closed, convex subset of $\mathbb{R}^N$, and suppose that g is a continuous, concave, non-negative function on C. If $(E, \mathcal{B}, \mu)$ is a probability space and $\mathbf{F} : E \longrightarrow C$ a measurable function on $(E, \mathcal{B})$ with the property that $|\mathbf{F}| \in L^1(\mu; \mathbb{R})$, then*

$$\int_E \mathbf{F}\, d\mu \equiv \begin{pmatrix} \int_E F_1\, d\mu \\ \vdots \\ \int_E F_N\, d\mu \end{pmatrix} \in C$$

and

$$\int_E g \circ \mathbf{F}\, d\mu \leq g\bigg(\int_E \mathbf{F}\, d\mu\bigg).$$

(See Exercise 6.1.5 for another derivation.)

Proof. First assume that $\mathbf{F}$ is simple. Then $\mathbf{F} = \sum_{k=0}^n y_k \mathbf{1}_{\Gamma_k}$ for some $n \in \mathbb{Z}^+$, $y_0, \ldots, y_n \in C$, and cover $\{\Gamma_0, \ldots, \Gamma_n\}$ of E by mutually disjoint elements of $\mathcal{B}$. Thus, since $\sum_0^n \mu(\Gamma_k) = 1$ and C is convex, $\int_E \mathbf{F}\, d\mu = \sum_0^n y_k \mu(\Gamma_k) \in C$ and, because g is concave,

$$g\Big(\int_E \mathbf{F}\,d\mu\Big) = g\Big(\sum_{k=0}^{n} y_k\mu(\Gamma_k)\Big) \ge \sum_{k=0}^{n} g(y_k)\mu(\Gamma_k) = \int_E g\circ\mathbf{F}\,d\mu.$$

Now let $\mathbf{F}$ be general. The idea is to approximate $\mathbf{F}$ by C-valued simple functions. For this purpose, choose and fix some element y_0 of C, and let $\{y_k : k \ge 1\}$ be a dense sequence in C. Given $m \in \mathbb{Z}^+$, choose $R_m > 0$ and $n_m \in \mathbb{Z}^+$ for which

$$\int_{\{|\mathbf{F}|\ge R_m\}} \big(|\mathbf{F}| + |y_0|\big)\,d\mu \le \frac{1}{m} \quad\text{and}\quad C \cap \overline{B(0,R_m)} \subseteq \bigcup_{k=1}^{n_m} B\left(y_k, \tfrac{1}{m}\right).$$

Next, set $\Gamma_{m,0} = \big\{\xi \in E : |\mathbf{F}(\xi)| \ge R_m\big\}$, and use induction to define

$$\Gamma_{m,\ell} = \left\{\xi \in E \setminus \bigcup_{k=0}^{\ell-1} \Gamma_{m,k} : \mathbf{F}(\xi) \in B\left(y_\ell, \tfrac{1}{m}\right)\right\}$$

for $1 \le \ell \le n_m$. Finally, set $\mathbf{F}_m = \sum_{k=0}^{n_m} y_k \mathbf{1}_{\Gamma_{m,k}}$.

By construction, the $\mathbf{F}_m$'s are simple and C-valued. Hence, by the preceding,

$$\int_E \mathbf{F}_m\,d\mu \in C \quad\text{and}\quad g\Big(\int_E \mathbf{F}_m\,d\mu\Big) \ge \int_E g\circ\mathbf{F}_m\,d\mu$$

for each $m \in \mathbb{Z}^+$. Moreover, since $|\mathbf{F} - \mathbf{F}_m| \le \frac{1}{m}$ on $\bigcup_{\ell=1}^{n_m} \Gamma_{m,\ell} = E \setminus \Gamma_{m,0}$,

$$\int_E |\mathbf{F} - \mathbf{F}_m|\,d\mu = \sum_{\ell=0}^{n_m} \int_{\Gamma_{m,\ell}} |\mathbf{F} - \mathbf{F}_m|\,d\mu \le \frac{1}{m} + \int_{\Gamma_{m,0}} \big(|\mathbf{F}| + |y_0|\big)\,d\mu \le \frac{2}{m}.$$

Thus, $\big\|\,|\mathbf{F}_m - \mathbf{F}|\,\big\|_{L^1(\mu;\mathbb{R})} \longrightarrow 0$ as $m \to \infty$; and so, because C is closed, we now see that $\int_E \mathbf{F}\,d\mu \in C$. At the same time, because (cf. Exercise 3.2.2) g is continuous, $g\circ\mathbf{F}_m \longrightarrow g\circ\mathbf{F}$ in μ-measure as $m \to \infty$. Hence, by the version of Fatou's lemma in Theorem 3.2.8,

$$\int_E g\circ\mathbf{F}\,d\mu \le \varlimsup_{m\to\infty} \int_E g\circ\mathbf{F}_m\,d\mu \le \varlimsup_{m\to\infty} g\Big(\int_E \mathbf{F}_m\,d\mu\Big) = g\Big(\int_E \mathbf{F}\,d\mu\Big).$$

□

A function f on a convex set C is said to be **convex** if $-f$ is concave. That is, if

$$f\big((1-t)y_1 + ty_2\big) \le (1-t)f(y_1) + tf(y_2) \text{ for } y_1,\, y_2 \in C \text{ and } t \in [0,1].$$

Corollary 6.1.2 *Let I be a closed, possibly infinite, interval with non-empty interior and $(E, \mathcal{B}, \mu)$ a probability space. If f a continuous, convex function on I and $\varphi : E \longrightarrow I$ is a $\mathcal{B}$-measurable, μ-integrable function, then $(f\circ\varphi)^-$ is μ-integrable and*

$$f\left(\int \varphi\, d\mu\right) \leq \int f \circ \varphi\, d\mu.$$

Proof. If $f \leq M$ for some constant $M < \infty$, then, by Theorem 6.1.1 applied to $g = M - f$,

$$M - \int f \circ \varphi\, d\mu \leq M - f\left(\int \varphi\, d\mu\right),$$

and from this the asserted inequality is trivial. Next assume that f is monotone, choose some $c \in \mathring{I}$, and define

$$\varphi_n = \begin{cases} \varphi \vee (c-n) & \text{if } f \text{ is non-increasing} \\ \varphi \wedge (c+n) & \text{if } f \text{ is non-decreasing.} \end{cases}$$

In either case, $f \circ \varphi_n$ is bounded, $|\varphi_0| \leq |c| \vee |\varphi|$, and $f \circ \varphi_n \leq f \circ \varphi_{n+1} \leq f \circ \varphi$. In particular, $f \circ \varphi_0$ and therefore $(f \circ \varphi)^-$ are μ-integrable. In addition,

$$f\left(\int \varphi_n\, d\mu\right) \leq \int f \circ \varphi_n\, d\mu \leq \int f \circ \varphi\, d\mu$$

for all $n \geq 1$, and so, by Lebesgue's dominated convergence theorem and continuity, the result follows after $n \to \infty$. Finally, suppose that f is not monotone. Then, because (cf. Exercise 6.1.1) f is convex, there exists a $c \in \mathring{I}$ such that $f \restriction I \cap [c, \infty)$ is non-decreasing and $f \restriction I \cap (-\infty, c]$ is non-increasing. Thus, if $\varphi_n = \big(\varphi \vee (c-n)\big) \wedge (c+n)$, then the preceding line of reasons leads to the desired result. □

In order to apply Jensen's inequality, one wants to develop a criterion for recognizing when a function is concave. Such a criterion is contained in the next theorem. Recall that the **Hessian matrix** $H_g(x)$ of a function g that is twice continuously differentiable at x is the symmetric matrix given by

$$H_g(x) \equiv \left(\left(\frac{\partial^2 g}{\partial x_i \partial x_j}(x)\right)\right)_{1 \leq i,j \leq N}.$$

Also, a symmetric, real $N \times N$ matrix A is said to be non-positive definite if all of its eigenvalues are non-positive, or, equivalently, if $\big(\xi, A\xi\big)_{\mathbb{R}^N} \leq 0$ for all $\xi \in \mathbb{R}^N$.

Lemma 6.1.3 *Suppose that U is an open, convex subset of $\mathbb{R}^N$, and set $C = \overline{U}$. Then C is also convex. Moreover, if $g : C \longrightarrow \mathbb{R}$ is continuous and $g \restriction U$ is concave, then g is concave on all of C. Finally, if $g : C \longrightarrow \mathbb{R}$ is continuous and $g \restriction U$ is twice continuously differentiable, then g is concave on C if and only if its Hessian matrix is non-positive definite for each $x \in U$.*

Proof. The convexity of C is obvious. In addition, if $g \restriction U$ is concave, the concavity of g on C follows trivially by continuity. Thus, what remains to show is that if $g : U \longrightarrow \mathbb{R}$ is twice continuously differentiable, then g is concave on U if and only if its Hessian is non-positive definite at each $x \in U$.

In order to prove that g is concave on U if $H_g(x)$ is non-positive definite at every $x \in U$, we will use the following simple result about functions on the interval [0,1]. Namely, suppose that $u \in C^2([0,1];\mathbb{R})$, $u(0) = 0 = u(1)$, and $u'' \le 0$. Then $u \ge 0$. To see this (cf. Exercise 6.1.6 for another approach), let $\epsilon > 0$ be given, and consider the function $u_\epsilon \equiv u+\epsilon t(1-t)$. Clearly it is enough to show that $u_\epsilon \ge 0$ on $[0,1]$ for every $\epsilon > 0$. Note that $u_\epsilon(0) = u_\epsilon(1) = 0$ and $u_\epsilon''(t) < 0$ for every $t \in [0,1]$. On the other hand, if $u_\epsilon(t) < 0$ for some $t \in [0,1]$, then there is an $s \in (0,1)$ at which u_ϵ achieves its global minimum. But this is impossible, since then, by the second derivative test, we would have that $u_\epsilon''(s) \ge 0$.

Now assume that $H_g(x)$ is non-positive definite for every $x \in U$. Given $x, y \in U$, define $u(t) = g((1-t)x + ty) - (1-t)g(x) - tg(y)$ for $t \in [0,1]$. Then $u(0) = u(1) = 0$ and

$$u''(t) = \Big(y - x, H_g((1-t)x + ty)(y-x)\Big)_{\mathbb{R}^N} \le 0$$

for every $t \in [0,1]$. Hence, by the preceding paragraph, $u \ge 0$ on $[0,1]$; and so $g((1-t)x + ty) \ge (1-t)g(x) + tg(y)$ for all $t \in [0,1]$. In other words, g is concave on U and therefore on C.

To complete the proof, suppose that $H_g(x)$ has a positive eigenvalue for some $x \in U$. We could then find an $\mathbf{e} \in \mathbb{S}^{N-1}$ and an $\epsilon > 0$ such that $\big(\mathbf{e}, H_g(x)\mathbf{e}\big)_{\mathbb{R}^N} > 0$ and $x + t\mathbf{e} \in U$ for all $t \in (-\epsilon, \epsilon)$. Set $u(t) = g(x + t\mathbf{e})$ for $t \in (-\epsilon, \epsilon)$. Then $u''(0) = \big(\mathbf{e}, H_g(x)\mathbf{e}\big)_{\mathbb{R}^N} > 0$. On the other hand,

$$u''(0) = \lim_{t \to 0} \frac{u(t) + u(-t) - 2u(0)}{t^2},$$

and, if g were concave,

$$\begin{aligned} 2u(0) = 2u\left(\frac{t-t}{2}\right) &= 2g\Big(\tfrac{1}{2}(x + t\mathbf{e}) + \tfrac{1}{2}(x - t\mathbf{e})\Big) \\ &\ge g(x + t\mathbf{e}) + g(x - t\mathbf{e}) = u(t) + u(-t), \end{aligned}$$

from which we would get the contradiction $0 < u''(0) \le 0$. □

Of course, when I is a closed interval with non-empty interior and $f : I \longrightarrow \mathbb{R}$ is a continuous function on I that is twice continuously differentiable on $\mathring{I}$, then Lemma 6.1.3 implies that f is convex on I if and only if $f'' \ge 0$.

When $N = 2$, the following lemma provides a useful test for non-positive definiteness.

Lemma 6.1.4 *Let $A = \left(\begin{smallmatrix} a & b \\ b & c \end{smallmatrix}\right)$ be a real symmetric matrix. Then A is non-positive definite if and only if both $a + c \le 0$ and $ac \ge b^2$. In particular, for each $\alpha \in (0,1)$, the functions $(x,y) \in [0,\infty)^2 \longmapsto x^\alpha y^{1-\alpha}$ and $(x,y) \in [0,\infty)^2 \longmapsto (x^\alpha + y^\alpha)^{\frac{1}{\alpha}}$ are continuous and concave.*

Proof. In view of Lemma 6.1.3, it suffices to check the first assertion and then apply it to Hessians of the specified functions. To this end, let $T = a + c$ be the trace and $D = ac - b^2$ the determinant of A. Also, let λ and μ denote the eigenvalues of A. Then, $T = \lambda + \mu$ and $D = \lambda\mu$.

If A is non-positive definite and therefore $\lambda \vee \mu \le 0$, then it is obvious that $T \le 0$ and that $D \ge 0$. Conversely, If $D > 0$, then either both λ and μ are positive or both are negative. Hence if, in addition, $T \le 0$, then λ and μ are negative. Finally, if $D = 0$ and $T \le 0$, then either $\lambda = 0$ and $\mu = T \le 0$ or $\mu = 0$ and $\lambda = T \le 0$. □

My first application of these considerations provides a generalization, known as **Minkowski's inequality**, of the triangle inequality.

Theorem 6.1.5 (Minkowski's inequality) *Let f_1 and f_2 be non-negative, measurable functions on the measure space $(E, \mathcal{B}, \mu)$. Then, for every $p \in [1, \infty)$,*

$$\left(\int_E (f_1 + f_2)^p \, d\mu\right)^{\frac{1}{p}} \le \left(\int_E f_1^p \, d\mu\right)^{\frac{1}{p}} + \left(\int_E f_2^p \, d\mu\right)^{\frac{1}{p}}.$$

Proof. The case $p = 1$ follows from (3.1.3), and so we will assume that $p \in (1, \infty)$. Also, without loss of generality, we will assume that f_1^p and f_2^p are μ-integrable and that f_1 and f_2 are $[0, \infty)$-valued.

Let $p \in (1, \infty)$ be given. If we assume that $\mu(E) = 1$ and take $\alpha = \frac{1}{p}$, then, by Lemma 6.1.4 and Jensen's inequality,

$$\begin{aligned}
\int_E (f_1 + f_2)^p \, d\mu &= \int_E \left[(f_1^p)^\alpha + (f_2^p)^\alpha\right]^{\frac{1}{\alpha}} d\mu \\
&\le \left[\left(\int_E f_1^p \, d\mu\right)^\alpha + \left(\int_E f_2^p \, d\mu\right)^\alpha\right]^{\frac{1}{\alpha}} \\
&= \left[\left(\int_E f_1^p \, d\mu\right)^{\frac{1}{p}} + \left(\int_E f_2^p \, d\mu\right)^{\frac{1}{p}}\right]^p.
\end{aligned}$$

More generally, if $\mu(E) = 0$ there is nothing to do, and if $0 < \mu(E) < \infty$ we can replace μ by $\frac{\mu}{\mu(E)}$ and apply the preceding. Hence, all that remains is the case $\mu(E) = \infty$. But if $\mu(E) = \infty$, take $E_n = \{f_1 \vee f_2 \ge \frac{1}{n}\}$, note that $\mu(E_n) \le n^p \int f_1^p \, d\mu + n^p \int f_2^p \, d\mu < \infty$, apply the preceding to f_1, f_2, and μ all restricted to E_n, and let $n \to \infty$. □

The next application, which is known as **Hölder's inequality**, gives a generalization of the inner product inequality $|(\xi, \eta)_{\mathbb{R}^N}| \le |\xi||\eta|$ for ξ, $\eta \in \mathbb{R}^N$. In the Euclidean context, this inequality can be seen as an application of the *law of the cosine*, which says that the inner product of vectors is the product of their lengths times the cosine of the angle between them.

Theorem 6.1.6 (Hölder's inequality) *Given $p \in (1,\infty)$, define the* **Hölder conjugate** *p' of $p \in (1,\infty)$ by the equation $\frac{1}{p} + \frac{1}{p'} = 1$ (i.e., $p' = \frac{p}{p-1}$). Then, for every pair of non-negative, measurable functions f_1 and f_2 on a measure space $(E, \mathcal{B}, \mu)$,*

$$\int_E f_1 f_2 \, d\mu \le \left(\int_E f_1^p \, d\mu\right)^{\frac{1}{p}} \left(\int_E f_2^{p'} \, d\mu\right)^{\frac{1}{p'}}$$

for every $p \in (1,\infty)$.

Proof. First note that if either factor on the right-hand side of the above inequality is 0, then $f_1 f_2 = 0$ (a.e., μ), and so the left-hand side is also 0. Thus we will assume that both factors on the right are strictly positive, in which case, we may and will assume in addition that both f_1^p and $f_2^{p'}$ are μ-integrable and that f_1 and f_2 are both $[0,\infty)$-valued. Also, just as in the proof of Minkowski's inequality, we can reduce everything to the case $\mu(E) = 1$. But then we can apply Jensen's inequality and Lemma 6.1.4 with $\alpha = \frac{1}{p}$ to see that

$$\begin{aligned}\int_E f_1 f_2 \, d\mu &= \int_E \left(f_1^p\right)^{\alpha} \left(f_2^{p'}\right)^{1-\alpha} d\mu \le \left(\int_E f_1^p \, d\mu\right)^{\alpha} \left(\int_E f_2^{p'} \, d\mu\right)^{1-\alpha} \\ &= \left(\int_E f_1^p \, d\mu\right)^{\frac{1}{p}} \left(\int_E f_2^{p'} \, d\mu\right)^{\frac{1}{p'}}.\end{aligned}$$

□

6.1.1 Exercises for § 6.1

Exercise 6.1.1 Let f be a convex function on a closed interval I with non-empty interior.

(i) Given points $x < y < z$ in I, show that

$$\frac{f(y) - f(x)}{y - x} \le \frac{f(z) - f(y)}{z - y}.$$

(ii) Set

$$c = \inf\{x \in I : \forall y \in I \cap [x,\infty) \; f(y) \ge f(x)\}.$$

Show that if $x \in I \cap (-\infty, c)$, then $f(x) \ge y$ for all $y \in I \cap [x, c]$, and conclude that $c \in \mathring{I}$ if and only if f is not monotone.

(iii) Assuming that $c \in \mathring{I}$, show that f is non-increasing on $I \cap (-\infty, c]$ and non-decreasing on $I \cap [c, \infty)$.

(iv) Given a measurable function φ on a probability space $(E, \mathcal{B}, \mu)$, show that

$$\int \varphi \, d\mu \le \frac{1}{\alpha} \log \left(\int e^{\alpha\varphi} \, d\mu \right)$$

for all $\alpha > 0$.

Exercise 6.1.2 As a consequence of the result in Exercise 6.1.1, we know that a convex function f on an interval $I = [a, b]$ is either monotone throughout or non-increasing at first and then non-decreasing. Thus, by Theorem 3.3.11, we know that f is differentiable at $\lambda_{\mathbb{R}}$-almost every point. In this exercise you are to show that more can be said.

(**i**) Show that if $a < x < y < z < b$, then

$$\frac{f(x) - f(a)}{x - a} \le \frac{f(y) - f(x)}{y - x} \le \frac{f(z) - f(x)}{z - a} \le \frac{f(b) - f(z)}{b - z}.$$

(**ii**) Use (**i**) to show that for $x \in (a, b)$

$$y \in (a, x) \longmapsto \frac{f(y) - f(x)}{y - x} \in \mathbb{R} \text{ and } y \in (x, b) \longmapsto \frac{f(y) - f(x)}{y - x}$$

are both non-decreasing and that the first of these is bounded above and the second bounded below. From this, show that

$$D^- f(x) = \lim_{y \nearrow x} \frac{f(y) - f(x)}{y - x} \text{ and } D^+ f(x) = \lim_{y \searrow x} \frac{f(y) - f(x)}{y - x}$$

exist in $\mathbb{R}$. In particular, conclude that f is continuous on (a, b).

(**iii**) Show that $D^+ f$ and $D^- f$ are both non-decreasing and that $D^+ f(y) \le D^- f(x) \le D^+(x) \le D^- f(z)$ for $a < y < x < z < b$, and conclude that $D^- f(x) = D^+ f(x)$ at points x of continuity for either $D^- f$ or $D^+ f$. Therefore f is differentiable at all but at most a countable numbers of $x \in (a, b)$.

Exercise 6.1.3 Here are a couple of easy applications of the results in this section.

(**i**) Show that log is continuous and concave on every interval $[\epsilon, \infty)$ with $\epsilon > 0$. Use this together with Jensen's inequality to show that for every $n \in \mathbb{Z}^+$, $\mu_1, \dots, \mu_n \in (0, 1)$ satisfying $\sum_{m=1}^n \mu_m = 1$, and $a_1, \dots, a_n \in [0, \infty)$,

$$\prod_{m=1}^{n} a_m^{\mu_m} \le \sum_{m=1}^{n} \mu_m a_m.$$

In particular, when $\mu_m = \frac{1}{n}$ for every $1 \le m \le n$, this yields $(a_1 \cdots a_n)^{\frac{1}{n}} \le \frac{1}{n} \sum_{m=1}^n a_m$, which is the statement that *the arithmetic mean dominates the geometric mean.*

(**ii**) Let $n \in \mathbb{Z}^+$, and suppose that $f_1, \dots, f_n$ are non-negative, measurable functions on the measure space $(E, \mathcal{B}, \mu)$. Given $p_1, \dots, p_n \in (1, \infty)$ satisfying $\sum_{m=1}^n \frac{1}{p_m} = 1$, show that

$$\int_E f_1 \cdots f_n \, d\mu \le \prod_{m=1}^{n} \left(\int_E f_m^{p_m} \, d\mu \right)^{\frac{1}{p_m}}.$$

Exercise 6.1.4 When $p = 2$, Minkowski's and Hölder's inequalities are intimately related and are both very simple to prove. Indeed, let f_1 and f_2 be bounded, measurable functions on the finite measure space $(E, \mathcal{B}, \mu)$. Given any $\alpha \neq 0$, observe that

$$0 \le \int_E \left(\alpha f_1 \pm \frac{1}{\alpha} f_2 \right)^2 d\mu = \alpha^2 \int_E f_1^2 \, d\mu \pm 2 \int_E f_1 f_2 \, d\mu + \frac{1}{\alpha^2} \int_E f_2^2 \, d\mu,$$

from which it follows that

$$2 \left| \int_E f_1 f_2 \, d\mu \right| \le t \int_E f_1^2 \, d\mu + \frac{1}{t} \int_E f_2^2 \, d\mu$$

for every $t > 0$. If either integral on the right vanishes, show from the preceding that $\int_E f_1 f_2 \, d\mu = 0$. On the other hand, if neither integral vanishes, choose $t > 0$ so that the preceding yields

$$\left| \int_E f_1 f_2 \, d\mu \right| \le \left(\int_E f_1^2 \, d\mu \right)^{\frac{1}{2}} \left(\int_E f_2^2 \, d\mu \right)^{\frac{1}{2}}. \qquad (6.1.1)$$

Hence, in any case, (6.1.1) holds. Finally, argue that one can remove the restriction that f_1 and f_2 be bounded, and then remove the condition that $\mu(E) < \infty$. In particular, even if they are not bounded, as long as f_1^2 and f_2^2 are μ-integrable, conclude that $f_1 f_2$ must be μ-integrable and that (6.1.1) continues to hold.

Clearly (6.1.1) is the special case of Hölder's inequality when $p = 2$. Because it is a particularly significant case, it is often referred to by a different name and is called **Schwarz's inequality**. Assuming that both f_1^2 and f_2^2 are μ-integrable, show that the inequality in Schwarz's inequality is an equality if and only if there exist $(\alpha, \beta) \in \mathbb{R}^2 \setminus \{\mathbf{0}\}$ such that $\alpha f_1 + \beta f_2 = 0$ (a.e., μ).

Finally, use Schwarz's inequality to obtain Minkowski's inequality for the case $p = 2$. Notice the similarity between the development here and that of the classical *triangle inequality* for the Euclidean metric on $\mathbb{R}^N$.

Exercise 6.1.5 A geometric proof of Jensen's inequality can be based on the following. Given a closed, convex subset C of $\mathbb{R}^N$, show that $q \notin C$ if and only if there is a $\mathbf{e}_q \in \mathbb{S}^{N-1}$ such that $\left(\mathbf{e}_q, q - x\right)_{\mathbb{R}^N} > 0$ for all $x \in C$. Next, given a probability space $(E, \mathcal{B}, \mu)$ and a μ-integrable $\mathbf{F} : E \longrightarrow C$, use the preceding to show that $p \equiv \int \mathbf{F} \, d\mu \in C$. Finally, let $g : C \longrightarrow [0, \infty)$ be a continuous, concave function, and use the first part to prove Jensen's inequality. Here are some steps that you might want follow in proving Jensen's inequality.

(i) Show that if g_1 and g_2 are continuous, concave functions on C, then so is $g_1 \wedge g_2$. In particular, if g is a non-negative, continuous, concave function, then $g \wedge n$ is also, and use this to reduce the proof of Jensen's inequality to the case in which g is bounded.

(ii) Assume that $g : C \longrightarrow [0,\infty)$ is a bounded, continuous, concave function, and set $\hat{C} = \{(x,t) \in \mathbb{R}^N \times \mathbb{R} : x \in C \text{ and } t \le g(x)\}$. Show that $\hat{C}$ is a closed, convex subset of $\mathbb{R}^{N+1}$. Next, define $\hat{\mathbf{F}} : E \longrightarrow \hat{C}$ by $\hat{F} = \binom{\mathbf{F}}{g\circ\mathbf{F}}$, note that $\hat{\mathbf{F}}$ is μ-integrable, and apply the first part to see that its μ-integral is an element of $\hat{C}$. Finally, notice that

$$\int \hat{\mathbf{F}}\, d\mu \in \hat{C} \implies \int g\circ\mathbf{F}\, d\mu \le g\left(\int \mathbf{F}\, d\mu\right).$$

Exercise 6.1.6 Suppose that $u \in C^2\big([0,1];\mathbb{R}\big)$ satisfies $u(0) = 0 = u(1)$. The goal of this exercise is to show that

$$u(t) = -\int_{[0,1]} (s \wedge t - st)u''(s)\, ds \quad \text{for } t \in [0,1]. \tag{$*$}$$

In particular, if $u'' \le 0$, then $u \ge 0$.

(i) Use integration by parts to show that

$$u(t) = tu'(0) + \int_{[0,t]} (t-s)u''(s)\, ds \quad \text{for } t \in [0,1].$$

(ii) Using **(i)**, show that $u'(0) = -\int_{[0,1]}(1-s)u''(s)\, ds$ and therefore that $(*)$ holds.

6.2 The Lebesgue Spaces

In the first part of this section I will introduce and briefly discuss the standard Lebesgue spaces $L^p(\mu;\mathbb{R})$. In the second part, I will introduce mixed Lebesgue spaces, one of the many useful variations on the standard ones.

6.2.1 The L^p-Spaces

Given a measure space $(E,\mathcal{B},\mu)$ and a $p \in [1,\infty)$, define

$$\|f\|_{L^p(\mu;\mathbb{R})} = \left(\int_E |f|^p\, d\mu\right)^{\frac{1}{p}}$$

for measurable functions f on $(E,\mathcal{B})$. Also, if f is a measurable function on $(E,\mathcal{B})$ define

$$\|f\|_{L^\infty(\mu;\mathbb{R})} = \inf\{M \in [0,\infty] : |f| \le M \text{ (a.e., } \mu)\}.$$

Although information about f can be gleaned from a study of $\|f\|_{L^p(\mu;\mathbb{R})}$ as p changes (for example, *spikes* in f will be emphasized by taking p to be large), all these quantities share the same flaw as $\|f\|_{L^1(\mu;\mathbb{R})}$: they cannot detect properties of f that occur on sets having μ-measure 0. Thus, before one can hope to use any of them to put a metric on measurable functions, one must invoke the same subterfuge that I introduced in § 3.1.2 in connection with the space $L^1(\mu;\mathbb{R})$. That is, for $p \in [1,\infty]$, denote by $L^p(\mu;\mathbb{R}) = L^p(E,\mathcal{B},\mu;\mathbb{R})$ the collection of equivalence classes $[f]^{\overset{\mu}{\sim}}$ of $\mathbb{R}$-valued, measurable functions f satisfying $\|f\|_{L^p(\mu;\mathbb{R})} < \infty$. Once again, I will consistently abuse notation by using f to denote its own equivalence class $[f]^{\overset{\mu}{\sim}}$.

Obviously $\|\alpha f\|_{L^p(\mu;\mathbb{R})} = |\alpha|\|f\|_{L^p(\mu;\mathbb{R})}$ for all $p \in [1,\infty)$, $\alpha \in \mathbb{R}$, and $\mathcal{B}$-measurable f's. Also, by (3.1.3) and Minkowski's inequality, we have the triangle inequality

$$\|f_1 + f_2\|_{L^p(\mu;\mathbb{R})} \le \|f_1\|_{L^p(\mu;\mathbb{R})} + \|f_2\|_{L^p(\mu;\mathbb{R})}$$

for all $p \in [1,\infty)$ and $f_1, f_2 \in L^p(\mu;\mathbb{R})$. Moreover, it is a simple matter to check that these relations hold equally well when $p = \infty$. Thus, each of the spaces $L^p(\mu;\mathbb{R})$ is a vector space. In addition, because of our convention and Markov's inequality (Theorem 3.1.5), $\|f\|_{L^p(\mu;\mathbb{R})} = 0$ if and only if $f = 0$ as an element of $L^p(\mu;\mathbb{R})$. Hence, $\|f_2 - f_1\|_{L^p(\mu;\mathbb{R})}$ determines a metric on $L^p(\mu;\mathbb{R})$, and I will write $f_n \longrightarrow f$ in $L^p(\mu;\mathbb{R})$ when $\{f_n : n \ge 1\} \cup \{f\} \subseteq L^p(\mu;\mathbb{R})$ and $\|f_n - f\|_{L^p(\mu;\mathbb{R})} \longrightarrow 0$.

The following theorem simply summarizes obvious applications of the results in §§ 3.1 and 3.2 to the present context. The reader should verify that each of the assertions here follows from the relevant result there.

Theorem 6.2.1 *Let $(E,\mathcal{B},\mu)$ be a measure space. Then, for any $p \in [1,\infty]$ and $f,\, g \in L^p(\mu;\mathbb{R})$,*

$$\big|\, \|g\|_{L^p(\mu;\mathbb{R})} - \|f\|_{L^p(\mu;\mathbb{R})} \big| \le \|g - f\|_{L^p(\mu;\mathbb{R})}.$$

Next suppose that $\{f_n : n \ge 1\} \subseteq L^p(\mu;\mathbb{R})$ for some $p \in [1,\infty]$ and that f is an $\mathbb{R}$-valued measurable function on $(E,\mathcal{B})$.

(i) *If $p \in [1,\infty)$ and $f_n \longrightarrow f$ in $L^p(\mu;\mathbb{R})$, then $f_n \longrightarrow f$ in μ-measure. If $f_n \longrightarrow f$ in $L^\infty(\mu;\mathbb{R})$, then $f_n \longrightarrow f$ uniformly off of a set of μ-measure 0.*

(ii) *If $p \in [1,\infty]$ and $f_n \longrightarrow f$ in μ-measure or* (a.e., μ), *then $\|f\|_{L^p(\mu;\mathbb{R})} \le \underline{\lim}_{n\to\infty} \|f_n\|_{L^p(\mu;\mathbb{R})}$. Moreover, if $p \in [1,\infty)$ and, in addition, there is a $g \in L^p(\mu;\mathbb{R})$ such that $|f_n| \le g$* (a.e., μ) *for each $n \in \mathbb{Z}^+$, then $f_n \longrightarrow f$ in $L^p(\mu;\mathbb{R})$.*

(iii) *If $p \in [1,\infty]$ and $\lim_{m\to\infty} \sup_{n\ge m} \|f_n - f_m\|_{L^p(\mu;\mathbb{R})} = 0$, then there is an $f \in L^p(\mu;\mathbb{R})$ such that $f_n \longrightarrow f$ in $L^p(\mu;\mathbb{R})$. In other words, the space $L^p(\mu;\mathbb{R})$ is complete with respect to the metric determined by $\|\cdot\|_{L^p(\mu;\mathbb{R})}$.*

Finally, we have the following variants of Theorem 3.2.10 and Corollary 3.2.11.

(iv) *Assume that* $\mu(E) < \infty$ *and that* $p, q \in [1, \infty)$. *Referring to Theorem 3.2.10, define* $\mathcal{S}$ *as in that theorem. Then, for each* $f \in L^p(\mu;\mathbb{R}) \cap L^q(\mu;\mathbb{R})$, *there is a sequence* $\{\varphi_n : n \geq 1\} \subseteq \mathcal{S}$ *such that* $\varphi_n \longrightarrow f$ *both in* $L^p(\mu;\mathbb{R})$ *and in* $L^q(\mu;\mathbb{R})$. *In particular, if* μ *is* σ*-finite and* $\mathcal{B}$ *is generated by a countable collection* $\mathcal{C}$, *then each of the spaces* $L^p(\mu;\mathbb{R})$, $p \in [1, \infty)$, *is separable.*

(v) *Let* (E, ρ) *be a metric space, and suppose that* μ *is a measure on* $(E, \mathcal{B}_E)$ *for which there exists a non-decreasing sequence of open sets* $E_n \nearrow E$ *satisfying* $\mu(E_n) < \infty$ *for each* $n \geq 1$. *Then, for each pair* $p, q \in [1, \infty)$ *and* $f \in L^p(\mu;\mathbb{R}) \cap L^q(\mu;\mathbb{R})$, *there is a sequence* $\{\varphi_n : n \geq 1\}$ *of bounded,* ρ*-uniformly continuous functions such that* $\varphi_n \equiv 0$ *off of* E_n *and* $\varphi_n \longrightarrow f$ *both in* $L^p(\mu;\mathbb{R})$ *and in* $L^q(\mu;\mathbb{R})$.

The version of the Brezis–Lieb variation on Fatou's lemma for L^p-spaces with $p \neq 1$ is not as easy as the assertions in Theorem 6.2.1. To prove it we will use the following lemma.

Lemma 6.2.2 *Let* $p \in (1, \infty)$, *and suppose that* $\{f_n : n \geq 1\} \subseteq L^p(\mu;\mathbb{R})$ *satisfies* $\sup_{n\geq 1} \|f_n\|_{L^p(\mu;\mathbb{R})} < \infty$ *and that* $f_n \longrightarrow 0$ *either in* μ*-measure or* (a.e., μ). *Then, for every* $g \in L^p(\mu;\mathbb{R})$,

$$\lim_{n\to\infty} \int |f_n|^{p-1}|g|\, d\mu = 0 = \lim_{n\to\infty} \int |f_n|\, |g|^{p-1}\, d\mu.$$

Proof. Without loss of generality, we will assume that all of the f_n's as well as g are non-negative. Given $\delta > 0$, we have that

$$\begin{aligned}\int f_n^{p-1} g\, d\mu &= \int_{\{f_n \leq \delta g\}} f_n^{p-1} g\, d\mu + \int_{\{f_n > \delta g\}} f_n^{p-1} g\, d\mu \\ &\leq \delta^{p-1}\|g\|_{L^p(\mu;\mathbb{R})}^p + \int_{\{f_n \geq \delta^2\}} f_n^{p-1} g\, d\mu + \int_{\{g \leq \delta\}} f_n^{p-1} g\, d\mu.\end{aligned}$$

Applying Hölder's inequality to each of the last two terms, we obtain

$$\begin{aligned}\int f_n^{p-1} g\, d\mu \leq &\delta^{p-1}\|g\|_{L^p(\mu;\mathbb{R})}^p \\ &+ \|f_n\|_{L^p(\mu;\mathbb{R})}^{p-1}\left[\left(\int_{\{f_n \geq \delta^2\}} g^p\, d\mu\right)^{\frac{1}{p}} + \left(\int_{\{g \leq \delta\}} g^p\, d\mu\right)^{\frac{1}{p}}\right].\end{aligned}$$

Since, by Lebesgue's dominated convergence theorem, the first term in the final brackets tends to 0 as $n \to 0$, we conclude that

$$\varlimsup_{n\to\infty} \int f_n^{p-1} g\, d\mu \leq \delta^{p-1}\|g\|_{L^p(\mu;\mathbb{R})}^p + \sup_{n\geq 1}\|f_n\|_{L^p(\mu;\mathbb{R})}^{p-1}\, \|\mathbf{1}_{\{g\leq\delta\}} g\|_{L^p(\mu;\mathbb{R})}$$

for every $\delta > 0$. Thus, after another application of Lebesgue's dominated convergence theorem, we get the first equality upon letting $\delta \searrow 0$.

To derive the other equality, apply the preceding with $f_n^{\frac{1}{p-1}}$ and g^{p-1} replacing, respectively, f_n and g and with $p' = \frac{p}{p-1}$ in place of p. □

Theorem 6.2.3 (Brezis–Lieb) *Let $(E, \mathcal{B}, \mu)$ be a measure space, $p \in [1, \infty)$, and $\{f_n : n \geq 1\} \cup \{f\} \subseteq L^p(\mu; \mathbb{R})$. If $\sup_{n \geq 1} \|f_n\|_{L^p(\mu;\mathbb{R})} < \infty$ and $f_n \longrightarrow f$ in μ-measure or* (a.e., μ), *then*

$$\lim_{n \to \infty} \int \left| |f_n|^p - |f|^p - |f_n - f|^p \right| d\mu = 0;$$

and therefore $\|f_n - f\|_{L^p(\mu;\mathbb{R})} \longrightarrow 0$ if $\|f_n\|_{L^p(\mu;\mathbb{R})} \longrightarrow \|f\|_{L^p(\mu;\mathbb{R})}$.

Proof. The case $p = 1$ is covered by Theorem 3.2.5, and so we will assume that $p \in (1, \infty)$. Given such a p, we will first check that there is a $K_p < \infty$ such that

$$\left| |b|^p - |a|^p - |b - a|^p \right| \leq K_p \left(|b - a|^{p-1} |a| + |a|^{p-1} |b - a| \right), \quad a, b \in \mathbb{R}. \quad (*)$$

Since it is clear that $(*)$ holds for all $a, b \in \mathbb{R}$ if it does for all $a \in \mathbb{R} \setminus \{0\}$ and $b \in \mathbb{R}$, we can assume that $a \neq 0$ and divide both sides by $|a|^p$, thereby showing that $(*)$ is equivalent to

$$\left| |c|^p - 1 - |c - 1|^p \right| \leq K_p \left(|c - 1|^{p-1} + |c - 1| \right), \quad c \in \mathbb{R}.$$

Finally, the existence of a $K_p < \infty$ for which this inequality holds can be easily verified with elementary consideration of what happens when c is near 1 and when $|c|$ is near infinity.

Applying $(*)$ with $a = f_n(x)$ and $b = f(x)$, we see that

$$\left| |f_n|^p - |f|^p - |f_n - f|^p \right| \leq K_p \left(|f_n - f|^{p-1} |f| + |f_n - f| |f|^{p-1} \right)$$

pointwise. Thus, by Lemma 6.2.2 with f_n and g there replaced by, respectively, $f_n - f$ and f here, our result follows. □

Before applying Hölder's inequality to the L^p-spaces, it makes sense to complete the definition of the Hölder conjugate p' that was given in Theorem 6.1.6 only for $p \in (1, \infty)$. Namely, I will take the Hölder conjugate of 1 to be ∞ and that of ∞ to be 1. Notice that this is completely consistent with the equation $\frac{1}{p} + \frac{1}{p'} = 1$ used before.

Theorem 6.2.4 *Let $(E, \mathcal{B}, \mu)$ be a measure space.*

(**i**) *If f and g are measurable functions on $(E, \mathcal{B})$, then for every $p \in [1, \infty]$,*

$$\|fg\|_{L^1(\mu;\mathbb{R})} \leq \|f\|_{L^p(\mu;\mathbb{R})} \|g\|_{L^{p'}(\mu;\mathbb{R})}.$$

In particular, if $f \in L^p(\mu; \mathbb{R})$ and $g \in L^{p'}(\mu; \mathbb{R})$, then $fg \in L^1(\mu; \mathbb{R})$.

(**ii**) *If $p \in [1,\infty)$ and $f \in L^p(\mu;\mathbb{R})$, then*

$$\|f\|_{L^p(\mu;\mathbb{R})} = \sup\bigl\{\|fg\|_{L^1(\mu;\mathbb{R})} : g \in L^{p'}(\mu;\mathbb{R}) \text{ and } \|g\|_{L^{p'}(\mu;\mathbb{R})} \le 1\bigr\}.$$

In fact, if $\|f\|_{L^p(\mu;\mathbb{R})} > 0$, then the supremum on the right is achieved by the function

$$g = \frac{|f|^{p-1}}{\|f\|^{p-1}_{L^p(\mu;\mathbb{R})}}.$$

(**iii**) *More generally, for any f that is measurable on $(E,\mathcal{B})$,*

$$\|f\|_{L^p(\mu;\mathbb{R})} \ge \sup\bigl\{\|fg\|_{L^1(\mu;\mathbb{R})} : g \in L^{p'}(\mu;\mathbb{R}) \text{ and } \|g\|_{L^{p'}(\mu;\mathbb{R})} \le 1\bigr\},$$

and equality holds if $p = 1$ or $p \in (1,\infty)$ and either $\mu(|f| \ge \delta) < \infty$ for every $\delta > 0$ or μ is σ-finite.

(**iv**) *If $f : E \longrightarrow \mathbb{R}$ is measurable and $\mu(|f| \ge R) \in (0,\infty)$ for some $R \ge 0$, then*

$$\|f\|_{L^\infty(\mu;\mathbb{R})} = \sup\bigl\{\|fg\|_{L^1(\mu;\mathbb{R})} : g \in L^1(\mu;\mathbb{R}) \text{ and } \|g\|_{L^1(\mu;\mathbb{R})} \le 1\bigr\}.$$

Proof. Part (**i**) is an immediate consequence of Hölder's inequality when $p \in (1,\infty)$. At the same time, when $p \in \{1,\infty\}$, the conclusion is clear without any further comment. Given (**i**), (**ii**) as well as the inequality in (**iii**) are obvious.

When $p = 1$, equality in (**iii**) is trivial, since one can take $g = \mathbf{1}$. Moreover, in view of (**ii**), the proof of equality in (**iii**) for $p \in (1,\infty)$ reduces to showing that, under either one of the stated conditions, $\|f\|_{L^p(\mu;\mathbb{R})} = \infty$ implies that the supremum on right-hand side is infinite. To this end, first suppose that $\mu(|f| \ge \delta) < \infty$ for every $\delta > 0$. Then, for each $n \ge 1$, the function given by

$$\psi_n \equiv |f|^{p-1}\Bigl(\mathbf{1}_{[\frac{1}{n},n]} \circ |f|\Bigr) + n\mathbf{1}_{\{\infty\}} \circ f$$

is an element of $L^{p'}(\mu;\mathbb{R})$. Moreover, if $\|f\|_{L^p(\mu;\mathbb{R})} = \infty$, then, by the monotone convergence theorem, $\|\psi_n\|_{L^{p'}(\mu;\mathbb{R})} \longrightarrow \infty$. Thus, since $\|f\psi_n\|_{L^1(\mu;\mathbb{R})} \ge \|\psi_n\|^{p'}_{L^{p'}(\mu;\mathbb{R})}$, we see that

$$\|fg_n\|_{L^1(\mu;\mathbb{R})} \longrightarrow \infty \text{ if } \|f\|_{L^p(\mu;\mathbb{R})} = \infty \text{ and } g_n \equiv \frac{\psi_n}{1 + \|\psi_n\|_{L^{p'}(\mu;\mathbb{R})}}.$$

To handle the case μ is σ-finite and $\mu(|f| \ge \delta) = \infty$ for some $\delta > 0$, choose $\{E_n : n \ge 1\} \subseteq \mathcal{B}$ such that $E_n \nearrow E$ and $\mu(E_n) < \infty$ for every $n \ge 1$. Then it is easy to see that $\lim_{n\to\infty} \|fg_n\|_{L^1(\mu;\mathbb{R})} = \infty$ when

$$g_n \equiv \frac{\mathbf{1}_{\Gamma_n}}{\bigl(1 + \mu(\Gamma_n)\bigr)^{1-\frac{1}{p}}} \quad \text{with} \quad \Gamma_n = E_n \cap \{|f| \ge \delta\}.$$

Since $\|g_n\|_{L^{p'}(\mu;\mathbb{R})} \le 1$, this completes the proof of **(iii)**.

Finally, to check **(iv)**, first note that the right side is dominated by the left. To get the opposite inequality, define $g_M = \frac{\mathbf{1}_{[M,\infty]}\circ f}{\mu(|f|\ge M)}$ for $M \in [0,\infty)$ satisfying $\mu(|f| \ge M) \in (0,\infty)$. Obviously, $\|g_M\|_{L^1(\mu;\mathbb{R})} = 1$ and $\|fg_M\|_{L^1(\mu;\mathbb{R})} \ge M$. If $R = \|f\|_{L^\infty(\mu;\mathbb{R})}$, take $M = R$ to get $\|fg_R\|_{L^1(\mu;\mathbb{R})} \ge \|f\|_{L^\infty(\mu;\mathbb{R})}$. If $R < \|f\|_{L^\infty(\mu;\mathbb{R})}$, get the same conclusion by considering $M \in \big[R, \|f\|_{L^\infty(\mu;\mathbb{R})}\big)$. Thus, the left side is dominated by the right. □

6.2.2 Mixed Lebesgue Spaces

For reasons that will become clearer in the next section, it is sometimes useful to consider a slight variation on the basic L^p-spaces. Namely, let $(E_1, \mathcal{B}_1, \mu_1)$ and $(E_2, \mathcal{B}_2, \mu_2)$ be a pair of σ-finite measure spaces and let $p_1, p_2 \in [1,\infty)$. Given a measurable function f on $(E_1 \times E_2, \mathcal{B}_1 \times \mathcal{B}_2)$, define

$$\|f\|_{L^{(p_1,p_2)}((\mu_1,\mu_2);\mathbb{R})} \equiv \left[\int_{E_2}\left(\int_{E_1} |f(x_1,x_2)|^{p_1}\,\mu_1(dx_1)\right)^{\frac{p_2}{p_1}} \mu_2(dx_2)\right]^{\frac{1}{p_2}},$$

and let $L^{(p_1,p_2)}\big((\mu_1,\mu_2);\mathbb{R}\big)$ denote the **mixed Lebesgue space** of $\mathbb{R}$-valued, $\mathcal{B}_1\times\mathcal{B}_2$-measurable f's for which $\|f\|_{L^{(p_1,p_2)}((\mu_1,\mu_2);\mathbb{R})} < \infty$. Obviously, when $p_1 = p = p_2$, $\|f\|_{L^{(p_1,p_2)}((\mu_1,\mu_2);\mathbb{R})} = \|f\|_{L^p(\mu_1\times\mu_2;\mathbb{R})}$ and $L^{(p_1,p_2)}\big((\mu_1,\mu_2);\mathbb{R}\big) = L^p(\mu_1\times\mu_2;\mathbb{R})$.

The goal of this subsection is to show that when $p_1 \le p_2$ the $L^{(p_1,p_2)}$-norm of a function is dominated by its $L^{(p_2,p_1)}$-norm, and the following lemma will play a crucial role in the proof.

Lemma 6.2.5 *For all f and g that are measurable on $(E_1 \times E_2, \mathcal{B}_1 \times \mathcal{B}_2)$,*

$$\begin{aligned}\|f+g\|_{L^{(p_1,p_2)}((\mu_1,\mu_2);\mathbb{R})} &\le \|f\|_{L^{(p_1,p_2)}((\mu_1,\mu_2);\mathbb{R})} + \|g\|_{L^{(p_1,p_2)}((\mu_1,\mu_2);\mathbb{R})},\\ \|fg\|_{L^1((\mu_1\times\mu_2);\mathbb{R})} &\le \|f\|_{L^{(p_1,p_2)}((\mu_1,\mu_2);\mathbb{R})}\,\|g\|_{L^{(p_1',p_2')}((\mu_1,\mu_2);\mathbb{R})}.\end{aligned} \tag{6.2.1}$$

Now assume $\{f_n : n \ge 1\} \cup \{f\} \subseteq L^{(p_1,p_2)}\big((\mu_1,\mu_2);\mathbb{R}\big)$ with $|f_n| \le g$ (a.e., $\mu_1 \times \mu_2$) for each $n \ge 1$ and some $g \in L^{(p_1,p_2)}\big((\mu_1,\mu_2);\mathbb{R}\big)$. If $f_n \longrightarrow f$ (a.e., $\mu_1 \times \mu_2$), then $\|f_n - f\|_{L^{(p_1,p_2)}((\mu_1,\mu_2);\mathbb{R})} \longrightarrow 0$. Finally, if μ_1 and μ_2 are finite and $\mathcal{G}$ denotes the class of all ψ's on $E_1 \times E_2$ having the form $\sum_{m=1}^n \mathbf{1}_{\Gamma_{1,m}}(\,\cdot_1)\varphi_m(\,\cdot_2)$ for some $n \ge 1$,

$$\{\varphi_m : 1 \le m \le n\} \subseteq L^\infty(\mu_2;\mathbb{R}),$$

and mutually disjoint $\Gamma_{1,1},\dots,\Gamma_{1,n} \in \mathcal{B}_1$, then, for each $f \in L^{(p_1,p_2)}\big((\mu_1,\mu_2);\mathbb{R}\big)$ and $\epsilon > 0$, there is a $\psi \in \mathcal{G}$ for which $\|f - \psi\|_{L^{(p_1,p_2)}((\mu_1,\mu_2);\mathbb{R})} < \epsilon$.

Proof. Note that

$$\|f\|_{L^{(p_1,p_2)}((\mu_1,\mu_2);\mathbb{R})} = \Big\| \|f(\,\cdot\,_1,\,\cdot\,_2)\|_{L^{p_1}(\mu_1;\mathbb{R})} \Big\|_{L^{p_2}(\mu_2;\mathbb{R})}. \tag{$*$}$$

Hence the assertions in (6.2.1) are consequences of repeated application of Minkowski's and Hölder's inequalities, respectively. Moreover, to prove the second statement, observe (cf. Exercise 4.1.3) that, for μ_2-almost every $x_2 \in E_2$, $f_n(\,\cdot\,,x_2) \longrightarrow f(\,\cdot\,,x_2)$ (a.e., μ_1), $|f_n(\,\cdot\,,x_2)| \le g(\,\cdot\,,x_2)$ (a.e., μ_1), and $g(\,\cdot\,,x_2) \in L^{p_1}(\mu_1;\mathbb{R})$. Thus, by part **(ii)** of Theorem 6.2.1, for μ_2-almost every $x_2 \in E_2$, $\|f_n(\,\cdot\,,x_2) - f(\,\cdot\,,x_2)\|_{L^{p_1}(\mu_1;\mathbb{R})} \longrightarrow 0$. In addition,

$$\|f_n(\,\cdot\,,x_2) - f(\,\cdot\,,x_2)\|_{L^{p_1}(\mu_1;\mathbb{R})} \le 2\|g(\,\cdot\,,x_2)\|_{L^{p_1}(\mu_1;\mathbb{R})}$$

for μ_2-almost every $x_2 \in E_2$ and, by $(*)$ with g replacing f,

$$\Big\| \|g(\,\cdot\,_1,\,\cdot\,_2)\|_{L^{p_1}(\mu_1;\mathbb{R})} \Big\|_{L^{p_2}(\mu_2;\mathbb{R})} < \infty.$$

Hence the required result follows after a second application of **(ii)** in Theorem 6.2.1.

We turn now to the final part of the lemma, in which both the measures μ_1 and μ_2 are assumed to be finite. In fact, without loss of generality, we will assume that they are probability measures. In addition, by the preceding, it is clear that, for each $f \in L^{(p_1,p_2)}\big((\mu_1,\mu_2);\mathbb{R}\big)$,

$$\|f - f_n\|_{L^{(p_1,p_2)}((\mu_1,\mu_2);\mathbb{R})} \longrightarrow 0 \quad \text{where } f_n \equiv f\mathbf{1}_{[-n,n]} \circ f.$$

Thus, we only need consider f's that are bounded. Finally, because $\mu_1 \times \mu_2$ is also a probability measure, part **(i)** of Exercise 6.2.1 together with $(*)$ imply that

$$\|f - \psi\|_{L^{(p_1,p_2)}(\mu_1,\mu_2)} \le \|f - \psi\|_{L^q(\mu_1\times\mu_2)} \quad \text{where} \quad q = p_1 \vee p_2.$$

Hence, all that remains is to show that, for every bounded, $\mathcal{B}_1\times\mathcal{B}_2$-measurable $f : E_1 \times E_2 \longrightarrow \mathbb{R}$ and $\epsilon > 0$, there is a $\psi \in \mathcal{G}$ for which $\|f - \psi\|_{L^q(\mu_1\times\mu_2)} < \epsilon$. But, by part **(iv)** of Theorem 6.2.1, the class of simple functions having the form

$$\psi = \sum_{m=1}^{n} a_m \mathbf{1}_{\Gamma_{1,m}\times\Gamma_{2,m}}$$

with $\Gamma_{i,m} \in \mathcal{B}_i$ is dense in $L^q(\mu_1 \times \mu_2;\mathbb{R})$. Thus, we will be done once we check that such a ψ is an element of $\mathcal{G}$. To this end, set $\mathcal{I} = \{0,1\}^n$ and, for $\eta \in \mathcal{I}$, define $\Gamma_{1,\eta} = \bigcap_{m=1}^{n} \Gamma_{1,m}^{(\eta_m)}$ where $\Gamma^{(0)} \equiv \Gamma^{\complement}$ and $\Gamma^{(1)} \equiv \Gamma$. Then

$$\psi(x_1,x_2) = \sum_{m=1}^{n} a_m \left(\sum_{\eta\in\mathcal{I}} \eta_m \mathbf{1}_{\Gamma_{1,\eta}}(x_1) \right) \mathbf{1}_{\Gamma_{2,m}}(x_2) = \sum_{\eta\in\mathcal{I}} \mathbf{1}_{\Gamma_{1,\eta}}(x_1)\,\varphi_\eta(x_2),$$

where

$$\varphi_\eta = \sum_{m=1}^{n} \eta_m a_m \mathbf{1}_{\Gamma_{2,m}}.$$

Since the $\Gamma_{1,\eta}$'s are mutually disjoint, this completes the proof. □

We can now prove the following **continuous version of Minkowski's inequality**.

Theorem 6.2.6 *Let $(E_i, \mathcal{B}_i, \mu_i)$, $i \in \{1,2\}$, be σ-finite measure spaces. Then, for any measurable function f on $(E_1 \times E_2, \mathcal{B}_1 \times \mathcal{B}_2)$,*

$$\|f\|_{L^{(p_1,p_2)}((\mu_1,\mu_2);\mathbb{R})} \le \|f\|_{L^{(p_2,p_1)}((\mu_2,\mu_1);\mathbb{R})} \quad \text{if } 1 \le p_1 \le p_2 < \infty.$$

Proof. Since it is easy to reduce the general case to the one in which both μ_1 and μ_2 are finite, we will take them to be probability measures from the outset.

Let $\mathcal{G}$ be the class described in the last part of Lemma 6.2.5. Given $\psi = \sum_1^n \mathbf{1}_{\Gamma_{1,m}}(\cdot_1)\,\varphi_m(\cdot_2)$ from $\mathcal{G}$, note that, since the $\Gamma_{1,m}$'s are mutually disjoint, $\left|\sum_1^n a_m \mathbf{1}_{\Gamma_{1,m}}\right|^r = \sum_1^n |a_m|^r \mathbf{1}_{\Gamma_{1,m}}$ for any $r \in [0,\infty)$ and $a_1, \dots, a_n \in \mathbb{R}$. Hence, by Minkowski's inequality with $p = \frac{p_2}{p_1}$,

$$\begin{aligned}
\|\psi\|_{L^{(p_1,p_2)}((\mu_1,\mu_2);\mathbb{R})} &= \left[\int_{E_2} \left(\sum_{m=1}^{n} \mu_1(\Gamma_{1,m})|\varphi_m(x_2)|^{p_1}\right)^{\frac{p_2}{p_1}} \mu_2(dx_2)\right]^{\frac{1}{p_2}} \\
&= \left\|\sum_{m=1}^{n} \mu_1(\Gamma_{1,m})|\varphi_m(\cdot_2)|^{p_1}\right\|_{L^{\frac{p_2}{p_1}}(\mu_2;\mathbb{R})}^{\frac{1}{p_1}} \\
&\le \left[\sum_{m=1}^{n} \mu_1(\Gamma_{1,m}) \left\| |\varphi_m|^{p_1} \right\|_{L^{\frac{p_2}{p_1}}(\mu_2;\mathbb{R})}\right]^{\frac{1}{p_1}} = \left[\sum_1^n \mu_1(\Gamma_{1,m}) \|\varphi_m\|_{L^{p_2}(\mu_2;\mathbb{R})}^{p_1}\right]^{\frac{1}{p_1}} \\
&= \left[\int_{E_1} \sum_{m=1}^{n} \mathbf{1}_{\Gamma_{1,m}}(x_1) \|\varphi_m\|_{L^{p_2}(\mu_2;\mathbb{R})}^{p_1} \mu_1(dx_1)\right]^{\frac{1}{p_1}} \\
&= \left[\int_{E_1} \left(\sum_{m=1}^{n} \mathbf{1}_{\Gamma_{1,m}}(x_1) \|\varphi_m\|_{L^{p_2}(\mu_2;\mathbb{R})}^{p_2}\right)^{\frac{p_1}{p_2}} \mu_1(dx_1)\right]^{\frac{1}{p_1}} \\
&= \left[\int_{E_1} \left(\int_{E_2} \left|\sum_1^n \mathbf{1}_{\Gamma_{1,m}}(x_1)\varphi_m(x_2)\right|^{p_2} \mu_2(dx_2)\right)^{\frac{p_1}{p_2}} \mu_1(dx_1)\right]^{\frac{1}{p_1}} \\
&= \|\psi\|_{L^{(p_2,p_1)}((\mu_2,\mu_1);R)}.
\end{aligned}$$

Therefore we are done when the function f is an element of $\mathcal{G}$.

To complete the proof, let f be a $\mathcal{B}_1 \times \mathcal{B}_2$-measurable function on $E_1 \times E_2$. Clearly we may assume that $\|f\|_{L^{(p_2,p_1)}((\mu_2,\mu_1);\mathbb{R})} < \infty$. Using the last part of Lemma 6.2.5, choose $\{\psi_n : n \geq 1\} \subseteq \mathcal{G}$ such that

$$\|\psi_n - f\|_{L^{(p_2,p_1)}((\mu_2,\mu_1);\mathbb{R})} \longrightarrow 0.$$

Then, by Jensen's inequality (cf. (i) in Exercise 6.2.1), one has that $\|\psi_n - f\|_{L^1((\mu_1 \times \mu_2);\mathbb{R})} \longrightarrow 0$, and therefore that $\psi_n \longrightarrow f$ in $\mu_1 \times \mu_2$-measure. Hence, without loss of generality, assume that $\psi_n \longrightarrow f$ (a.e., $\mu_1 \times \mu_2$). In particular, by Fatou's lemma and Exercise 4.1.3, this means that

$$\int_{E_1} |f(x_1, x_2)|^{p_1}\, \mu_1(dx_1) \leq \varliminf_{n\to\infty} \int_{E_1} |\psi_n(x_1, x_2)|^{p_1}\, \mu_1(dx_1)$$

for μ_2-almost every $x_2 \in E_2$; and so, by the result for elements of $\mathcal{G}$ and another application of Fatou's lemma, the required result follows for f. □

6.2.3 *Exercises for § 6.2*

Exercise 6.2.1 Let $(E, \mathcal{B}, \mu)$ be a measure space and $f : E \longrightarrow \mathbb{R}$ a $\mathcal{B}$-measurable function.

(i) If μ is finite, show that $\|f\|_{L^p(\mu;\mathbb{R})} \leq \mu(E)^{\frac{1}{p}-\frac{1}{q}} \|f\|_{L^q(\mu;\mathbb{R})}$ for $1 \leq p < q \leq \infty$. In particular, when μ is a probability measure, this means that $\|f\|_{L^p(\mu;\mathbb{R})}$ is non-decreasing as a function of p.

(ii) When E is countable, $\mathcal{B} = \mathcal{P}(E)$, and μ is the counting measure on E (i.e., $\mu(\{x\}) = 1$ for each $x \in E$), show that $\|f\|_{L^p(\mu;\mathbb{R})}$ is a non-increasing function of $p \in [1, \infty]$.

Hint: First reduce to the case when $E = \{1, \ldots, n\}$ for some $n \in \mathbb{Z}^+$. Second, show that it suffices to prove that $\sum_{m=1}^n a_m^p \leq 1$ for every $p \in [1, \infty)$ and $\{a_m : 1 \leq m \leq n\} \subseteq [0, 1]$ with $\sum_{m=1}^n a_m = 1$, and apply elementary calculus to check this.

(iii) If μ is finite or f is μ-integrable, show that, as $p \to \infty$, $\|f\|_{L^p(\mu;\mathbb{R})} \longrightarrow \|f\|_{L^\infty(\mu;\mathbb{R})}$ for any $\mathcal{B}$-measurable $f : E \longrightarrow \mathbb{R}$.

(iv) Let $(E_1, \mathcal{B}_1)$ and $(E_2, \mathcal{B}_2)$ be a pair of measurable spaces, and let μ_2 be a σ-finite measure on $(E_2, \mathcal{B}_2)$. If $f : E_1 \times E_2 \longrightarrow \mathbb{R}$ is $\mathcal{B}_1 \times \mathcal{B}_2$-measurable, use **(iii)** to show that $x_1 \in E_1 \longmapsto \|f(x_1, \cdot)\|_{L^\infty(\mu_2;\mathbb{R})} \in [0, \infty]$ is $\mathcal{B}_1$-measurable. Hence, we could have defined $L^{(p_1,p_2)}\big((\mu_1, \mu_2); \mathbb{R}\big)$ for all p_1, $p_2 \in [1, \infty]$.

Exercise 6.2.2 Let a measure space $(E, \mathcal{B}, \mu)$ and $1 \leq q_0 \leq q_1 \leq \infty$ be given. If $f \in L^{q_0}(\mu; \mathbb{R}) \cap L^{q_1}(\mu; \mathbb{R})$, show that for every $t \in (0, 1)$

$$\|f\|_{L^{q_t}(\mu;\mathbb{R})} \leq \|f\|^t_{L^{q_0}(\mu;\mathbb{R})} \|f\|^{1-t}_{L^{q_1}(\mu;\mathbb{R})} \quad \text{where } \frac{1}{q_t} = \frac{t}{q_0} + \frac{1-t}{q_1}.$$

Note that this can be summarized by saying that $p \rightsquigarrow -\log \|f\|_{L^p(\mu;\mathbb{R})}$ is a concave function of $\frac{1}{p}$.

Exercise 6.2.3 If $(E_1, \mathcal{B}_1, \mu_1)$ and $(E_2, \mathcal{B}_2, \mu_2)$ are a pair of σ-finite measure spaces and $p \in [1, \infty)$, show that the set of functions that can be written in the form $\sum_{m=1}^{n} f_{1,m}(x_1) f_{2,m}(x_2)$, where $n \geq 1$, $\{f_{1,m} : 1 \leq m \leq n\} \subseteq L^p(\mu_1; \mathbb{R})$, and $\{f_{2,m} : 1 \leq m \leq n\} \subseteq L^p(\mu_2; \mathbb{R})$, is dense in $L^p(\mu_1 \times \mu_2; \mathbb{R})$.

Hint: First reduce to the case in which μ_1 and μ_2 are finite, and then apply part **(iv)** of Theorem 6.2.1 to handle this case.

Exercise 6.2.4 Let $(E, \mathcal{B}, \mu)$ be a measure space, g a non-negative element of $L^p(\mu; \mathbb{R})$ for some $p \in (1, \infty)$, and f a non-negative, $\mathcal{B}$-measurable function for which there exists a $C \in (0, \infty)$ such that

$$\mu(f \geq t) \leq \frac{C}{t} \int_{\{f \geq t\}} g \, d\mu, \quad t \in (0, \infty). \tag{$*$}$$

The purpose of this exercise is to show that $(*)$ allows one to estimate the L^p-norm of f in terms of that of g when $p > 1$. The result here is a very special case of a general result known as Marcinkiewicz's Interpolation Theorem.

(i) Set $\nu(\Gamma) = \int_\Gamma g \, d\mu$ for $\Gamma \in \mathcal{B}$, note that $(*)$ is equivalent to

$$\mu(f > t) \leq \frac{C}{t} \nu(f > t), \quad t \in (0, \infty),$$

and use (5.1.5) or Exercise 5.1.2 to justify

$$\begin{aligned} \|f\|_{L^p(\mu;\mathbb{R})}^p &= p \int_{(0,\infty)} t^{p-1} \mu(f > t) \, dt \\ &\leq Cp \int_{(0,\infty)} t^{p-2} \nu(f > t) \, dt = \frac{Cp}{p-1} \int f^{p-1} \, d\nu. \end{aligned}$$

Finally, note that $\int f^{p-1} \, d\nu = \int f^{p-1} g \, d\mu$, and apply Hölder's inequality to conclude that

$$\|f\|_{L^p(\mu;\mathbb{R})}^p \leq \frac{Cp}{p-1} \|f\|_{L^p(\mu;\mathbb{R})}^{p-1} \|g\|_{L^p(\mu;\mathbb{R})}. \tag{$**$}$$

(ii) Under the condition that $\|f\|_{L^p(\mu;\mathbb{R})} < \infty$, it is clear that $(**)$ implies

$$\|f\|_{L^p(\mu;\mathbb{R})} \leq \frac{Cp}{p-1} \|g\|_{L^p(\mu;\mathbb{R})}. \tag{6.2.2}$$

Now suppose that $\mu(E) < \infty$. After checking that $(*)$ for f implies $(*)$ for $f_R \equiv f \wedge R$, conclude that (6.2.2) holds first with f_R replacing f and then, after $R \nearrow \infty$, for f itself. In other words, when μ is finite, $(*)$ always implies (6.2.2).

(iii) Even if μ is not finite, show that $(*)$ implies (6.2.2) if, for every $\epsilon > 0$, $\mu(f > \epsilon) < \infty$.
Hint: Given $\epsilon > 0$, consider $\mu_\epsilon = \mu \restriction \mathcal{B}\big[\{f > \epsilon\}\big]$, note that $(*)$ with μ implies itself with μ_ϵ, and use **(ii)** to conclude that (6.2.2) holds with μ_ϵ in place of μ. Finally, let $\epsilon \searrow 0$.

Exercise 6.2.5 In § 3.3.2 I derived the Hardy–Littlewood inequality (3.3.1) and pointed out the one cannot pass from such an estimate to an estimate on the L^1-norm. Nonetheless, use (3.3.1) together with Exercise 6.2.4 to show that

$$\|Mf\|_{L^p(\lambda_{\mathbb{R}};\mathbb{R})} \leq \frac{2p}{p-1}\|f\|_{L^p(\lambda_{\mathbb{R}};\mathbb{R})} \quad \text{for all } p \in (1,\infty).$$

6.3 Some Elementary Transformations on Lebesgue Spaces

The most natural transformations on a vector are linear, and in this section I will discuss some linear transformations on the Lebesgue spaces.

Linear transformations on finite-dimensional vector spaces are represented in terms of matrices. To think about $\mathbb{R}^n$ in measure-theoretic terms, one first identifies it with the space of $\mathbb{R}$-valued functions on $\{1,\dots,n\}$ and then gives it any one of the topologies that comes from putting a measure μ on $E = \{1,\dots,n\}$ and introducing the norm $\|\cdot\|_{L^p(\mu;\mathbb{R})}$ for some $p \in [1,\infty]$. (Of course, in order for the resulting topology to be Hausdorff, one has to choose a μ that assigns positive measure to each element of E.) Next suppose that $\mathcal{K}$ is a linear transformation on $\mathbb{R}^n$. Then the measure-theoretic analog of the matrix representation of $\mathcal{K}$ takes the form $\mathcal{K}f(x) = \int_E K(x,y)f(y)\,\mu(dy)$. That is, summation is replaced by integration with respect to μ, and the matrix is replaced by a function, called a **kernel**, on $E \times E$.

As long as one is working in a finite-dimensional context, the choice of norm is not crucial. Indeed, every norm on a finite-dimensional vector space is commensurate with every other norm on that space, and every linear operator is continuous no matter what norm is used. However, in infinite dimensions, the choice of norm is critical. In fact, for a given kernel, in general the linear operator that it represents will be well-defined only when the norm is judiciously chosen. In the following subsection, I will derive a useful criterion for deciding when a kernel determines a continuous map between Lebesgue spaces, and in the subsequent subsection I will apply that criterion.

6.3.1 A General Estimate for Linear Transformations

In the following, $(E_1, \mathcal{B}_1, \mu_1)$ and $(E_2, \mathcal{B}_2, \mu_2)$ are a pair of σ-finite measure spaces and K is a measurable function on $(E_1 \times E_2, \mathcal{B}_1 \times \mathcal{B}_2)$.

Lemma 6.3.1 *Assume that*

$$M_1 \equiv \sup_{x_2 \in E_2} \|K(\,\cdot\,, x_2)\|_{L^q(\mu_1;\mathbb{R})} < \infty \text{ and } M_2 \equiv \sup_{x_1 \in E_1} \|K(x_1, \,\cdot\,)\|_{L^q(\mu_2;\mathbb{R})} < \infty$$

for some $q \in [1, \infty)$, *and define*

$$\mathcal{K}f(x_1) = \int_{E_2} K(x_1, x_2) f(x_2)\, \mu_2(dx_2) \quad \text{for } x_1 \in E_1 \text{ and } f \in L^{q'}(\mu_2; \mathbb{R}).$$

Then for each $p \in [1, \infty]$ *satisfying* $\frac{1}{r} \equiv \frac{1}{p} + \frac{1}{q} - 1 \geq 0$,

$$\|\mathcal{K}f\|_{L^r(\mu_1;\mathbb{R})} \leq M_1^{\frac{q}{r}} M_2^{1-\frac{q}{r}} \|f\|_{L^p(\mu_2;\mathbb{R})}, \quad f \in L^{q'}(\mu_2; \mathbb{R}).$$

Proof. First suppose that $r = \infty$ and therefore that $p = q'$. Then, by part (**i**) of Theorem 6.2.4,

$$\big|\mathcal{K}f(x_1)\big| \leq \big\|K(x_1, \cdot)\big\|_{L^q(\mu_2;\mathbb{R})} \|f\|_{L^p(\mu_2;\mathbb{R})} \quad \text{for all } x_1 \in E_1;$$

and so there is nothing more to do in this case.

Next, suppose that $p = 1$ and therefore that $q = r$. Noting that $\|\mathcal{K}f\|_{L^r(\mu_1;\mathbb{R})} \leq \|Kf\|_{L^{(1,r)}((\mu_2,\mu_1);\mathbb{R})}$, we can apply Theorem 6.2.6 to obtain

$$\begin{aligned}
\|\mathcal{K}f\|_{L^r(\mu_1;\mathbb{R})} &\leq \|Kf\|_{L^{(1,r)}((\mu_2,\mu_1);\mathbb{R})} \leq \|Kf\|_{L^{(r,1)}((\mu_1,\mu_2);\mathbb{R})} \\
&= \int_{E_2} \left(\int_{E_1} |K(x_1, x_2) f(x_2)|^r \, \mu_1(dx_1) \right)^{\frac{1}{r}} \mu_2(dx) \\
&= \int_{E_2} \|K(\,\cdot\,, x_2)\|_{L^r(\mu_1;\mathbb{R})} |f(x_2)|\, \mu_2(dx_2) \leq M_1 \|f\|_{L^1(\mu_2;\mathbb{R})}.
\end{aligned}$$

Finally, the only case remaining is that in which $r \in [1, \infty)$ and $p \in (1, \infty)$. Noting that $r \in (q, \infty)$, set $\alpha = \frac{q}{r}$. Then, $\alpha \in (0, 1)$ and $(1 - \alpha)p' = q$. Given $g \in L^{r'}(\mu_1; \mathbb{R})$, we have, by the second inequality in (6.2.1), that

$$\begin{aligned}
\|g\,\mathcal{K}f\|_{L^1(\mu_1;\mathbb{R})} &\leq \|g\,Kf\|_{L^1(\mu_1 \times \mu_2;\mathbb{R})} \\
&\leq \big\|\, |K|^\alpha f \big\|_{L^{(r,p)}((\mu_1,\mu_2);\mathbb{R})} \big\|g\,|K|^{1-\alpha}\big\|_{L^{(r',p')}((\mu_1,\mu_2);\mathbb{R})}.
\end{aligned}$$

Next, observe that

$$\big\| |K|^\alpha f \big\|_{L^{(r,p)}((\mu_1,\mu_2);\mathbb{R})}$$

$$= \left[\int_{E_2} \left(\int_{E_1} |K(x_1,x_2)|^{\alpha r} |f(x_2)|^r \, \mu_1(dx_1) \right)^{\frac{p}{r}} \mu_2(dx_2) \right]^{\frac{1}{p}} \le M_1^\alpha \|f\|_{L^p(\mu_2;\mathbb{R})}.$$

At the same time, since $p \le r$ and therefore $r' \le p'$, one can apply Theorem 6.2.6 to see that $\|g\,|K|^{1-\alpha}\|_{L^{(r',p')}((\mu_1,\mu_2);\mathbb{R})} \le \|g\,|K|^{1-\alpha}\|_{L^{(p',r')}((\mu_2,\mu_1);\mathbb{R})}$. By the same reasoning as we just applied to $\||K|^\alpha f\|_{L^{(r,p)}((\mu_1,\mu_2);\mathbb{R})}$, we find that $\|g\,|K|^{1-\alpha}\|_{L^{(r',p')}((\mu_1,\mu_2);\mathbb{R})} \le M_2^{1-\alpha}\|g\|_{L^{r'}(\mu_2;\mathbb{R})}$. Combining these two, we arrive at $\|g\,\mathcal{K}f\|_{L^1(\mu_1;\mathbb{R})} \le M_1^\alpha M_2^{1-\alpha}\|f\|_{L^p(\mu_1;\mathbb{R})}$ for all $g \in L^{r'}(\mu_1;\mathbb{R})$ with $\|g\|_{L^{r'}(\mu_1;\mathbb{R})} \le 1$; and so the asserted estimate now follows from part **(iii)** of Theorem 6.2.4 . □

In Lemma 6.3.1, $\mathcal{K}f$ was defined only when $f \in L^{q'}(\mu_2;\mathbb{R})$. However, the estimate in that lemma says that, as long as $\frac{1}{p} + \frac{1}{q} \ge 1$, the definition can be extended as a map from $L^p(\mu_2;\mathbb{R})$ into $L^r(\mu_1;\mathbb{R})$, where $\frac{1}{r} = \frac{1}{p} + \frac{1}{q} - 1$. A precise description of this extension is the content of the following.

Theorem 6.3.2 *Let everything be as in Lemma 6.3.1. For measurable $f : E_2 \longrightarrow \mathbb{R}$, define*

$$\Lambda_K(f) = \left\{ x_1 \in E_1 : \int_{E_2} \big|K(x_1,x_2)\big|\,\big|f(x_2)\big|\,\mu_2(dx_2) < \infty \right\}$$

and

$$\overline{\mathcal{K}}f(x_1) = \begin{cases} \int_{E_2} K(x_1,x_2)\,f(x_2)\,\mu_2(dx_2) & \text{if } x_1 \in \Lambda_K(f) \\ 0 & \text{otherwise.} \end{cases}$$

Next, let $p \in [1,\infty]$ satisfying $\frac{1}{p} + \frac{1}{q} \ge 1$ be given, and define $r \in [1,\infty]$ by $\frac{1}{r} = \frac{1}{p} + \frac{1}{q} - 1$. Then

$$\mu_1\left(\Lambda_K(f)^\complement\right) = 0 \quad \text{and} \quad \big\|\overline{\mathcal{K}}f\big\|_{L^r(\mu_1;\mathbb{R})} \le M_1^{\frac{q}{r}} M_2^{1-\frac{q}{r}} \|f\|_{L^p(\mu_2;\mathbb{R})} \tag{6.3.1}$$

for $f \in L^p(\mu_2;\mathbb{R})$. In particular, as a map from $L^p(\mu_2;\mathbb{R})$ into $L^r(\mu_1;\mathbb{R})$, $\overline{\mathcal{K}}$ is linear. In fact, $f \in L^p(\mu_2;\mathbb{R}) \longmapsto \overline{\mathcal{K}}f \in L^r(\mu_1;\mathbb{R})$ is the unique continuous mapping from $L^p(\mu_2;\mathbb{R})$ into $L^r(\mu_1;\mathbb{R})$ that coincides with $\mathcal{K}$ on $L^p(\mu_2;\mathbb{R}) \cap L^{q'}(\mu_2;\mathbb{R})$.

Proof. If $r = \infty$, and therefore $p = q'$, there is nothing to do. Thus, we will assume that r and therefore p are finite.

Let $f \in L^p(\mu_2;\mathbb{R})$ be given, and set $f_n = f\mathbf{1}_{[-n,n]} \circ f$ for $n \in \mathbb{Z}^+$. Because $p < q'$ and $f_n \in L^p(\mu_2;\mathbb{R}) \cap L^\infty(\mu_2;\mathbb{R})$, $f_n \in L^p(\mu_2;\mathbb{R}) \cap L^{q'}(\mu_2;\mathbb{R})$. Hence, by the estimate in Lemma 6.3.1 applied to $|K|$ and $|f_n|$,

$$\int_{E_1}\left(\int_{\{x_2:|f(x_2)|\le n\}}|K(x_1,x_2)|\,|f_n(x_2)|\,\mu_2(dx_2)\right)^r\mu_1(dx_1)$$
$$\le M_1^q\,M_2^{r-q}\|f_n\|^r_{L^p(\mu_2;\mathbb{R})}\le M_1^q\,M_2^{r-q}\|f\|^r_{L^p(\mu_2;\mathbb{R})}.$$

In particular, by the monotone convergence theorem, this proves both parts of (6.3.1). Furthermore, if $f,\,g\in L^p(\mu_2;\mathbb{R})$ and $\alpha,\,\beta\in\mathbb{R}$, then

$$\overline{\mathcal{K}}(\alpha f+\beta g)=\alpha\overline{\mathcal{K}}f+\beta\overline{\mathcal{K}}g\quad\text{on}\quad\Lambda_K(f)\cap\Lambda_K(g).$$

Thus, since both $\Lambda_K(f)^\complement$ and $\Lambda_K(g)^\complement$ have μ_1-measure 0, we now see that, as a mapping from $L^p(\mu;\mathbb{R})$ into $L^r(\mu_1;\mathbb{R})$, $\overline{\mathcal{K}}$ is linear. Finally, it is obvious that $\overline{\mathcal{K}}f=\mathcal{K}f$ for $f\in L^p(\mu_2;\mathbb{R})\cap L^{q'}(\mu_2;\mathbb{R})$. Hence, if $\mathcal{K}'$ is any extension of $\mathcal{K}\restriction L^p(\mu_2;\mathbb{R})\cap L^{q'}(\mu_2;\mathbb{R})$ as a continuous, linear mapping from $L^p(\mu_2;\mathbb{R})$ to $L^r(\mu_1;\mathbb{R})$, then (with the same choice of $\{f_n:n\ge 1\}$ as above)

$$\left\|\overline{\mathcal{K}}f-\mathcal{K}'f\right\|_{L^r(\mu_1;\mathbb{R})}\le\varliminf_{n\to\infty}\left\|\overline{\mathcal{K}}f-\mathcal{K}f_n\right\|_{L^r(\mu_1;\mathbb{R})}$$
$$=\varliminf_{n\to\infty}\left\|\overline{\mathcal{K}}(f-f_n)\right\|_{L^r(\mu_1;\mathbb{R})}\le M_1^{\frac{q}{r}}\,M_2^{1-\frac{q}{r}}\varliminf_{n\to\infty}\|f-f_n\|_{L^p(\mu_1;\mathbb{R})}=0.$$

□

6.3.2 Convolutions and Young's inequality

One of the many applications of Theorem 6.3.2 is to the multiplication operation known as convolution. To be precise, given $p\in[1,\infty]$, define the **convolution** $f*g$ of a function $f\in L^p(\lambda_{\mathbb{R}^N};\mathbb{R})$ with a function $g\in L^{p'}(\lambda_{\mathbb{R}^N};\mathbb{R})$ by

$$f*g(x)=\int_{\mathbb{R}^N}f(y)g(x-y)\,dy. \tag{6.3.2}$$

Clearly $f*g$ is well-defined and measurable for f and g from Hölder complementary Lebesgue spaces. In addition, for such functions, it is obvious that $f*g=g*f$, $f*g$ is bounded and that $\|f*g\|_{\mathrm{u}}\le\|f\|_{L^p(\lambda_{\mathbb{R}^N};\mathbb{R})}\|g\|_{L^{p'}(\lambda_{\mathbb{R}^N};\mathbb{R})}$. However, it is less obvious whether convolution can be extended to functions that are not in complementary Hölder spaces or, if it can, in what space the resulting $f*g$ will land. For this reason, the following is interesting.

Theorem 6.3.3 (Young's inequality) *Let p and q from $[1,\infty]$ satisfying $\frac{1}{p}+\frac{1}{q}\ge 1$ be given, and define $r\in[1,\infty]$ by $\frac{1}{r}=\frac{1}{p}+\frac{1}{q}-1$. Then, for each $f\in L^p(\lambda_{\mathbb{R}^N};\mathbb{R})$ and $g\in L^q(\lambda_{\mathbb{R}^N};\mathbb{R})$, the complement of the set*

$$\Lambda(f,g)\equiv\left\{x\in\mathbb{R}^N:\int_{\mathbb{R}^N}|f(x-y)|\,|g(y)|\,dy<\infty\right\}$$

has Lebesgue measure 0*. Furthermore, if*

$$f * g(x) \equiv \begin{cases} \int_{\mathbb{R}^N} f(x-y)\, g(y)\, dy & \text{when } x \in \Lambda(f,g) \\ 0 & \text{otherwise,} \end{cases}$$

then $f * g = g * f$ *and*

$$\|f * g\|_{L^r(\lambda_{\mathbb{R}^N};\mathbb{R})} \le \|f\|_{L^p(\lambda_{\mathbb{R}^N};\mathbb{R})} \|g\|_{L^q(\lambda_{\mathbb{R}^N};\mathbb{R})}. \tag{6.3.3}$$

Finally, the mapping $(f,g) \in L^p(\lambda_{\mathbb{R}^N};\mathbb{R}) \times L^q(\lambda_{\mathbb{R}^N};\mathbb{R}) \longmapsto f*g \in L^r(\lambda_{\mathbb{R}^N};\mathbb{R})$ *is bilinear.*

Proof. As was observed above, there is nothing to do when $r = \infty$ and therefore $q = p'$. Thus, we will assume throughout that r and therefore also p and q are all finite. Next, using the translation invariance of Lebesgue measure, first note that $\Lambda(f,g) = \Lambda(g,f)$ and then conclude that $f*g = g*f$. Finally, given $q \in [1,\infty)$ and $g \in L^q(\mathbb{R}^N;\mathbb{R})$, set $K(x,y) = g(x-y)$ for $x,\ y \in \mathbb{R}^N$. Obviously,

$$\sup_{y \in \mathbb{R}^N} \|K(\,\cdot\,,y)\|_{L^q(\lambda_{\mathbb{R}^N};\mathbb{R})} = \sup_{x \in \mathbb{R}^N} \|K(x,\cdot\,)\|_{L^q(\lambda_{\mathbb{R}^N};\mathbb{R})} = \|g\|_{L^q(\lambda_{\mathbb{R}^N};\mathbb{R})} < \infty;$$

and, in the notation of Theorem 6.3.2, $\Lambda(f,g) = \Lambda_K(f)$ and $f * g = \overline{\mathcal{K}}f$. In particular, for each $f \in L^p(\lambda_{\mathbb{R}^N};\mathbb{R})$, $\Lambda(f,g)^{\complement}$ has Lebesgue measure 0 and (6.3.3) holds. In addition, $f \in L^p(\lambda_{\mathbb{R}^N};\mathbb{R}) \longmapsto f * g \in L^r(\lambda_{\mathbb{R}^N};\mathbb{R})$ is linear for each $g \in L^q(\lambda_{\mathbb{R}^N};\mathbb{R})$; and therefore the bilinearity assertion follows after one reverses the roles of f and g. □

The most frequent applications of these results are to situations in which $f \in L^p(\lambda_{\mathbb{R}^N};\mathbb{R})$ and $g \in L^q(\lambda_{\mathbb{R}^N};\mathbb{R})$ where either $p = q'$ (and therefore $r = \infty$) or $p = 1$ (and therefore $r = q$). To get more information about the case $p = q'$, we will need the following.

Lemma 6.3.4 *Given* $h \in \mathbb{R}^N$*, define* $\tau_h f$ *for functions* f *on* $\mathbb{R}^N$ *by* $\tau_h f = f \circ T_h = f(\cdot + h)$*. Then* τ_h *maps* $L^p(\lambda_{\mathbb{R}^N};\mathbb{R})$ *isometrically onto itself for every* $h \in \mathbb{R}^N$ *and* $p \in [1,\infty]$*. Moreover, if* $p \in [1,\infty)$ *and* $f \in L^p(\lambda_{\mathbb{R}^N};\mathbb{R})$*, then*

$$\lim_{h \to 0} \|\tau_h f - f\|_{L^p(\lambda_{\mathbb{R}^N};\mathbb{R})} = 0. \tag{6.3.4}$$

Proof. The first assertion is an immediate consequence of the translation invariance of Lebesgue measure.

Next, let $p \in [1,\infty)$ be given. If $\mathcal{G}$ denotes the class of $f \in L^p(\lambda_{\mathbb{R}^N};\mathbb{R})$ for which (6.3.4) holds, it is clear that $C_c(\mathbb{R}^N;\mathbb{R}) \subseteq \mathcal{G}$. Hence, by **(v)** in Theorem 6.2.1, we will know that $\mathcal{G} = L^p(\lambda_{\mathbb{R}^N};\mathbb{R})$ as soon as we show that $\mathcal{G}$ is closed in $L^p(\lambda_{\mathbb{R}^N};\mathbb{R})$. To this end, let $\{f_n : n \ge 1\} \subseteq \mathcal{G}$ and suppose that $f_n \longrightarrow f$ in $L^p(\lambda_{\mathbb{R}^N};\mathbb{R})$. Then

$$\begin{aligned}&\overline{\lim_{h\to 0}}\,\|\tau_h f - f\|_{L^p(\lambda_{\mathbb{R}^N};\mathbb{R})}\\&\le \overline{\lim_{h\to 0}}\,\|\tau_h(f-f_n)\|_{L^p(\lambda_{\mathbb{R}^N};\mathbb{R})} + \overline{\lim_{h\to 0}}\,\|\tau_h f_n - f_n\|_{L^p(\lambda_{\mathbb{R}^N};\mathbb{R})} + \|f_n - f\|_{L^p(\lambda_{\mathbb{R}^N};\mathbb{R})}\\&\quad = 2\|f_n - f\|_{L^p(\lambda_{\mathbb{R}^N};\mathbb{R})} \longrightarrow 0 \quad \text{as } n\to\infty.\end{aligned}$$

□

Theorem 6.3.5 *Let $p \in [1,\infty]$, $f \in L^p(\lambda_{\mathbb{R}^N};\mathbb{R})$, and $g \in L^{p'}(\lambda_{\mathbb{R}^N};\mathbb{R})$. Then*

$$\tau_h(f*g) = (\tau_h f)*g = f*(\tau_h g) \quad \textit{for all } h\in\mathbb{R}^N.$$

Moreover, $f*g$ is uniformly continuous on $\mathbb{R}^N$ and

$$\|f*g\|_{\mathrm{u}} \le \|f\|_{L^p(\lambda_{\mathbb{R}^N};\mathbb{R})}\,\|g\|_{L^{p'}(\lambda_{\mathbb{R}^N};\mathbb{R})}.$$

Finally, if $p\in(1,\infty)$, then $\lim_{|x|\to\infty} f*g(x) = 0$.

Proof. The first assertion is again just an expression of translation invariance for $\lambda_{\mathbb{R}^N}$, and the estimate on $\|f*g\|_{\mathrm{u}}$ is a simple application of either Hölder's inequality or Theorem 6.3.3. To see that $f*g$ is uniformly continuous, first suppose that $p\in[1,\infty)$. Then, by the preceding estimate, Hölder's inequality, and (6.3.4),

$$\|\tau_h(f*g) - f*g\|_{\mathrm{u}} = \|(\tau_h f - f)*g\|_{\mathrm{u}} \le \|\tau_h f - f\|_{L^p(\lambda_{\mathbb{R}^N};\mathbb{R})}\|g\|_{L^{p'}(\mathbb{R}^N)} \longrightarrow 0$$

as $|h|\to 0$. When $p=\infty$, reverse the roles of f and g in this argument.

To prove the final assertion, first let $f\in C_{\mathrm{c}}(\mathbb{R}^N;\mathbb{R})$ be given, and define $\mathcal{G}_f$ to be the class of $g\in L^{p'}(\lambda_{\mathbb{R}^N};\mathbb{R})$ for which the assertion holds. Then it is easy to check that $C_{\mathrm{c}}(\mathbb{R}^N;\mathbb{R})\subseteq\mathcal{G}_f$. Moreover, from our estimate on $\|f*g\|_{\mathrm{u}}$ one sees that $\mathcal{G}_f$ is closed in $L^{p'}(\lambda_{\mathbb{R}^N};\mathbb{R})$. Hence, just as in the final step of the proof of Lemma 6.3.4, we conclude that $\mathcal{G}_f = L^{p'}(\lambda_{\mathbb{R}^N};\mathbb{R})$. To complete the proof, let $g\in L^{p'}(\lambda_{\mathbb{R}^N};\mathbb{R})$ be given, and define $\mathcal{H}_g$ to be the class of $f\in L^p(\lambda_{\mathbb{R}^N};\mathbb{R})$ for which the assertion holds. By the preceding, we know that $C_{\mathrm{c}}(\lambda_{\mathbb{R}^N};\mathbb{R})\subseteq\mathcal{H}_g$. Moreover, just as before, $\mathcal{H}_g$ is closed in $L^p(\lambda_{\mathbb{R}^N};\mathbb{R})$, and therefore $\mathcal{H}_g = L^p(\lambda_{\mathbb{R}^N};\mathbb{R})$. □

Both Theorems 6.3.3 and 6.3.5 tell us that the convolution product of two functions is often more regular than one or both of its factors. An application of this fact is given in Exercise 6.3.2 below, where it used to give an elegant derivation of Lemma 2.2.14. What follows can be seen as a further development of the same basic fact.

Lemma 6.3.6 *Let $p\in[1,\infty]$ and $g\in C^1(\mathbb{R}^N;\mathbb{R})$ be given. If g as well as $\partial_{x_i}g$ for each $1\le i\le N$ are elements of $L^{p'}(\lambda_{\mathbb{R}^N};\mathbb{R})$, then, for every $f\in L^p(\lambda_{\mathbb{R}^N};\mathbb{R})$, $f*g\in C^1(\mathbb{R}^N;\mathbb{R})$ and $\partial_{x_i}(f*g) = f*(\partial_{x_i}g)$.*

Proof. Let $\mathbf{e} \in \mathbb{S}^{N-1}$ be given. By Theorem 6.3.5,

$$\tau_{t\mathbf{e}}(f * g) - f * g = f * \big(\tau_{t\mathbf{e}}g - g\big)$$

for every $t \in \mathbb{R}$. Since

$$\frac{\tau_{t\mathbf{e}}g(y) - g(y)}{t} = \int_{[0,1]} \big(\mathbf{e}, \nabla g(y + st\mathbf{e})\big)_{\mathbb{R}^N}\, ds,$$

and, when $p' \in [1, \infty)$, Theorem 6.2.6 together with Lemma 6.3.4 implies that

$$\begin{aligned} &\left\| \int_{[0,1]} \big(\mathbf{e}, \nabla g(\,\cdot\, + st\mathbf{e}) - \nabla g(\,\cdot\,)\big)_{\mathbb{R}^N}\, ds \right\|_{L^{p'}(\lambda_{\mathbb{R}^N};\mathbb{R})} \\ &\qquad \le \int_{[0,1]} \left\| \tau_{st\mathbf{e}}\big(\mathbf{e}, \nabla g\big)_{\mathbb{R}^N} - \big(\mathbf{e}, \nabla g\big)_{\mathbb{R}^N} \right\|_{L^{p'}(\lambda_{\mathbb{R}^N};\mathbb{R})} ds \longrightarrow 0 \end{aligned}$$

as $t \to 0$, the required result follows from (6.3.3). On the other hand, if $p' = \infty$, then, for $\lambda_{\mathbb{R}^N}$-almost every y,

$$\frac{\tau_{t\mathbf{e}}g(y) - g(y)}{t} = \int_{[0,1]} \big(\mathbf{e}, \nabla g(y + st\mathbf{e})\big)_{\mathbb{R}^N}\, ds \longrightarrow \big(\mathbf{e}, \nabla g(y)\big)_{\mathbb{R}^N}$$

boundedly and pointwise, and therefore the result follows in this case from Lebesgue's dominated convergence theorem. □

6.3.3 Friedrichs Mollifiers

The results in the preceding subsection lead immediately to the conclusion that the smoother g is, the smoother $f * g$ is. Therefore convolution provides a flexible procedure for transforming integrable functions into smooth ones. The systematic development of this procedure seems to been made first by K.O. Friedrichs.

For a given a $\alpha = (\alpha_1, \dots, \alpha_N) \in \mathbb{N}^N$, set $\|\alpha\| = \sum_1^N \alpha_i$ and

$$\partial^\alpha = \partial_{x_1}^{\alpha_1} \cdots \partial_{x_N}^{\alpha_N}.$$

One of the two key ingredients in Friedrichs's procedure is the following immediate corollary of Lemma 6.3.6. Namely, if $g \in C^\infty(\lambda_{\mathbb{R}^N}; \mathbb{R})$ and $\partial^\alpha g \in L^{p'}(\lambda_{\mathbb{R}^N}; \mathbb{R})$ for some $p \in [1, \infty]$ and all α's, then

$$f * g \in C^\infty(\mathbb{R}^N; \mathbb{R}) \quad \text{and} \quad \partial^\alpha(f * g) = f * (\partial^\alpha g) \tag{6.3.5}$$

for every $f \in L^p(\lambda_{\mathbb{R}^N}; \mathbb{R})$.

The second key ingredient is the following general method for constructing what are called approximate identities.

Theorem 6.3.7 *Given* $g \in L^1(\lambda_{\mathbb{R}^N};\mathbb{R})$*, define* $g_t(\,\cdot\,) = t^{-N}g(t^{-1}\,\cdot\,)$ *for* $t > 0$*. Then* $g_t \in L^1(\lambda_{\mathbb{R}^N};\mathbb{R})$ *and* $\int g_t\,dx = \int g\,dx$*. In addition, if* $\int g\,dx = 1$*, then for every* $f \in C_{\mathrm{b}}(\mathbb{R}^N;\mathbb{C})$*,* $f * g_t \longrightarrow f$ *uniformly on compacts, and, for each* $p \in [1,\infty)$ *and* $f \in L^p(\lambda_{\mathbb{R}^N};\mathbb{R})$*,*

$$\lim_{t\searrow 0}\big\|f * g_t - f\big\|_{L^p(\lambda_{\mathbb{R}^N};\mathbb{R})} = 0.$$

Proof. The initial assertions are trivial.

Now assume that $\int g\,dx = 1$. Given $f \in C_{\mathrm{b}}(\mathbb{R}^N;\mathbb{C})$, note that, for $\lambda_{\mathbb{R}^N}$-almost every $x \in \mathbb{R}^N$,

$$f*g_t(x)-f(x) = \int_{\mathbb{R}^N}\big(f(x-y)-f(x)\big)g_t(y)\,dy = \int_{\mathbb{R}^N}\big(f(x-ty)-f(x)\big)g(y)\,dy.$$

Hence, if f is continuous, then, by Lebesgue's dominated convergence theorem, $f * g_t \longrightarrow f$ uniformly on compacts. To handle the case when $f \in L^p(\mu;\mathbb{C})$ for some $p \in [1,\infty)$, set $\Psi^t(x,y) = \big(f(x-ty) - f(x)\big)g(y)$. Then, by Theorem 6.2.6,

$$\begin{aligned}\|f * g_t - f\|_{L^p(\lambda_{\mathbb{R}^N};\mathbb{R})} &\le \|\Psi^t\|_{L^{(1,p)}((\lambda_{\mathbb{R}^N},\lambda_{\mathbb{R}^N});\mathbb{R})}\\ &\le \|\Psi^t\|_{L^{(p,1)}((\lambda_{\mathbb{R}^N},\lambda_{\mathbb{R}^N});\mathbb{R})} = \int_{\mathbb{R}^N}\big\|\tau_{-ty}f - f\big\|_{L^p(\lambda_{\mathbb{R}^N};\mathbb{R})}|g(y)|\,dy.\end{aligned}$$

Since $\|\tau_{-ty}f - f\|_{L^p(\lambda_{\mathbb{R}^N};\mathbb{R})} \le 2\|f\|_{L^p(\lambda_{\mathbb{R}^N};\mathbb{R})}$, the result follows from the above combined with Lebesgue's dominated convergence theorem and (6.3.4). □

Theorem 6.3.7 should make it clear why, when $g \in L^1(\lambda_{\mathbb{R}}^N;\mathbb{R})$ and $\int g\,dx = 1$, the corresponding family $\{g_t : t > 0\}$ is called an **approximate identity**. To understand how an approximate identity actually produces *an approximation of the identity,* consider a g which is non-negative and vanishes off of $B(0,1)$. Then *the volume under the graph* of g_t continues to be 1 as $t \searrow 0$, whereas the base of the graph is restricted to $B(0,t)$. Hence, all the *mass* is getting concentrated over the origin.

Combining Theorem 6.3.5 and (6.3.5), we arrive at Friedrichs's main result.

Corollary 6.3.8 *Let* $g \in C^\infty(\mathbb{R}^N;\mathbb{R}) \cap L^1(\lambda_{\mathbb{R}^N};\mathbb{R})$ *with* $\int_{\mathbb{R}^N} g(x)\,dx = 1$ *be given. In addition, let* $p \in [1,\infty)$*, and assume that* $\partial^\alpha g \in L^{p'}(\lambda_{\mathbb{R}^N};\mathbb{R})$ *for all* $\alpha \in \mathbb{N}^N$*. Then, for each* $f \in L^p(\lambda_{\mathbb{R}^N};\mathbb{R})$*,* $f * g_t \longrightarrow f$ *in* $L^p(\lambda_{\mathbb{R}^N};\mathbb{R})$ *as* $t \searrow 0$*. Finally, for each* $t > 0$*,* $f * g_t$ *has bounded, continuous derivatives of all orders and* $\partial^\alpha\big(f * g_t\big) = f * \big(\partial^\alpha g_t\big)$*,* $\alpha \in \mathbb{N}^N$*.*

Because it is usually employed to transform rough functions into smooth ones, the procedure described in Corollary 6.3.8 is called a **mollification**

procedure, and, when it is non-negative, smooth, and has compact support (i.e., vanishes off of a compact), the function g is called a **Friedrichs mollifier**. It takes a second to figure out how one might go about constructing a Friedrichs mollifier: it must be smooth, but it cannot be analytic. The standard construction is the following. Begin by considering the function $\psi : [0,\infty) \longrightarrow [0,\infty)$ that vanishes at 0 and is equal to $e^{-\frac{1}{\xi}}$ for $\xi > 0$. Because derivatives of ψ away from 0 are all polynomials in ξ^{-1} times ψ, it is easy to convince oneself that $\psi \in C^\infty\big([0,\infty);[0,\infty)\big)$. Hence, the function $x \in \mathbb{R}^N \longmapsto \psi\big((1-|x|^2)^+\big) \in [0,\infty)$ is a smooth function that vanishes off of $B(0,1)$. Now set

$$\rho(x) = k_N \begin{cases} \exp\big(-(1-|x|^2)^{-1}\big) & \text{if } x \in B(0,1) \\ 0 & \text{if } x \notin B(0,1), \end{cases} \tag{6.3.6}$$

where k_N is chosen to make ρ have total integral 1. Then ρ is a non-negative Friedrichs mollifier that is supported on $B(0,1)$.

An important application of Friedrichs mollifiers is to the construction of smoothed replacements, known as **bump functions**, for indicator functions. That is, given a $\Gamma \in \mathcal{B}_{\mathbb{R}^N}$ and an $\epsilon > 0$, then an associated bump function is a smooth function that is bounded below by $\mathbf{1}_\Gamma$ and above by $\mathbf{1}_{\Gamma^{(\epsilon)}}$. Obviously, if ρ is a Friedrichs mollifier supported on $B(0,1)$, then $\rho_{\frac{\epsilon}{2}} * \mathbf{1}_{\Gamma^{(\frac{\epsilon}{2})}}$ is an associated bump function. To appreciate how useful bump functions can be, recall the statement that I gave of the Divergence Theorem. Even though I assumed that the vector field V was globally defined and had uniformly bounded first derivatives, one should have suspected that only properties of V in a neighborhood of the closure of G ought to matter, and we can now show that to be the case. Namely, suppose that V is a twice continuously differentiable vector field defined on an open neighborhood of the bounded smooth region G, and choose an $\epsilon > 0$ for which $\overline{G^{(\epsilon)}}$ is a compact subset of that neighborhood. Next, take $\eta_\epsilon = \rho_{\frac{\epsilon}{2}} * \mathbf{1}_{G^{(\frac{\epsilon}{2})}}$, and define V_ϵ to be equal to $\eta_\epsilon V$ on $G^{(\epsilon)}$ and 0 off of $G^{(\epsilon)}$. Since the Divergence Theorem applies to V_ϵ and gives a conclusion that depends only on $V \restriction \bar{G}$, we now know that the global hypotheses in Theorem 5.3.1 can be replaced by local ones. Similarly, the hypotheses in Corollary 5.3.2 as well as those in Exercises 5.3.2–5.3.4 admit localization.

Finally, the following application of Friedrichs mollifiers shows that the smooth, compactly supported functions are dense in the Lebesgue spaces.

Theorem 6.3.9 *For each $p \in [1,\infty)$, the space $C_c^\infty(\mathbb{R}^N;\mathbb{R})$ of compactly supported, infinitely differentiable functions is dense in $L^p(\lambda_{\mathbb{R}^N};\mathbb{R})$. In fact, if $p,\ q \in [1,\infty)$ and $f \in L^p(\lambda_{\mathbb{R}^N};\mathbb{R}) \cap L^q(\lambda_{\mathbb{R}^N};\mathbb{R})$, then there exists a sequence $\{\varphi_n : n \geq 1\} \subseteq C_c^\infty(\mathbb{R}^N;\mathbb{R})$ for which $\varphi_n \longrightarrow f$ in both $L^p(\lambda_{\mathbb{R}^N};\mathbb{R})$ and $L^q(\lambda_{\mathbb{R}^N};\mathbb{R})$.*

Proof. Given $f \in L^p(\lambda_{\mathbb{R}^N};\mathbb{R}) \cap L^q(\lambda_{\mathbb{R}^N};\mathbb{R})$, Lebesgue's dominated convergence theorem implies that

$$\lim_{R\to\infty} \int_{|x|>R} \left(|f(x)|^p + |f(x)|^q\right) dx = 0.$$

Hence, it suffices to treat f's that have compact support. Furthermore, if $f \in L^p(\lambda_{\mathbb{R}^N};\mathbb{R}) \cap L^q(\lambda_{\mathbb{R}^N};\mathbb{R})$ has compact support and ρ is a Friedrichs mollifier, then $\rho_t * f \in C_c^\infty(\mathbb{R}^N;\mathbb{R})$ for each $t > 0$ and, as $t \searrow 0$, $\rho_t * f \longrightarrow f$ in both $L^p(\lambda_{\mathbb{R}^N};\mathbb{R})$ and $L^q(\lambda_{\mathbb{R}^N};\mathbb{R})$. □

6.3.4 Exercises for § 6.3

Exercise 6.3.1 Given $f,\, g \in L^1(\lambda_{\mathbb{R}^N};\mathbb{R})$, show that

$$\int f * g(x)\, dx = \int f(x)\, dx \int g(x)\, dx.$$

Exercise 6.3.2 Here is the promised alternative derivation of Lemma 2.2.14. Let $\Gamma \in \mathcal{B}_{\mathbb{R}^N}$ have finite, positive Lebesgue measure, and set $u = \mathbf{1}_{-\Gamma} * \mathbf{1}_{\Gamma}$. Show that $u(x) \le \lambda_{\mathbb{R}^N}(\Gamma)\mathbf{1}_{\Delta}$, where $\Delta = \Gamma - \Gamma \equiv \{y - x : x, y \in \Gamma\}$, and that $u(0) = \lambda_{\mathbb{R}^N}(\Gamma) > 0$. Use these observations, together with Theorem 6.3.5, to prove that Δ contains a ball $B(0,\delta)$ for some $\delta > 0$. Thus, not only Lemma 2.2.14 but its $\mathbb{R}^N$-analog of hold.

Exercise 6.3.3 Given a family $\{f_\alpha : \alpha \in (0,\infty)\} \subseteq L^1(\lambda_{\mathbb{R}^N};\mathbb{R})$, one says that the family is a **convolution semigroup** if $f_{\alpha+\beta} = f_\alpha * f_\beta$ for all $\alpha,\, \beta \in (0,\infty)$. Here are four famous examples of convolution semigroups, three of which are obtained by scaling.

(i) Define the **Gauss kernel** $\gamma(x) = (2\pi)^{-\frac{N}{2}} \exp\left(-\frac{|x|^2}{2}\right)$ for $x \in \mathbb{R}^N$. Using the result in part **(i)** Exercise 5.1.3, show that $\int_{\mathbb{R}^N} \gamma(x)\, dx = 1$ and that

$$\gamma_{\sqrt{s}} * \gamma_{\sqrt{t}} = \gamma_{\sqrt{s+t}} \quad \text{for } s,\, t \in (0,\infty).$$

Clearly this says that the approximate identity $\{\gamma_{\sqrt{t}} : t \in (0,\infty)\}$ is a convolution semigroup of functions. It is known as either the **heat flow semigroup** or the **Weierstrass semigroup**.

(ii) Define ν on $\mathbb{R}$ by

$$\nu(y) = \frac{\mathbf{1}_{(0,\infty)}(y)\, e^{-\frac{1}{y}}}{\pi^{\frac{1}{2}} y^{\frac{3}{2}}}.$$

Again using the notation in Theorem 6.3.7, show that $\int_{\mathbb{R}} \nu(y)\, dy = 1$ and that

$$\nu_{s^2} * \nu_{t^2} = \nu_{(s+t)^2} \quad \text{for } s,\, t \in (0,\infty).$$

Hence, here again, we have an approximate identity that is a convolution semigroup. The family $\{\nu_{t^2} : t > 0\}$, or, more precisely, the probability measures $A \rightsquigarrow \int_A \nu_{t^2}(y)\,dy$, play a role in probability theory, where they are called the **one-sided stable laws of order $\frac{1}{2}$**.

Hint: Note that for $x \in (0,\infty)$,

$$\nu_{s^2} * \nu_{t^2}(x) = \frac{st}{\pi}\int_{(0,x)} \frac{1}{\big(y(x-y)\big)^{\frac{3}{2}}}\exp\left[-\frac{s^2}{y} - \frac{t^2}{x-y}\right] dy,$$

try the change of variable $\xi = \Phi(y) = \frac{y}{x-y}$, and use part (**iv**) of Exercise 5.1.3.

(**iii**) Using part (**ii**) of Exercise 5.2.3, check that the function P on $\mathbb{R}^N$ given by

$$P(x) = \frac{2}{\omega_N}\left(1+|x|^2\right)^{-\frac{N+1}{2}}, \quad x \in \mathbb{R}^N,$$

has Lebesgue integral 1. Next prove the representation

$$P_t(x) = t^{-1}P(t^{-1}x) = \int_{(0,\infty)} \gamma_{\sqrt{\frac{y}{2}}}(x)\nu_{t^2}(y)\,dy.$$

Finally, using this together with the preceding parts of this exercise, show that

$$P_s * P_t = P_{s+t}, \quad s,t \in (0,\infty),$$

and therefore that $\{P_t : t > 0\}$ is a convolution semigroup. This semigroup is known as the **Poisson semigroup** among harmonic analysts and as the **Cauchy semigroup** in probability theory. The representation of $P(x)$ in terms of the Gauss kernel is an example of how to obtain one semigroup from another via the method of **subordination**.

(**iv**) For each $t > 0$, define $G_t : \mathbb{R} \longrightarrow [0,\infty)$ so that $g_t(x) = 0$ if $x \le 0$ and, for $t > 0$ (cf. (**ii**) in Exercise 5.1.3) $G_t(x) = \big(\Gamma(t)\big)^{-1}x^{t-1}e^{-x}$ for $x > 0$. Clearly, $\int_{\mathbb{R}} G_t(x)\,dx = 1$. Next, check that

$$G_s * G_t(x) = \left(\frac{\Gamma(s+t)}{\Gamma(s)\Gamma(t)}\int_{(0,1)} y^{s-1}(1-y)^{t-1}\,dt\right) g_{s+t}(x),$$

and use this, together with Exercise 6.3.1, to give another derivation of the formula in (**i**) of Exercise 5.2.3, for the beta function. Clearly, it also shows that $\{G_t : t > 0\}$ is yet another convolution semigroup, although this one is not obtained by rescaling.

Exercise 6.3.4 Show that if μ is a finite Borel measure on $\mathbb{R}^N$ and $p \in [1,\infty]$, then for all $f \in C_c(\mathbb{R}^N;\mathbb{R})$ the function $f * \mu$ given by

$$f * \mu(x) = \int_{\mathbb{R}^N} f(x-y)\,\mu(dy), \quad x \in \mathbb{R}^N,$$

is continuous and satisfies

$$\|f * \mu\|_{L^p(\lambda_{\mathbb{R}^N};\mathbb{R})} \le \mu(\mathbb{R}^N)\|f\|_{L^p(\lambda_{\mathbb{R}^N};\mathbb{R})}. \tag{$*$}$$

Next, use $(*)$ to show that for each $p \in [1,\infty)$ there is a unique continuous map $\overline{\mathcal{K}}_\mu : L^p(\lambda_{\mathbb{R}^N};\mathbb{R}) \longrightarrow L^p(\lambda_{\mathbb{R}^N};\mathbb{R})$ such that $\overline{\mathcal{K}}_\mu f = f * \mu$ for $f \in C_c(\mathbb{R}^N;\mathbb{R})$. Finally, note that $(*)$ continues to hold when $f * \mu$ is replaced by $\overline{\mathcal{K}}_\mu f$.

Exercise 6.3.5 Define the σ-finite measure μ on $\big((0,\infty),\mathcal{B}_{(0,\infty)}\big)$ by

$$\mu(\Gamma) = \int_\Gamma \frac{1}{x}\,dx \quad \text{for} \quad \Gamma \in \mathcal{B}_{(0,\infty)},$$

and show that μ is **invariant under the multiplicative group** in the sense that

$$\int_{(0,\infty)} f(\alpha x)\,\mu(dx) = \int_{(0,\infty)} f(x)\,\mu(dx), \qquad \alpha \in (0,\infty),$$

and

$$\int_{(0,\infty)} f\left(\frac{1}{x}\right)\mu(dx) = \int_{(0,\infty)} f(x)\,\mu(dx)$$

for every $\mathcal{B}_{(0,\infty)}$-measurable $f : (0,\infty) \longrightarrow [0,\infty]$. Next, for $\mathcal{B}_{(0,\infty)}$-measurable, $\mathbb{R}$-valued functions f and g, set

$$\Lambda_\mu(f,g) = \left\{x \in (0,\infty) : \int_{(0,\infty)} \left|f\left(\frac{x}{y}\right)\right| |g(y)|\,\mu(dy) < \infty\right\},$$

$$f \bullet g(x) = \begin{cases} \int_{(0,\infty)} f\left(\frac{x}{y}\right) g(y)\,\mu(dy) & \text{when} \quad x \in \Lambda_\mu(f,g) \\ 0 & \text{otherwise,} \end{cases}$$

and show that $f \bullet g = g \bullet f$. In addition, show that if $p,\, q \in [1,\infty]$ satisfy $\frac{1}{r} \equiv \frac{1}{p} + \frac{1}{q} - 1 \ge 0$, then

$$\mu\left(\Lambda_\mu(f,g)^{\complement}\right) = 0 \quad \text{and} \quad \|f \bullet g\|_{L^r(\mu;\mathbb{R})} \le \|f\|_{L^p(\mu;\mathbb{R})}\,\|g\|_{L^q(\mu;\mathbb{R})}$$

for all $f \in L^p(\mu;\mathbb{R})$ and $g \in L^q(\mu;\mathbb{R})$. Finally, use these considerations to prove the following one of G.H. Hardy's many inequalities:

$$\left[\int_{(0,\infty)} \frac{1}{x^{1+\alpha}} \left(\int_{(0,x)} \varphi(y)\,dy\right)^p dx\right]^{\frac{1}{p}} \le \frac{p}{\alpha}\left(\int_{(0,\infty)} \frac{\big(y\varphi(y)\big)^p}{y^{1+\alpha}}\,dy\right)^{\frac{1}{p}}$$

for all $\alpha \in (0,\infty)$, $p \in [1,\infty)$, and non-negative, $\mathcal{B}_{(0,\infty)}$-measurable φ's.

Hint: To prove everything except Hardy's inequality, one can simply repeat the arguments used in the proof of Young's inequality. Alternatively, one can map the present situation into the one for convolution of functions on $\mathbb{R}$ by making a logarithmic change of variables. To prove Hardy's result, take

$$f(x) = \left(\frac{1}{x}\right)^{\frac{\alpha}{p}} \mathbf{1}_{[1,\infty)}(x) \quad \text{and} \quad g(x) = x^{1-\frac{\alpha}{p}}\varphi(x),$$

and use $\|f \bullet g\|_{L^p(\mu;\mathbb{R})} \leq \|f\|_{L^1(\mu;\mathbb{R})} \|g\|_{L^p(\mu;\mathbb{R})}$.

Chapter 7
Hilbert Space and Elements of Fourier Analysis

The basic tools of analysis rely on the translation invariance properties of functions and operations on Euclidean space, and, at least for applications involving integration, Fourier analysis is one of the most powerful techniques for exploiting translation invariance.

This chapter provides an introduction to Fourier analysis, but the treatment here barely scratches the surface. For more complete accounts, the reader should consult any one of the many excellent books devoted to the subject. For example, if a copy can be located, H. Dym and H.P. McKean's book *Fourier Series and Integrals*, published by Academic Press, is among the best places to gain an appreciation for the myriad ways in which Fourier analysis has been applied. More readily available is the loving, if somewhat whimsical, treatment given by T.W. Körner in *Fourier Analysis*, published by Cambridge University Press. At a more technically sophisticated level, E.M. Stein and G. Weiss's book *Introduction to Fourier Analysis on Euclidean Spaces*, published by Princeton University Press, is a good place to start, and for those who want to delve into the origins of the subject, A. Zygmund's classic *Trigonometric Series*, now available in paperback from Cambridge University Press, is indispensable.

Before getting to Fourier analysis itself, I will begin by presenting a few facts about Hilbert space, facts that subsequently will play a crucial role in my development of Fourier analysis.

7.1 Hilbert Space

In Exercise 6.1.4 it was shown that, among the L^p-spaces, the space L^2 is the most closely related to familiar Euclidean geometry. In the present section, I will expand on this observation and give some applications of it.

D. W. Stroock, *Essentials of Integration Theory for Analysis*, Graduate Texts in Mathematics 262, https://doi.org/10.1007/978-3-030-58478-8_7

7.1.1 Elementary Theory of Hilbert Spaces

By part (**iii**) of Theorem 6.2.1, if $(E, \mathcal{B}, \mu)$ is a measure space, then $L^2(\mu;\mathbb{R})$ is a vector space that becomes a complete metric space when $\|g-f\|_{L^2(\mu;\mathbb{R})}$ is used to measure the distance between functions g and f. Moreover, if $\mathcal{B}$ is countably generated and μ is σ-finite, then (cf. (**iv**) in Theorem 6.2.1) $L^2(\mu;\mathbb{R})$ is separable.

Next, if one defines

$$(f,g) \in \left(L^2(\mu;\mathbb{R})\right)^2 \longmapsto \left(f,g\right)_{L^2(\mu;\mathbb{R})} \equiv \int_E fg\,d\mu \in \mathbb{R},$$

then $\left(f,g\right)_{L^2(\mu;\mathbb{R})}$ is bilinear (i.e., it is linear as a function of each of its entries), and (cf. part (**ii**) of Theorem 6.2.4), for $f \in L^2(\mu;\mathbb{R})$,

$$\begin{aligned}\|f\|_{L^2(\mu;\mathbb{R})} &= \sqrt{(f,f)_{L^2(\mu;\mathbb{R})}}\\ &= \sup\Big\{\left(f,g\right)_{L^2(\mu;\mathbb{R})} : g \in L^2(\mu;\mathbb{R}) \text{ with } \|g\|_{L^2(\mu;\mathbb{R})} \le 1\Big\}.\end{aligned}$$

Thus, $\left(f,g\right)_{L^2(\mu;\mathbb{R})}$ plays the same role for $L^2(\mu;\mathbb{R})$ that the Euclidean inner product plays in $\mathbb{R}^N$. That is, by Schwarz's inequality (cf. Exercise 6.1.4),

$$\left|(f,g)_{L^2(\mu;\mathbb{R})}\right| \le \|f\|_{L^2(\mu;\mathbb{R})}\|g\|_{L^2(\mu;\mathbb{R})},$$

and

$$\arccos\left(\frac{(f,g)_{L^2(\mu;\mathbb{R})}}{\|f\|_{L^2(\mu;\mathbb{R})}\|g\|_{L^2(\mu;\mathbb{R})}}\right)$$

can be thought of as the *angle between* f *and* g in the plane that they span. For this reason, I will say that $f \in L^2(\mu;\mathbb{R})$ is **orthogonal** or **perpendicular** to $S \subseteq L^2(\mu;\mathbb{R})$ and write $f \perp S$ if $\left(f,g\right)_{L^2(\mu;\mathbb{R})} = 0$ for every $g \in S$ and will write $f \perp g$ when $S = \{g\}$.

For ease of presentation, as well as conceptual clarity, it is helpful to abstract these properties and to call a vector space that has them a **real Hilbert space**. More precisely, I will say that H is a **Hilbert space over the field** $\mathbb{R}$ if it is a vector space over $\mathbb{R}$ that possesses a symmetric bilinear map, known as the **inner product**, $(x,y) \in H^2 \longmapsto (x,y)_H \in \mathbb{R}$ with the properties that:

(**a**) $(x,x)_H \ge 0$ for all $x \in H$ and $\|x\|_H \equiv \sqrt{(x,x)_H} = 0$ if and only if $x = 0$ in H.

(**b**) H is complete as a metric space when the metric assigns $\|y-x\|_H$ as the distance between y and x.

A look at the reasoning suggested in Exercise 6.1.4 should be enough to convince one that bilinearity and symmetry combined with (**a**) implies the abstract *Schwarz's inequality* $|(x,y)_H| \le \|x\|_H\|y\|_H$. Thus, once again, it is natural to interpret

$$\arccos\left(\frac{(x,y)_H}{\|x\|_H\|y\|_H}\right)$$

as the angle between x and y. In particular, if $S \subseteq H$ and $x \in H$, I will say that x is **orthogonal** to S and will write $x \perp S$ if $(x,y)_H = 0$ for all $y \in S$ and $x \perp y$ when $S = \{y\}$. Finally, once one knows that Schwarz's inequality holds, the reasoning in Exercise 6.1.4 shows that $\|\cdot\|_H$ satisfies the triangle inequality $\|x+y\|_H \le \|x\|_H + \|y\|_H$, which, together with (**a**), means that $(x,y) \in H^2 \longrightarrow \|x-y\|_H \in [0,\infty)$ is a metric. In addition, because $|(x,z)_H - (y,z)_H| = |(x-y,z)_H| \le \|x-y\|_H\|z\|_H$, it should be clear that $(x,y) \in H^2 \longmapsto (x,y)_H \in \mathbb{R}$ is continuous.

A further abstraction, and one that will prove important when I discuss Fourier analysis, entails replacing the field $\mathbb{R}$ of real numbers by the field $\mathbb{C}$ of complex numbers and considering a **complex Hilbert space**. That is, H is a Hilbert space over $\mathbb{C}$ if it is a vector space over C that possesses a **Hermitian inner product** (property (**a**) below) $(z,\zeta) \in H^2 \longmapsto (z,\zeta)_H \in \mathbb{C}$ with the properties that

(**a**) $\overline{(z,\zeta)_H} = (\zeta,z)_H$ for all $z,\ \zeta \in H$,
(**b**) for each $\zeta \in H$, $z \in H \longmapsto (z,\zeta)_H \in \mathbb{C}$ is (complex) linear,
(**c**) $(z,z)_H \ge 0$ and $\|z\|_H \equiv \sqrt{(z,z)_H} = 0$ if and only if $z = 0$ in H,
(**d**) H is complete as a metric space when the metric assigns $\|z-\zeta\|_H$ as the distance between z and ζ.

Once again, the reasoning in Exercise 6.1.4 allows one to pass from (**a**), (**b**), and (**c**) to Schwarz's inequality $|(z,\zeta)_H| \le \|z\|_H\|\zeta\|_H$ and thence to the triangle inequality and the conclusion that $(z,\zeta) \in H^2 \longrightarrow \|z-\zeta\|_H \in [0,\infty)$ is a metric. Perhaps the easiest way to see that the argument survives the introduction of complex numbers is to begin with the observation that it suffices to handle the case in which $(z,\zeta)_H \ge 0$. Indeed, if this is not already the case, one can achieve it by multiplying z by a $\theta \in \mathbb{C}$ of modulus (i.e., absolute value) 1. Clearly, none of the quantities entering the inequality is altered by this multiplication, and when $(x,\zeta)_H \ge 0$ the reasoning given in Exercise 6.1.4 applies without change. Having verified Schwarz's inequality, there is good reason to continue thinking of

$$\arccos\left(\frac{|(z,\zeta)_H|}{\|z\|_H\|\zeta\|_H}\right)$$

as measuring the size of the angle between z and ζ. In particular, I will continue to say that z is orthogonal to S and write $z \perp S$ ($z \perp \zeta$) if $(z,\zeta)_H = 0$ for all $\zeta \in S$ (when $S = \{\zeta\}$), and, just as before, $(z,\zeta) \in H^2 \longmapsto (z,\zeta)_H \in \mathbb{C}$ is continuous.

Notice that any Hilbert space H over $\mathbb{R}$ can be *complexified.* For example, the complexification of $L^2(\mu;\mathbb{R})$ is $L^2(\mu;\mathbb{C})$, the space of measurable, $\mathbb{C}$-

valued functions whose absolute values are μ-square integrable, and with inner product given by

$$(f,g)_{L^2(\mu;\mathbb{C})} = \int f(x)\overline{g(x)}\,dx.$$

More generally, if H is a Hilbert space over $\mathbb{R}$, its complexification H_c is the complex vector space whose elements are of the form $x+iy$, where $x,\ y \in H$. The associated inner product is given by

$$\big(x+iy,\xi+i\eta\big)_{H_c} = \big(x,\xi\big)_H + \big(y,\eta\big)_H + i\Big(\big(y,\xi\big)_H - \big(x,\eta\big)_H\Big),$$

and so $\|x+iy\|_{H_c}^2 = \|x\|_H^2 + \|y\|_H^2$. Thus, for example, the complexification of $\mathbb{R}$ is $\mathbb{C}$ with the inner product of $z,\ \zeta \in \mathbb{C}$ given by multiplying z by the complex conjugate $\bar{\zeta}$ of ζ.

It is interesting to observe that, as distinguished from the preceding construction of a complex from a real Hilbert space, the passage from a complex Hilbert space to a real one of which it is the complexification is not canonical. See Exercise 7.1.2 for more details.

7.1.2 Orthogonal Projection and Bases

Given a closed linear subspace L of a real or complex Hilbert space, I want to show that for each $x \in H$ there is a unique $\Pi_L x$ with the properties that

$$\Pi_L x \in L \quad \text{and} \quad \|x - \Pi_L x\|_H = \inf\{\|x-y\|_H : y \in L\}. \tag{7.1.1}$$

An alternative characterization of $\Pi_L x$ is as the unique element of $y \in L$ for which $x - y \perp L$. That is, $\Pi_L x$ is uniquely determined by the properties that

$$\Pi_L x \in L \quad \text{and} \quad x - \Pi_L x \perp L. \tag{7.1.2}$$

To see that (7.1.1) is equivalent to (7.1.2), first suppose that y_0 is an element of L for which $\|x-y\|_H \ge \|x-y_0\|_H$ whenever $y \in L$. Then, for any $y \in L$, the quadratic function

$$t \in \mathbb{R} \longmapsto \|x-y_0+ty\|_H^2 = \|x-y_0\|_H^2 + 2t\mathfrak{Re}\big((x-y_0,y)_H\big) + t^2\|y\|_H^2$$

achieves its minimum at $t=0$. Hence, by the first derivative test, the real part of $(x-y_0,y)_H$ vanishes. When H is a real Hilbert space, this proves that $x-y_0 \perp L$. When H is complex, choose $\theta \in \mathbb{C}$ of modulus 1 to make $(x-y_0,\theta y)_H = \bar{\theta}(x-y_0,y)_H \ge 0$, and apply the preceding with θy to conclude that $|(x-y_0,y)_H| = (x-y_0,\theta y)_H = 0$. Conversely, if $y_0 \in L$ and $x - y_0 \perp L$, then, for every $y \in L$,

$$\|x-y\|_H^2 = \|x-y_0\|_H^2 + 2\mathfrak{Re}\big((x-y_0, y_0-y)_H\big) + \|y_0-y\|_H^2 = \|x-y_0\|_H^2 + \|y_0-y\|_H^2,$$

from which is clear that $\|x-y_0\|_H$ is the minimum value of $\|x-y\|_H$ as y runs over L. Hence, (7.1.1) and (7.1.2) are indeed equivalent descriptions of $\Pi_L x$. Moreover, $\Pi_L x$ is uniquely determined by (7.1.2), since if y_1, $y_2 \in L$ and both $x - y_1$ and $x - y_2$ are orthogonal to L, then $y_2 - y_1$ is also orthogonal to L and therefore to itself.

What remains unanswered in the preceding discussion is the question of existence. That is, we know that there is at most one choice of $\Pi_L x$ that satisfies either (7.1.1) or (7.1.2), but I have yet to show that such a $\Pi_L x$ exists. If H is finite-dimensional, then existence is easy. Namely, one chooses a minimizing sequence $\{y_n : n \geq 1\} \subseteq L$, one for which

$$\|x - y_n\|_H \longrightarrow \delta \equiv \inf\{\|x-y\|_H : y \in L\}.$$

Since $\{\|y_n\|_H : n \geq 1\}$ is bounded and bounded subsets of a finite dimensional Hilbert space are relatively compact, it follows that there exists a subsequence of $\{y_n : n \geq 1\}$ has a limit y_0, and it is obvious that any such limit $y_0 \in L$ will satisfy $\|x - y_0\|_H = \delta$.

However, when H is infinite-dimensional, one has to find another argument. It simply is not true that every bounded subset of an infinite-dimensional space is relatively compact. For example, take $\ell^2(\mathbb{N};\mathbb{R})$ to be $L^2(\mu;\mathbb{R})$ where μ is the counting measure on $\mathbb{N}$ (i.e., $\mu(\{n\}) = 1$ for each $n \in \mathbb{N}$), and let $x_n = \mathbf{1}_{\{n\}}$ be the element of $\ell^2(\mathbb{N};\mathbb{R})$ that is 1 at n and 0 elsewhere. Clearly, $\|x_n\|_{\ell^2(\mathbb{N};\mathbb{R})} = 1$ for all $n \in \mathbb{N}$, and therefore $\{x_n : n \in \mathbb{N}\}$ is bounded. On the other hand, $\|x_n - x_m\|_{\ell^2(\mathbb{N};\mathbb{R})} = \sqrt{2}$ for all $n \neq m$, and therefore $\{x_n : n \in \mathbb{N}\}$ can have no limit point in $\ell^2(\mathbb{N};\mathbb{R})$.

As the preceding makes clear, in infinite dimensions one has to base the proof of existence of $\Pi_L x$ on something other than compactness, and so it is fortunate that completeness comes to the rescue. Namely, I will show that every minimizing sequence $\{y_n : n \geq 1\}$ is Cauchy convergent. The key to doing so is the *parallelogram equality*, which says that the sum of the squares of the lengths of the diagonals in a parallelogram is equal to the sum to the squares of the lengths of its sides. That is, for any a, $b \in H$, $\|a+b\|_H^2 + \|a-b\|_H^2 = 2\|a\|_H^2 + 2\|b\|_H^2$, an equation that is easy to check by expanding the terms on the left-hand side. Applying this when $a = x - y_n$ and $b = x - y_m$, one gets

$$4\left\|x - \frac{y_n + y_m}{2}\right\|_H^2 + \|y_n - y_m\|_H^2 = 2\|x - y_n\|_H^2 + 2\|x - y_m\|_H^2,$$

and therefore, since $\frac{y_n + y_m}{2} \in L$, that

$$\|y_n - y_m\|_H^2 \leq 2\|x - y_n\|_H^2 + 2\|x - y_m\|_H^2 - 4\delta^2,$$

where $\delta = \inf\{\|x-y\|_H : y \in L\}$. Thus, the Cauchy convergence of $\{y_n : n \geq 1\}$ follows from the convergence of $\{\|x-y_n\|_H : n \geq 1\}$ to δ.

Before moving on, I will summarize our progress thus far in the following theorem, and for this purpose it is helpful to introduce a little additional terminology. A map Φ taking H into itself is said to be **idempotent** if $\Phi \circ \Phi = \Phi$, it is called a **contraction** if $\|\Phi(x)\|_H \leq \|x\|_H$ for all $x \in H$, and it is said to be **symmetric** (or sometimes, in the complex case, **Hermitian**) if $\big(\Phi(x), y\big)_H = \big(x, \Phi(y)\big)_H$ for all $x, y \in H$.

Theorem 7.1.1 *Let L be a closed, linear subspace of the real or complex Hilbert space H. Then, for each $x \in H$ there is a unique $\Pi_L x \in L$ for which* (7.1.1) *holds. Moreover, $\Pi_L x$ is the unique element of L for which* (7.1.2) *holds. Finally, the map $x \rightsquigarrow \Pi_L x$ is a linear, idempotent, symmetric contraction.*

Proof. Only the final assertions need comment. However, linearity follows from the obvious fact that if, depending on whether H is real or complex, α_1 and α_2 are elements of $\mathbb{R}$ or $\mathbb{C}$, then for any $x_1, x_2 \in H$, $\alpha_1 \Pi_L x_1 + \alpha_2 \Pi_L x_2 \in L$ and $\alpha_1 x_1 + \alpha_2 x_2 - \alpha_1 \Pi_L x_1 - \alpha_2 \Pi_L x_2 \perp L$. As for idempotency, use (7.1.1) to check that $\Pi_L x = x$ if $x \in L$. To see that Π_L is symmetric, observe that, because $x - \Pi_L x \perp L$ and $y - \Pi_L y \perp L$,

$$\big(x, \Pi_L y\big)_H = \big(\Pi_L x, \Pi_L y\big)_H = \big(\Pi_L x, y\big)_H$$

for all $x, y \in H$. Finally, because $x - \Pi_L x \perp L$,

$$\|x\|_H^2 = \|\Pi_L x\|_H^2 + \|x - \Pi_L x\|_H^2,$$

and so $\|\Pi_L x\|_H \leq \|x\|_H$. □

In view of its properties, especially (7.1.2), it should be clear why the map Π_L is called the **orthogonal projection operator** from H onto L.

An immediate corollary of Theorem 7.1.1 is the following useful criterion. In its statement, and elsewhere, $S^\perp$ denotes the **orthogonal complement** a subset of S of H. That is, $x \in S^\perp$ if and only if $x \perp S$. Clearly, for any $S \subseteq H$, $S^\perp$ is a closed linear subspace. Also, the **span**, denoted by span(S), of S is the smallest linear subspace containing S. Thus, span(S) is the set of vectors $\sum_{m=1}^n \alpha_m x_m$, where $n \in \mathbb{Z}^+$, $\{x_m : 1 \leq m \leq n\} \subseteq S$, and, depending on whether H is real or complex, $\{\alpha_m : 1 \leq m \leq n\}$ is a subset of $\mathbb{R}$ or $\mathbb{C}$.

Corollary 7.1.2 *If S is a subset of a real or complex Hilbert space H, then S spans a dense subset of H if and only if $S^\perp = \{0\}$.*

Proof. Let L denote the closure of the subspace of H spanned by S, and note that $S^\perp = L^\perp$. Hence, without loss in generality, we will assume that $S = L$ and must show that $L = H$ if and only if $L^\perp = \{0\}$. But if $L = H$ and $x \perp L$, then $x \perp x$ and therefore $\|x\|_H^2 = 0$. Conversely, if $L \neq H$, then there exists an $x \notin L$, and so $x - \Pi_L x$ is a non-zero element of $L^\perp$. □

A second important corollary of Theorem 7.1.1 is the easiest of F. Riesz's renowned representation theorems.

Theorem 7.1.3 *If H is a real or complex Hilbert space and Λ is a linear map with values in $\mathbb{R}$ or $\mathbb{C}$, then Λ is continuous if and only if there exists a $g \in H$ such that $\Lambda(h) = (h, g)_H$ for all $h \in H$, in which case there is only one such g.*

Proof. Since, for any $g \in H$, it is clear that $h \rightsquigarrow (h, g)_H$ is continuous, linear, and uniquely determines g, it remains only to prove the existence part of the "only if" assertion. Thus, assume that Λ is continuous, and take L to be the null space $\{h \in H : \Lambda(h) = 0\}$ of Λ. By linearity, L is a linear subspace, and, by continuity, L is closed. If $L = H$, then one can take $g = 0$, and so we will assume now that $L \neq H$. Then, by Corollary 7.1.2, one can find an $f \in L^\perp$ with $\|f\|_H = 1$, in which case $\Lambda(f) \neq 0$ and, for any $h \in H$, $\Lambda\left(h - \frac{\Lambda(h)}{\Lambda(f)}f\right) = 0$. Equivalently, $h - \frac{\Lambda(h)}{\Lambda(f)}f \in L$ and therefore $(h, f)_H = \frac{\Lambda(h)}{\Lambda(f)}$ for all $h \in H$. In other words, when $L \neq H$ one can take $g = \overline{\Lambda(f)}f$. □

Elements of a subset S are said to be **linearly independent** if, for each finite collection F of distinct elements from S, the only linear combination $\sum_{x \in F} \alpha_x x$ that is 0 is the one for which $\alpha_x = 0$ for each $x \in F$. A **basis** in H is a subset S of H whose elements are linearly independent and whose span is dense in H. A Hilbert space H is **infinite-dimensional** if it admits no finite basis.

Lemma 7.1.4 *Assume that H is an infinite-dimensional, separable Hilbert space. Then there exists a countable basis in H. Moreover, if $\{x_n : n \geq 0\}$ is linearly independent in H, then there exists a sequence $\{e_n : n \geq 0\}$ such that $(e_m, e_n)_H = \delta_{m,n}$ for all $m, n \in \mathbb{N}$ and*

$$\operatorname{span}(\{x_0, \dots, x_n\}) = \operatorname{span}(\{e_0, \dots, e_n\}) \quad \text{for each } n \in \mathbb{N}.$$

In particular, if $\{x_n : n \geq 0\}$ is a basis for H, then $\{e_n : n \geq 0\}$ is also.

Proof. To produce a countable basis, start with a sequence $\{y_n : n \geq 0\} \subseteq H \setminus \{0\}$ that is dense in H, and filter out its linearly dependent elements. That is, take $n_0 = 0$, and, proceeding by induction, take n_{m+1} to be the smallest n for which $y_n \notin \operatorname{span}(\{y_0, \dots, y_{n-1}\})$. If $x_m = y_{n_m}$, then it is clear that the x_m's are linearly independent and that they span the same subspace as the y_n's do. Hence, $\{x_m : m \geq 0\}$ is a countable basis for H.

Now suppose that $\{x_n : n \geq 0\}$ is linearly independent, and set $L_n = \operatorname{span}(\{x_0, \dots, x_n\})$. Then L_n is finite-dimensional and therefore (cf. part **(ii)** in Exercise 7.1.1) closed. Because $x_{n+1} - \Pi_{L_n} x_{n+1} \neq 0$ for any $n \geq 0$, we can take

$$e_0 = \frac{x_0}{\|x_0\|_H} \quad \text{and} \quad e_{n+1} = \frac{x_{n+1} - \Pi_{L_n} x_{n+1}}{\|x_{n+1} - \Pi_{L_n} x_{n+1}\|_H}.$$

If we do so, then it is obvious that $\mathrm{span}(\{e_0,\dots,e_n\})$ is contained in L_n for each $n \ge 0$. To see that it is equal to L_n, one can work by induction. Obviously, there is nothing to do when $n=0$, and if $L_n = \mathrm{span}(\{e_0,\dots,e_n\})$, then, because $x_{n+1} - \|x_{n+1} - \Pi_{L_n} x_{n+1}\|_H e_{n+1} \in L_n$, the same is true for $n+1$. Finally, because, by construction, $\|e_n\|_H = 1$ and $e_{n+1} \perp L_n$ for all $n \in \mathbb{N}$, $(e_m, e_n)_H = \delta_{m,n}$. □

A sequence $\{e_n : n \ge 0\} \subseteq H$ is said to be **orthonormal** if $(e_m, e_n)_H = \delta_{m,n}$, and the preceding construction of an orthonormal sequence from a linearly independent one is called the **Gram–Schmidt orthonormalization procedure**.

Lemma 7.1.5 *Suppose that $\{e_n : n \ge 0\}$ is an orthonormal sequence in H. Then, depending on whether H is real or complex, for each $\{\alpha_m : m \ge 0\} \in \ell^2(\mathbb{N};\mathbb{R})$ or $\{\alpha_m : m \ge 0\} \in \ell^2(\mathbb{N};\mathbb{C})$, the series $\sum_{n=0}^\infty \alpha_n e_n$ converges in H. That is, the limit*

$$\sum_{n=0}^{\infty} \alpha_n e_n \equiv \lim_{N\to\infty} \sum_{n=0}^{N} \alpha_n e_n \quad \textit{exists in } H.$$

Moreover,

$$\alpha_m = \left(\sum_{n=0}^{\infty} \alpha_n e_n, e_m\right)_H \quad \textit{for all } m \in \mathbb{N} \quad \textit{and} \quad \left\|\sum_{n=0}^{\infty} \alpha_n e_n\right\|_H^2 = \sum_{n=0}^{\infty} |\alpha_n|^2.$$

Finally, the closed linear span[1] *L of $\{e_n : n \ge 0\}$ in H coincides with the set of sums $\sum_{n=1}^\infty \alpha_n e_n$ as $\{\alpha_n : n \ge 1\}$ runs over $\ell^2(\mathbb{N};\mathbb{R})$ or $\ell^2(\mathbb{N};\mathbb{C})$. In fact, if $x \in H$, then, depending or whether H is real or complex, $\{(x,e_n)_H : n \ge 0\}$ is an element of $\ell^2(\mathbb{N};\mathbb{R})$ or $\ell^2(\mathbb{N};\mathbb{C})$, and*

$$\Pi_L x = \sum_{n=0}^{\infty} (x, e_n)_H e_n.$$

In particular,

$$\sum_{n=0}^{\infty} \left|(x,e_n)_H\right|^2 = \|\Pi_L x\|_H^2 \le \|x\|_H^2 \quad \textit{for all } x \in H,$$

and

$$(\Pi_L x, y)_H = \sum_{n=0}^{\infty} (x,e_n)_H \overline{(y,e_n)_H} \quad \textit{for all } x,\ y \in H,$$

where the series on the right is absolutely convergent.

[1] The closed linear span of a set is the closure of the subspace spanned by that set. It is easy to check that an equivalent description is as the smallest closed linear subspace containing the set.

Proof. To prove that $\sum_{n=0}^{\infty} \alpha_n e_n$ converges, note that, because $(e_m, e_n)_H = \delta_{m,n}$,

$$\left\| \sum_{n=0}^{N} \alpha_n e_n - \sum_{n=0}^{M} \alpha_n e_n \right\|_H^2 = \left\| \sum_{n=M+1}^{N} \alpha_n e_n \right\|_H^2 = \sum_{n=M+1}^{N} |\alpha_n|^2$$

for all $M < N$. Hence, since $\sum_{n=0}^{\infty} |\alpha_n|^2 < \infty$, $\left\{\sum_{n=0}^{N} \alpha_n e_n : N \geq 0\right\}$ satisfies Cauchy's convergence criterion. Furthermore,

$$\alpha_m = \lim_{N\to\infty} \left(\sum_{n=0}^{N} \alpha_n e_n, e_m \right)_H = \left(\sum_{n=0}^{\infty} \alpha_n e_n, e_m \right)_H,$$

and

$$\left\| \sum_{n=0}^{\infty} \alpha_n e_n \right\|_H^2 = \lim_{N\to\infty} \left\| \sum_{n=0}^{N} \alpha_n e_n \right\|_H^2 = \lim_{N\to\infty} \sum_{n=0}^{N} |\alpha_n|^2 = \sum_{n=0}^{\infty} |\alpha_n|^2.$$

Obviously, $\sum_{n=0}^{\infty} \alpha_n e_n \in L$ for all $\{\alpha_n : n \geq 0\}$ from $\ell^2(\mathbb{N};\mathbb{R})$ or $\ell^2(\mathbb{N};\mathbb{C})$. Conversely, if $L_N = \mathrm{span}(\{e_0, \dots, e_N\})$, then, because

$$\left(\sum_{n=0}^{N} (x, e_n)_H e_n, e_m \right)_H = (x, e_m)_H \quad \text{for all } 0 \leq m \leq N,$$

it is easy to see that $x - \sum_{n=0}^{N} (x, e_n)_H e_n \perp L_N$ and therefore that $\Pi_{L_N} x = \sum_{n=0}^{N} (x, e_n)_H e_n$ for all $N \geq 0$ and $x \in H$. Hence,

$$\sum_{n=0}^{N} \left| (x, e_n)_H \right|^2 = \left\| \sum_{n=0}^{N} (x, e_n)_H e_n \right\|_H^2 = \left\| \Pi_{L_N} x \right\|_H^2 \leq \|x\|_H^2,$$

and therefore

$$\sum_{n=0}^{\infty} \left| (x, e_n)_H \right|^2 \leq \|x\|_H^2 < \infty.$$

In particular, the series $\{(x, e_n)_H : n \geq 0\}$ is an element of $\ell^2(\mathbb{N};\mathbb{R})$ or $\ell^2(\mathbb{N};\mathbb{C})$, $\sum_{n=0}^{\infty} (x, e_n)_H e_n$ converges in H to an element of L, and

$$(x, e_m)_H = \left(\sum_{n=0}^{\infty} (x, e_n)_H e_n, e_m \right)_H \quad \text{for all } m \in \mathbb{N}.$$

Since this means that $x - \sum_{n=0}^{\infty} (x, e_n)_H e_n$ is orthogonal to $\{e_m : m \geq 0\}$, and therefore to L, we now know that $\Pi_L x = \sum_{n=0}^{\infty} (x, e_n)_H e_n$ for $x \in H$. Finally, if $x, y \in H$, then

$$
\begin{aligned}
(\Pi_L x, y)_H &= \lim_{N\to\infty} \left(\sum_{n=0}^{N} (x, e_n)_H e_n, y \right)_H \\
&= \lim_{N\to\infty} \sum_{n=0}^{N} (x, e_n)_H (e_n, y)_H = \sum_{n=0}^{\infty} (x, e_n)_H \overline{(y, e_n)_H},
\end{aligned}
$$

and the absolute convergence follows from the Schwarz's inequality for $\ell^2(\mathbb{N};\mathbb{R})$ or $\ell^2(\mathbb{N};\mathbb{C})$. □

The inequality $\sum_{n=0}^{\infty} |(x, e_n)_H|^2 \le \|x\|_H^2$ in Lemma 7.1.5 is often called **Bessel's inequality**.

By combining Lemmas 7.1.4 and 7.1.5, one arrives at a structure theorem for infinite-dimensional, separable Hilbert spaces.

Theorem 7.1.6 *Let H be an infinite-dimensional, separable Hilbert space. Then, there exists a linear isometry Φ from H onto, depending on whether H is real or complex, $\ell^2(\mathbb{N};\mathbb{R})$ or $\ell^2(\mathbb{N};\mathbb{C})$.*

Proof. Choose, via Lemma 7.1.4, an orthonormal basis $\{e_n : n \ge 0\}$ for H, and define $\Phi(x) = \{(x, e_n)_H : n \ge 0\}$. Now apply Lemma 7.1.5 to check that Φ has the required properties. □

I will finish this survey of Hilbert spaces with a procedure for constructing a basis out of other bases. In order to avoid the introduction of the general notion of tensor products, I will state and prove this result only for L^2-spaces.

If $f_1 : E_1 \longrightarrow \mathbb{C}$ and $f_2 : E_2 \longrightarrow \mathbb{C}$, then $f_1 \otimes f_2$ is the $\mathbb{C}$-valued function on $E_1 \times E_2$ given by $f_1 \otimes f_2(x_1, x_2) = f_1(x_1) f_2(x_2)$. Of course, if f_1 and f_2 are $\mathbb{R}$-valued, then so is $f_1 \otimes f_2$.

Theorem 7.1.7 *Let $(E_1, \mathcal{B}_1, \mu_1)$ and $(E_2, \mathcal{B}_2, \mu_2)$ be a pair of σ-finite measure spaces whose σ-algebras are countably generated. If $\{e_{1,n} : n \ge 0\}$ and $\{e_{2,n} : n \ge 0\}$ are orthonormal bases for $L^2(\mu_1;\mathbb{R})$ (or $L^2(\mu_1;\mathbb{C})$) and $L^2(\mu_2;\mathbb{R})$ (or $L^2(\mu_2;\mathbb{C})$) respectively, then $\{e_{1,n_1} \otimes e_{2,n_2} : (n_1, n_2) \in \mathbb{N}^2\}$ is an orthonormal basis for $L^2(\mu_1 \times \mu_2;\mathbb{R})$ (or $L^2(\mu_1 \otimes \mu_2;\mathbb{C})$).*

Proof. Since the proof is the same in both cases, we will deal only with the $\mathbb{R}$-valued case.

By Fubini's Theorem,

$$
\begin{aligned}
&\big(e_{1,m_1} \otimes e_{2,m_2}, e_{1,n_1} \otimes e_{2,n_2}\big)_{L^2(\mu_1 \times \mu_2;\mathbb{R})} \\
&\qquad = \big(e_{1,m_1}, e_{1,n_1}\big)_{L^2(\mu_1;\mathbb{R})} \big(e_{2,m_2}, e_{2,n_2}\big)_{L^2(\mu_1;\mathbb{R})},
\end{aligned}
$$

and so it is clear that $\{e_{1,n_1} \otimes e_{2,n_2} : (n_1, n_2) \in \mathbb{N}^2\}$ is orthonormal. Thus, it suffices to check that $L = \operatorname{span}\big(\{e_{1,m_1} \otimes e_{2,m_2} : (m_1, m_2) \in \mathbb{N}^2\}\big)$ is dense in $L^2(\mu_1 \times \mu_2;\mathbb{R})$. Since, for $f_1 \in L^2(\mu_1;\mathbb{R})$ and $f_2 \in L^2(\mu_2;\mathbb{R})$, one has

$$\lim_{M\to\infty}\left(\sum_{m_1=0}^{M}(f_1,e_{1,m_1})_{L^2(\mu_1;\mathbb{R})}e_{1,m_1}\right)\otimes\left(\sum_{m_2=0}^{M}(f_2,e_{2,m_2})_{L^2(\mu_2;\mathbb{R})}e_{2,m_2}\right)$$

$$=\lim_{M\to\infty}\sum_{m_1,m_2=0}^{M}(f_1,e_{1,m_1})_{L^2(\mu_1;\mathbb{R})}(f_2,e_{2,m_2})_{L^2(\mu_2;\mathbb{R})}e_{1,m_1}\otimes e_{2,m_2},$$

where the convergence is in $L^2(\mu_1\times\mu_2;\mathbb{R})$, $f_1\otimes f_2\in\overline{L}$. Hence, $\overline{L}$ contains the linear span of such functions, and, by Exercise 6.2.3, that span is dense in $L^2(\mu_1\times\mu_2;\mathbb{R})$. □

7.1.3 Exercises for § 7.1

Exercise 7.1.1 Let H be a real or complex Hilbert space, and note that every closed subspace of H becomes a Hilbert space with the inner product obtained by restriction. Here are a few more simple facts about subspaces of a Hilbert space.

(i) Show that for any $S\subseteq H$, $S^\perp$ is always a closed linear subspace of H. In addition, show that $(S^\perp)^\perp$ is the closed, linear subspace spanned by S.

(ii) The proof of Lemma 7.1.4 used the fact that the subspaces L_n there are closed. Of course, once one knows that L_n admits an orthonormal basis $\{e_m : 0\le m\le n\}$, this can be easily checked by noting that $x\in L_n \implies x=\sum_{m=0}^{n}(x,e_m)_He_m$, and therefore if $\{x_k : k\ge 1\}\subseteq L_n$ and $x_k\longrightarrow x$ in H, then

$$\left\|x-\sum_{m=0}^{n}(x,e_m)_He_m\right\|_H=\lim_{k\to\infty}\left\|x_k-\sum_{m=0}^{n}(x_k,e_m)_He_m\right\|_H=0.$$

More generally, without using the existence of orthonormal bases, show that any finite dimensional subspace L of H is closed.

Hint: If $L=\{0\}$ there is nothing to do. Otherwise, choose a basis $\{b_1,\dots,b_\ell\}$ for L, and show that $\left\|\sum_{k=1}^{\ell}\alpha_kb_k\right\|_H\ge\epsilon\left(\sum_{k=1}^{\ell}|\alpha_k|^2\right)^{\frac12}$ for some $\epsilon>0$. Conclude from this that if $\{x_n : n\ge 1\}\subseteq L$ converges to x, then $x=\sum_{k=1}^{\ell}\alpha_kb_k\in L$ for some real or complex coefficients $\{\alpha_k : 1\le k\le\ell\}$.

(iii) Show that $C([0,1];\mathbb{R})$ is a non-closed subspace of $L^2(\lambda_{[0,1]};\mathbb{R})$. Hence, when dealing with infinite-dimensional subspaces, closedness is something that requires checking.

Exercise 7.1.2 Given a complex, infinite-dimensional, separable Hilbert space H, there are myriad ways to produce a real Hilbert space of which H is the complexification. To wit, choose an orthonormal basis $\{e_n : n\ge 0\}$ for H, and let L be the set of $x \in H$ such that $(x, e_n)_H \in \mathbb{R}$ for all $n \geq 0$. Show that L becomes a real Hilbert space when one takes the restriction of $(\cdot\,,\,\cdot)_H$ to L to be its inner product. In addition, show that H is the complexification of L.

Exercise 7.1.3 Suppose that Π is a linear map from the Hilbert space H into itself. Show that Π is the orthogonal projection operator onto the closed subspace L if and only if $L = \text{Range}(\Pi)$ and Π is idempotent and symmetric (i.e., $\Pi^2 = \Pi$ and $(\Pi x, y)_H = (x, \Pi y)_H$ for all $x,\ y \in H$). Also, show that if L is a closed, linear subspace of H, then $\Pi_{L^\perp} = I - \Pi_L$, where I is the identity map.

Exercise 7.1.4 It may be reassuring to know that, in some sense, the dimension of a separable, infinite-dimensional Hilbert space is well-defined and equal to the cardinality of the integers. To see this, show that if E is an infinite subset of H that is orthonormal in the sense that

$$(e, f)_H = \begin{cases} 1 & \text{if } f = e \\ 0 & \text{otherwise} \end{cases} \quad \text{for } e,\ f \in E,$$

then the elements of E are in one-to-one correspondence with the integers.

Exercise 7.1.5 Assume that H is a separable, infinite-dimensional, real or complex Hilbert space, and let L be a closed, linear subspace of H. Show that L is also a separable Hilbert space and that *every orthonormal basis for L can be extended to an orthonormal basis for H*. That is, if E is an orthonormal basis for L, then there is an orthonormal basis $\tilde{E}$ for H with $E \subseteq \tilde{E}$. In fact, show that $\tilde{E} = E \cup E'$, where E' is an orthonormal basis for $L^\perp$.

7.2 Fourier Series

Much of analysis depends on the clever selection of the "best" orthonormal basis for a particular task. Unfortunately, the "best" choice is often unavailable in any practical sense, and one has to settle for a choice that represents a compromise between what would be ideal and what is available. For example, when dealing with a situation in which it is important to exploit features that derive from translation invariance, a reasonable choice is a basis whose elements transform nicely under translations, namely, exponentials of linear functions. This is the choice made by Fourier, and it is still the basis of choice in a large variety of applications.

7.2.1 The Fourier Basis

For each $n \in \mathbb{Z}$, set $\mathfrak{e}_n(x) = e^{i2\pi nx}$. Obviously, $\{\mathfrak{e}_n : n \in \mathbb{Z}\}$ is an orthonormal sequence in $L^2(\lambda_{[0,1]};\mathbb{C})$, and my goal in this subsection is to show that it is an orthonormal basis there. There are many ways to prove this result. For example, one can make a change of venue and replace $[0,1)$ by the unit circle $\mathbb{S}^1$, thought of as a subset of the complex plane $\mathbb{C}$. That is, from Exercise 5.2.1, one sees that $\lambda_{\mathbb{S}^1} = 2\pi(\mathfrak{e}_1)_*\lambda_{[0,1)}$, and so the problem becomes one of showing that the sequence $\{(2\pi)^{-\frac{1}{2}}z^n : n \in \mathbb{Z}\}$ is an orthonormal basis in $L^2(\lambda_{\mathbb{S}^1};\mathbb{C})$. Again orthonormality is obvious. Moreover, by the complex version of the Stone–Weierstrass Theorem (cf. W. Rudin's *Principles of Mathematical Analysis* published by McGraw-Hill), the span of the functions $\{z^n : n \in \mathbb{Z}\}$ is dense in $C(\mathbb{S}^1;\mathbb{C})$. Hence, if $f \perp \{z^n : n \in \mathbb{Z}\}$ in $L^2(\lambda_{\mathbb{S}^1};\mathbb{C})$, then $f \perp C(\mathbb{S}^1;\mathbb{C})$, which, since $C(\mathbb{S}^1;\mathbb{C})$ is dense in $L^2(\lambda_{\mathbb{S}^1};\mathbb{C})$, is possible only if $f = 0$ as an element of $L^2(\mathbb{S}^1;\mathbb{C})$. Now apply Corollary 7.1.2 to conclude that the span of $\{z^n : n \in \mathbb{Z}\}$ is dense in $L^2(\lambda_{\mathbb{S}^1};\mathbb{C})$ and therefore that $\{\mathfrak{e}_n : n \in \mathbb{Z}\}$ is a basis in $L^2(\lambda_{[0,1]};\mathbb{C})$.

Elegant as the preceding approach is, its reliance on Stone–Weierstrass is unfortunate from a measure-theoretic standpoint. For this reason, I will give a second proof, one that seems to me more consistent with the content of this book. Again the idea is to prove that there are enough functions in the span of $\{\mathfrak{e}_n : n \in \mathbb{Z}\}$ to apply Corollary 7.1.2. For this purpose, consider the function given by

$$P(r,x) = \sum_{n\in\mathbb{Z}} r^{|n|}\mathfrak{e}_n(x) \quad \text{for } (r,x) \in [0,1) \times \mathbb{R}.$$

For each $r \in [0,1)$, it is clear that $P(r,\,\cdot\,)$ is a smooth function of period 1, each of whose translates is an element of $L \equiv \overline{\text{span}(\{\mathfrak{e}_n : n \in \mathbb{Z}\})}$. Now let $C^0_1([0,1];\mathbb{C})$ be the space of continuous functions $f : [0,1] \longrightarrow \mathbb{C}$ which are periodic in the sense that $f(0) = f(1)$, and define

$$u_f(r,x) = \int_{[0,1]} P(r,x-y)f(y)\,dy \quad \text{for } f \in C^0_1([0,1];\mathbb{C}).$$

Then $u_f(r,\,\cdot\,) \in L$ for each $r \in [0,1)$. Therefore, if I show that, as $r \nearrow 1$, $u(r,\,\cdot\,) \longrightarrow f$ in $L^2(\lambda_{[0,1]};\mathbb{C})$, then we will know that $C^0_1([0,1];\mathbb{C}) \subseteq L$. Finally, because (cf. Corollary 3.2.11) it is clear that $C^0_1([0,1];\mathbb{C})$ is dense in $L^2(\lambda_{[0,1]};\mathbb{C})$, we will have a second proof that $\{\mathfrak{e}_n : n \in \mathbb{Z}\}$ is a basis for $L^2(\lambda_{[0,1]};\mathbb{C})$.

To carry out this strategy, break the summands defining $P(r,\,\cdot\,)$ into two groups: those with $n \geq 0$ and those with $n < 0$. The resulting sums are geometric series that add up to, respectively,

$$\frac{1}{1 - r\mathfrak{e}_1(x)} \quad \text{and} \quad \frac{r\mathfrak{e}_{-1}(x)}{1 - r\mathfrak{e}_{-1}(x)}.$$

Hence, another expression for $P(r, \cdot)$ is

$$P(r, x) = \frac{1 - r^2}{|r\mathfrak{e}_1(x) - 1|^2}. \tag{7.2.1}$$

Obviously, $P(r, \cdot) > 0$. Secondly, by using the initial expression for $P(r, \cdot)$, one sees that

$$\int_{[0,1]} P(r, x)\, dx = \sum_{n \in \mathbb{Z}} r^{|n|} \big(\mathfrak{e}_n, \mathfrak{e}_0\big)_{L^2(\lambda_{[0,1]};\mathbb{C})} = 1.$$

In particular, after combining these, one finds that $\|u_f\|_{\mathrm{u}} \le \|f\|_{\mathrm{u}}$. Finally, if $f \in C_1^0\big([0,1];\mathbb{C}\big)$ and, for $\delta > 0$, $\omega_f(\delta) = \sup\{|f(y) - f(x)| : (x, y) \in S(\delta)\}$, where

$$S(\delta) = \{(x, y) \in [0,1]^2 : |y - x| \wedge |1 - y + x| \wedge |1 - x + y| \le \delta\},$$

then

$$\begin{aligned} \big|u_f(r, x) - f(x)\big| &= \left| \int_{[0,1]} P(r, x - y)\big(f(y) - f(x)\big)\, dy \right| \\ &\le \int_{[0,1]} P(r, y - x)\big|f(y) - f(x)\big|\, dy \\ &\le \omega_f(\delta) + 2\|f\|_{\mathrm{u}} \int_{\{y:\, (x,y) \notin S(\delta)\}} P(r, x - y)\, dy, \end{aligned}$$

Hence, since for $\delta \in \big(0, \frac{1}{4}\big)$

$$P(r, x - y) \le \frac{2(1 - r)}{1 - \cos(2\pi\delta)} \quad \text{when } (x, y) \notin S(\delta),$$

we see that $\overline{\lim}_{r \nearrow 1} \big|u_f(r, x) - f(x)\big| \le \omega_f(\delta)$ for all $\delta \in \big(0, \frac{1}{4}\big)$. That is, I have shown that, as $r \nearrow 1$, $u_f(r, x) \longrightarrow f(x)$ for each $x \in [0,1]$, and so, since u_f is bounded, one can use Lebesgue's dominated convergence theorem to complete the proof that $\lim_{r \nearrow 1} u_f(r, \cdot) = f$ in $L^2(\lambda_{[0,1]};\mathbb{C})$ for each $f \in C_1^0\big([0,1];\mathbb{C}\big)$. As a consequence, we now have two verifications of Fourier's basic idea.[2]

Theorem 7.2.1 *The sequence* $\{\mathfrak{e}_n : n \in \mathbb{Z}\}$ *is an orthonormal basis for* $L^2(\lambda_{[0,1]};\mathbb{C})$. *Hence, for every* $f \in L^2(\lambda_{[0,1]};\mathbb{C})$, *the Fourier series*

[2] Fourier himself introduced the idea of using trigonometric series to approximate more general functions, but he did not have the machinery to carry the program to completion. Indeed, the statement here would not be possible without Lebesgue's integration theory.

$$\sum_{n\in\mathbb{Z}}(f,\mathfrak{e}_n)_{L^2(\lambda_{[0,1]};\mathbb{C})}\mathfrak{e}_n$$

converges in $L^2(\lambda_{[0,1]};\mathbb{C})$ *to* f.

Theorem 7.2.1 has several more or less immediate corollaries, of which the following is perhaps the most familiar.

Corollary 7.2.2 *The family*

$$\{\sqrt{2}\cos(2\pi nx) : n\in\mathbb{N}\}\cup\{\sqrt{2}\sin(2\pi nx) : n\in\mathbb{Z}^+\}$$

is an orthonormal basis for $L^2(\lambda_{[0,1]};\mathbb{R})$.

Proof. There is no doubt that this family is orthonormal. To prove that it is a basis, suppose that $f\in L^2(\lambda_{\mathbb{R}};\mathbb{R})$ is orthogonal to all its members. Then, as an element of $L^2(\lambda_{[0,1]};\mathbb{C})$, $f\perp\mathfrak{e}_n$ for all $n\in\mathbb{N}$, and so, by Theorem 7.2.1, $f=0$. Now apply Corollary 7.1.2. □

Although, as we have just seen, the Fourier series for an $f\in L^2(\lambda_{[0,1]};\mathbb{C})$ converges to f in $L^2(\lambda_{[0,1]};\mathbb{C})$, for many years it was unknown whether the Fourier series converges to f $\lambda_{[0,1]}$-almost everywhere. That is, given an $f\in L^2(\lambda_{[0,1]};\mathbb{C})$, we know that

$$\sum_{n\in\mathbb{Z}}\big(f,\mathfrak{e}_n\big)_{L^2(\lambda_{[0,1]};\mathbb{C})}\mathfrak{e}_n \text{ converges to } f \text{ in } L^2(\lambda_{[0,1]};\mathbb{C}),$$

but does it do so almost everywhere? An affirmative answer to this question was given by L. Carleson in an article[3] that remains one of the outstanding Twentieth Century contributions to analysis. Beautiful as Carleson's theorem is, in many circumstances one knows enough about the function f that one can check almost everywhere convergence, and more, without recourse to his profound result. For example, if, for $\ell\in\mathbb{Z}^+$, $C_1^\ell([0,1];\mathbb{C})$ is the space of ℓ-times, continuously differentiable functions that, together with their first $\ell-1$ derivatives, take the same value at 0 and 1, then for $f\in C_1^\ell([0,1];\mathbb{C})$ one can use integration by parts to check that

$$\big(f,\mathfrak{e}_n\big)_{L^2(\lambda_{[0,1]};\mathbb{C})}=(i2\pi n)^{-\ell}\big(f^{(\ell)},\mathfrak{e}_n\big)_{L^2(\lambda_{[0,1]};\mathbb{C})}\quad\text{for } n\neq 0, \tag{7.2.2}$$

where $f^{(\ell)}$ denotes the ℓth derivative of f. Starting from (7.2.2), it is easy to prove the following corollary to Theorem 7.2.1.

Corollary 7.2.3 *If* $\ell\in\mathbb{Z}^+$ *and* $f\in C_1^\ell([0,1];\mathbb{C})$, *then* (7.2.2) *holds and so the series* $\sum_{n\in\mathbb{Z}}(f,\mathfrak{e}_n)_{L^2(\lambda_{[0,1]};\mathbb{C})}\mathfrak{e}_n$ *converges uniformly and absolutely to* f. *Moreover, for each* $1\le k<\ell$,

---

[3] *On convergence and growth of partial sums of Fourier series,* Acta Math. 116, Nos. 1-2, 135–157 (1966).

$$f^{(k)} = \sum_{n\in\mathbb{Z}} (i2\pi n)^k \big(f, \mathfrak{e}_n\big)_{L^2(\lambda_{[0,1]};\mathbb{C})} \mathfrak{e}_n,$$

where again the series is absolutely and uniformly convergent.

Proof. Because we already know that $\sum_{n\in\mathbb{Z}^+}(f,\mathfrak{e}_n)_{L^2(\lambda_{[0,1]};\mathbb{C})}\mathfrak{e}_n$ converges to f in $L^2(\lambda_{[0,1]};\mathbb{C})$ and that (7.2.2) holds, the first assertion comes down to the fact that $\sum_{n=1}^{\infty} n^{-2\ell} < \infty$. Indeed, by Schwarz's and Bessel's inequalities,

$$\begin{aligned}\left(\sum_{|n|\ge N} \left|\big(f,\mathfrak{e}_n\big)_{L^2(\lambda_{[0,1]};\mathbb{C})}\mathfrak{e}_n\right|\right)^2 &\le \sum_{|n|\ge N} \left|\big(f^{(\ell)},\mathfrak{e}_n\big)_{L^2(\lambda_{[0,1]};\mathbb{C})}\right|^2 \sum_{|n|\ge N} |n|^{-2\ell} \\ &\le \|f^{(\ell)}\|^2_{L^2(\lambda_{[0,1]};\mathbb{C})} \sum_{|n|\ge N} |n|^{-2\ell},\end{aligned}$$

and so there is nothing more to do.

Turning to the second part, assume that $1 \le k < \ell$, and apply the first part to $f^{(k)}$ to see that

$$f^{(k)} = \sum_{n\in\mathbb{Z}^+\setminus\{0\}} \big(f^{(k)}, \mathfrak{e}_n\big)_{L^2(\lambda_{[0,1]};\mathbb{C})} \mathfrak{e}_n,$$

where the convergence is absolute and uniform, and then use (7.2.2) to show that

$$\big(f^{(k)}, \mathfrak{e}_n\big)_{L^2(\lambda_{[0,1]};\mathbb{C})} = (i2\pi n)^k \big(f, \mathfrak{e}_n\big)_{L^2(\lambda_{[0,1]};\mathbb{C})}$$

for each $n \ne 0$. □

7.2.2 An Application to Euler–Maclaurin

In this subsection I will use the considerations in § 7.2.1 to give another derivation of (1.3.3) and to evaluate the numbers b_ℓ in (1.3.7).

Let $f \in C^\ell_1([0,1];\mathbb{C})$ for some $\ell \ge 1$. By the first part of Corollary 7.2.3 and 7.2.2, we know that

$$f = \int_{[0,1]} f(x)\,dx + (i2\pi)^{-\ell} \sum_{j\in\mathbb{Z}\setminus\{0\}} j^{-\ell}\big(f^{(\ell)}, \mathfrak{e}_j\big)_{L^2(\lambda_{[0,1]};\mathbb{C})}\mathfrak{e}_j, \tag{7.2.3}$$

where the convergence is absolute and uniform. Next, observe that if $n \in \mathbb{Z}^+$, then (cf. the notation in (1.3.2))

$$\mathcal{R}_n(\mathfrak{e}_j) = \frac{1}{n}\sum_{m=1}^{n} \mathfrak{e}_j\left(\tfrac{m}{n}\right) = \begin{cases} 1 & \text{if } n \mid j \\ 0 & \text{if } n \nmid j, \end{cases}$$

where $n \mid j$ means that $\frac{j}{n}$ is an integer. Hence,

$$\int_{[0,1]} f(x)\,dx - \mathcal{R}_n(f) = -(i2\pi n)^{-\ell} \sum_{j\neq 0} j^{-\ell} \big(f^{(\ell)}, \mathfrak{e}_{nj}\big)_{L^2(\lambda_{[0,1]};\mathbb{C})},$$

and so

$$\left| \int_{[0,1]} f(x)\,dx - \mathcal{R}_n(f) \right| \leq \frac{\|f^{(\ell)}\|_{L^2(\lambda_{[0,1]};\mathbb{C})}}{(2\pi n)^{\ell}} \left(2\sum_{j=1}^{\infty} j^{-2\ell} \right)^{\frac{1}{2}}, \tag{7.2.4}$$

which represents a slight improvement of the result obtained in (1.3.3) of § 1.3.1.

More interesting than the re-derivation of (1.3.3) is the fact that we can now compute the numbers b_ℓ for $\ell \geq 2$. To see how this is done, let

$$P_\ell(x) = \sum_{k=0}^{\ell} (-1)^k \frac{b_{\ell-k} x^k}{k!}$$

be the polynomial introduced in (1.3.17). By (1.3.18), we know that $P_\ell' = P_{\ell-1}$ for $\ell \geq 1$ and that $P_\ell(1) = P_\ell(0)$ for $\ell \geq 2$. Thus,

$$\big(P_\ell, \mathfrak{e}_0\big)_{L^2(\lambda_{[0,1)};\mathbb{C})} = P_{\ell+1}(0) - P_{\ell+1}(1) = 0 \quad \text{for } \ell \geq 1,$$

and, by (7.2.2),

$$\big(P_\ell, \mathfrak{e}_n\big)_{L^2(\lambda_{[0,1)};\mathbb{C})} = (-1)^{\ell-1} \big(i2\pi n\big)^{1-\ell} \big(P_1, \mathfrak{e}_n\big)_{L^2(\lambda_{[0,1)};\mathbb{C})}$$
$$\text{for } \ell \geq 2 \text{ and } n \neq 0.$$

Hence, since $P_1(x) = \frac{1}{2} - x$ and therefore

$$\big(P_1, \mathfrak{e}_n\big)_{L^2(\lambda_{[0,1)};\mathbb{C})} = -\int_{[0,1)} x\mathfrak{e}_n(-x)\,dx = \big(i2\pi n\big)^{-1},$$

we now know that

$$P_\ell(x) = -\left(\frac{i}{2\pi}\right)^{\ell} \sum_{n\neq 0} \frac{\mathfrak{e}_n(x)}{n^{\ell}} \quad \text{for } \ell \geq 2 \text{ and } x \in [0,1]. \tag{7.2.5}$$

Starting from (7.2.5), the calculation of the b_ℓ's is easy. Namely, $b_0 = 1$ by definition, $b_1 = \frac{1}{2}$ follows from (1.3.7), and, by evaluating both sides of (7.2.5) at 0,

$$b_\ell = -\left(\frac{i}{2\pi}\right)^{\ell} \sum_{n\neq 0} n^{-\ell} \quad \text{for } \ell \geq 2.$$

Thus, if $\zeta(s) \equiv \sum_{n=1}^{\infty} n^{-s}$ is the **Riemann zeta function** for $s > 1$, then

$$b_\ell = \begin{cases} (-1)^{\frac{\ell}{2}+1} 2(2\pi)^{-\ell}\zeta(\ell) & \text{for even } \ell \geq 2 \\ 0 & \text{for odd } \ell \geq 2. \end{cases} \tag{7.2.6}$$

Alternatively, one can combine (7.2.6) with (1.3.7) to evaluate $\zeta(2m)$ for $m \in \mathbb{Z}^+$. Namely, (7.2.6) says that

$$\zeta(2m) = \frac{(-1)^{m+1}(2\pi)^{2m} b_{2m}}{2} \quad \text{for } m \in \mathbb{Z}^+, \tag{7.2.7}$$

and from (1.3.7) and the fact that the $b_\ell = 0$ when $\ell \geq 2$ is odd, we know that

$$b_0 = 1,\, b_1 = \frac{1}{2}, \quad \text{and } b_{2m} = \frac{m - \frac{1}{2}}{(2m+1)!} - \sum_{1 \leq k < m} \frac{b_{2k}}{(2m-2k+1)!}. \tag{7.2.8}$$

Hence, $\zeta(2) = \frac{\pi^2}{6}$, $\zeta(4) = \frac{\pi^4}{90}$, $\zeta(6) = \frac{\pi^6}{945}$, etc.

Before closing, I would be remiss if I did not point out that the polynomials P_ℓ in (1.3.17) have a long history. Indeed, $P_\ell = (-1)^\ell \ell! B_\ell$, where B_ℓ is the ℓth **Bernoulli polynomial**.[4] To see this, one can use the characterization of the B_ℓ's as the unique functions that satisfy $B_0 \equiv 1$, $B_\ell' = \ell B_\ell$ for $\ell \geq 1$, and $B_\ell(0) = B_\ell(1)$ for $\ell \geq 2$. Hence, by Exercise 1.3.3, it is clear that $P_\ell = \ell! B_\ell$. In the literature, $B_\ell(0)$ is known as the ℓth **Bernoulli number**, and so we now know that *b_ℓ is $(-1)^\ell \ell!$ times the ℓth Bernoulli number.*

7.2.3 Exercises for § 7.2

Exercise 7.2.1 Here are a few easy consequences of Theorem 7.2.1.

(i) For $\mathbf{n} = (n_1, \ldots, n_N) \in \mathbb{Z}^N$ and $x = (x_1, \ldots, x_N) \in \mathbb{R}^N$, set $\mathfrak{e}_{\mathbf{n}}(x) = \prod_{j=1}^N \mathfrak{e}_{n_j}(x_j)$ and show that $\{\mathfrak{e}_{\mathbf{n}} : \mathbf{n} \in \mathbb{Z}^N\}$ is an orthonormal basis for $L^2(\lambda_{[0,1]^N}; \mathbb{C})$.

(ii) Given $a < b$, set

$$\mathfrak{e}_{[a,b],n}(x) = (b-a)^{-\frac{1}{2}} \mathfrak{e}_n\left(\frac{x}{b-a}\right) \quad \text{for } n \in \mathbb{Z} \text{ and } x \in [a,b],$$

and show that $\{\mathfrak{e}_{[a,b],n} : n \in \mathbb{Z}\}$ is an orthonormal basis for $L^2(\lambda_{[a,b]}; \mathbb{C})$.

(iii) Show that both

[4] For an account of the Bernoulli polynomials, see Chapter VIII of G.H. Hardy's *Divergent Series*, now available from the AMS in its Chelsea Series.

$$\{1\} \cup \{2^{\frac{1}{2}} \cos(\pi n\,\cdot\,) : n \in \mathbb{Z}^+\} \quad \text{and} \quad \{2^{\frac{1}{2}} \sin(\pi n\,\cdot\,) : n \in \mathbb{Z}^+\}$$

are orthonormal bases for $L^2(\lambda_{[0,1]}; \mathbb{R})$.

Hint: In doing part (**iii**), consider what happens when one extends a real-valued function on $[0,1]$ as an even or an odd function on $[-1,1]$.

Exercise 7.2.2 The evaluation of $\zeta(2) = \sum_{n=1}^{\infty} n^{-2}$ can be done much more easily than that of $\zeta(2m)$ when $m \geq 2$. Namely, take $f(x) = x$, show that

$$\left(f, \mathfrak{e}_n\right)_{L^2(\lambda_{[0,1)};\mathbb{C})} = \begin{cases} \frac{i}{2\pi n} & \text{if } n \neq 0 \\ \frac{1}{2} & \text{if} n = 0, \end{cases}$$

and derive $\zeta(2) = \frac{\pi^2}{6}$ from $\|f\|^2_{L^2(\lambda_{[0,1)};\mathbb{C})} = \sum_{n\in\mathbb{Z}} \left|\left(f, \mathfrak{e}_n\right)_{L^2(\lambda_{[0,1)};\mathbb{C})}\right|^2$.

Exercise 7.2.3 Let $\varphi \in C\big([0,1]; \mathbb{C}\big)$, and assume that

$$\sum_{n\in\mathbb{Z}} \left|(\varphi, \mathfrak{e}_n)_{L^2([0,1];\mathbb{C})}\right| < \infty.$$

Show that $\sum_{n\in\mathbb{Z}} (\varphi, \mathfrak{e}_n)_{L^2([0,1];\mathbb{C})} \mathfrak{e}_n$ converges uniformly to φ. In particular, $\varphi(1) = \varphi(0) = \sum_{n\in\mathbb{Z}} (\varphi, \mathfrak{e}_n)_{L^2([0,1];\mathbb{C})}$.

Exercise 7.2.4 One should ask whether there is a theorem about Fourier series for $f \in L^1(\lambda_{[0,1]}; \mathbb{C})$. That is, given $f \in L^1(\lambda_{[0,1]}; \mathbb{C})$, set $a_n(f) = \int_{[0,1]} f(x)\mathfrak{e}_{-n}(x)\, dx$ for $n \in \mathbb{Z}$. The question is what one can say about the series $\sum_{n\in\mathbb{Z}} a_n(f)\mathfrak{e}_n$.

(**i**) Show that $\sup_{n\in\mathbb{Z}} |a_n(f)| \leq \|f\|_{L^1(\lambda_{[0,1]};\mathbb{C})}$. Next, show that $a_n(f) \longrightarrow 0$ as $|n| \to \infty$, first for bounded and then for general $f \in L^1(\lambda_{[0,1]}; \mathbb{C})$. This fact is known as the **Riemann–Lebesgue Lemma**.

(**ii**) Define $P(r,x)$ as in (7.2.1), and set $u_f(r,x) = \int_{[0,1]} P(r, x-y) f(y)\, dy$ for $r \in [0,1)$. Show that $\|u_f(r,\,\cdot\,)\|_{L^1(\lambda_{[0,1]};\mathbb{C})} \leq \|f\|_{L^1(\lambda_{[0,1]};\mathbb{C})}$ and that $u_f(r,\,\cdot\,) \longrightarrow f$ in $L^1(\lambda_{[0,1]}; \mathbb{C})$ as $r \nearrow 1$, first for $f \in C^0_1\big([0,1]; \mathbb{C}\big)$ and then for all $f \in L^1(\lambda_{[0,1]}; \mathbb{C})$.

(**iii**) Continuing (**ii**), show that

$$u_f(r,x) = \sum_{n\in\mathbb{Z}} r^{|n|} a_n(f)\mathfrak{e}_n(x),$$

and conclude that the series on the right converges to f in $L^1(\lambda_{[0,1]}; \mathbb{C})$ as $r \nearrow 1$.

7.3 The Fourier Transform

Let f be a smooth, compactly supported, complex-valued function on $\mathbb{R}$. Then as soon as $L > 0$ is large enough that $[-L, L]$ contains its support, we know from part **(ii)** of Exercise 7.2.1 that

$$f(x) = \frac{1}{2L} \sum_{m \in \mathbb{Z}} \int_{[-L,L]} f(y) e^{\frac{i2\pi m(x-y)}{2L}}\, dy.$$

Thus, if one closes ones eyes, suspends ones disbelief, lets $L \to \infty$, interprets the sum as a Riemann approximation, and indulges in a certain amount of unjustified re-arrangement, one is led to guess that

$$f(x) = \int_{\mathbb{R}} e^{i2\pi\xi x} \left(\int_{\mathbb{R}} f(y)\, e^{-i2\pi\xi y}\, dy \right) d\xi. \tag{7.3.1}$$

In this section, I will give two justifications for this guess. The first justification is for functions $f \in L^1(\lambda_{\mathbb{R}}; \mathbb{C})$, and the second, which is the more intricate but more satisfactory one, is for $f \in L^2(\lambda_{\mathbb{R}}; \mathbb{C})$.

7.3.1 L^1-Theory of the Fourier Transform

Set $\mathfrak{e}_\xi(x) = e^{i2\pi(\xi,x)_{\mathbb{R}^N}}$ for $(x, \xi) \in \mathbb{R}^N \times \mathbb{R}^N$, and define

$$\left.\begin{aligned} \hat{f}(\xi) &= \int_{\mathbb{R}^N} f(x)\mathfrak{e}_\xi(x)\, dx, \\ \check{f}(x) &= \int_{\mathbb{R}^N} f(\xi)\mathfrak{e}_{-x}(\xi)\, d\xi \end{aligned}\right\} \quad \text{for } f \in L^1(\lambda_{\mathbb{R}^N}; \mathbb{C}). \tag{7.3.2}$$

The following lemma is the analog in this setting of part **(i)** in Exercise 7.2.4.

Lemma 7.3.1 *Let $C_0(\mathbb{R}^N; \mathbb{C})$ be the space of continuous, $\mathbb{C}$-valued functions on $\mathbb{R}^N$ that tend to 0 at infinity, and give $C_0(\mathbb{R}^N; \mathbb{C})$ the uniform topology, the one corresponding to uniform convergence on $\mathbb{R}^N$. Then $\hat{f} \in C_0(\mathbb{R}^N; \mathbb{C})$ for each $f \in L^1(\lambda_{\mathbb{R}^N}; \mathbb{C})$ and the map $f \in L^1(\mathbb{R}^N; \mathbb{C}) \longmapsto \hat{f} \in C_0(\mathbb{R}^N; \mathbb{C})$ is a linear contraction. Furthermore, if $f, g \in L^1(\mathbb{R}^N; \mathbb{C})$, then*

$$\int_{\mathbb{R}^N} f(x)\, \overline{\check{g}(x)}\, dx = \int_{\mathbb{R}^N} \hat{f}(\xi)\, \overline{g(\xi)}\, d\xi. \tag{7.3.3}$$

Proof. Since, by Lebesgue's dominated convergence theorem, it is obvious that $\hat{f} \in C_b(\mathbb{R}^N; \mathbb{C})$ with $\|\hat{f}\|_u \le \|f\|_{L^1(\mathbb{R}^N;\mathbb{C})}$ and that $f \rightsquigarrow \hat{f}$ is linear, the only part of the first assertion that requires comment is the statement that $\hat{f}$ tends to 0 at infinity. However, $C_0(\mathbb{R}^N;\mathbb{C})$ is closed under uniform limits, and so, by continuity, Theorem 6.3.9 tells us that it is enough to check $\hat{f} \in C_0(\mathbb{R}^N;\mathbb{C})$ when $f \in C_c^\infty(\mathbb{R}^N;\mathbb{C})$. But if $f \in C_c^\infty(\mathbb{R}^N;\mathbb{C})$, then we can use Green's Formula (cf. Exercise 5.3.2) to see that, because f has compact support,

$$-|2\pi\xi|^2\hat{f}(\xi) = \int_{\mathbb{R}^N} f(x)\Delta\mathfrak{e}_\xi(x)\,dx = \int_{\mathbb{R}^N} \Delta f(x)\mathfrak{e}_\xi(x)\,dx = \widehat{\Delta f}(\xi).$$

Thus $|\hat{f}(\xi)| \le |2\pi\xi|^{-2}\|\Delta f\|_{L^1(\mathbb{R}^N;\mathbb{C})} \longrightarrow 0$ as $|\xi| \to \infty$.

Finally, to prove (7.3.3), apply Fubini's Theorem to justify

$$\begin{aligned}\int_{\mathbb{R}^N} \hat{f}(\xi)\overline{g(\xi)}\,d\xi &= \iint\limits_{\mathbb{R}^N\times\mathbb{R}^N} f(x)\mathfrak{e}_\xi(x)\overline{g(\xi)}\,dx\,d\xi \\ &= \iint\limits_{\mathbb{R}^N\times\mathbb{R}^N} f(x)\overline{\mathfrak{e}_x(-\xi)g(\xi)}\,dx\,d\xi = \int_{\mathbb{R}^N} f(x)\overline{\check{g}(x)}\,dx.\end{aligned}$$

□

The function $\hat{f}$ is called the **Fourier transform** of the function f. The fact that $\hat{f}$ tends to 0 at infinity was observed originally (in the context of Fourier series) by Riemann and is usually called the **Riemann–Lebesgue Lemma**. In this connection, the reader should not be deluded into thinking that $\{\hat{f} : f \in L^1(\mathbb{R}^N;\mathbb{C})\} = C_0(\mathbb{R}^N;\mathbb{C})$; it is not! In fact, I know of no simple, satisfactory characterization of $\{\hat{f} : f \in L^1(\mathbb{R}^N;\mathbb{C})\}$.

To complete my justification of (7.3.1) for $f \in L^1(\mathbb{R}^N;\mathbb{C})$, I will need the computation contained in the next lemma.

Lemma 7.3.2 *Given $t \in (0,\infty)$, define $g_t : \mathbb{R}^N \longrightarrow (0,\infty)$ so that*

$$g_t(x) = t^{-\frac{N}{2}} \exp\left(-\frac{\pi|x|^2}{t}\right), \quad x \in \mathbb{R}^N. \tag{7.3.4}$$

Then, for all $t > 0$, g_t has total integral 1. In fact, for all $\zeta \in \mathbb{C}^N$,

$$\int_{\mathbb{R}^N} e^{2\pi(\zeta,x)_{\mathbb{C}^N}} g_t(x)\,dx = \exp\left(t\pi\sum_{j=1}^N \zeta_j^2\right). \tag{7.3.5}$$

In particular,

$$\widehat{g_t}(\xi) = e^{-t\pi|\xi|^2} \quad \text{and} \quad (\widehat{g_t})^\vee = g_t. \tag{7.3.6}$$

(See part (**i**) *of Exercise* 8.2.7 *for another derivation of* (7.3.6)*).*

Proof. The first part of (7.3.6) is an easy application of (7.3.5), and, given the first part, the second part is another application of (7.3.5). Thus we need only prove (7.3.5). First note that, by Fubini's Theorem, it is enough to handle the case $N = 1$. That is, we have to check that

$$t^{-\frac{1}{2}} \int_{\mathbb{R}} \exp\left(-\frac{\pi x^2}{t} + 2\pi\zeta x\right) dx = e^{t\pi\zeta^2} \qquad (*)$$

for all $t > 0$ and $\zeta \in \mathbb{C}$. To this end, observe that, for any given $t > 0$, both sides of $(*)$ are analytic functions of ζ in the entire complex plane $\mathbb{C}$. Hence, for each $t > 0$, $(*)$ will hold for all $\zeta \in \mathbb{C}$ as soon as it holds for $\zeta \in \mathbb{R}$. Furthermore, given $\zeta \in \mathbb{R}$, a change of variables shows that $(*)$ for some $t > 0$ implies $(*)$ for all $t > 0$. Thus, we need only prove $(*)$ for $\zeta \in \mathbb{R}$ and $t = 2\pi$. But, for $\zeta \in \mathbb{R}$,

$$\begin{aligned}\frac{1}{(2\pi)^{\frac{1}{2}}} \int_{\mathbb{R}} e^{-\frac{x^2}{2}+2\pi\zeta x}\, dx &= \frac{1}{(2\pi)^{\frac{1}{2}}} e^{2\pi^2\zeta^2} \int_{\mathbb{R}} e^{-\frac{1}{2}(x-2\pi\zeta)^2}\, dx \\ &= \frac{1}{(2\pi)^{\frac{1}{2}}} e^{2\pi^2\zeta^2} \int_{\mathbb{R}} e^{-\frac{x^2}{2}}\, dx = e^{2\pi^2\zeta^2},\end{aligned}$$

where, in the last equation, the computation made in part **(i)** of Exercise 5.3.2 was used. □

We can now prove for the Fourier transform the analog of part **(iii)** in Exercise 7.2.4.

Theorem 7.3.3 *Given $f \in L^1(\mathbb{R}^N;\mathbb{C})$ and $t \in (0,\infty)$, (cf. (7.3.4))*

$$f * g_t(x) = \int_{\mathbb{R}^N} e^{-t\pi|\xi|^2} \mathfrak{e}_x(-\xi)\hat{f}(\xi)\, d\xi,$$

and

$$\int_{\mathbb{R}^N} e^{-t\pi|\xi|^2} \mathfrak{e}_x(-\xi)\hat{f}(\xi)\, d\xi \longrightarrow f(x) \quad \text{in } L^1(\mathbb{R}^N;\mathbb{C}) \text{ as } t \searrow 0,$$

and so

$$f(x) = \int_{\mathbb{R}^N} \mathfrak{e}_{-x}(\xi)\hat{f}(\xi)\, d\xi$$

if $\hat{f}$ as well as f are $\lambda_{\mathbb{R}^N}$-integrable. In particular, if $f \in L^1(\mathbb{R}^N;\mathbb{C})$ and $\hat{f} = 0$, then $f = 0$ $\lambda_{\mathbb{R}^N}$-almost everywhere.

Proof. By Theorem 6.3.7,[5] it suffices to prove the initial equation. To this end, observe that (recall that $\tau_x f = f \circ T_x$)

$$\widehat{\tau_x f}(\xi) = \mathfrak{e}_\xi(-x)\hat{f}(\xi). \qquad (7.3.7)$$

Hence, because, by (7.3.6), $g_t(-y) = g_t(y) = (\widehat{g_t})^\vee(y)$, the desired equality follows from (7.3.3) applied to $\tau_x f$ and g_t. □

[5] In the notation of that theorem, g_t here would have been $g_{\sqrt{t}}$ there.

The preceding proof turns on a general principle that, because of its interpretation in quantum mechanics, is known as the **uncertainty principle**. Crudely stated, this principle says that the more *localized* a function f is, the more *delocalized* its Fourier transform $\hat{f}$ will be. Thus, because g_t gets more and more concentrated near 0 as $t \searrow 0$, $\widehat{g_t}$ gets more and more evenly spread. Exercise 7.3.1 below presents this phenomenon in a more general context.

Before closing this discussion, I describe a couple of the properties that indicate why the Fourier transform is a powerful tool for the analysis of operations based on translation should be pointed out.

Theorem 7.3.4 *If $f \in C^1(\mathbb{R}^N;\mathbb{C})$ and both f and $|\nabla f|$ are $\lambda_{\mathbb{R}^N}$-integrable, then*

$$\widehat{\partial_{x_j} f}(\xi) = -i2\pi\xi_j \hat{f}(\xi) \quad \text{for each } 1 \le j \le N.$$

In particular, if $f \in C^{N+1}(\mathbb{R}^N;\mathbb{C})$ and $\partial^\alpha f \in L^1(\mathbb{R}^N;\mathbb{C})$ for $\|\alpha\| \le N+1$, then $\hat{f} \in L^1(\mathbb{R}^N;\mathbb{C})$. Also, if $f,\ g \in L^1(\lambda_{\mathbb{R}^N};\mathbb{C})$ then $\widehat{f * g} = \hat{f}\hat{g}$.

Proof. The proofs of both these facts turn on (7.3.7). Indeed, since by Lebesgue's dominated convergence theorem,

$$\widehat{\partial_{x_j} f}(\xi) = \lim_{t \searrow 0} \frac{\widehat{\tau_{te_j} f}(\xi) - \hat{f}(\xi)}{t},$$

the first assertion follows. As for the second, write $f * g(x)$ as $\int \tau_{-y} f(x) g(y)\, dy$ and apply Fubini's Theorem and (7.3.7) to conclude that

$$\widehat{f * g}(\xi) = \hat{f}(\xi) \int \mathfrak{e}_\xi(y) g(y)\, dy = \hat{f}(\xi)\hat{g}(\xi).$$

Finally, to prove the concluding assertion, use the preceding to see that there is a $C < \infty$ for which $|\hat{f}(\xi)| \le C(1+|\xi|^2)^{-\frac{N+1}{2}}$ and therefore that

$$\int |\hat{f}(\xi)|\, d\xi \le C \int (1+|\xi|^2)^{-\frac{N+1}{2}}\, d\xi = C\Omega_N \int_{(0,\infty)} \frac{r^{N-1}}{(1+r^2)^{\frac{N+1}{2}}}\, dr < \infty.$$

$\square$

7.3.2 The Hermite Functions

My treatment of the L^2-theory of the Fourier transform will require the introduction of a special orthonormal basis for $L^2(\lambda_{\mathbb{R}^N};\mathbb{C})$, and because the basis for $L^2(\lambda_{\mathbb{R}^N};\mathbb{C})$ is the one constructed from the basis for $L^2(\lambda_{\mathbb{R}};\mathbb{C})$ as an application of Theorem 7.1.7, it suffices to deal with the case $N = 1$.

Define

$$h_n(x) = e^{-\pi x^2} H_n(x) \quad \text{where } H_n(x) = (-1)^n e^{2\pi x^2} \partial_x^n \left(e^{-2\pi x^2}\right) \text{ for } n \in \mathbb{N}.$$

These functions were introduced by C. Hermite, and I will call h_n and H_n, respectively, the nth **unnormalized Hermite function** and **Hermite polynomial**. Because $H_0 = \mathbf{1}$ and $H_n(x) = 4\pi x H_{n-1}(x) - H'_{n-1}(x)$ for $n \geq 1$, it should be clear that H_n is an nth order polynomial whose leading coefficient is $(4\pi)^n$. In particular, for each $n \geq 0$, the span of $\{H_m : 0 \leq m \leq n\}$ is the same as that of $\{x^m : 0 \leq m \leq n\}$.

In order to analyze the Hermite functions, it is helpful to introduce the **raising operator** $a_+ = 2\pi x - \partial_x$ and **lowering operator** $a_- = 2\pi x + \partial_x$. Alternatively,

$$a_+\varphi(x) = -e^{\pi x^2} \partial_x\left(e^{-\pi x^2}\varphi(x)\right) \text{ and } a_-\varphi(x) = e^{-\pi x^2}\partial_x\left(e^{\pi x^2}\varphi(x)\right).$$

Using this second expression, it is easy to check that $a_+ h_n = h_{n+1}$, which explains the origin of its name. At the same time, $a_- h_n(x) = e^{-\pi x^2}\partial_x H_n(x)$, which shows that a_- is lowering in the sense that it annihilates h_0 and, for $n \geq 1$, takes h_n into the span of $\{h_m : 0 \leq m < n\}$. Arguing by induction, one can easily pass from these to

$$a_+^m h_n = h_{m+n} \quad \text{and} \quad a_-^m h_n = e^{-\pi x^2}\partial_x^m H_n \quad \text{for all } m, n \in \mathbb{N}, \tag{7.3.8}$$

and, taking into account the earlier remark about the structure of H_n, one can use the second of these to see that

$$a_-^m h_n = 0 \text{ if } m > n \quad \text{and} \quad a_-^n h_n = (4\pi)^n n! h_0. \tag{7.3.9}$$

Lemma 7.3.5 *If $\varphi, \psi \in C^1(\mathbb{R};\mathbb{C})$ and $\varphi\psi$, $(a_+\varphi)\psi$, and $\varphi(a_-\psi)$ are all in $L^1(\mathbb{R};\mathbb{C})$, then $\int (a_+\varphi)\psi d\lambda_{\mathbb{R}} = \int \varphi(a_-\psi)\, d\lambda_{\mathbb{R}}$.*

Proof. When at least one of the functions φ and ψ has compact support, the equation is an easy application of integration by parts. To handle the general case, choose a bump function $\eta \in C^\infty\big(\mathbb{R};[0,1]\big)$ that is identically 1 on $[-1,1]$ and vanishes off of $[-2,2]$, and set $\eta_R(x) = \eta(R^{-1}x)$ for $R > 0$. Then, for each $R > 0$,

$$\int (a_+\varphi)\eta_R\psi d\lambda_{\mathbb{R}} = \int \varphi a_-(\eta_R\psi)\, d\lambda_{\mathbb{R}} = \int (\eta_R\varphi)a_-\psi\, d\lambda_{\mathbb{R}} + \int \eta'_R\varphi\psi\, d\lambda_{\mathbb{R}}.$$

Hence, since $\|\eta_R\|_{\mathrm{u}} \leq 1$ and $\|\eta'_R\|_{\mathrm{u}} = R^{-1}\|\eta'\|_{\mathrm{u}}$, one can use Lebesgue's dominated convergence theorem to get the desired result after letting $R \to \infty$. □

By combining the first part of (7.3.8), (7.3.9), and Lemma (7.3.5), one sees that

$$\big(h_m, h_n\big)_{L^2(\lambda_{\mathbb{R}};\mathbb{C})} = \big(h_0, a_-^m h_n\big)_{L^2(\lambda_{\mathbb{R}};\mathbb{C})} = \begin{cases} (4\pi)^n n! \|h_0\|^2_{L^2(\lambda_{\mathbb{R}};\mathbb{C})} & \text{if } m = n \\ 0 & \text{if } m > n. \end{cases}$$

Hence, since $\|h_0\|^2_{L^2(\lambda_{\mathbb{R}};\mathbb{C})} = 2^{-\frac{1}{2}}$,

$$\big(h_m, h_n\big)_{L^2(\lambda_{\mathbb{R}};\mathbb{C})} = 2^{-\frac{1}{2}}(4\pi)^m m!\delta_{m,n}. \tag{7.3.10}$$

This proves that the h_n's are mutually orthogonal in $L^2(\lambda_{\mathbb{R}};\mathbb{C})$, and, among other things, this fact also allows us to compute a_-h_n. We already know that $a_-h_0 = 0$. Further, as I noted earlier, when $n \geq 1$, a_-h_n is in the span of $\{h_m : 0 \leq m < n\}$, and so $a_-h_n = \sum_{m=0}^{n-1} \alpha_m h_m$ for some choice of α_m's. In fact, using Lemma 7.3.5 and (7.3.10), one finds that

$$2^{-\frac{1}{2}}(4\pi)^n n!\delta_{n,m+1} = \big(a_-h_n, h_m\big)_{L^2(\lambda_{\mathbb{R}};\mathbb{C})} = 2^{-\frac{1}{2}}(4\pi)^m m!\alpha_m \quad \text{for } 0 \leq m < n,$$

from which it follows that

$$a_-h_0 = 0 \quad \text{and} \quad a_-h_n = 4\pi n h_{n-1} \text{ for } n \geq 1. \tag{7.3.11}$$

Noting that $4\pi x = a_+ + a_-$ and adding the first part of (7.3.8) with $m = 1$ to (7.3.11), one arrives at

$$4\pi x h_n(x) = h_{n+1}(x) + 4\pi n h_{n-1}(x) \quad \text{for } n \geq 1. \tag{7.3.12}$$

Similarly, since $-2\partial = a_+ - a_-$,

$$-2\partial h_n = h_{n+1} - 4\pi n h_{n-1}. \tag{7.3.13}$$

We now have all the machinery needed to prove the basic properties of the Hermite functions, properties that will be used in the next subsection to develop the L^2-theory of the Fourier transform.

Theorem 7.3.6 *Set*

$$\tilde{h}_n = \frac{2^{\frac{1}{4}}}{\sqrt{(4\pi)^n n!}} h_n \quad \textit{for } n \in \mathbb{N}.$$

Then $\{\tilde{h}_n : n \in \mathbb{N}\}$ is an orthonormal basis for $L^2(\lambda_{\mathbb{R}};\mathbb{C})$, and

$$(2\pi x)^2 \tilde{h}_n - \tilde{h}_n''(x) = 2\pi(2n+1)\tilde{h}_n \quad \textit{for each } n \in \mathbb{N}. \tag{7.3.14}$$

Furthermore, $\|\tilde{h}_0\|_{\mathrm{u}} \leq \|\tilde{h}_0\|_{L^1(\lambda_{\mathbb{R}};\mathbb{C})} = 2^{\frac{1}{4}}$ and

$$\|\tilde{h}_n\|_{\mathrm{u}} \leq \|\tilde{h}_n\|_{L^1(\lambda_{\mathbb{R}};\mathbb{C})} \leq \frac{1 + 4\pi + 2n^{\frac{1}{2}}}{2} \quad \textit{for } n \geq 1. \tag{7.3.15}$$

Finally (cf. the second part of Remark 7.3.1 below),

$$\widehat{\tilde{h}_n} = i^n \tilde{h}_n \textit{ for all } n \in \mathbb{N}. \tag{7.3.16}$$

Proof. The orthonormality assertion is simply a restatement of (7.3.10). To check that $\{\tilde{h}_n : n \in \mathbb{N}\}$ is a basis, Corollary 7.1.2 says that it suffices to check that the only $f \in L^2(\lambda_{\mathbb{R}};\mathbb{C})$ satisfying $(f, \tilde{h}_n)_{L^2(\lambda_{\mathbb{R}};\mathbb{C})} = 0$ for all $n \in \mathbb{N}$ vanishes $\lambda_{\mathbb{R}}$-almost everywhere. To this end, let f be given and set $\varphi = fh_0$. Then, by (7.3.5), for any $\xi \in \mathbb{R}$,

$$\left(\int e^{2\pi\xi x}|\varphi(x)|\,dx\right)^2 \leq \|f\|^2_{L^2(\lambda_{\mathbb{R}};\mathbb{C})}\int e^{2\pi(2\xi x - x^2)}\,dx = 2^{-\frac{1}{2}}\|f\|^2_{L^2(\lambda_{\mathbb{R}};\mathbb{C})}e^{2\pi\xi^2}.$$

Hence, since

$$\sum_{m=0}^{n}\left|\frac{(i2\pi\xi x)^m}{m!}\varphi(x)\right| \leq \left(e^{2\pi\xi x} + e^{-2\pi\xi x}\right)|\varphi(x)|,$$

Lebesgue's dominated convergence theorem says that

$$\sum_{m=0}^{n}\frac{(i2\pi\xi x)^m}{m!}\varphi(x) \longrightarrow \mathfrak{e}_\xi(x)\varphi(x) \quad \text{in } L^1(\lambda_{\mathbb{R}};\mathbb{C}).$$

Now suppose that f is orthogonal to $\{\tilde{h}_n : n \in \mathbb{N}\}$. Since, for all $n \in \mathbb{N}$, the sets $\{H_m : 0 \leq m \leq n\}$ and $\{x^m : 0 \leq m \leq n\}$ span the same subspace, $\int x^m\varphi(x)\,dx = 0$ for all $m \in \mathbb{N}$ and therefore, by the preceding, $\hat{\varphi}(\xi) = 0$ for all $\xi \in \mathbb{R}$. Now apply the last part of Theorem 7.3.3 to conclude first that $\varphi = 0$ and then that $f = 0$ $\lambda_{\mathbb{R}}$-almost everywhere.

To derive (7.3.14), combine the first part of (7.3.8) with (7.3.11) to see that $\left((2\pi x)^2 - \partial_x^2 - 2\pi\right)h_n = a_+a_-h_n = 4\pi n h_n$.

To prove the L^1-estimates, first note that $\|\tilde{h}_0\|_{L^1(\lambda_{\mathbb{R}};\mathbb{C})} = 2^{\frac{1}{4}}\|h_0\|_{L^1(\lambda_{\mathbb{R}};\mathbb{C})} = 2^{\frac{1}{4}}$. When $n \geq 1$, use (7.3.12) and (7.3.10) to see that

$$16\pi^2\int x^2 h_n(x)^2\,dx = 2^{-\frac{1}{2}}(4\pi)^{n+1}n!(2n+1),$$

or, equivalently, that $\int x^2\tilde{h}_n(x)^2\,dx = (4\pi)^{-1}(2n+1)$. Thus, since

$$\|\tilde{h}_n\|^2_{L^1(\lambda_{\mathbb{R}};\mathbb{C})} = \left(\int_{\mathbb{R}}\frac{(1+x^2)^{\frac{1}{2}}}{(1+x^2)^{\frac{1}{2}}}|\tilde{h}_n(x)|\,dx\right)^2 \leq \pi\int_{\mathbb{R}}(1+x^2)\tilde{h}_n(x)^2\,dx,$$

the required estimate follows. Moreover, $\|\tilde{h}_n\|_{\mathrm{u}} \leq \|\tilde{h}_n\|_{L^1(\lambda_R;\mathbb{C})}$ will follow as soon as we know that $\widehat{\tilde{h}_n} = i^n\tilde{h}_n$ and therefore that $\|\widehat{\tilde{h}_n}\|_{L^1(\lambda_{\mathbb{R}};\mathbb{C})} = \|\tilde{h}_n\|_{L^1(\lambda_{\mathbb{R}};\mathbb{C})}$.

The proof of (7.3.16) comes down to another application of Lemma 7.3.5 and the first part of (7.3.8). Namely, begin with the observation that $\hat{h}_0 = h_0$. Next, assume that $\widehat{h_n} = i^n h_n$, and apply Lemma 7.3.5, (7.3.8), and the identity $(a_-)_x\mathfrak{e}_\xi(x) = i(a_+)_\xi\mathfrak{e}_\xi(x)$, where the subscript indicates on which variable the operator is acting, to justify

$$\widehat{h_{n+1}}(\xi) = \int \mathfrak{e}_\xi(x)(a_+ h_n)(x)\,dx = \int (a_-)_x \mathfrak{e}_\xi(x) h_n(x)\,dx$$
$$= ia_+\hat{h}_n(\xi) = i^{n+1}h_{n+1}(\xi).$$

□

The functions $\tilde{h}_n$ in Theorem 7.3.6 are called the **normalized Hermite functions**.

The following corollary is an easy consequence of Theorems 7.3.6 and 7.1.7. In its statement, $\|\mathbf{n}\| = \sum_{j=1}^N n_j$ for $\mathbf{n} \in \mathbb{N}^N$.

Corollary 7.3.7 *For each* $\mathbf{n} = (n_1, \dots, n_N) \in \mathbb{N}^N$, *define* $\tilde{h}_\mathbf{n} : \mathbb{R}^N \longrightarrow \mathbb{R}$ *by*

$$\tilde{h}_\mathbf{n}(x) = \prod_{j=1}^N \tilde{h}_{n_j}(x_j) \quad \text{for } x = (x_1, \dots, x_N) \in \mathbb{R}^N.$$

Then $\{\tilde{h}_\mathbf{n} : \mathbf{n} \in \mathbb{N}^N\}$ *is an orthonormal basis in* $L^2(\lambda_{\mathbb{R}^N};\mathbb{C})$ *and*

$$(2\pi|x|)^2\tilde{h}_\mathbf{n}(x) - \Delta\tilde{h}_\mathbf{n}(x) = 2\pi(2\|\mathbf{n}\| + 1)\tilde{h}_\mathbf{n}(x) \quad \text{for all } \mathbf{n} \in \mathbb{N}^N.$$

Furthermore, $\|\tilde{h}_\mathbf{n}\|_\mathrm{u} \le \|\tilde{h}_\mathbf{n}\|_{L^1(\lambda_{\mathbb{R}^N};\mathbb{C})} \le C_N(1 + \|\mathbf{n}\|)^{\frac{N}{2}}$ *for some* $C_N < \infty$, *and* $\widehat{\tilde{h}_\mathbf{n}} = i^{\|\mathbf{n}\|}\tilde{h}_\mathbf{n}$ *for all* $\mathbf{n} \in \mathbb{N}^N$.

7.3.3 L^2-Theory of the Fourier Transform

Referring to Corollary 7.3.7, define the **Fourier operator** $\mathcal{F}: L^2(\lambda_{\mathbb{R}^N};\mathbb{C}) \longrightarrow L^2(\lambda_{\mathbb{R}^N};\mathbb{C})$ by

$$\mathcal{F}f = \sum_{\mathbf{n}\in\mathbb{Z}^N} i^{\|\mathbf{n}\|}\big(f, \tilde{h}_\mathbf{n}\big)_{L^2(\lambda_{\mathbb{R}^N};\mathbb{C})}\tilde{h}_\mathbf{n}.$$

Clearly $\mathcal{F}$ is a linear isometry from $L^2(\lambda_{\mathbb{R}^N};\mathbb{C})$ onto itself, and its inverse is given by

$$\mathcal{F}^{-1}f = \sum_{\mathbf{n}\in\mathbb{Z}^N} (-i)^{\|\mathbf{n}\|}\big(f, \tilde{h}_\mathbf{n}\big)_{L^2(\lambda_{\mathbb{R}^N};\mathbb{C})}\tilde{h}_\mathbf{n} = \overline{\mathcal{F}\bar{f}}.$$

In addition, as an invertible isometry, $\mathcal{F}$ is **unitary** in the sense that

$$\big(\mathcal{F}f, g\big)_{L^2(\lambda_{\mathbb{R}^N};\mathbb{C})} = \big(f, \mathcal{F}^{-1}g\big)_{L^2(\lambda_{\mathbb{R}^N};\mathbb{C})}. \tag{7.3.17}$$

Indeed, simply note that, because it is an isometry,

$$\big(f, \mathcal{F}^{-1}g\big)_{L^2(\lambda_{\mathbb{R}^N};\mathbb{C})} = \big(\mathcal{F}f, \mathcal{F}\circ\mathcal{F}^{-1}g\big)_{L^2(\lambda_{\mathbb{R}^N};\mathbb{C})} = \big(\mathcal{F}f, g\big)_{L^2(\lambda_{\mathbb{R}^N};\mathbb{C})}.$$

Theorem 7.3.8 *For every $f \in L^1(\lambda_{\mathbb{R}^N};\mathbb{C}) \cap L^2(\lambda_{\mathbb{R}^N};\mathbb{C})$, $\mathcal{F}f = \hat{f}$ and $\overline{\mathcal{F}^{-1}f} = \widehat{\bar{f}}$ $\lambda_{\mathbb{R}^N}$-almost everywhere. In particular, for every $f \in L^2(\lambda_{\mathbb{R}^N};\mathbb{C})$,*

$$\left.\begin{aligned}\int_{|x|\le R} \mathfrak{e}_\xi(x) f(x)\,dx &\longrightarrow \mathcal{F}f(\xi)\\ \int_{|\xi|\le R} \mathfrak{e}_{-x}(\xi) f(\xi)\,d\xi &\longrightarrow \mathcal{F}^{-1}f(x)\end{aligned}\right\} \quad \text{in } L^2(\lambda_{\mathbb{R}^N};\mathbb{C}) \text{ as } R\to\infty.$$

Proof. It suffices to prove the first assertion when $f \in C_c^\infty(\mathbb{R}^N;\mathbb{C})$. Indeed, given any $f \in L^1(\lambda_{\mathbb{R}^N};\mathbb{C}) \cap L^2(\lambda_{\mathbb{R}^N};\mathbb{C})$, one can apply Theorem 6.3.9 to its real and imaginary parts to produce a sequence $\{f_n : n \ge 1\} \subseteq C_c^\infty(\mathbb{R}^N;\mathbb{C})$ for which $f_n \longrightarrow f$ in both $L^1(\lambda_{\mathbb{R}^N};\mathbb{C})$ and $L^2(\lambda_{\mathbb{R}^N};\mathbb{C})$. Hence, if we knew that $\mathcal{F}f_n = \widehat{f_n}$ for each n, then, because $\widehat{f_n} \longrightarrow \hat{f}$ uniformly and $\mathcal{F}f_n \longrightarrow \mathcal{F}f$ in $L^2(\lambda_{\mathbb{R}^N};\mathbb{C})$, we would know that $\mathcal{F}f = \hat{f}$ $\lambda_{\mathbb{R}^N}$-almost everywhere.

Now assume that $f \in C_c^\infty(\mathbb{R}^N;\mathbb{C})$, and set

$$\varphi_m = \sum_{\|\mathbf{n}\|\le m} \big(f, \tilde{h}_{\mathbf{n}}\big)_{L^2(\lambda_{\mathbb{R}^N};\mathbb{C})} \tilde{h}_{\mathbf{n}} \quad \text{for } m \ge 0.$$

We know that $\varphi_m \longrightarrow f$ and therefore that $\mathcal{F}\varphi_m \longrightarrow \mathcal{F}f$ in $L^2(\lambda_{\mathbb{R}^N};\mathbb{C})$. Furthermore, by Corollary 7.3.7, $\varphi_m \in L^1(\lambda_{\mathbb{R}^N};\mathbb{C})$ and $\widehat{\varphi_m} = \mathcal{F}\varphi_m$ for each $m \ge 0$. Hence, if we can show that $\varphi_m \longrightarrow f$ in $L^1(\lambda_{\mathbb{R}^N};\mathbb{C})$, then we will know that $\hat{f} = \mathcal{F}f$ $\lambda_{\mathbb{R}^N}$-almost everywhere. But, since $\varphi_m \longrightarrow f$ in $\lambda_{\mathbb{R}^N}$-measure, Fatou's lemma says that

$$\begin{aligned}\|f - \varphi_m\|_{L^1(\lambda_{\mathbb{R}^N};\mathbb{C})} &\le \varliminf_{M\to\infty} \|\varphi_M - \varphi_m\|_{L^1(\lambda_{\mathbb{R}^N};\mathbb{C})}\\ &\le \sum_{\|\mathbf{n}\|>m} \big|\big(f, \tilde{h}_{\mathbf{n}}\big)_{L^2(\lambda_{\mathbb{R}^N};\mathbb{C})}\big| \|\tilde{h}_{\mathbf{n}}\|_{L^1(\lambda_{\mathbb{R}^N};\mathbb{C})}.\end{aligned}$$

Furthermore, by Corollary 7.3.7, $\|\tilde{h}_{\mathbf{n}}\|_{L^1(\lambda_{\mathbb{R}^N};\mathbb{C})} \le C_N(1+\|\mathbf{n}\|)^{\frac{N}{2}}$ and

$$\begin{aligned}&\big(2\pi(2\|\mathbf{n}\|+1)\big)^{N+1}\big(f, \tilde{h}_{\mathbf{n}}\big)_{L^2(\lambda_{\mathbb{R}^N};\mathbb{C})}\\ &\quad = \big(f, \mathcal{H}^{N+1}\tilde{h}_{\mathbf{n}}\big)_{L^2(\lambda_{\mathbb{R}^N};\mathbb{C})} = \big(\mathcal{H}^{N+1}f, \tilde{h}_{\mathbf{n}}\big)_{L^2(\lambda_{\mathbb{R}^N};\mathbb{C})},\end{aligned}$$

where $\mathcal{H}$ is the operator $(2\pi|x|)^2 - \Delta$ and Green's formula was applied to get the final equality. Thus, we now have that

$$\begin{aligned}\|f - \varphi_m\|_{L^1(\lambda_{\mathbb{R}^N};\mathbb{C})} &\le C_N \sum_{\|\mathbf{n}\|>m} \frac{(1+\|\mathbf{n}\|)^{\frac{N}{2}}}{(2\pi(2\|\mathbf{n}\|+1))^{N+1}} \big|\big(\mathcal{H}^{N+1}f, \tilde{h}_{\mathbf{n}}\big)_{L^2(\lambda_{\mathbb{R}^N};\mathbb{C})}\big|\\ &\le C_N \left(\sum_{\|\mathbf{n}\|>m} \frac{(1+\|\mathbf{n}\|)^N}{(2\pi(2\|\mathbf{n}\|+1))^{2N+2}}\right)^{\frac{1}{2}} \|\mathcal{H}^{N+1}f\|_{L^2(\lambda_{\mathbb{R}^N};\mathbb{C})} \longrightarrow 0 \text{ as } m\to\infty.\end{aligned}$$

Knowing that $\mathcal{F}f = \hat{f}$ for $f \in L^1(\lambda_{\mathbb{R}^N};\mathbb{C}) \cap L^2(\lambda_{\mathbb{R}^N};\mathbb{C})$ and noting that $\mathcal{F}^{-1}f = \overline{\mathcal{F}\bar{f}}$, one sees that $\overline{\mathcal{F}^{-1}f} = \widehat{\bar{f}}$ for $f \in L^1(\lambda_{\mathbb{R}^N};\mathbb{C}) \cap L^2(\lambda_{\mathbb{R}^N};\mathbb{C})$.

Given the first part, the second part is easy. Namely, if $f \in L^2(\lambda_{\mathbb{R}^N};\mathbb{C})$, then $f_R = \mathbf{1}_{\overline{B(0,R)}} f \in L^1(\lambda_{\mathbb{R}^N};\mathbb{C}) \cap L^2(\lambda_{\mathbb{R}^N};\mathbb{C})$, and therefore

$$\mathcal{F}f_R(\xi) = \widehat{f_R}(\xi) = \int_{|x|\le R} \mathfrak{e}_\xi(x) f(x)\, dx,$$

$$\mathcal{F}^{-1}f_R(x) = \overline{\widehat{\overline{f_R}}}(x) = \int_{|\xi|\le R} \mathfrak{e}_{-x}(\xi) f(\xi)\, d\xi.$$

Hence, since $f_R \longrightarrow f$ and therefore $\mathcal{F}f_R \longrightarrow \mathcal{F}f$ and $\mathcal{F}^{-1}f_R \longrightarrow \mathcal{F}^{-1}f$ in $L^2(\lambda_{\mathbb{R}^N};\mathbb{C})$, we are done. □

The fact that $\mathcal{F}$ is unitary is so important that two of its consequences have names of their own. The formula

$$f = \lim_{R\to\infty} \int_{|\xi|\le R} \mathfrak{e}_{-x}(\xi)\mathcal{F}f(\xi)\, d\xi \quad \text{in } L^2(\lambda_{\mathbb{R}^N};\mathbb{C}) \tag{7.3.18}$$

is called the **Fourier inversion formula** and

$$\big(\mathcal{F}f, \mathcal{F}g\big)_{L^2(\lambda_{\mathbb{R}^N};\mathbb{C})} = \big(f, g\big)_{L^2(\lambda_{\mathbb{R}^N};\mathbb{C})} \tag{7.3.19}$$

is known as **Parseval's identity**.

Remark 7.3.1 As Theorem 7.3.8 shows, $\mathcal{F}$ on $L^2(\lambda_{\mathbb{R}^N};\mathbb{C})$ is the continuous extension from $L^1(\lambda_{\mathbb{R}^N};\mathbb{C}) \cap L^2(\lambda_{\mathbb{R}^N};\mathbb{C})$ of the L^1-Fourier transform, and it is conventional to continue to use $\hat{f}$ and $\check{f}$ to denote $\mathcal{F}f$ and $\mathcal{F}^{-1}f$. This convention is a little misleading since $\hat{f}$ for $f \in L^1(\lambda_{\mathbb{R}^N};\mathbb{C})$ is defined pointwise by a well-defined Lebesgue integral whereas $\mathcal{F}f$ is, in general, defined only up to a set of $\lambda_{\mathbb{R}^N}$-measure 0 and is given as a limit in $L^2(\lambda_{\mathbb{R}^N};\mathbb{C})$ of Lebesgue integrals.

It should also be pointed out that there is a general principle that should have made one suspect that the Fourier transform takes a Hermite function into a scalar multiple of itself. Namely, (7.3.14) says that h_n is a eigenfunction of the operator $\mathcal{H} = (2\pi x)^2 - \partial^2$, and it is easy that check that $\widehat{\mathcal{H}\varphi} = \mathcal{H}\hat{\varphi}$ for smooth functions φ that, together with their derivates, vanish sufficiently rapidly at infinity. In other words, the Fourier transform commutes with $\mathcal{H}$ and therefore should share eigenfunctions with it.

The L^2-version of Theorem 7.3.4 is the following.

Theorem 7.3.9 *If $f \in C^1(\mathbb{R}^N;\mathbb{C})$ and both f and $|\nabla f|$ are in $L^2(\lambda_{\mathbb{R}^N};\mathbb{C})$, then, for each $1 \le j \le N$,*

$$\mathcal{F}(\partial_{x_j} f)(\xi) = -i2\pi\xi_j \mathcal{F}f(\xi) \quad \textit{for } \lambda_{\mathbb{R}^N}\textit{-almost every } \xi \in \mathbb{R}^N,$$

and therefore $|\xi|\mathcal{F}f(\xi)$ *is in* $L^2(\lambda_{\mathbb{R}^N};\mathbb{C})$. *If* $f \in L^1(\lambda_{\mathbb{R}^N};\mathbb{C})$ *and* $g \in L^2(\lambda_{\mathbb{R}^N};\mathbb{C})$, *then* $\mathcal{F}(f * g) = \hat{f}\mathcal{F}g$.

Proof. Choose a bump function $\eta \in C^\infty(\mathbb{R}^N;[0,1])$ for which $\eta = 1$ on $B(0,1)$ and 0 off $B(0,2)$, and set $\eta_R(x) = \eta(R^{-1}x)$. Given an f with the properties in the first assertion, take $f_R = \eta_R f$. Then, by Theorems 7.3.8 and 7.3.4, $\mathcal{F}(\partial_{x_j} f_R)(\xi) = -i2\pi\xi_j \mathcal{F}f_R(\xi)$ for $\lambda_{\mathbb{R}^N}$-almost every $\xi \in \mathbb{R}^N$. Furthermore, as $R \to \infty$, $f_R \longrightarrow f$ and

$$\partial_{x_j} f_R(x) = R^{-1}\big((\partial_{x_j}\eta)(R^{-1}x)\big)f(x) + \eta_R(x)\partial_{x_j} f(x) \longrightarrow \partial_{x_j} f(x)$$

both pointwise and in $L^2(\lambda_{\mathbb{R}^N};\mathbb{C})$. Hence, $\mathcal{F}f_R \longrightarrow \mathcal{F}f$ and $\mathcal{F}(\partial_{x_j} f_R) \longrightarrow \mathcal{F}(\partial_{x_j} f)$ in $L^2(\lambda_{\mathbb{R}^N};\mathbb{C})$, and so we can find $R_n \nearrow \infty$ such that

$$-i2\pi\xi_j \mathcal{F}f(\xi) = -\lim_{n\to\infty} i2\pi\xi_j \widehat{f_{R_n}}(\xi) = \lim_{n\to\infty} \widehat{\partial_{x_j} f_{R_n}}(\xi) = \big[\mathcal{F}(\partial_{x_j} f)\big](\xi)$$

for $\lambda_{\mathbb{R}^N}$-almost every $\xi \in \mathbb{R}^N$.

To prove the second part, take η_R as above and $g_R = \eta_R g$. By Theorems 7.3.8 and 7.3.4, $\mathcal{F}(f * g_R) = \hat{f}\mathcal{F}g_R$. Moreover, $g_R \longrightarrow g$ and, by Young's inequality, $f * g_R \longrightarrow f * g$ in $L^2(\lambda_{\mathbb{R}^N};\mathbb{C})$. Hence, the second part is also proved. □

7.3.4 Schwartz Test Function Space and Tempered Distributions

Even though it is not really a topic in measure theory, having introduced them, it seems appropriate to develop the relationship between Hermite functions and Laurent Schwartz's theory of tempered distributions. [6] His theory is an attempt to put into a general context ideas going back to Boole and Heaviside in connection with applications of the Laplace transform to ordinary differential equations, and later developed by Sobolev for partial differential equations.

The theory of distributions deals with quantities that one would like to think of as functions but defy an interpretation as traditional functions. The most famous such quantity is the *Dirac delta function*, the "function" δ with the property that

$$\int \delta(x)\varphi(x)\,dx = \varphi(0)$$

for $\varphi \in C(\mathbb{R};\mathbb{C})$. (See Exercise 7.3.9 for further information.) Of course, such a function does not exist if one insists that a function is a quantity that can

[6] There are many books in which Schwartz's theory is presented, but his own original treatment in "Théorie des distributions, I" published in 1950 by Hermann, Paris remains one of the best accounts.

be defined pointwise. On the other hand, if one thinks in measure theoretic terms, then one can interpret Dirac's delta function as the non-existent function for which the unit point mass $\delta_0(dx) = \delta(x)\,\lambda_{\mathbb{R}}(dx)$. That is, although one cannot evaluate δ point-wise, one knows how to evaluate integrals in which the "integrand" is the product of δ with a continuous function. Similarly, the derivative δ' of δ is even less amenable to interpretation as a traditional function, but nonetheless one knows that its "integral" against a $\varphi \in C^1(\mathbb{R};\mathbb{C})$ should be equal to $-\varphi'(0)$.

Schwartz's idea was to use functional analysis to embed quantities like Dirac's delta function and its derivative in a much larger space, one that arises as the dual space (i.e., the space of continuous, linear functions) of a topological vector space of functions. That is, instead of trying to evaluate these quantities as functions at points in a space like $\mathbb{R}^N$, he replaces evaluation at points with evaluation at a "test function," and the more stringent the requirements that one imposes on the test functions, the larger the space of quantities that can act on them. Thus, in the most general version of his theory, $C_c^\infty(\mathbb{R}^N;\mathbb{C})$ is the space of test functions, but, unfortunately, the appropriate topology on $C_c^\infty(\mathbb{R}^N;\mathbb{C})$ has several unpleasant features. For example, there are no countable neighborhood bases, and so it admits no metric. Nonetheless, if one is willing to make a compromise between generality and approachability, there is a closely related space of test functions whose dual includes most of the quantities that are needed in applications, and it is in this space of test functions that the Hermite functions play a central role.

The test function space alluded to above is denoted by $\mathscr{S}(\mathbb{R}^N;\mathbb{C})$ and consists of functions $\varphi \in C^\infty(\mathbb{R}^N;\mathbb{C})$ with the property that $x \rightsquigarrow |x|^j \partial^\alpha \varphi(x)$ is bounded for all $m \geq 0$ and $\alpha \in \mathbb{N}^N$. Obviously, $\mathscr{S}(\mathbb{R}^N;\mathbb{C})$ is a vector space. In addition, it is closed under differentiation as well as products with smooth functions which, together with all their derivatives, have polynomial growth (i.e., grow no faster than some power of $(1+|x|^2)$). Thus the Hermite functions are all in $\mathscr{S}(\mathbb{R};\mathbb{C})$. Finally, since, for $\varphi \in \mathscr{S}(\mathbb{R}^N;\mathbb{C})$,

$$\int_{\mathbb{R}^N} |\varphi(x)|\,dx \leq \|(1+|x|^2)^{N+1}\varphi\|_{\mathrm{u}} \int_{\mathbb{R}^N} (1+|x|^2)^{-N-1}\,dx,$$

$\mathscr{S}(\mathbb{R}^N;\mathbb{C}) \subseteq \bigcap_{p\in[1,\infty]} L^p(\lambda_{\mathbb{R}^N};\mathbb{C})$.

One of the goals here will be to show that $\mathscr{S}(\mathbb{R}^N;\mathbb{C})$ shares many characteristics with a Hilbert space in which the Hermite functions play the role of an orthonormal basis. In order to avoid cumbersome notation, I will carry out this program only in the case when $N = 1$, but it should be clear how one can modify the arguments to handle arbitrary $N \in \mathbb{Z}^+$.

There is an obvious notion of convergence for sequences in $\mathscr{S}(\mathbb{R};\mathbb{C})$. Namely, define the norms

$$\|\varphi\|_{\mathrm{u}}^{(j,\ell)} = \|x^j \partial^\ell \varphi\|_{\mathrm{u}}$$

for $j,\ \ell \in \mathbb{N}$, and say that $\varphi_k \longrightarrow \varphi$ in $\mathscr{S}(\mathbb{R};\mathbb{C})$ if $\lim_{k\to\infty} \|\varphi_k - \varphi\|_{\mathrm{u}}^{(j,\ell)} = 0$ for all $j,\ \ell \in \mathbb{N}$. The corresponding topology is the one for which G is open if and only if for each $\varphi \in G$ there an $m \in \mathbb{N}$ and $r > 0$ such that

$$\{\psi : \|\psi - \varphi\|_{\mathrm{u}}^{(m)} < r\} \subseteq G,$$

where

$$\|\bullet\|_{\mathrm{u}}^{(m)} \equiv \sum_{\substack{j,\ell\in\mathbb{N}\\ j+\ell\le m}} \|\bullet\|_{\mathrm{u}}^{(j,\ell)}.$$

Finally, define the operator $\mathcal{H}$ on $\mathscr{S}(\mathbb{R};\mathbb{C})$ into itself by

$$\mathcal{H}\varphi = (2\pi x)^2\varphi - \partial^2\varphi.$$

Also, set $\mu_n = 2\pi(2n+1)$, and note that, by (7.3.14), $\mathcal{H}\tilde{h}_n = \mu_n\tilde{h}_n$ for all $n \ge 0$.

Theorem 7.3.10 *Set*

$$\|\varphi\|_{\mathscr{S}}^{(m)} = \left(\sum_{n=0}^{\infty} \mu_n^m \left|(\varphi, \tilde{h}_n)_{L^2(\mathbb{R};\mathbb{C})}\right|^2\right)^{\frac{1}{2}}$$

for $m \in \mathbb{N}$ and $\varphi \in \mathscr{S}(\mathbb{R};\mathbb{C})$. For each $m \in \mathbb{N}$ there exists a $K_m < \infty$ such that

$$\|\varphi\|_{\mathscr{S}}^{(m)} \le K_m \|\varphi\|_{\mathrm{u}}^{(m+1)} \tag{7.3.20}$$

and

$$\|\varphi\|_{\mathrm{u}}^{(m)} \le K_m \|\varphi\|_{\mathscr{S}}^{(m+5)}. \tag{7.3.21}$$

for all $\varphi \in \mathscr{S}(\mathbb{R};\mathbb{C})$. In particular, $\varphi_k \longrightarrow \varphi$ in $\mathscr{S}(\mathbb{R};\mathbb{C})$ if and only if

$$\lim_{k\to\infty} \|\varphi_k - \varphi\|_{\mathscr{S}}^{(m)} = 0$$

for all $m \in \mathbb{N}$. Finally,

$$\sum_{n=0}^{\infty} \left|(\varphi, \tilde{h}_n)_{L^2(\lambda_{\mathbb{R}};\mathbb{C})}\right| \|\tilde{h}_n\|_{\mathrm{u}} < \infty$$

and so

$$\sum_{n=0}^{\infty} (\varphi, \tilde{h}_n)_{L^2(\lambda_{\mathbb{R}};\mathbb{C})} \tilde{h}_n$$

converges to φ in $\mathscr{S}(\mathbb{R};\mathbb{C})$.

Proof. To prove (7.3.20), begin by using integration by parts to show that

$$\big(\mathcal{H}\varphi, \psi\big)_{L^2(\lambda_{\mathbb{R}};\mathbb{C})} = \big(\varphi, \mathcal{H}\psi\big)_{L^2(\lambda_{\mathbb{R}};\mathbb{C})}$$

for all $\varphi, \psi \in \mathscr{S}(\mathbb{R};\mathbb{C})$. By combining this with (7.3.14), one sees that

$$\left(\mathcal{H}^m\varphi, \tilde{h}_n\right)_{L^2(\lambda_\mathbb{R};\mathbb{C})} = \mu_n^m\left(\varphi, \tilde{h}_n\right)_{L^2(\lambda_R;\mathbb{C})}$$

and therefore that

$$\begin{aligned}\left(\varphi, \mathcal{H}^m\varphi\right)_{L^2(\lambda_\mathbb{R};\mathbb{C})} &= \sum_{n=0}^{\infty}(\varphi, \tilde{h}_n)_{L^2(\lambda_\mathbb{R};\mathbb{C})}\overline{\left(\mathcal{H}^m\varphi, \tilde{h}_n\right)_{L^2(\lambda_\mathbb{R};\mathbb{C})}} \\ &= \sum_{n=0}^{\infty}\mu_n^m\left|(\varphi, \tilde{h}_n)_{L^2(\lambda_\mathbb{R};\mathbb{C})}\right|^2.\end{aligned}$$

for all $m \in \mathbb{N}$. Next observe that there exist constants $c_{j,\ell}^{(m)} \in \mathbb{R}$ such that

$$\left((2\pi x)^2 - \partial^2\right)^m\varphi = \sum_{\substack{j,\ell\in\mathbb{N}\\ j+\ell\le 2m}} c_{j.\ell}^{(m)}x^j\partial^\ell\varphi,$$

and use integration by parts to see that

$$\left(\varphi, x^j\partial^\ell\varphi\right)_{L^2(\lambda_\mathbb{R};\mathbb{C})} = (-1)^{\ell'}\left(\partial^{\ell'}(x^{j'}\varphi), x^{j-j'}\partial^{\ell-\ell'}\varphi\right)_{L^2(\lambda_\mathbb{R};\mathbb{C})},$$

where

$$j' = \begin{cases}\frac{j}{2} & \text{if } j \text{ is even}\\ \frac{j-1}{2} & \text{if } j \text{ is odd}\end{cases} \quad \text{and} \quad \ell' = \begin{cases}\frac{\ell}{2} & \text{if } \ell \text{ is even}\\ \frac{\ell+1}{2} & \text{if } \ell \text{ is odd.}\end{cases}$$

Hence there exist constants $b_{(j_1,\ell_1),(j_2,\ell_2)}^{(m)} \in \mathbb{R}$ such that

$$\begin{aligned}\left(\varphi, \mathcal{H}^m\varphi\right)_{L^2(\lambda_\mathbb{R};\mathbb{C})} &\le \sum_{\substack{(j_1,\ell_1),(j_2,\ell_2)\in\mathbb{N}^2\\ (j_1+\ell_1)\vee(j_2+\ell_2)\le m}} \left|b_{(j_1,\ell_1),(j_2,\ell_2)}^{(m)}\left(x^{j_1}\partial^{\ell_1}\varphi, x^{j_2}\partial^{\ell_2}\varphi\right)_{L^2(\lambda_\mathbb{R};\mathbb{C})}\right| \\ &\le \sum_{\substack{(j_1,\ell_1),(j_2,\ell_2)\in\mathbb{N}^2\\ (j_1+\ell_1)\vee(j_2+\ell_2)\le m}} \left|b_{(j_1,\ell_1),(j_2,\ell_2)}^{(m)}\right|\left\|x^{j_1}\partial^{\ell_1}\varphi\right\|_{L^2(\lambda_\mathbb{R};\mathbb{C})}\left\|x^{j_2}\partial^{\ell_2}\varphi\right\|_{L^2(\lambda_\mathbb{R};\mathbb{C})}.\end{aligned}$$

Finally, observe that

$$\begin{aligned}\|x^j\partial^\ell\varphi\|_{L^2(\lambda_\mathbb{R};\mathbb{C})}^2 &= \int (1+x^2)^{-1}\left|(1+x^2)^{\frac{1}{2}}x^j\partial^\ell\varphi(x)\right|^2\,dx \\ &\le 2\pi\left(\|x^j\partial^\ell\varphi\|_{\mathrm{u}}^2 + \|x^{j+1}\partial^\ell\varphi\|_{\mathrm{u}}^2\right),\end{aligned}$$

and combine this with the preceding to arrive at (7.3.20).

To prove (7.3.21), begin by observing that, by (7.3.20), $\|\varphi\|_{\mathscr{S}}^{(m)} < \infty$ for all $m \in \mathbb{N}$ and, by (7.3.15), $\|\tilde{h}_n\|_{\mathrm{u}} \le \mu_n^{\frac{1}{2}}$. Therefore, since $|(\varphi, \tilde{h}_n)_{L^2(\lambda_\mathbb{R};\mathbb{C})}| \le \mu_n^{-\frac{m}{2}}\|\varphi\|_{\mathscr{S}}^{(m)}$,

$$\big|(\varphi, \tilde{h}_n)_{L^2(\lambda_\mathbb{R};\mathbb{C})}\big| \|\tilde{h}_n\|_\mathrm{u} \le \mu_n^{-\frac{m-1}{2}} \|\varphi\|_{\mathscr{S}}^{(m)}. \tag{*}$$

Hence,

$$\sum_{n=0}^{\infty} \big|(\varphi, \tilde{h}_n)_{L^2(\lambda_\mathbb{R};\mathbb{C})}\big| \|\tilde{h}_n\|_\mathrm{u} < \infty$$

and so the series

$$\sum_{n=0}^{\infty} (\varphi, \tilde{h}_n)_{L^2(\lambda_\mathbb{R};\mathbb{C})} \tilde{h}_n$$

converges uniformly to φ. The next step is to apply the preceding and integration by parts to see that

$$\|x^j \partial^\ell \varphi\|_\mathrm{u} \le \sum_{n=0}^{\infty} \mu_n^{\frac{1}{2}} \big|\big(x^j \partial^\ell \varphi, \tilde{h}_n\big)_{L^2(\lambda_\mathbb{R};\mathbb{C})}\big| = \sum_{n=0}^{\infty} \mu_n^{\frac{1}{2}} \big|\big(\varphi, \partial^\ell(x^j \tilde{h}_n)\big)_{L^2(\lambda_\mathbb{R};\mathbb{C})}\big|.$$

By repeated applications of equations (7.3.12) and (7.3.13) (cf. Exercise 7.3.6), one sees that there exist constants $a_{n,k}^{(j,\ell)} \in \mathbb{R}$ such that

$$\partial^\ell(x^j \tilde{h}_n) = \mu_n^{\frac{j+\ell}{2}} \sum_{k=-(j+\ell)\wedge n}^{j+\ell} a_{n,k}^{(j,\ell)} \tilde{h}_{n+k}.$$

and therefore, by $(*)$, there exists a $C_{(j,\ell)} < \infty$ such that

$$\|x^j \partial^\ell \varphi\|_\mathrm{u} \le C_{(j,\ell)} \|\varphi\|_{\mathscr{S}}^{(m)} \sum_{n=0}^{\infty} \mu_n^{\frac{j+\ell}{2}} \sum_{k=-(j+\ell)\wedge n}^{j+\ell} \mu_{n+k}^{\frac{-m+1}{2}}.$$

Since the preceding double sum converges when $m \ge j+\ell+5$, (7.3.21) follows. □

Corollary 7.3.11 *Let $\mathfrak{s}(\mathbb{R};\mathbb{C})$ be the space of functions $a : \mathbb{N} \longrightarrow \mathbb{C}$ such that*

$$\|a\|_{\mathfrak{s}}^{(m)} = \left(\sum_{n=0}^{\infty} \mu_n^m |a(n)|^2\right)^{\frac{1}{2}} < \infty$$

for all $m \in \mathbb{N}$. Then, for each $\varphi \in \mathscr{S}(\mathbb{R};\mathbb{C})$, the function $n \in \mathbb{N} \longmapsto (\varphi, \tilde{h}_n)_{L^2(\lambda_\mathbb{R};\mathbb{C})} \in \mathbb{C}$ is an element of $\mathfrak{s}(\mathbb{R};\mathbb{C})$. Conversely, if $a \in \mathfrak{s}(\mathbb{R};\mathbb{C})$, then the series $\sum_{n=0}^{\infty} a(n)\tilde{h}_n$ converges in $\mathscr{S}(\mathbb{R};\mathbb{C})$. Thus, the map $\Psi : \mathscr{S}(\mathbb{R};\mathbb{C}) \longrightarrow \mathfrak{s}(\mathbb{R};\mathbb{C})$ given by

$$\big(\Psi(\varphi)\big)(n) = \big(\varphi, \tilde{h}_n\big)_{L^2(\lambda_\mathbb{R};\mathbb{C})}$$

is one-to-one and onto. In addition, $\|\Phi(\varphi)\|_{\mathfrak{s}}^{(m)} = \|\varphi\|_{\mathscr{S}}^{(m)}$ for all $m \in \mathbb{N}$.

Proof. The fact that, for each $\varphi \in \mathscr{S}(\mathbb{R};\mathbb{C})$, $n \rightsquigarrow (\varphi, \tilde{h}_n)_{L^2(\lambda_{\mathbb{R}};\mathbb{C})}$ is an element of $\mathfrak{s}(\mathbb{R};\mathbb{C})$ follows from (7.3.20), and, by definition, its $\| \bullet \|_{\mathfrak{s}}^{(m)}$-norm equals $\|\varphi\|_{\mathscr{S}}^{(m)}$. Conversely, given $a \in \mathfrak{s}(\mathbb{R};\mathbb{C})$, set $\varphi_N = \sum_{n=0}^{N} a(n)\tilde{h}_n$. Then

$$\sup_{N\geq M} \|\varphi_N - \varphi_M\|_{\mathscr{S}}^{(m)} \leq \left(\sum_{n=M+1}^{\infty} \mu_n^m |a_n|^2 \right)^{\frac{1}{2}}.$$

and so, by (7.3.21), $\lim_{M\to\infty} \sup_{N\geq M} \|\varphi_N - \varphi_M\|_{\mathrm{u}}^{(m)} = 0$ for all $m \in \mathbb{N}$. From this it follows that φ_N and all its derivatives converge to an element of $\varphi \in C^\infty(\mathbb{R};\mathbb{C})$ and that $(\varphi, \tilde{h}_n)_{L^2(\lambda_{\mathbb{R}};\mathbb{C})} = a(n)$ for all $n \in \mathbb{N}$. Hence $\varphi \in \mathscr{S}(\mathbb{R};\mathbb{C})$.

Given the preceding, it is clear that Ψ is a one-to-one map of $\mathscr{S}(\mathbb{R};\mathbb{C})$ onto $\mathfrak{s}(\mathbb{R};\mathbb{C})$, that $\Phi^{-1}(a) = \sum_{n=0}^{\infty} a(n)\tilde{h}_n$, and that $\|\Phi(\varphi)\|_{\mathfrak{s}}^{(m)} = \|\varphi\|_{\mathscr{S}}^{(m)}$. □

As a consequence of Theorem 7.3.10 and Corollary 7.3.11, we know that, for any function $g : [0,\infty) \longrightarrow \mathbb{C}$ with at most polynomial growth, the series

$$\sum_{n=0}^{\infty} g(\mu_n)(\varphi, \tilde{h}_n)_{L^2(\lambda_{\mathbb{R}};\mathbb{C})}\tilde{h}_n$$

converges in $\mathscr{S}(\mathbb{R};\mathbb{C})$ and determines a continuous, linear operator $g(\mathcal{H})$ on $\mathscr{S}(\mathbb{R};\mathbb{C})$ into itself. Of particular importance here is the case when $g(t) = t^s$ for some $s \in \mathbb{R}$, in which case

$$\mathcal{H}^s\varphi = \sum_{n=0}^{\infty} \mu_n^s(\varphi, \tilde{h}_n)_{L^2(\lambda_{\mathbb{R}};\mathbb{C})}\tilde{h}_n.$$

Of course, $\mathcal{H}^0$ is the identify map, and, for $m \in \mathbb{Z}^+$, $\mathcal{H}^m = \left((2\pi x)^n - \partial^2\right)^m$. Observe that $\mathcal{H}^s \circ \mathcal{H}^t = \mathcal{H}^{s+t}$ and that

$$\|\varphi\|_{\mathscr{S}}^{(m)} = \left\|\mathcal{H}^{\frac{m}{2}}\varphi\right\|_{L^2(\lambda_{\mathbb{R}};\mathbb{C})}. \tag{7.3.22}$$

We now have the machinery to discuss the role of $\mathscr{S}(\mathbb{R};\mathbb{C})$ as a test function space for distributions. Recall that the dual space X^* of a topological vector space X over $\mathbb{C}$ is the space to $\mathbb{C}$-valued, continuous, linear functions on X. When X is a Hilbert space, Theorem 7.1.3 says that X^* can be identified with X via the inner product. Although $\mathscr{S}(\mathbb{R};\mathbb{C})$ is not itself a Hilbert space, Theorem 7.3.10 and (7.3.22) show that it is the intersection of the Hilbert spaces

$$H_m = \left\{\varphi \in L^2(\mathbb{R};\mathbb{C}) : \|\mathcal{H}^{\frac{m}{2}}\varphi\|_{L^2(\mathbb{R};\mathbb{C})} < \infty\right\}.$$

Thus, it should come as no surprise that $\mathscr{S}(\mathbb{R};\mathbb{C})^*$ is intimately related to the dual spaces associated with these Hilbert spaces. However, as the following theorem makes precise, the identification is not via the inner products on the spaces H_m but instead via the inner product on $L^2(\lambda_{\mathbb{R}};\mathbb{C})$.

In the distribution theory literature, an element of $\mathscr{S}(\mathbb{R};\mathbb{C})^*$ is called a **tempered distribution** to distinguish it from a more general distribution, an element of $C_c^\infty(\mathbb{R};\mathbb{C})^*$.

Theorem 7.3.12 *If $\Lambda \in \mathscr{S}(\mathbb{R};\mathbb{C})^*$, then there is an $m \in \mathbb{N}$ and $C < \infty$ such that $|\Lambda(\varphi)| \le C\|\varphi\|_{\mathscr{S}}^{(m)}$. Thus, $\Lambda \in \mathscr{S}(\mathbb{R};\mathbb{C})^*$ if and only if*

$$\Lambda(\varphi) = \left(\mathcal{H}^{\frac{m}{2}}\varphi, \bar{f}\right)_{L^2(\lambda_{\mathbb{R}};\mathbb{C})}$$

for some $m \ge 0$ and $f \in L^2(\lambda_{\mathbb{R}};\mathbb{C})$.

Proof. By the estimates in Theorem 7.3.10, sets of the form

$$\{\psi : \|\psi\|_{\mathscr{S}}^{(m)} < r\} \text{ for } m \in \mathbb{N} \text{ and } r > 0$$

constitute a neighborhood basis at 0. Thus, if $\Lambda \in \mathscr{S}(\mathbb{R};\mathbb{C})^*$, then there exists an $m \ge 0$ and $r > 0$ such that $|\Lambda(\varphi)| \le 1$ if $\|\varphi\|_{\mathscr{S}}^{(m)} \le r$, which means that

$$\frac{r}{\|\varphi\|_{\mathscr{S}}^{(m)}}|\Lambda(\varphi)| = \left|\Lambda\left(\frac{r\varphi}{\|\varphi\|_{\mathscr{S}}^{(m)}}\right)\right| \le 1.$$

If $f \in L^2(\lambda_{\mathbb{R}};\mathbb{C})$ and $m \ge 0$, then it is clear that $\varphi \rightsquigarrow \left(\mathcal{H}^{\frac{m}{2}}\varphi, \bar{f}\right)_{L^2(\lambda_{\mathbb{R}};\mathbb{C})}$ is a continuous, linear function on $\mathscr{S}(\mathbb{R};\mathbb{C})$. Conversely, if $\Lambda \in \mathscr{S}(\mathbb{R};\mathbb{C})^*$ and $|\Lambda(\varphi)| \le C\|\varphi\|_{\mathscr{S}}^{(m)}$, then $\left|\Lambda\left(\mathcal{H}^{-\frac{m}{2}}\varphi\right)\right| \le C\|\varphi\|_{L^2(\lambda_{\mathbb{R}};\mathbb{C})}$, and so $\Lambda \circ \mathcal{H}^{-\frac{m}{2}}$ extends to a continuous linear function on $L^2(\lambda_{\mathbb{R}};\mathbb{C})$. Hence, by Theorem 7.1.3, there exists an $f \in L^2(\lambda_{\mathbb{R}};\mathbb{C})$ such that $\Lambda(\mathcal{H}^{-\frac{m}{2}}\varphi) = (\varphi, \bar{f})_{L^2(\lambda_{\mathbb{R}};\mathbb{C})}$, which means that $\Lambda(\varphi) = \left(\mathcal{H}^{\frac{m}{2}}\varphi, \bar{f}\right)_{L^2(\lambda_{\mathbb{R}};\mathbb{C})}$. □

As was said in the introduction to this subsection, the reason why Schwartz introduced $\mathscr{S}(\mathbb{R};\mathbb{C})$ and its dual is that he wanted to have a mathematically rigorous description of quantities that are not conventional functions but one would like to use as if they were. In particular, one would like to apply operations like differentiation to these quantities even though the operation is not classically defined. For this reason, distributions are often called *generalized functions.* This term is justified by the observation that any Borel measurable $f : \mathbb{R} \longrightarrow \mathbb{C}$ with moderate growth, in the sense that $(1+x^2)^{-m} f \in L^1(\lambda_{\mathbb{R}};\mathbb{C})$ for some $m \in \mathbb{N}$, gives rise to a tempered distribution Λ_f defined

$$\Lambda_f(\varphi) = \int \varphi(x)f(x)\,dx = \left(\varphi, \bar{f}\right)_{L^2(\lambda_{\mathbb{R}};\mathbb{C})}.$$

Moreover, Λ_f determines f up to a set of $\lambda_{\mathbb{R}}$-measure 0, and so Borel measurable functions with moderate growth can be seen as a subspace of $\mathscr{S}(\mathbb{R};\mathbb{C})^*$.

The preceding observation motivates the extension of many familiar operations on functions to $\mathscr{S}(\mathbb{R};\mathbb{C})^*$. For example, given $\Lambda \in \mathscr{S}^*(\mathbb{R};\mathbb{C})$, the map

$\varphi \rightsquigarrow -\Lambda(\partial\varphi)$ is an element of $\mathscr{S}(\mathbb{R};\mathbb{C})^*$. Furthermore, if f is an element of $C^1(\mathbb{R};\mathbb{C})$ whose derivative has polynomial growth, then f also has polynomial growth and $-\Lambda_f(\partial\varphi) = \Lambda_{\partial f}(\varphi)$. Therefore, it is reasonable to use the notation $\partial\Lambda$ to denote the map $\varphi \rightsquigarrow -\Lambda(\partial\varphi)$ and to think of this as the extension of differentiation to tempered distributions.

More generally, if L is a continuous, linear map of $\mathscr{S}(\mathbb{R};\mathbb{C})$ into itself, then the map $\Lambda \rightsquigarrow L^\top\Lambda$ given by $L^\top\Lambda(\varphi) = \Lambda(L\varphi)$ is a linear map of $\mathscr{S}(\mathbb{R};\mathbb{C})^*$ into itself. Furthermore, if there is a continuous linear operator $L^\top$ on $\mathscr{S}(\mathbb{R};\mathbb{C})$ into itself such that

$$\int \psi L\varphi \, d\lambda_{\mathbb{R}} = \int \varphi L^\top \psi \, d\lambda_{\mathbb{R}}$$

for all $\varphi, \psi \in \mathscr{S}(\mathbb{R};\mathbb{C})$, then $L = (L^\top)^\top$ and so $L\Lambda(\varphi) = \Lambda(L^\top\varphi)$ and $L\Lambda_f = \Lambda_{Lf}$ if $f \in \mathscr{S}(\mathbb{R};\mathbb{C})$. That is, when $L^\top$ exists, L admits a natural extention to $\mathscr{S}(\mathbb{R};\mathbb{C})$. In particular, the concluding result in Theorem 7.3.12 can be summarized as the statement that

$$\mathscr{S}(\mathbb{R};\mathbb{C})^* = \big\{\mathcal{H}^{\frac{m}{2}}\Lambda_f : m \geq 0 \,\&\, f \in L^2(\lambda_{\mathbb{R}};\mathbb{C})\big\}.$$

Finally, it should be said that in most applications, the role of distributions is in large part psychological. For example, in applications to differential equations, one often begins by showing that the equation is satisfied by a distribution. Such a solution is hardly ever satisfactory, but at least it provides a starting point. The hard work comes later, when one tries to show that the distributional solution is better than just an abstract quantity and may even be a classical solution. Exercise 8.2.4 contains an elementary example of the sort of reasoning that is used in showing that a distribution is often a more tractable quantity than it appears at first.

7.3.5 Exercises for § 7.3

Exercise 7.3.1 Let $g \in L^1(\lambda_{\mathbb{R}^N};\mathbb{R})$ with $\int_{\mathbb{R}} g(x)\,dx = 1$ be given, and define $g_t(x) = t^{-N}g(t^{-1}x)$. Clearly, as $t \searrow 0$, g_t gets more and more localized at 0 in a sense made precise by Theorem 6.3.7. Show that, at the same time, $\widehat{g_t}$ is becoming delocalized in the sense that $\widehat{g_t}(\xi) = \hat{g}(t\xi) \longrightarrow 1$ uniformly on compacts. As I said earlier, this property is a manifestation of the uncertainty principle.[7]

Exercise 7.3.2 In this exercise you are to compute the Fourier transforms of some of the functions in Exercise 6.3.3.

[7] For the reader who wants to learn more about the "uncertainty principle", G. Folland's *Harmonic Analysis in Phase Space*, published by Princeton Univ. Press, is a good place to start.

(i) Define γ_t as in part **(i)** of Exercise 6.3.3, and use (7.3.6) to see that

$$\widehat{\gamma_t}(\xi) = e^{-t^2 2\pi^2 |\xi|^2}.$$

(ii) Let $\{\nu_t : t > 0\}$ be the functions in part **(ii)** of Exercise 6.3.3. To compute their Fourier transforms, begin by observing that $\widehat{\nu_t}(\xi) = \hat{\nu}(t^{\frac{1}{2}}\xi)$ and therefore that it suffices to treat the case when $t = 1$.

As was the case in the derivation of (7.3.6), analytic continuation provides a way to carry this out. Namely, consider the function

$$f(\zeta) = \pi^{-\frac{1}{2}} \int_{(0,\infty)} y^{-\frac{3}{2}} e^{-\frac{1}{y}} e^{\zeta y}\, dy$$

for $\zeta \in \bar{\Omega}$ where $\Omega \equiv \{z \in \mathbb{C} : \mathfrak{Re}\, z < 0\}$. Show that this function is analytic on Ω and continuous on $\bar{\Omega}$. Next, use part **(iv)** of Exercise 5.1.3 to see that $f(\zeta) = e^{-2\sqrt{-\zeta}}$ if $\zeta \in (-\infty, 0)$. To show how to extend $(-\zeta)^{\frac{1}{2}}$ as an analytic function on Ω, take log to be the principal branch of the logarithm function on $\mathbb{C} \setminus (-\infty, 0]$. That is, if $z \in \mathbb{C} \setminus (-\infty, 0]$, then $\log z = \log|z| + i\theta$ where θ is the element to $(-\pi, \pi)$ such that $z = |z|e^{i\theta}$. Show that if $\zeta = re^{i\theta}$ where $r > 0$ and $\theta \in [-\pi, \pi] \setminus \{0\}$, then

$$\log(-\zeta) = \begin{cases} \log r + i(\theta - \pi) & \text{if } \theta \in (0, \pi] \\ \log r + i(\theta + \pi) & \text{if } \theta \in [-\pi, 0). \end{cases}$$

Now define $(-\zeta)^{\frac{1}{2}} = e^{\frac{1}{2}\log(-\zeta)}$ for $\zeta \in \mathbb{C} \setminus (-\infty, 0]$, and check that, with this definition, $(-\zeta)^{\frac{1}{2}}$ is an analytic function of $\zeta \in \mathbb{C} \setminus [0, \infty)$ and that $(-\zeta)^{\frac{1}{2}} = \sqrt{-\zeta}$ for $\zeta \in (-\infty, 0)$. Thus, by analytic continuation, $f(\zeta) = e^{-(\zeta)^{\frac{1}{2}}}$ for $\zeta \in \Omega$ and therefore, by continuity, for $\zeta \in \bar{\Omega}$. Finally, check that

$$(i\xi)^{\frac{1}{2}} = \begin{cases} \sqrt{|\xi|}e^{-i\frac{\pi}{4}} & \text{if } \xi \geq 0 \\ \sqrt{|\xi|}e^{i\frac{\pi}{4}} & \text{if } \xi \leq 0, \end{cases}$$

Conclude that

$$\hat{\nu}(\xi) = \exp\bigl(-\sqrt{\pi|\xi|}(1 - \operatorname{sgn}(\xi)\, i)\bigr) \text{ for } \xi \in \mathbb{R}.$$

Finally, define $\check{\nu}(x) = \nu(-x)$, and show that

$$\widehat{\nu * \check{\nu}}(\xi) = e^{-2\sqrt{\pi|\xi|}}.$$

(iii) Let $\{P_t : t > 0\}$ be the functions in part **(iii)** of Exercise 6.3.3, and use their representation given there together with part **(i)** of this exercise to show that

$$\widehat{P_t}(\xi) = e^{-2\pi t|\xi|}.$$

(iv) Let $G(x) = \mathbf{1}_{[0,\infty)}(x)e^{-x}$ be the function G_1 in **(iv)** of Exercise 6.3.3, and set

$$f(\zeta) = \int_{(0,\infty)} e^{\zeta} e^{-x}\, dx.$$

Observe that f is analytic on $\{z \in \mathbb{C} : \mathfrak{Re}\, z < 1\}$ and that $f(\zeta) = (1-\zeta)^{-1}$ if $\zeta \in (-\infty, 1)$, and conclude that $\hat{G}(\xi) = (1 - i2\pi\xi)^{-1}$. In part **(iv)** of Exercise 8.2.7 below, you will show that

$$\widehat{G_t}(\xi) = \exp\bigl(-t\log(1 - i2\pi\xi)\bigr) = \bigl(1 + (2\pi\xi)^2\bigr)^{-\frac{t}{2}} e^{it\arctan(2\pi\xi)},$$

where log is again the principal branch of the logarithm on $\mathbb{C} \setminus (-\infty, 0]$, and arctan is the branch of $\tan^{-1}$ which vanishes at 0.

Exercise 7.3.3 There are a few formulas whose inherent beauty brings tears to the eyes of even the most jaded mathematicians, and among these is the formula to be derived in this exercise. Let $f \in L^1(\lambda_{\mathbb{R}}; \mathbb{C}) \cap C(\mathbb{R}; \mathbb{C})$, and assume that

$$\Biggl(\sum_{n\in\mathbb{Z}} \sup_{x\in[0,1]} |f(x+n)|\Biggr) \vee \Biggl(\sum_{n\in\mathbb{Z}} |\hat{f}(n)|\Biggr) < \infty.$$

The **Poisson summation formula** is the intriguing statement that

$$\sum_{n\in\mathbb{Z}} \hat{f}(n) = \sum_{n\in\mathbb{Z}} f(n). \tag{7.3.23}$$

To prove (7.3.23), define $\tilde{f}(x)$ for $x \in [0,1]$ by $\tilde{f}(x) = \sum_{n\in\mathbb{Z}} f(x+n)$, and show that $\tilde{f} \in C_1^0([0,1]; \mathbb{C})$ and that $(\tilde{f}, \mathfrak{e}_k)_{L^2([0,1];\mathbb{C})} = \hat{f}(-k)$. Now apply Exercise 7.2.3 with $\varphi = \tilde{f}$ to arrive at (7.3.23).

Exercise 7.3.4 Using (7.3.23) and (7.3.6), show that

$$\sum_{n\in\mathbb{Z}} e^{-\frac{\pi n^2}{t}} = t^{\frac{1}{2}} \sum_{n\in\mathbb{Z}} e^{-t\pi n^2}, \quad t > 0.$$

Similarly, using the calculation in part **(iii)** of Exercise 7.3.2, show that

$$\sum_{n\in\mathbb{Z}} \frac{1}{t^2 + n^2} = \frac{\pi}{t}\frac{1 + e^{-2\pi t}}{1 - e^{-2\pi t}} = \frac{\pi}{t}\coth(\pi t), \quad t > 0.$$

Starting from the second of these, give another proof that $\sum_{n\geq 1} \frac{1}{n^2} = \frac{\pi^2}{6}$. In addition, show that it leads to

$$\sinh(\pi t) = \pi t \prod_{n=1}^{\infty} \bigl(1 + \tfrac{t^2}{n^2}\bigr),$$

which, after analytic continuation, yields **Euler's product formula**

$$\sin(\pi t) = \pi t \prod_{n=1}^{\infty} \left(1 - \tfrac{t^2}{n^2}\right).$$

Exercise 7.3.5 Let γ be the Borel measure on $\mathbb{R}$ determined by $\gamma(\Gamma) = \int_\Gamma e^{-2\pi x^2}\,dx$.

(i) Set $\tilde{H}_n = 2^{\frac{1}{4}}\left(\sqrt{(4\pi)^n n!}\right)^{-1} H_n$, and show that $\{\tilde{H}_n : n \in \mathbb{N}\}$ is an orthonormal basis for $L^2(\gamma;\mathbb{C})$.

(ii) Show that

$$e^{4\pi\zeta x - 2\pi\zeta^2} = \sum_{n=0}^{\infty} \frac{\zeta^n}{n!} H_n(x) \quad \text{for } (\zeta, x) \in \mathbb{C}\times\mathbb{R},$$

where the convergence is uniform on compact subsets of $\mathbb{C}\times\mathbb{R}$. In addition, show that, for each $\zeta \in \mathbb{C}$, the convergence is in $L^2(\gamma;\mathbb{C})$ and that the rate of this convergence is uniform in ζ from compact subsets of $\mathbb{C}$.

(iii) Use part **(ii)** to give another derivation of the fact that $\hat{h}_n = i^n h_n$.

(iv) Show that $h_n(0) = i^n \int h(x)\,dx$, and conclude that

$$h_n(0) = \begin{cases} (-1)^{\frac{n}{2}} \int h_n(x)\,dx & \text{if } n \text{ is even} \\ 0 & \text{if } n \text{ is odd.} \end{cases}$$

Exercise 7.3.6 Using (7.3.12) and (7.3.13), show that

$$2\pi x\tilde{h}_n = \pi^{\frac{1}{2}}\left((n+1)^{\frac{1}{2}}\tilde{h}_{n+1} + n^{\frac{1}{2}}\tilde{h}_{n-1}\right) \text{ and } -\partial\tilde{h}_n = \pi^{\frac{1}{2}}\left((n+1)^{\frac{1}{2}}\tilde{h}_{n+1} - n^{\frac{1}{2}}\tilde{h}_{n-1}\right).$$

Starting from these, show that, for each $m \in \mathbb{Z}^+$, there exist constants $\{a^{(m)}_{n,k} : n \geq 0 \;\&\; -m\wedge n \leq k \leq m\}$ and $\{b^{(m)}_{n,k} : n \geq 0 \;\&\; -m\wedge n \leq k \leq m\}$ such that $|a^{(m)}_{n,k}| \vee |b^{(m)}_{n,k}| \leq C_m(n+1)^{\frac{m}{2}}$ for some $C_m < \infty$ and

$$(2\pi x)^m \tilde{h}_n = \sum_{k=-m\wedge n}^{m} a^{(m)}_{n,k}\tilde{h}_{n+k} \text{ and } \partial^m\tilde{h}_n = \sum_{k=-m\wedge n}^{m} b^{(m)}_{n,k}\tilde{h}_{n+k}.$$

If one is sufficiently careful, one sees that $|a^{(m)}_{n,k}| = |b^{m)}_{n,k}|$ and thereby concludes that

$$(2\pi)^m \int_{\mathbb{R}} x^{2m} h_n(x)^2\,dx = \int_{\mathbb{R}} \partial^m h_n(x)^2\,dx.$$

Give another derivation of this conclusion using the facts that $\partial^m_\xi \widehat{h_n}(\xi)$ is the Fourier transform of $(i2\pi x)^m h_n(x)$ and that $\widehat{h_n} = i^n h_n$.

Exercise 7.3.7 Referring to Corollary 7.3.11, define

$$\rho_{\mathscr{S}}(\varphi, \psi) = \sum_{m=0}^{\infty} \frac{1}{2^m} \frac{\|\varphi - \psi\|_{\mathscr{S}}^{(m)}}{1 + \|\varphi - \psi\|_{\mathscr{S}}^{(m)}}$$

for φ, $\psi \in \mathscr{S}(\mathbb{R}; \mathbb{C})$ and

$$\rho_{\mathfrak{s}}(a, b) = \sum_{m=0}^{\infty} \frac{1}{2^m} \frac{\|a - b\|_{\mathfrak{s}}^{(m)}}{1 + \|a - b\|_{\mathfrak{s}}^{(m)}}$$

for a, $b \in \mathfrak{s}(\mathbb{R}; \mathbb{C})$. Show that $\rho_{\mathscr{S}}$ and ρ_s are metrics on $\mathscr{S}(\mathbb{R}; \mathbb{C})$ and $\mathfrak{s}(\mathbb{R}; \mathbb{C}$, respectively, and that the topology determined by $\rho_{\mathscr{S}}$ is the same as the topology on $\mathscr{S}(\mathbb{R}; \mathbb{C})$. In addition, show that the map Ψ is isometric. Finally, show that $\big(\mathscr{S}(\mathbb{R}; \mathbb{C}), \rho_{\mathscr{S}}\big)$ an $\big(\mathfrak{s}(\mathbb{R}; \mathbb{C}), \rho_{\mathfrak{s}}\big)$ are complete, separable metric spaces.

Exercise 7.3.8 Show that $\|\hat{\varphi}\|_{\mathscr{S}}^{(m)} = \|\varphi\|_{\mathscr{S}}^{(m)}$ for all $m \geq 0$, and conclude that the Fourier transform is an isometric map of $\big(\mathscr{S}(\mathbb{R}; \mathbb{C}), \rho_{\mathscr{S}}\big)$ onto itself and therefore that $\hat{\Lambda}$ given by $\hat{\Lambda}(\varphi) = \Lambda(\hat{\varphi})$ is a tempered distribution for all $\Lambda \in \mathscr{S}(\mathbb{R}; \mathbb{C})^*$. Next note that

$$\widehat{\Lambda_f}(\varphi) = \int f(\xi)\hat{\varphi}(\xi)\, d\xi$$

when f is a Borel measurable function f for which there is an $m \geq 0$ such that $(1 + x^2)^{-m} f \in L^1(\lambda_{\mathbb{R}}; \mathbb{C})$, and conclude from this that if f is the polynomial $\sum_{k=0}^{n} c_k x^k$, then

$$\widehat{\Lambda_f}(\varphi) = \sum_{k=0}^{n} \frac{i^k c_k}{(2\pi)^k} \partial^k \varphi(0).$$

Exercise 7.3.9 Show that $\partial \Lambda_{\mathbf{1}_{[0,\infty)}}(\varphi) = \varphi(0)$. Hence, $\partial \Lambda_{\mathbf{1}_{[0,\infty)}}$ can be identified with the unit, point measure δ_0. Next, show that

$$f = \sum_{n=0}^{\infty} \mu_n^{-\frac{3}{2}} \tilde{h}_n(0)\tilde{h}_n \in L^2(\lambda_{\mathbb{R}}; \mathbb{C})$$

and that $\varphi(0) = \big(\mathcal{H}^{\frac{3}{2}}\varphi, f\big)_{L^2(\lambda_{\mathbb{R}};\mathbb{C})}$. Conclude that $\partial \Lambda_{\mathbf{1}_{[0,\infty)}} = \mathcal{H}^{\frac{3}{2}} \Lambda_f$.

Exercise 7.3.10 Here are a couple more examples of tempered distributions.

(i) Given $\varphi \in \mathscr{S}(\mathbb{R}; \mathbb{C})$, show that

$$\lim_{r \searrow 0} \int_{|x| \geq r} \varphi(x) \frac{1}{x}\, dx = -\int \partial\varphi(x) \log|x|\, dx,$$

and conclude that

$$\partial \Lambda_{\log|x|}(\varphi) = \lim_{r \searrow 0} \int_{|x| \geq r} \varphi(x) \frac{1}{x}\, dx.$$

Limits of the sort on the right hand side of the above are known as *principal value* integrals.

(ii) Use $\Lambda_{\frac{1}{x}}$ to denote the representation obtained in **(i)** of $\partial\Lambda_{\log|x|}$, and show that

$$\Lambda_{\frac{1}{x}}(\varphi) = \lim_{R\to\infty}\lim_{r\searrow 0}\int_{r\le|x|\le R}\frac{1}{x}\varphi(x)\,dx.$$

Next, show that

$$\lim_{r\searrow 0}\int_{r\le|x|\le R}\frac{1}{x}\hat{\varphi}(x)\,dx = 2i\int\varphi(\xi)\left(\int_{(0,R]}\frac{\sin(2\pi\xi x)}{x}\,dx\right)d\xi$$

$$= 2i\int\varphi(\xi)\mathrm{sgn}(\xi)g_R(\xi)\,d\xi \text{ where } g_R(\xi) = \int_{(0,2\pi|\xi|R]}\frac{\sin x}{x}\,dx.$$

Finally, show that, as $R\to\infty$, $\mathbf{1}_{\mathbb{R}\setminus\{0\}}g_R$ converges boundedly and pointwise to a constant $c\in(0,\infty)$ and therefore that $\widehat{\Lambda_{\frac{1}{x}}} = \Lambda_{i2c\mathrm{sgn}}$. In part **(iii)** of Exercise 8.2.7, it is shown that $c = \frac{\pi}{2}$.

7.4 The Fourier Transform of Probability Measures

Given a Borel probability measure μ on $\mathbb{R}^N$, define its Fourier transform $\hat{\mu}:\mathbb{R}^N\longrightarrow\mathbb{C}$ by

$$\hat{\mu}(\xi) = \int_{\mathbb{R}^N}\mathfrak{e}_\xi(x)\,\mu(dx).$$

Clearly, $\hat{\mu}$ is a continuous function, and $\|\hat{\mu}\|_{\mathrm{u}} = 1 = \hat{\mu}(0)$.

7.4.1 Parseval for Measures

The following is the analog for probability measures of Parseval's identity for functions.

Theorem 7.4.1 *If* $\varphi\in C_{\mathrm{b}}(\mathbb{R}^N;\mathbb{C})\cap L^1(\lambda_{\mathbb{R}^N};\mathbb{C})$ *and* $\hat{\varphi}\in L^1(\lambda_{\mathbb{R}^N};\mathbb{C})$, *then*

$$\int_{\mathbb{R}^N}\varphi(x)\,\mu(dx) = \int_{\mathbb{R}^N}\hat{\varphi}(\xi)\overline{\hat{\mu}(\xi)}\,d\xi.$$

Proof. The idea is to get this conclusion as an application of (7.3.19). To this end, choose an even function $\rho\in C_{\mathrm{c}}^\infty\big(\mathbb{R}^N;[0,\infty)\big)$ with integral 1, and set $\rho_\epsilon(x) = \epsilon^{-N}\rho(\epsilon^{-1}x)$ for $\epsilon>0$. Then it is easy to check that $\widehat{\rho_\epsilon}(\xi) = \hat{\rho}(\epsilon\xi)$. In addition, as an application of the last part of Theorem 7.3.4, one sees that $\hat{\rho}\in L^1(\lambda_{\mathbb{R}^N};\mathbb{C})$. and therefore that $\widehat{\rho_\epsilon}\in L^1(\lambda_{\mathbb{R}^N};\mathbb{R})$ for all $\epsilon>0$. Now set

$$\psi_\epsilon(x) = \int_{\mathbb{R}^N} \rho_\epsilon(x-y)\,\mu(dy).$$

By Theorem 4.1.5, one knows that $\|\psi_\epsilon\|_{L^1(\lambda_{\mathbb{R}^N};\mathbb{R})} = 1$ and $\widehat{\psi_\epsilon}(\xi) = \widehat{\rho_\epsilon}(\xi)\hat{\mu}(\xi)$.

Let φ be given, and set $\varphi_\epsilon = \rho_\epsilon * \varphi$. Then $\varphi_\epsilon \in C_{\mathrm{b}}^\infty(\mathbb{R}^N;\mathbb{C})$, and, again by Theorem 4.1.5, it and all its derivatives are in $L^1(\lambda_{\mathbb{R}^N};\mathbb{C})$, which, by Theorem 7.3.4, means that $\widehat{\varphi_\epsilon} \in L^1(\lambda_{\mathbb{R}^N};\mathbb{C}) \cap C_0(\mathbb{R}^N;\mathbb{C})$. Also, since φ is bounded and continuous, Theorem 6.3.7 says that, as $\epsilon \searrow 0$, $\varphi_\epsilon \longrightarrow \varphi$ boundedly and uniformly on compacts. Further, because ρ is even, yet another application of Theorem 4.1.5 shows that

$$\int_{\mathbb{R}^N} \varphi(x)\psi_\epsilon(x)\,dx = \int_{\mathbb{R}^N} \varphi_\epsilon(x)\,\mu(dx).$$

Thus, by (7.3.19),

$$\begin{aligned}\int_{\mathbb{R}^N} \varphi(x)\,\mu(dx) &= \lim_{\epsilon\searrow 0}\int_{\mathbb{R}^N} \varphi_\epsilon(x)\,\mu(dx) = \lim_{\epsilon\searrow 0}\int_{\mathbb{R}^N} \varphi(x)\psi_\epsilon(x)\,dx \\ &= \lim_{\epsilon\searrow 0}\int_{\mathbb{R}^N} \hat{\varphi}(\xi)\overline{\widehat{\psi_\epsilon}(\xi)}\,d\xi = \lim_{\epsilon\searrow 0}\int_{\mathbb{R}^N} \hat{\varphi}(\xi)\widehat{\rho_\epsilon}(\xi)\overline{\hat{\mu}(\xi)}\,d\xi = \int_{\mathbb{R}^N} \hat{\varphi}(\xi)\overline{\hat{\mu}(\xi)}\,d\xi\end{aligned}$$

since $\widehat{\rho_\epsilon}(\xi) = \hat{\rho}(\epsilon\xi) \longrightarrow 1$ boundedly and pointwise. □

A trivial consequence of Theorem 7.4.1 is that two probability measures on $\mathbb{R}^N$ are equal if and only if their Fourier transforms are equal. In particular, it shows that $\nu = \mu$ if and only if, for each $\mathbf{e} \in \mathbb{S}^{N-1}$ and all $\Gamma \in \mathcal{B}_{\mathbb{R}}$, $\nu(\{x \in \mathbb{R}^N : (x,\mathbf{e})_{\mathbb{R}^N} \in \Gamma\}) = \mu(\{x \in \mathbb{R}^N : (x,\mathbf{e})_{\mathbb{R}^N} \in \Gamma\})$. However, as is shown in Theorem 7.4.3 below, much more is true.

7.4.2 Weak Convergence of Probability Measures

Given a sequence $\{\mu_n : n \ge 1\}$ of probability measures on $\mathbb{R}^N$, one says that $\{\mu_n : n \ge 1\}$ **converges weakly** to the probability measure μ and writes $\mu_n \xrightarrow{\mathrm{w}} \mu$ if

$$\lim_{n\to\infty}\int_{\mathbb{R}^N} \varphi(x)\,\mu_n(dx) = \int_{\mathbb{R}^N} \varphi(x)\,\mu(dx) \tag{7.4.1}$$

for all $\varphi \in C_{\mathrm{b}}(\mathbb{R}^N;\mathbb{C})$. The goal here is to show that $\{\mu_n : n \ge 1\}$ converges weakly to μ if and only if $\{\widehat{\mu_n} : n \ge 1\}$ converges pointwise to $\hat{\mu}$, in which case this convergence is uniform on compacts.

Lemma 7.4.2 *If* (7.4.1) *holds for all* $\varphi \in C_{\mathrm{c}}^\infty(\mathbb{R}^N;\mathbb{R})$, *then* $\mu_n \xrightarrow{\mathrm{w}} \mu$. *Furthermore, if* $\mu_n \xrightarrow{\mathrm{w}} \mu$ *and* $\{\varphi_n : n \ge 1\}$ *is a uniformly bounded sequence of continuous functions which converge uniformly on compacts to* φ, *then*

$$\lim_{n\to\infty}\int_{\mathbb{R}^N}\varphi_n(x)\,\mu_n(dx)=\int_{\mathbb{R}^N}\varphi(x)\,\mu(dx).$$

Proof. Assume that (7.4.1) holds for all $\varphi\in C_{\mathrm{c}}^{\infty}(\mathbb{R}^N;\mathbb{R})$ and therefore for all $\varphi\in C_{\mathrm{c}}^{\infty}(\mathbb{R}^N;\mathbb{C})$. Choose $\eta\in C_{\mathrm{c}}^{\infty}\big(\mathbb{R}^N;[0,1]\big)$ so that $\eta(x)=1$ for $x\in\overline{B_{\mathbb{R}^N}(0,1)}$, and set $\eta_R(x)=\eta(R^{-1}x)$ for $R>0$. Given $\varphi\in C_{\mathrm{b}}^{\infty}(\mathbb{R}^N;\mathbb{C})$, define $\varphi_R(x)=\eta_R(x)\varphi(x)$. Then $\varphi_R\in C_{\mathrm{c}}^{\infty}(\mathbb{R}^N;\mathbb{C})$ and $\|\varphi_R\|_{\mathrm{u}}\le\|\varphi\|_{\mathrm{u}}$. Thus, for each $R>0$,

$$\begin{aligned}&\overline{\lim_{n\to\infty}}\left|\int_{\mathbb{R}^N}\varphi\,d\mu_n-\int_{\mathbb{R}^N}\varphi\,d\mu\right|\\&\qquad\le\overline{\lim_{n\to\infty}}\left|\int_{\mathbb{R}^N}(\varphi-\varphi_R)\,d\mu_n\right|+\overline{\lim_{n\to\infty}}\left|\int_{\mathbb{R}^N}(\varphi-\varphi_R)\,d\mu\right|.\end{aligned}$$

Clearly

$$\left|\int_{\mathbb{R}^N}(\varphi-\varphi_R)\,d\mu_n\right|\le\|\varphi\|_{\mathrm{u}}\int_{\mathbb{R}^N}(1-\eta_R)\,d\mu_n=\|\varphi\|_{\mathrm{u}}\left(1-\int_{\mathbb{R}^N}\eta_R\,d\mu_n\right),$$

and, since (7.4.1) holds when $\varphi=\eta_R$, this means that

$$\overline{\lim_{n\to\infty}}\left|\int_{\mathbb{R}^N}(\varphi-\varphi_R)\,d\mu_n\right|\le\|\varphi\|_{\mathrm{u}}\left(1-\int_{\mathbb{R}^N}\eta_R\,d\mu\right).$$

Therefore, for all $R>0$,

$$\overline{\lim_{n\to\infty}}\left|\int_{\mathbb{R}^N}\varphi\,d\mu_n-\int_{\mathbb{R}^N}\varphi\,d\mu\right|\le2\|\varphi\|_{\mathrm{u}}\int_{\mathbb{R}^N}(1-\eta_R)\,d\mu,$$

and so, since $\eta_R\longrightarrow1$ as $R\to\infty$, we have proved that (7.4.1) holds when $\varphi\in C_{\mathrm{b}}^{\infty}(\mathbb{R}^N;\mathbb{C})$.

Now choose ρ as in the proof of Theorem 7.4.1, and define ρ_ϵ accordingly. Given $\varphi\in C_{\mathrm{b}}(\mathbb{R}^N;\mathbb{C})$, set $\varphi_\epsilon=\rho_\epsilon*\varphi$. Then $\varphi_\epsilon\in C_{\mathrm{b}}^{\infty}(\mathbb{R}^N;\mathbb{C})$, $\|\varphi_\epsilon\|_{\mathrm{u}}\le\|\varphi\|_{\mathrm{u}}$, and $\varphi_\epsilon\longrightarrow\varphi$ uniformly on compacts. Thus, for each $\epsilon>0$,

$$\overline{\lim_{n\to\infty}}\left|\int_{\mathbb{R}^N}\varphi\,d\mu_n-\int_{\mathbb{R}^N}\varphi\,d\mu\right|\le\overline{\lim_{n\to\infty}}\left|\int_{\mathbb{R}^N}(\varphi-\varphi_\epsilon)\,d\mu_n\right|+\left|\int_{\mathbb{R}^N}(\varphi-\varphi_\epsilon)\,d\mu\right|,$$

and, therefore, by the same reasoning as was used in the preceding, for each $R>0$,

$$\begin{aligned}&\overline{\lim_{n\to\infty}}\left|\int_{\mathbb{R}^N}(\varphi-\varphi_\epsilon)\,d\mu_n\right|+\left|\int_{\mathbb{R}^N}(\varphi-\varphi_\epsilon)\,d\mu\right|\\&\qquad\le2\sup_{x\in B_{\mathbb{R}^N}(0,R)}|\varphi(x)-\varphi_\epsilon(x)|+2\|\varphi\|_{\mathrm{u}}\int_{\mathbb{R}^N}(1-\eta_R)\,d\mu.\end{aligned}$$

Finally, let $\epsilon\searrow0$, and conclude that

$$\varlimsup_{n\to\infty}\left|\int_{\mathbb{R}^N}\varphi\,d\mu_n-\int_{\mathbb{R}^N}\varphi\,d\mu\right|\le 2\|\varphi\|_{\mathrm{u}}\int_{\mathbb{R}^N}(1-\eta_R)\,d\mu,$$

which tends to 0 as $R\to\infty$. Thus, the first assertion has been proved.

The proof of the second assertion is essentially the same. Namely, one begins with

$$\left|\int_{\mathbb{R}^N}\varphi_n\,d\mu_n-\int_{\mathbb{R}^N}\varphi\,d\mu\right|\le\left|\int_{\mathbb{R}^N}(\varphi_n-\varphi)\,d\mu_n\right|+\left|\int_{\mathbb{R}^N}\varphi\,d\mu_n-\int_{\mathbb{R}^N}\varphi\,d\mu\right|.$$

As $n\to\infty$, the second term on the right tends to 0, and, just as above, for each $R>0$,

$$\varlimsup_{n\to\infty}\left|\int_{\mathbb{R}^N}(\varphi_n-\varphi)\,d\mu_n\right|\le 2\sup_{n\ge1}\|\varphi_n\|_{\mathrm{u}}\int_{\mathbb{R}^N}(1-\eta_R)\,d\mu.$$

□

Theorem 7.4.3 *Let $\{\mu_n : n\ge1\}$ be a sequence of Borel probability measures on $\mathbb{R}^N$. If μ is a Borel probability measure which is the weak limit of $\{\mu_n : n\ge1\}$, then $\widehat{\mu_n}$ converges uniformly on compacts to $\hat\mu$. Conversely, if $\{\widehat{\mu_n} : n\ge1\}$ converges pointwise to $\hat\mu$, then $\mu_n\xrightarrow{\mathrm{w}}\mu$, and therefore $\widehat{\mu_n}\longrightarrow\hat\mu$ uniformly on compacts.*

Proof. Because $\mathfrak{e}_{\xi_n}\longrightarrow\mathfrak{e}_\xi$ uniformly on compacts if $\xi_n\longrightarrow\xi$, the first assertion follows immediately from Lemma 7.4.2.

Now assume that $\widehat{\mu_n}\longrightarrow\hat\mu$ pointwise, and let $\varphi\in C_{\mathrm{c}}^\infty(\mathbb{R}^N;\mathbb{C})$ be given. Then, $\hat\varphi\in L^1(\lambda_{\mathbb{R}^N};\mathbb{C})$, and, by Theorem 7.4.1,

$$\int_{\mathbb{R}^N}\varphi\,d\mu_n=\int_{\mathbb{R}^N}\varphi(\xi)\overline{\widehat{\mu_n}(\xi)}\,d\xi\longrightarrow\int_{\mathbb{R}^N}\varphi(\xi)\overline{\hat\mu(\xi)}\,d\xi=\int_{\mathbb{R}^N}\varphi\,d\mu.$$

Hence, by Lemma 7.4.2, $\mu_n\xrightarrow{\mathrm{w}}\mu$. □

7.4.3 Convolutions and The Central Limit Theorem

A renowned corollary of Theorem 7.4.3 is a result that probabilists call **The Central Limit Theorem** and abbreviate as the CLT. To state the CLT, one needs to introduce the notion of convolution for probability measures. Namely, given Borel probability measure μ and ν on $\mathbb{R}^N$, define their convolution $\mu*\nu$ on $\mathcal{B}_{\mathbb{R}^N}$ by

$$\mu*\nu(\Gamma)=\int_{\mathbb{R}^N\times\mathbb{R}^N}\mathbf{1}_\Gamma(x+y)\,\mu\times\nu(dx\times dy).$$

It is easy to check that $\mu * \nu$ is again a Borel probability measure and that, if $d\mu = \varphi d\lambda_{\mathbb{R}^N}$ and $d\nu = \psi d\lambda_{\mathbb{R}^N}$, then $d(\mu * \nu) = \varphi * \psi d\lambda_{\mathbb{R}^N}$. In addition, by Theorem 4.1.5, one sees that $\widehat{\mu * \nu} = \hat{\mu}\hat{\nu}$.

Given $n \in \mathbb{Z}^+$ and μ, set

$$\mu^{*n} = \underbrace{\mu * \cdots * \mu}_{n \text{ times}},$$

the n-fold convolution of μ with itself, let $F_n : \mathbb{R}^N \longrightarrow \mathbb{R}^N$ be the map $F_n(x) = \frac{x}{n^{\frac{1}{2}}}$, and take μ_n to be the pushforward $(F_n)_*(\mu^{*n})$ of μ^{*n}. Equivalently, μ_n is the Borel probability with the property that, for all bounded, Borel measurable functions φ,

$$\int_{\mathbb{R}^N} \varphi \, d\mu_n = \int_{(\mathbb{R}^N)^n} \varphi\left(\frac{x_1 + \cdots x_n}{n^{\frac{1}{2}}}\right) \mu^{\times n}(dx_1 \times \cdots \times dx_n),$$

where

$$\mu^{\times n} = \underbrace{\mu \times \cdots \times \mu}_{n \text{ times}}.$$

Finally, let γ be the Borel probability measure determined by

$$\gamma(\Gamma) = (2\pi)^{-\frac{N}{2}} \int_\Gamma e^{-\frac{|x|^2}{2}} \, dx$$

for $\Gamma \in \mathcal{B}_{\mathbb{R}^N}$. By (7.3.6),

$$\hat{\gamma}(\xi) = e^{-2\pi^2|\xi|^2}. \tag{7.4.2}$$

Theorem 7.4.4 (Central Limit Theorem) *Let μ be a Borel probability measure on $\mathbb{R}^N$ with the properties that $x \rightsquigarrow |x|^2$ is μ-integrable and, for all $1 \le i, j \le N$,*

$$\int_{\mathbb{R}^N} x_i \, d\mu = 0 \text{ and } \int_{\mathbb{R}^N} x_i x_j \, d\mu = \delta_{i,j}.$$

Then $\mu_n \xrightarrow{\text{w}} \gamma$.

Proof. By Theorem 7.4.3 and (7.4.2), it suffices to show that $\widehat{\mu_n}(\xi) \longrightarrow e^{-2\pi^2|\xi|^2}$ for each $\xi \in \mathbb{R}^N$. Equivalently, what needs to be shown is that for each $\mathbf{e} \in \mathbb{S}^{N-1}$ and $t \in \mathbb{R}$,

$$\widehat{\mu_n}(t\mathbf{e}) \longrightarrow e^{-2\pi^2 t^2}.$$

To this end, note that

$$\int_{\mathbb{R}^N} (x, \mathbf{e})_{\mathbb{R}^N} \, \mu(dx) = 0 \text{ and } \int_{\mathbb{R}^N} (x, \mathbf{e})_{\mathbb{R}^N}^2 \, \mu(dx) = 1.$$

By Taylor's theorem,

$$\exp\left(\frac{i2\pi t\xi}{n^{\frac{1}{2}}}\right) = 1 + \frac{i2\pi t\xi}{n^{\frac{1}{2}}} - \frac{2\pi^2 t^2}{n} + \frac{E_n(t.\xi)}{n},$$

where

$$E_n(t,\xi) = 4\pi^2 t^2 \int_{[0,1]} (1-s)\left(1 - \exp\left(\frac{i2\pi st\xi}{n^{\frac{1}{2}}}\right)\right) ds.$$

Hence,

$$\hat{\mu}\left(\frac{t\mathbf{e}}{n^{\frac{1}{2}}}\right) = 1 - \frac{2\pi^2 t^2}{n} + \frac{1}{n}\int E_n\left(t, (x,\mathbf{e})_{\mathbb{R}^N}\right)\mu(dx),$$

and so, since $|E_n(t,\xi)| \leq C\xi^2$ and $E_n(t,\xi) \longrightarrow 0$ pointwise as $n \to \infty$, the Lebesgue dominated convergence theorem proves that

$$\hat{\mu}\left(\frac{t\mathbf{e}}{n^{\frac{1}{2}}}\right) = 1 - \frac{2\pi^2 t^2}{n} + o(n^{-1}).$$

Since

$$\widehat{\mu_n}(\xi) = \hat{\mu}\left(\frac{\xi}{n^{\frac{1}{2}}}\right)^n,$$

it follows that

$$\widehat{\mu_n}(t\mathbf{e}) = \left(1 - \frac{2\pi^2 t^2}{n} + o(n^{-1})\right)^n \longrightarrow e^{-2\pi^2 t^2}.$$

□

7.4.4 Exercises for § 7.4

Exercise 7.4.1 Assume that $\mu_n \xrightarrow{\text{w}} \mu$. In this exercise, you are to show that if $\Gamma \in \mathcal{B}_{\mathbb{R}^N}$ and $\mu(\partial\Gamma) = 0$, then $\mu_n(\Gamma) \longrightarrow \mu(\Gamma)$.

(i) Show that there exists a sequence $\{\varphi_k : k \geq 1\}$ and $\{\psi_k : k \geq 1\}$ in $C(\mathbb{R}^N; [0,1])$ such that $\varphi_k \nearrow \mathbf{1}_{\mathring{\Gamma}}$ and $\psi_k \searrow \mathbf{1}_{\bar{\Gamma}}$ as $k \to \infty$.

(ii) Referring to the preceding, show that, for each $k \geq 1$,

$$\varliminf_{n\to\infty} \mu_n(\Gamma) \geq \lim_{n\to\infty} \int \varphi_k \, d\mu_n = \int \varphi_k \, d\mu,$$

and therefore $\varliminf_{n\to\infty} \mu_n(\Gamma) \geq \mu(\mathring{\Gamma})$. Similarly, using $\{\psi_k : k \geq 1\}$, show that $\varlimsup_{n\to\infty} \mu_n(\Gamma) \leq \mu(\bar{\Gamma})$, and thereby conclude that $\lim_{n\to\infty} \mu_n(\Gamma) = \mu(\Gamma)$ if $\mu(\partial\Gamma) = 0$.

(iii) Take $N = 1$ in Theorem 7.4.4, and suppose that $\{\mu_n : n \geq 1\}$ is the sequence there. Show that

$$\lim_{n\to\infty} \mu_n\big((a,b]\big) = \frac{1}{\sqrt{2\pi}} \int_{(a,b]} e^{-\frac{x^2}{2}} \, dx$$

for $-\infty \le a < b \le \infty$. This is the standard statement of the Central Limit Theorem found in elementary probability texts.

Exercise 7.4.2 The goal here is to use Theorem 7.4.4 to show that the constant C in part (**iv**) of Exercise 1.2.2 is 2π.

(**i**) Let ν be the Borel measure on $\mathbb{R}$ such that

$$\nu(\Gamma) = \int_{[0,\infty)\cap\Gamma} e^{-t}\,dt \text{ for } \Gamma \in \mathcal{B}_{\mathbb{R}},$$

and set $\mu = \delta_{-1} * \nu$. That is,

$$\int \varphi\,d\mu = \int \varphi(t-1)\,d\nu = \int_{[0,\infty)} \varphi(t-1)e^{-t}\,dt$$

for bounded $\mathcal{B}_{\mathbb{R}}$-measurable functions φ. Show that μ satisfies the conditions in Theorem 7.4.4 and that $\mu^{*n} = \delta_{-n} * \nu^{*n}$.

(**ii**) Using part (**iv**) of Exercise 6.3.3, show that $\nu^{*n}(dt) = \frac{1}{(n-1)!}t^{n-1}e^{-t}\,dt$ and therefore that

$$\int \varphi\,d\mu_{n+1} = \frac{1}{n!}\int_{[0,\infty)} \varphi\left(\frac{t}{(n+1)^{\frac{1}{2}}} - (n+1)^{\frac{1}{2}}\right) t^n e^{-t}\,dt,$$

where $\{\mu_n : n \ge 1\}$ is defined from μ by the prescription in Theorem 7.4.4.

(**iii**) Making the change of variables $t = n^{\frac{1}{2}}s + n$ in the final expression in (**ii**), show that

$$\int \varphi\,d\mu_{n+1} = \frac{n^{n+\frac{1}{2}}e^{-n}}{n!} \int\limits_{[-n^{\frac{1}{2}},\infty)} \varphi\left(\sqrt{\frac{n}{n+1}}s - (n+1)^{-\frac{1}{2}}\right)\left(1+n^{-\frac{1}{2}}s\right)^n e^{-n^{\frac{1}{2}}s}\,ds.$$

Now choose $\varphi \in C(\mathbb{R};[0,1])$ so that $\varphi = 1$ on $\left[\frac{1}{4},\frac{3}{4}\right]$ and $\varphi = 0$ off $(0,1)$, and show that

$$\int\limits_{[-n^{\frac{1}{2}},\infty)} \varphi\left(\sqrt{\frac{n}{n+1}}s - (n+1)^{-\frac{1}{2}}\right)\left(1+n^{-\frac{1}{2}}s\right)^n e^{-n^{\frac{1}{2}}s}\,ds \longrightarrow \int \varphi(x)e^{-\frac{x^2}{2}}\,dx.$$

Hence, since, by Theorem 7.4.4,

$$\lim_{n\to\infty}\int \varphi\,d\mu_n = \frac{1}{\sqrt{2\pi}}\int \varphi(x)e^{-\frac{x^2}{2}}\,dx,$$

conclude that

$$\lim_{n\to\infty} \frac{n^{n+\frac{1}{2}}e^{-n}}{n!} = \frac{1}{\sqrt{2\pi}}$$

and therefore that $C = 2\pi$.

Chapter 8
Radon–Nikodym, Hahn, Daniell, and Carathéodory

In this concluding chapter I will deal with several matters that are of a rather abstract nature. The first of these is the famous theorem of Radon and Nikodym, which can be viewed as a generalization of the results in § 3.3 and plays an important role in the proof of the Hahn decomposition theorem, which is an abstraction of the results in § 1.2.2. The second is the abstraction of Lebesgue's integration theory that results from thinking about integrals as linear functions. For reasons that I do not understand, this theory is called Daniell integration even though all the key ideas seem to have been F. Riesz's. Be that as it may, Daniell's theory is an interesting interpretation of integration theory and leads to a powerful procedure for constructing measures. However, the finiteness hypotheses under which this construction procedure works are too restrictive to handle the construction of measures, like those of Hausdorff, that are wildly non-finite. For this reason, I will close in § 8.3.4 with a construction procedure, invented by C. Carathéodory, which can be used to construct such measures, and I then will apply his procedure to the construction of Hausdorff measures. Finally, I will prove the relationship between Hausdorff measure and surface measure on hypersurfaces.

8.1 The Radon–Nikodym and Hahn Decomposition Theorems

Given a pair of measures μ and ν on the measurable space $(E, \mathcal{B})$, recall (cf. Exercises 2.1.8 and (2.1.9)) that μ is **absolutely continuous** with respect to ν, abbreviated by $\mu \ll \nu$, if $\mu(\Gamma) = 0$ whenever Γ is an element of $\mathcal{B}$ for which $\nu(\Gamma) = 0$ and that μ is **singular** to ν, abbreviated by $\mu \perp \nu$, if there is a $\Gamma \in \mathcal{B}$ for which $\mu(\Gamma^{\complement}) = 0 = \nu(\Gamma)$. A purely measure-theoretic statement of Theorem 3.3.8 is that each finite Borel measure μ on $\mathbb{R}$ has a unique decomposition into a part μ_{a} that is absolutely continuous with respect to $\lambda_{\mathbb{R}}$ and a part μ_{s} that is singular to $\lambda_{\mathbb{R}}$. In addition, Theorem 3.3.6

D. W. Stroock, *Essentials of Integration Theory for Analysis*, Graduate Texts in Mathematics 262, https://doi.org/10.1007/978-3-030-58478-8_8

says that there is a unique $f \in L^1(\lambda_{\mathbb{R}};\mathbb{R})$ for which $\mu_{\mathrm{a}}(\Gamma) = \int_\Gamma f\,d\lambda_{\mathbb{R}}$. The theorem of Radon and Nikodym, which is the topic of this section, shows that both these statements admit vast generalizations which can be used to prove results like the one in § 8.1.2.

8.1.1 The Radon-Nikodym Theorem

The proof that I will give of the Radon–Nikodym Theorem is due to J. von Neumann. Von Neumann's idea was that the absolutely continuous part of μ should obtained by some sort of projection of $L^2(\nu;\mathbb{R})$ onto $L^2(\mu,\mathbb{R})$, and he implemented this idea as an application of Theorem 7.1.3.

The first step is to prove an analog of Lemma 3.3.4 in this context. That is, I will prove the Radon–Nikodym Theorem in the special case of measures satisfying $\mu \le \nu$ in the sense that $\mu(\Gamma) \le \nu(\Gamma)$ for all $\Gamma \in \mathcal{B}$. Notice that, by starting with simple functions and taking monotone limits, one can show that this condition implies $\int \varphi\,d\mu \le \int \varphi\,d\nu$ for every non-negative, $\mathcal{B}$-measurable φ.

Lemma 8.1.1 *Suppose that $(E,\mathcal{B},\nu)$ is a σ-finite measure space and that μ is a finite measure on $(E,\mathcal{B})$ with the property that $\mu \le \nu$. Then there is a unique $\varphi \in L^1(\nu;\mathbb{R})$ such that $\mu(\Gamma) = \int_\Gamma \varphi\,d\nu$ for all $\Gamma \in \mathcal{B}$. Moreover, φ can be chosen to take its values in $[0,1]$.*

Proof. Since E can be written as the countable union of disjoint elements of $\mathcal{B}$ on each of which ν is finite, we will, without loss of generality, assume that ν is finite.

Noting that $L^2(\nu;\mathbb{R}) \subseteq L^2(\mu;\mathbb{R}) \subseteq L^1(\mu;\mathbb{R})$, define $\Lambda(h) = \int h\,d\mu$ for $h \in L^2(\nu;\mathbb{R})$, and observe that $h \rightsquigarrow \Lambda(h)$ is linear and satisfies

$$|\Lambda(h)| \le \|h\|_{L^1(\mu;\mathbb{R})} \le \mu(E)^{\frac{1}{2}}\|h\|_{L^2(\mu;\mathbb{R})} \le \mu(E)^{\frac{1}{2}}\|h\|_{L^2(\nu;\mathbb{R})}.$$

Hence, $|\Lambda(h') - \Lambda(h)| = |\Lambda(h'-h)| \le \mu(E)^{\frac{1}{2}}\|h'-h\|_{L^2(\nu;\mathbb{R})}$, which means that Λ is continuous on $L^2(\nu;\mathbb{R})$. In particular, by Theorem 7.1.3, there exists a $\varphi \in L^2(\nu;\mathbb{R})$ such that $\int h\,d\mu = \Lambda(h) = \int h\varphi\,d\nu$ for all $h \in L^2(\nu;\mathbb{R})$, and clearly this means that $\mu(\Gamma) = \int_\Gamma \varphi\,d\nu$ for all $\Gamma \in \mathcal{B}$. Finally, since $\int_\Gamma \varphi\,d\nu = \mu(\Gamma) \in \big[0,\nu(E)\big]$ for all $\Gamma \in \mathcal{B}$, we know (cf. Exercise 3.1.6) that φ is uniquely determined up to a set of ν-measure 0 and that it takes its values in $[0,1]$ ν-almost everywhere. □

The first part of the next theorem is called **Lebesgue's Decomposition Theorem**, and the second part is the **Radon–Nikodym Theorem**.

Theorem 8.1.2 *Suppose that $(E,\mathcal{B},\nu)$ is a σ-finite measure space and that μ is a finite measure on $(E,\mathcal{B})$. Then there is one and only one way of writing μ as the sum of a measure $\mu_{\mathrm{a}} \ll \nu$ and a measure $\mu_{\mathrm{s}} \perp \nu$. Moreover, there is* *a $B \in \mathcal{B}$ such that $\mu_s(\Gamma) = \mu(\Gamma \cap B)$ and therefore $\mu_a(\Gamma) = \mu(\Gamma \cap B^{\complement})$ for all $\Gamma \in \mathcal{B}$. Finally, there is a unique $f \in L^1(\nu;\mathbb{R})$ such that $\mu_a(\Gamma) = \int_\Gamma f\,d\nu$ for all $\Gamma \in \mathcal{B}$, and f can be chosen to be non-negative. In particular, $\mu \ll \nu$ if and only if $\mu(\Gamma) = \int_\Gamma f\,d\nu$, $\Gamma \in \mathcal{B}$, for some $f \in L^1(\nu;\mathbb{R})$.*

Proof. First note that if $\mu(\Gamma) = \int_\Gamma f\,d\nu$ for some $f \in L^1(\nu;\mathbb{R})$ and all $\Gamma \in \mathcal{B}$, then, by Exercise 3.1.5, $\mu \ll \nu$. In addition, by Exercise 3.1.6, f is uniquely determined up to a set of ν-measure 0 and is non-negative (a.e., ν).

Next we will show that there is at most one choice of μ_a and therefore of μ_s. To this end, suppose that $\mu = \mu_a + \mu_s = \mu'_a + \mu'_s$, where μ_a and μ'_a are both absolutely continuous with respect to ν and both μ_s and μ'_s are singular to ν. Choose B, $B' \in \mathcal{B}$ such that

$$\nu(B) = \nu(B') = 0 \quad \text{and} \quad \mu_s(B^{\complement}) = 0 = \mu'_s\big((B')^{\complement}\big),$$

and set $A = B \cup B'$. Then $\nu(A) = 0$ and $\mu_s(A^{\complement}) = 0 = \mu'_s(A^{\complement})$, and so, for any $\Gamma \in \mathcal{B}$,

$$\mu_a(\Gamma) = \mu_a(\Gamma \cap A^{\complement}) = \mu(\Gamma \cap A^{\complement}) = \mu'_a(\Gamma \cap A^{\complement}) = \mu'_a(\Gamma).$$

In other words, $\mu_a = \mu'_a$.

To prove the existence statements, first use Lemma 8.1.1, applied to μ and $\mu + \nu$, to find a $\mathcal{B}$-measurable $\varphi : E \longrightarrow [0,1]$ with the property that

$$\mu(\Gamma) = \int_\Gamma \varphi\,d\mu + \int_\Gamma \varphi\,d\nu \quad \text{for all } \Gamma \in \mathcal{B}.$$

Proceeding by way of simple functions and monotone limits, one sees that

$$\int g\,d\mu = \int g\varphi\,d\mu + \int g\varphi\,d\nu$$

for all non-negative, $\mathcal{B}$-measurable g's, from which it is clear that

$$\int g(1-\varphi)\,d\mu = \int g\varphi\,d\nu,$$

first for all $g \in L^1(\mu;\mathbb{R}) \cap L^1(\nu;\mathbb{R})$ and then for all non-negative, $\mathcal{B}$-measurable g's. Now set $B = \{\varphi = 1\}$, and define $\mu_a(\Gamma) = \mu(\Gamma \cap B^{\complement})$ and $\mu_s(\Gamma) = \mu(\Gamma \cap B)$ for all $\Gamma \in \mathcal{B}$. Since

$$\nu(B) = \nu(\{\varphi = 1\}) = \int_{\{\varphi=1\}} \varphi\,d\nu = \int_{\{\varphi=1\}} (1-\varphi)\,d\mu = 0,$$

it is clear that μ_s is singular to ν. At the same time, if $f \equiv (1-\varphi)^{-1}\varphi \mathbf{1}_{B^{\complement}}$, then, by taking $g = (1-\varphi)^{-1}\mathbf{1}_{\Gamma \cap B^{\complement}}$, one sees that

$$\mu_{\mathrm{a}}(\Gamma) = \mu(\Gamma \cap B^{\complement}) = \int g(1-\varphi)\,d\mu = \int_\Gamma f\,d\nu$$

for each $\Gamma \in \mathcal{B}$. □

Given a finite measure μ and a σ-finite measure ν, the corresponding measures μ_{a} and μ_{s} are called the **absolutely continuous** and **singular parts of** μ **with respect to** ν. Also, if μ is absolutely continuous with respect to ν, then the corresponding non-negative $f \in L^1(\nu;\mathbb{R})$ is called the **Radon–Nikodym derivative** of μ with respect to ν and is often denoted by $\frac{d\mu}{d\nu}$. The choice of this terminology and notation comes from the general fact, of which Theorem 3.3.6 is a special case, that the Radon–Nikodym derivative of μ with respect to ν is truly a derivative.[1]

8.1.2 Hahn Decomposition

Throughout, $(E,\mathcal{B})$ will be a measurable space. A function $\mu : \mathcal{B} \longrightarrow \mathbb{R}$ is called a **finite signed measure** if there exists a finite Borel measure ν such that $|\mu(\Gamma)| \le \nu(\Gamma)$ for all $\Gamma \in \mathcal{B}$ and is countably additive in the sense that

$$\mu\left(\bigcup_{n=1}^\infty \Gamma_n\right) = \sum_{n=1}^\infty \mu(\Gamma_n)$$

for any sequence $\{\Gamma_n : n \ge 1\} \subseteq \mathcal{B}$ of mutually disjoint sets. Clearly, if $\nu_\pm = \frac{\nu\pm\mu}{2}$, then $\nu_\pm$ are finite measures, and $\mu = \nu_+ - \nu_-$. Hence it is reasonable to define

$$\int \varphi\,d\mu = \int \varphi\,d\nu_+ - \int \varphi\,d\nu_-$$

for $\varphi \in L^1(\nu;\mathbb{R})$. However, before adopting this definition, one has to check that if ν' is a second finite measure for which $|\mu(\Gamma)| \le \nu'(\Gamma)$ for all $\Gamma \in \mathcal{B}$, then

$$\int \varphi\,d\nu'_+ - \int \varphi\,d\nu'_- = \int \varphi\,d\nu_+ - \int \varphi\,d\nu_-$$

for all $\varphi \in L^1(\nu;\mathbb{R}) \cap L^1(\nu';\mathbb{R})$. But clearly this is true for any function φ which is simple, and therefore, if $\{\varphi_n : n \ge 1\}$ is a sequence of simple functions that tend to φ in $L^1(\nu+\nu';\mathbb{R})$, then

$$\begin{aligned}\int \varphi\,d\nu'_+ - \int \varphi\,d\nu'_- &= \lim_{n\to\infty}\left(\int \varphi_n\,d\nu'_+ - \int \varphi_n\,d\nu'_-\right)\\ &= \lim_{n\to\infty}\left(\int \varphi_n\,d\nu_+ - \int \varphi_n\,d\nu_-\right) = \int \varphi\,d\nu_+ - \int \varphi\,d\nu_-.\end{aligned}$$

[1] For further information, see Theorem 5.2.20 in the second edition of my book *Probability Theory, an Analytic View*, published by Cambridge Univ. Press.

Because the preceding says how to integrate φ when φ is integrable with respect to ν, it is desirable to choose ν to be as small as possible. When E is countable and $\mathcal{B} = \mathcal{P}(E)$, it is clear that the minimal ν is the one which assigns each point $x \in E$ measure $|\mu(\{x\})|$, in which case

$$\nu_+(\Gamma) = \nu(\Gamma \cap A) \text{ and } \nu_-(\Gamma) = \nu(\Gamma \cap A^\complement),$$

where $A = \{x \in E : \mu(\{x\}) \geq 0\}$. The goal in this subsection is to prove the following theorem, proved originally by H. Hahn[2], which shows that the preceeding line of reasoning has a version that works in general. As Exercise 8.2.6 makes clear, the procedure used here is the analog for measures of the results in § 1.2.2 and Exercise 1.2.6 about functions of bounded variation. See Exercise 8.1.5 for more information.

Theorem 8.1.3 (Hahn Decomposition) *There is a unique measure $|\mu|$ with the property that $|\mu(\Gamma)| \leq |\mu|(\Gamma)$ for all $\Gamma \in \mathcal{B}$ and $|\mu| \leq \nu$ for any measure ν with these properties. Furthermore, if $\mu_\pm = \frac{|\mu| \pm \mu}{2}$, then there exists a set $A \in \mathcal{B}$, which is unique up to a set of $|\mu|$-measure 0, such that $\mu_+(\Gamma) = \mu(\Gamma \cap A)$ and $\mu_-(\Gamma) = \mu(\Gamma \cap A^\complement)$. Hence*

$$\int \varphi \, d\mu = \int_A \varphi \, d|\mu| - \int_{A^\complement} \varphi \, d|\mu|$$

for all $\varphi \in L^1(|\mu|; \mathbb{R})$.

I will divide the proof of Hahn's result into two steps. In the following, $\mathcal{C}$ will denote a generic, countable cover of E by mutually disjoint sets $A \in \mathcal{B}$, and, for $\Gamma \in \mathcal{B}$,

$$S(\Gamma; \mathcal{C}) = \sum_{A \in \mathcal{C}} |\mu(\Gamma \cap A)|.$$

Lemma 8.1.4 *If μ is a finite signed measure on $(E, \mathcal{B})$, then*

$$\sup_{\mathcal{C}} S(E; \mathcal{C}) < \infty.$$

Moreover, if $|\mu|(\Gamma)$ is defined for $\Gamma \in \mathcal{B}$ by

$$|\mu|(\Gamma) = \sup\{\{S(\Gamma; \mathcal{C}\} : \mathcal{C} \text{ a countable cover of } E \text{ by mutually disjoint } C \in \mathcal{B}\},$$

then $|\mu|$ is a finite measure on $(E, \mathcal{B})$, $|\mu(\Gamma)| \leq |\mu|(\Gamma)$ for all $\Gamma \in \mathcal{B}$, and $|\mu| \leq \nu$ for any measure ν satisfying $|\mu(\Gamma)| \leq \nu(\Gamma)$ for all $\Gamma \in \mathcal{B}$.

Proof. If ν is a finite measure for which $|\mu(\Gamma)| \leq \nu(\Gamma)$, then it is obvious that $S(E; \mathcal{C}) \leq \nu(E)$ for all $\mathcal{C}$ and therefore that $|\mu| \leq \nu$.

[2] Hahn's result is a sharpening of the Jordon decomposition theorem which states that a signed measure can be decomposed into the difference of two (non-negative) measures.

Next, let $|\mu|$ be defined as in the statement. It is easy to check that $|\mu|$ is countably subadditive. To see that it is countably additive, let $\Gamma = \bigcup_{n=1}^{\infty} \Gamma_n$, where $\{\Gamma_n : n \geq 1\} \subseteq \mathcal{B}$ is a sequence of mutually disjoint sets. Given $\epsilon > 0$, choose a sequence $\{\mathcal{C}_n : n \geq 1\}$ of countable covers so the $|\mu|(\Gamma_n) \leq S(\Gamma_n; \mathcal{C}_n) + 2^{-n}\epsilon$ for each $n \geq 1$, and take

$$\mathcal{C} = \{\Gamma^{\complement}\} \cup \bigcup \{A \cap \Gamma_n : n \geq 1 \;\&\; A \in \mathcal{C}_n\}.$$

Then $\mathcal{C}$ is a countable collection of mutually disjoint elements of $\mathcal{B}$, and $\{\Gamma_n \cap A : A \in \mathcal{C}\} = \{\Gamma_n \cap A : A \in \mathcal{C}_n\}$ for all $n \geq 1$. Hence,

$$|\mu|(\Gamma) \geq S(\Gamma; \mathcal{C}) = \sum_{n=1}^{\infty} S(\Gamma_n; \mathcal{C}) = \sum_{n=1}^{\infty} S(\Gamma_n; \mathcal{C}_n) \geq \sum_{n=1}^{\infty} |\mu|(\Gamma_n) - \epsilon.$$

Thus, $|\mu|$ is a measure, and it is clear that $|\mu(\Gamma)| \leq |\mu|(\Gamma)$ for all $\Gamma \in \mathcal{B}$ and that it is the smallest measure with that property. □

Proof of Theorem 8.1.3. Set $\mu_\pm = \frac{|\mu| \pm \mu}{2}$, and take $f_\pm = \frac{d\mu_\pm}{d|\mu|}$ and $f = f_+ - f_-$. Since $\mu_+ + \mu_- = |\mu|$, we can choose $f_\pm$ so that $f_+ + f_- = 1$. Also, since $\mu_+ - \mu_- = \mu$,

$$\int \varphi \, d\mu = \int \varphi f \, d|\mu|,$$

for all $\varphi \in L^1(|\mu|; \mathbb{R})$. Therefore, if $\nu(\Gamma) = \int_\Gamma |f| \, d|\mu|$ for $\Gamma \in \mathcal{B}$, then $\nu(\Gamma) \geq |\mu(\Gamma)|$ for all $\Gamma \in \mathcal{B}$, which implies that $\nu \geq |\mu|$. Thus we can choose f so that $|f| \geq 1$, which, because $1 = f_+ + f_- \geq |f_+ - f_-|$, means that $f_+ + f_- = 1 = |f_+ - f_-|$. Hence, if $A = \{f \geq 0\}$, then $f_+ = \mathbf{1}_A$ and $f_- = \mathbf{1}_{A^{\complement}}$.

Finally, to prove the uniqueness assertion, first observe that if $\nu_\pm$ are measures for which $\nu_+ - \nu_- = \mu$, then $\nu = \nu_+ + \nu_- \geq |\mu|$, and so

$$\nu_\pm = \frac{\nu \pm \mu}{2} \geq \frac{|\mu| \pm \mu}{2} = \mu_\pm.$$

Now suppose that $\varphi_\pm$ are non-negative, $\mathcal{B}$-measurable functions for which $\varphi_+ + \varphi_- = 1$ $|\mu|$-a.e. and $\mu(\Gamma) = \int_\Gamma (\varphi_+ - \varphi_-) \, d|\mu|$ for all $\Gamma \in \mathcal{B}$. Define $\nu_\pm$ so that $d\nu_\pm = \varphi_\pm d|\mu|$, and set $\nu = \nu_+ + \nu_-$. Then $\nu_+ - \nu_- = \mu$, and therefore $\nu_\pm \geq \mu_\pm$. Since this means that $\varphi_+ \geq \mathbf{1}_A$ and $\varphi_- \geq \mathbf{1} - \mathbf{1}_A$ at the same time as $\varphi_+ + \varphi_- = 1$ $|\mu|$-a.e., it follows that $\varphi_+ = \mathbf{1}_A$ and $\varphi_- = \mathbf{1}_{A^{\complement}}$ $|\mu|$-a.e. □

8.1.3 Exercises for § 8.1

Exercise 8.1.1 Let $(E, \mathcal{B}, \nu)$ and μ be as in Theorem 8.1.2, and show that

$$\mu_s(\Gamma) = \sup\{\mu(\Gamma \cap A) : A \in \mathcal{B} \text{ and } \nu(A) = 0\} \quad \text{for all } \Gamma \in \mathcal{B}.$$

In addition, show that there is a $B_0 \in \mathcal{B}$ such that $\nu(B_0) = 0$ and

$$\mu(B_0) = \kappa \equiv \sup\{\mu(A) : A \in \mathcal{B} \text{ and } \nu(A) = 0\}.$$

Finally, show that if $B \in \mathcal{B}$ and $\nu(B) = 0$, then $\mu_s(\Gamma) = \mu(\Gamma \cap B)$ for all $\Gamma \in \mathcal{B}$ if and only if $\mu(B) = \kappa$.

Exercise 8.1.2 Suppose that $\mathcal{P}$ is a countable partition of the non-empty set E, and use $\mathcal{B}$ to denote $\sigma(\mathcal{P})$.

(i) Show that $f : E \longrightarrow \overline{\mathbb{R}}$ is $\mathcal{B}$-measurable if and only if f is constant on each $A \in \mathcal{P}$. Also, show that a measure ν is σ-finite on $(E, \mathcal{B})$ if and only if $\nu(A) < \infty$ for every $A \in \mathcal{P}$. Finally, if μ is a second measure on $(E, \mathcal{B})$, show that $\mu \ll \nu$ if and only if $\mu(A) = 0$ for all $A \in \mathcal{P}$ satisfying $\nu(A) = 0$.

(ii) Given any measures μ and ν on $(E, \mathcal{B})$ and a $\mathcal{B}$-measurable, ν-integrable $f : E \longrightarrow [0, \infty)$, show that $\mu(A) = \int_A f\, d\nu$ for all $A \in \mathcal{B}$ implies that, for every $A \in \mathcal{P}$, $\nu(A) \in (0, \infty) \implies f \restriction A = \frac{\mu(A)}{\nu(A)}$ and $\nu(A) = \infty \implies f \restriction A = 0$.

(iii) Using the preceding, show that, in general, one cannot dispense with the assumption in Theorem 8.1.2 that ν is σ-finite.

Exercise 8.1.3 Let μ_1 and μ_2 be a pair of finite measures on the measurable space $(E, \mathcal{B})$. Given a σ-finite measure ν on $(E, \mathcal{B})$ for which $\mu_1 \ll \nu$ and $\mu_2 \ll \nu$, set

$$(\mu_1, \mu_2) = \int f_1^{\frac{1}{2}} f_2^{\frac{1}{2}}\, d\nu \quad \text{where } f_1 = \frac{d\mu_1}{d\nu} \text{ and } f_2 = \frac{d\mu_2}{d\nu}.$$

Show that the number (μ_1, μ_2) is independent of the choice of ν and that $(\mu_1, \mu_2) = 0$ if and only if $\mu_1 \perp \mu_2$. In particular, one can always take $\nu = \mu_1 + \mu_2$.

Exercise 8.1.4 Given a Banach space X over $\mathbb{R}$, in functional analysis an important role is played by its **dual space** X^*: the space of continuous linear maps $\Lambda : X \longrightarrow \mathbb{R}$. When X is a Hilbert space, Theorem 7.1.3 allows one to identify X^* with X itself via the inner product. In particular, if $X = L^2(\mu; \mathbb{R})$, then $\Lambda \in X^*$ if and only if there exists an $f \in L^2(\mu; \mathbb{R})$ such that $\Lambda(\varphi) = (\varphi, f)_{L^2(\mu;\mathbb{R})}$ for all $\varphi \in L^2(\mu; \mathbb{R})$. More generally, at least when μ is σ-finite, it is known[3] that, for any $p \in [1, \infty)$, a similar identification exists of $L^p(\mu; \mathbb{R})^*$ with $L^{p'}(\mu; \mathbb{R})$, where p' is the Hölder conjugate of p. Namely, $\Lambda \in L^p(\mu; \mathbb{R})^*$ if and only if there exists an $f \in L^{p'}(\mu; \mathbb{R})$ such that $\Lambda(\varphi) = \int \varphi f\, d\mu$ for all $\varphi \in L^p(\mu; \mathbb{R})$. The goal of this exercise is to prove this identification in a special case. Since, by **(i)** in Theorem 6.2.4, it is clear that $\varphi \in L^p(\mu; \mathbb{R}) \longmapsto \int \varphi f\, d\mu \in \mathbb{R}$ is a continuous, linear map for every

[3] If you can afford it, see, for example, IV.8 in *Linear Operators, I* by N. Dunford and J. Schwartz, published by Wiley-Interscience.

$f \in L^{p'}(\mu;\mathbb{R})$, what remains is to show that every $\Lambda \in L^p(\mu;\mathbb{R})$ arises in this way. See Exercise 8.2.5 for further information.

(i) Show that a linear map $\Lambda : X \longrightarrow \mathbb{R}$ is continuous if and only if there exists a $C < \infty$ such that $|\Lambda(x)| \le C\|x\|_X$ for all $x \in X$.

(ii) Given a measure space $(E, \mathcal{B}, \mu)$, a $p \in [1, 2]$, and an $f \in L^2(\mu;\mathbb{R})$, use Theorem 6.2.4 to show that $f \in L^{p'}(\mu;\mathbb{R})$ if there exists $C < \infty$ for which $\|\varphi f\|_{L^1(\mu;\mathbb{R})} \le C\|\varphi\|_{L^p(\mu;\mathbb{R})}$ for all $\varphi \in L^2(\mu;\mathbb{R})$.

(iii) Suppose that $(E, \mathcal{B}, \mu)$ is a finite measure space and that $p \in [1, 2]$. Given $\Lambda \in L^p(\mu;\mathbb{R})^*$, choose $C < \infty$ as in **(i)**, and show that $|\Lambda(\varphi)| \le C\mu(E)^{\frac{1}{p}-\frac{1}{2}}\|\varphi\|_{L^2(\mu;\mathbb{R})}$ for all $\varphi \in L^2(\mu;\mathbb{R})$. Next, apply Theorem 7.1.3 to produce an $f \in L^2(\mu;\mathbb{R})$ such that $\Lambda(\varphi) = (\varphi, f)_{L^2(\mu;\mathbb{R})}$ for $\varphi \in L^2(\mu;\mathbb{R})$, use **(ii)** to see that $f \in L^{p'}(\mu;\mathbb{R})$, and conclude that $\Lambda(\varphi) = \int \varphi f \, d\mu$ for all $\varphi \in L^p(\mu;\mathbb{R})$.

(iv) Show that the same conclusion holds when μ is σ-finite.

Exercise 8.1.5 Let $\psi : \mathbb{R} \longrightarrow \mathbb{R}$ be a right-continuous function which tends to 0 at $-\infty$ and for which $\sup\{\mathrm{Var}(\psi; J) : J \text{ a compact interval}\} < \infty$. By Exercise 1.2.6, we know that $t \in [a, b] \longmapsto \mathrm{Var}(\psi; [a, t])$ and $t \in [a, b] \longmapsto \mathrm{Var}_\pm(\psi; [a, t])$ are all right continuous functions for every $-\infty < a < b < \infty$. Define $F(t) = \lim_{n\to\infty} \mathrm{Var}(\psi; [t-n, t])$ and $F_\pm(t) = \lim_{n\to\infty} \mathrm{Var}_\pm(\psi; [t-n, t])$ for $t \in \mathbb{R}$.

(i) Using (1.2.4) and Lemma 1.2.4 together with Exercise 1.2.6, show that F and $F_\pm$ are all bounded, right continuous, non-decreasing functions which tend to 0 at $-\infty$. In addition, show that $F(b) - F(a) = \mathrm{Var}(\psi; [a, b])$ and $F_\pm(b) - F_\pm(a) = \mathrm{Var}_\pm(\psi; [a, b])$, and conclude that $F = F_+ + F_-$ and $\psi = F_+ - F_-$.

(ii) Let ν and $\nu_\pm$ be the finite Borel measures on $\mathbb{R}$ with, respectively, distribution functions F and $F_\pm$, and set $\mu = \nu_+ - \nu_-$. Show that $\nu = \nu_+ + \nu_-$ and that

$$(\mathrm{R})\int_{[a,b]} \varphi \, d\psi = \int_{(a,b]} \varphi \, d\mu$$

for all $-\infty < a < b < \infty$ and $\varphi \in C(\mathbb{R};\mathbb{R})$.

(iii) Since $|\mu(\Gamma)| \le \nu(\Gamma)$ for all $\Gamma \in \mathcal{B}_\mathbb{R}$, we know that $|\mu| \le \nu$. On the other hand, if $-\infty < a < b < \infty$, show that $\nu\big((a, b]\big) \le |\mu|\big((a, b]\big)$, and conclude from this and Lemma 2.1.4 that $\nu \le |\mu|$. Hence, $\nu = |\mu|$ and $\nu_\pm = \mu_\pm$.

(iv) Show that there exists an $A \in \mathcal{B}_\mathbb{R}$ such that

$$(\mathrm{R})\int_{[a,b]} \varphi d\psi = \int_{(a,b]\cap A} \varphi d|\mu| - \int_{(a,b]\cap A^\complement} \varphi d|\mu|$$

for all $-\infty < a < b < \infty$ and $\varphi \in C(\mathbb{R};\mathbb{R})$.

8.2 The Daniell Integral

Riesz's idea underlying Daniell's theory is that it is smarter to start with an abstract theory of integration and extract the measure from the integration theory rather than start with the measure. In other words, one should reverse the procedure that was adopted in Chapters 2 and 3. In this section, I will describe how this can be done, and I will begin by setting the stage.

Let E be a non-empty set. I will say that a subset $\mathbf{L}$ of the functions $f : E \longrightarrow \mathbb{R}$ is a **vector lattice** if $\mathbf{L}$ is a vector space and $f^+ = f \vee 0$ is in $\mathbf{L}$ whenever $f \in \mathbf{L}$. Because $f \vee g = f + (g - f)^+$, $f \vee g$ is in $\mathbf{L}$ and, because $f \wedge g = -\big((-f) \vee (-g)\big)$, $f \wedge g$ is also in $\mathbf{L}$ whenever both f and g are. In particular, f^- and $|f| \in \mathbf{L}$ if $f \in \mathbf{L}$.

Given a vector lattice $\mathbf{L}$, I will say that the map $I : \mathbf{L} \longrightarrow \mathbb{R}$ is an **integral on L** if

(a) I is linear,
(b) I is **non-negative** in the sense that $I(f) \geq 0$ for every non-negative $f \in \mathbf{L}$,
(c) $I(f_n) \searrow 0$ whenever $\{f_n : n \geq 1\} \subseteq \mathbf{L}$ is a non-increasing sequence that tends (pointwise) to 0.

Finally, say that a triple $\mathcal{I} = (E, \mathbf{L}, I)$ is an **integration theory** if $\mathbf{L}$ is a vector lattice of functions $f : E \longrightarrow \mathbb{R}$ and I is an integral on $\mathbf{L}$.

Example 8.2.1 Here are three situations from which the preceding notions are derived.

(i) The basic model for the preceding definitions is the one that comes from the integration theory for a measure space $(E, \mathcal{B}, \mu)$. Indeed, in that case, $\mathbf{L} = L^1(\mu; \mathbb{R})$ and $I(f) = \int f \, d\mu$.

(ii) A second basic source of integration theories is the one that comes from *finitely additive functions on an algebra*. That is, let (cf. Exercise 2.1.1) $\mathcal{A}$ be an algebra of subsets of E, and denote by $\mathbf{L}(\mathcal{A})$ the space of simple functions $f : E \longrightarrow \mathbb{R}$ with the property that $\{f = a\} \in \mathcal{A}$ for every $a \in \mathbb{R}$. It is then an easy matter to check that $\mathbf{L}(\mathcal{A})$ is a vector lattice. Now let $\mu : \mathcal{A} \longrightarrow [0, \infty)$ be **finitely additive** in the sense that

$$\mu(\Gamma_1 \cup \Gamma_2) = \mu(\Gamma_1) + \mu(\Gamma_2) \quad \text{for disjoint } \Gamma_1, \Gamma_2 \in \mathcal{A}.$$

Note that, since $\mu(\emptyset) = \mu(\emptyset \cup \emptyset) = 2\mu(\emptyset)$, $\mu(\emptyset)$ must be 0. Also, because only finite additivity was used in the proof of Lemma 3.1.2, one can use the same argument here to show that

$$f \in \mathbf{L}(\mathcal{A}) \longmapsto I(f) \equiv \sum_{a \in \mathrm{Range}(f)} a \, \mu(\{f = a\})$$

is linear and non-negative. Finally, observe that I cannot be an integral unless μ has the property that

$$\mu(\Gamma_n) \searrow 0 \quad \text{whenever } \{\Gamma_n : n \ge 1\} \subseteq \mathcal{A} \text{ decreases to } \emptyset. \tag{8.2.1}$$

On the other hand, if (8.2.1) holds and $\{f_n : n \ge 1\} \subseteq \mathbf{L}(\mathcal{A})$ is a non-increasing sequence that tends pointwise to 0, then for each $\epsilon > 0$,

$$\varlimsup_{n\to\infty} I(f_n) \le \epsilon I(\mathbf{1}) + \|f_1\|_{\mathrm{u}} \varlimsup_{n\to\infty} \mu(\{f_n > \epsilon\}) = \epsilon I(\mathbf{1}).$$

Thus, in this setting, (8.2.1) is equivalent to I being an integral.

(iii) A third important example of an integration theory is provided by the following abstraction of Riemann's theory. Namely, let E be a compact topological space, and check that $C(E;\mathbb{R})$ is a vector lattice. Next, suppose that $I : C(E;\mathbb{R}) \longrightarrow \mathbb{R}$ is a linear map that is non-negative. It is then clear that $|I(f)| \le K\|f\|_{\mathrm{u}}$, $f \in C(E;\mathbb{R})$, where $K = I(\mathbf{1})$. In particular, this means that $|I(f_n) - I(f)| \le K\|f_n - f\|_{\mathrm{u}} \longrightarrow 0$ if $f_n \longrightarrow f$ uniformly. Thus, to see that I is an integral, all that we have to do is use Dini's Lemma (cf. Lemma 8.2.8 below), which says that $f_n \longrightarrow 0$ uniformly on E if $\{f_n : n \ge 1\} \subseteq C(E;\mathbb{R})$ decreases pointwise to 0.

My main goal will be to show that, at least when $\mathbf{1} \in \mathbf{L}$, every integration theory is a sub-theory of the sort of theory described in **(i)** above. Thus, we must learn how to extract the *measure* μ from the *integral*. At least in case **(ii)** above, it is clear how one might begin such a procedure. Namely, $\mathcal{A} = \{\Gamma \subseteq E : \mathbf{1}_\Gamma \in \mathbf{L}(\mathcal{A})\}$ and $\mu(\Gamma) = I(\mathbf{1}_\Gamma)$ for $\Gamma \in \mathcal{A}$. Hence, what we are attempting to do in this case is tantamount to showing that μ can be extended as a measure to the σ-algebra $\sigma(\mathcal{A})$ generated by $\mathcal{A}$. On the other hand, it is not so immediately clear where to start looking for the measure μ in case **(iii)**; the procedure that got us started in case **(ii)** does not work here since there will seldom be many $\Gamma \subseteq E$ for which $\mathbf{1}_\Gamma \in C(E;\mathbb{R})$. More generally, we must learn first how to extend I to a larger class of functions $f : E \longrightarrow \mathbb{R}$ and only then look for μ.

8.2.1 Extending an Integration Theory

The extension procedure has two steps, the first of which is nothing but a rerun of what I did in § 3.1, and the second one is a minor variation on what I did in § 2.2.

Lemma 8.2.1 *Let $(E, \mathbf{L}, I)$ be an integration theory, and define $\mathbf{L}_u$ to be the class of $f : E \longrightarrow (-\infty, \infty]$ that can be written as the pointwise limit of a non-decreasing sequence $\{\varphi_n : n \ge 1\} \subseteq \mathbf{L}$. Then $f \vee g$ and $f \wedge g$ are in $\mathbf{L}_u$ if f and g are, and $\mathbf{L}_{\mathrm{u}}$ is closed under non-negative linear combinations and non-decreasing sequential limits (i.e., $\{f_n : n \ge 1\} \subseteq \mathbf{L}_u$ and $f_n \nearrow f$ implies that $f \in \mathbf{L}_u$). Moreover, I admits a unique extension to $\mathbf{L}_u$ in such a way that $I(f_n) \nearrow I(f)$ whenever f is the limit of a non-decreasing sequence*

$\{f_n : n \geq 1\} \subseteq \mathbf{L}_u$. In particular, for all $f, g \in \mathbf{L}_u$, $-\infty < I(f) \leq I(g)$ if $f \leq g$ and $I(\alpha f + \beta g) = \alpha I(f) + \beta I(g)$ for all $\alpha, \beta \in [0, \infty)$.

Proof. The closedness properties of $\mathbf{L}_u$ are obvious. Moreover, given that an extension of I with the stated properties exists, it is clear that that extension is unique, monotone, and linear under non-negative linear combinations.

Just as in the development in § 3.1 that eventually led to the monotone convergence theorem, the proof (cf. Lemma 3.1.3) that I extends to $\mathbf{L}_u$ is simply a matter of checking that the desired extension of I does not depend on the choice of approximating φ_n's. Thus, what we must show is that when $\psi \in \mathbf{L}$ and $\{\varphi_n : n \geq 1\} \subseteq \mathbf{L}$ is a non-decreasing sequence with the property that $\psi \leq \lim_{n\to\infty} \varphi_n$ pointwise, then $I(\psi) \leq \lim_{n\to\infty} I(\varphi_n)$. To this end, note that $\varphi_n = \psi - (\psi - \varphi_n) \geq \psi - (\psi - \varphi_n)^+$, $\mathbf{L} \ni (\psi - \varphi_n)^+ \searrow 0$, and therefore that

$$\lim_{n\to\infty} I(\varphi_n) \geq I(\psi) - \lim_{n\to\infty} I\big((\psi - \varphi_n)^+\big) = I(\psi).$$

As was said before, once one knows that I is consistently defined on $\mathbf{L}_u$, the rest of the proof differs in no way from the proofs of Lemma 3.1.3 and Theorem 3.2.2. □

Lemma 8.2.1 is the initial extension of I. What it provides is a rich class of functions that will be used to play the role that open sets played in my construction of measures in § 2.2. Thus, given any $f : E \longrightarrow \overline{\mathbb{R}}$, define

$$\overline{I}(f) = \inf\big\{I(\varphi) : \varphi \in \mathbf{L}_u \text{ and } \varphi \geq f\big\}.$$

Because of (2.2.1), it is clear that $\overline{I}(f)$ is the analog here of the set function $\tilde{\mu}$ introduced in § 2.2 following Lemma 2.2.1. At the same time, it will be convenient to have introduced

$$\underline{I}(f) = \sup\big\{-I(\varphi) : \varphi \in \mathbf{L}_u \text{ and } -\varphi \leq f\big\},$$

the analog of which in § 2.2 would have been $\sup\big\{\tilde{\mu}(F) : \mathfrak{F}(E) \ni F \subseteq \Gamma\big\}$.

Lemma 8.2.2 *For any $\overline{\mathbb{R}}$-valued function f on E, $\underline{I}(f) \leq \overline{I}(f)$, and*

$$\begin{aligned} \overline{I}(\alpha f) = \alpha \overline{I}(f) \text{ and } \underline{I}(\alpha f) = \alpha \underline{I}(f) \quad \text{if } \alpha \in [0, \infty), \\ \overline{I}(\alpha f) = \alpha \underline{I}(f) \text{ and } \underline{I}(\alpha f) = \alpha \overline{I}(f) \quad \text{if } \alpha \in (-\infty, 0]. \end{aligned} \tag{8.2.2}$$

Moreover, if f and g map E to $\overline{\mathbb{R}}$, then

$$f \leq g \implies \overline{I}(f) \leq \overline{I}(g) \text{ and } \underline{I}(f) \leq \underline{I}(g), \tag{8.2.3}$$

and, when (f, g) takes values in (cf. § 3.1.1) $\widehat{\mathbb{R}^2}$,

$$\begin{aligned}\big(\overline{I}(f),\overline{I}(g)\big)\in\widehat{\mathbb{R}^2} &\implies \overline{I}(f+g)\le\overline{I}(f)+\overline{I}(g)\\ \big(\underline{I}(f),\underline{I}(g)\big)\in\widehat{\mathbb{R}^2} &\implies \underline{I}(f+g)\ge\underline{I}(f)+\underline{I}(g).\end{aligned}\tag{8.2.4}$$

Finally,

$$f\in\mathbf{L}_u \implies \overline{I}(f)=I(f)=\underline{I}(f).\tag{8.2.5}$$

Proof. To prove the initial assertion, first note that it suffices to treat the case in which $\underline{I}(f)>-\infty$ and $\overline{I}(f)<\infty$, and then observe that, for any $(\varphi,\psi)\in\mathbf{L}_u^2$ with $-\varphi\le f\le\psi$, $\varphi+\psi\ge 0$ and therefore $I(\varphi)+I(\psi)=I(\varphi+\psi)\ge 0$.

Both (8.2.2) and (8.2.3) are obvious, and, because of (8.2.2), it suffices to prove only the first line in (8.2.4). Moreover, when $\overline{I}(f)\wedge\overline{I}(g)>-\infty$, the required result is easy. On the other hand, if $\overline{I}(f)=-\infty$ and $\overline{I}(g)<\infty$, then we can choose $\{\varphi_n : n\ge 1\}\cup\{\psi\}\subseteq\mathbf{L}_u$ such that $f\le\varphi_n$, $I(\varphi_n)\le -n$ for each $n\in\mathbb{Z}^+$, $g\le\psi$, and $I(\psi)<\infty$. In particular, $f+g\le\varphi_n+\psi$ for all $n\in\mathbb{Z}^+$, and so $\overline{I}(f+g)\le\lim_{n\to\infty}I(\varphi_n)+I(\psi)=-\infty$.

Finally, suppose that $f\in\mathbf{L}_u$. Obviously, $\overline{I}(f)\le I(f)$. At the same time, if $\{\varphi_n : n\ge 1\}\subseteq\mathbf{L}$ is chosen such that $\varphi_n\nearrow f$, then (because $-\varphi_n\in\mathbf{L}\subseteq\mathbf{L}_u$ and $-(-\varphi_n)=\varphi_n\le f$ for each $n\in\mathbb{Z}^+$)

$$\underline{I}(f)\ge\lim_{n\to\infty}-I\big(-\varphi_n\big)=\lim_{n\to\infty}I\big(\varphi_n\big)=I(f).$$

□

From now on, I will use $\mathfrak{M}(\mathcal{I})$ to denote the class of those $f:E\longrightarrow\overline{\mathbb{R}}$ for which $\overline{I}(f)=\underline{I}(f)$, and I will define $\tilde{I}:\mathfrak{M}(\mathcal{I})\longrightarrow\overline{\mathbb{R}}$ so that $\underline{I}(f)=\tilde{I}(f)=\overline{I}(f)$. Obviously (cf. (8.2.2)), for all $\alpha\in\mathbb{R}$,

$$f\in\mathfrak{M}(\mathcal{I})\implies\alpha f\in\mathfrak{M}(\mathcal{I})\text{ and }\tilde{I}(\alpha f)=\alpha\tilde{I}(f).$$

Finally, let $L^1(\mathcal{I};\mathbb{R})$ denote the class of $\mathbb{R}$-valued $f\in\mathfrak{M}(\mathcal{I})$ with $\tilde{I}(f)\in\mathbb{R}$.

Lemma 8.2.3 *If $f:E\longrightarrow\mathbb{R}$, then $f\in L^1(\mathcal{I};\mathbb{R})$ if and only if, for each $\epsilon>0$, there exist $\varphi,\psi\in\mathbf{L}_u$ for which $-\varphi\le f\le\psi$ and $I(\varphi)+I(\psi)<\epsilon$. Moreover, $L^1(\mathcal{I};\mathbb{R})$ is a vector lattice and $\tilde{I}$ is linear on $L^1(\mathcal{I};\mathbb{R})$. Finally, if $\{f_n : n\ge 1\}\subseteq L^1(\mathcal{I};\mathbb{R})$ and $0\le f_n\nearrow f$, then $f\in\mathfrak{M}(\mathcal{I})$ and $0\le\tilde{I}(f_n)\nearrow\tilde{I}(f)$.*

Proof. First suppose that $f\in L^1(\mathcal{I};\mathbb{R})$. Given $\epsilon>0$, there exists $(\varphi,\psi)\in\mathbf{L}_u^2$ such that $-\varphi\le f\le\psi$, $\tilde{I}(f)<-I(\varphi)+\frac{\epsilon}{2}$, and $I(\psi)<\tilde{I}(f)+\frac{\epsilon}{2}$, from which it is clear that $I(\varphi)+I(\psi)<\epsilon$. Conversely, suppose that $f:E\longrightarrow\mathbb{R}$ and that, for every $\epsilon\in(0,\infty)$, there exists $(\varphi,\psi)\in\mathbf{L}_u^2$ for which $-\varphi\le f\le\psi$ and $I(\varphi)+I(\psi)<\epsilon$. Because $I(\varphi)\wedge I(\psi)>-\infty$, $-\infty<I(\psi)<\epsilon-I(\varphi)<\infty$ and $-\infty<I(\varphi)<\epsilon-I(\psi)<\infty$. In addition, $-I(\varphi)\le\underline{I}(f)\le\overline{I}(f)\le I(\psi)$. Hence, not only are both $\underline{I}(f)$ and $\overline{I}(f)$ in $\mathbb{R}$, but also $\overline{I}(f)-\underline{I}(f)<\epsilon$, which completes the proof of the first assertion.

Next, suppose that $f \in L^1(\mathcal{I};\mathbb{R})$. To prove that $f^+ \in L^1(\mathcal{I};\mathbb{R})$, let $\epsilon \in (0,\infty)$ be given, and choose $(\varphi,\psi) \in \mathbf{L}_u^2$ such that $-\varphi \le f \le \psi$ and $I(\varphi) + I(\psi) < \epsilon$. Note that $-\varphi^- = \varphi\wedge 0 \in \mathbf{L}_u$, $\varphi^+ = \varphi\vee 0 \in \mathbf{L}_u$, and that $\varphi^- \le f^+ \le \psi^+$. Moreover, because $\varphi + \psi \ge 0$, it is easy to see that $-\varphi^- + \psi^+ \le \varphi + \psi$, and therefore that

$$I(-\varphi^-) + I(\psi^+) = I(-\varphi^- + \psi^+) \le I(\varphi+\psi) = I(\varphi) + I(\psi) < \epsilon.$$

To see that $L^1(\mathcal{I};\mathbb{R})$ is a vector space and that $\tilde{I}$ is linear there, simply apply (8.2.2) and (8.2.4). Finally, let $\{f_n : n \ge 1\}$ be a non-decreasing sequence of non-negative elements of $L^1(\mathcal{I};\mathbb{R})$, and set $f = \lim_{n\to\infty} f_n$. Obviously, $\lim_{n\to\infty} \tilde{I}(f_n) \le \underline{I}(f) \le \overline{I}(f)$. Thus, all that we have to do is prove that $\overline{I}(f) \le \lim_{n\to\infty} \tilde{I}(f_n)$. To this end, set $h_1 = f_1$, $h_n = f_n - f_{n-1}$ for $n \ge 2$, and note that each h_n is a non-negative element of $L^1(\mathcal{I};\mathbb{R})$. Next, given $\epsilon > 0$, choose, for each $m \in \mathbb{Z}^+$, $\psi_m \in \mathbf{L}_u$ such that $h_m \le \psi_m$ and $I(\psi_m) \le \tilde{I}(h_m) + 2^{-m}\epsilon$. Clearly, $\psi \equiv \sum_1^\infty \psi_m \in \mathbf{L}_u$ and $f \le \psi$. Thus, $\overline{I}(f) \le I(\psi)$. Moreover, by Lemma 8.2.1 and the linearity of $\tilde{I}$ on $L^1(\mathcal{I};\mathbb{R})$,

$$\begin{aligned} I(\psi) &= \lim_{n\to\infty} I\left(\sum_1^n \psi_m\right) = \lim_{n\to\infty} \sum_1^n I(\psi_m) \\ &\le \lim_{n\to\infty} \sum_1^n \tilde{I}(h_m) + \epsilon = \lim_{n\to\infty} \tilde{I}(f_n) + \epsilon. \end{aligned}$$

□

Theorem 8.2.4 (Daniell) *Let $\mathcal{I} = (E,\mathbf{L},I)$ be an integration theory. Then $\tilde{\mathcal{I}} = \big(E, L^1(\mathcal{I};\mathbb{R}), \tilde{I}\big)$ is again an integration theory, $\mathbf{L} \subseteq L^1(\mathcal{I};\mathbb{R})$, and $\tilde{I}$ agrees with I on $\mathbf{L}$. Moreover, if $\{f_n : n \ge 1\} \subseteq L^1(\mathcal{I};\mathbb{R})$ is non-decreasing and $f_n \nearrow f$, then $f \in L^1(\mathcal{I};\mathbb{R})$ if and only if $\sup_n f(x) < \infty$ for each $x \in E$ and $\sup_n \tilde{I}(f_n) < \infty$, in which case $\tilde{I}(f_n) \nearrow \tilde{I}(f)$.*

Proof. By Lemma 8.2.3, $L^1(\mathcal{I};\mathbb{R})$ is a vector lattice and $\tilde{I}$ is linear there. Moreover, by (8.2.5), $\mathbf{L} \subseteq L^1(\mathcal{I};\mathbb{R})$ and $\tilde{I} \restriction \mathbf{L} = I \restriction \mathbf{L}$; and, by (8.2.3), $\tilde{I}(f) \ge I(\mathbf{0}) = 0$ if $f \in L^1(\mathcal{I};\mathbb{R})$ is non-negative. In addition, if $f_n \nearrow f$ for some $\{f_n : n \ge 1\} \subseteq L^1(\mathcal{I};\mathbb{R})$, then, by the last part of Lemma 8.2.3 applied to $\{f_n - f_1 : n \ge 1\}$, $f - f_1 \in \mathfrak{M}(\mathcal{I})$ and

$$\tilde{I}(f_n) = \tilde{I}(f_1) + \tilde{I}(f_n - f_1) \nearrow \tilde{I}(f_1) + \tilde{I}(f - f_1).$$

Since $f = f_1 + (f - f_1) \in \mathfrak{M}(\mathcal{I})$ and, by (8.2.4), $\tilde{I}(f) = \tilde{I}(f_1) + \tilde{I}(f - f_1)$, the last assertion is now proved.

Finally, if $\{f_n : n \ge 1\} \subseteq L^1(\mathcal{I};\mathbb{R})$ and $f_n \searrow 0$, then $\tilde{I}(f_n) \searrow 0$ follows from the preceding applied to $\{-f_n : n \ge 1\}$, and this completes the proof that $\tilde{\mathcal{I}}$ is an integration theory. □

8.2.2 Identification of the Measure

I am now ready to return to the problem, raised in the discussion in the introduction to this section, of identifying the measure underlying a given integration theory $(E, \mathbf{L}, I)$. For this purpose, I introduce the notation $\sigma(\mathbf{L})$ to denote the smallest σ-algebra over E with respect to which all of the functions in the vector lattice $\mathbf{L}$ are measurable. Obviously, $\sigma(\mathbf{L})$ is generated by the sets $\{f > a\}$ as f runs over $\mathbf{L}$ and a runs over $\mathbb{R}$.

Theorem 8.2.5 (**Stone**) *Let $(E, \mathbf{L}, I)$ be an integration theory, and assume that $\mathbf{1} \in \mathbf{L}$. Then*

$$\sigma\big(L^1(\mathcal{I};\mathbb{R})\big) = \big\{\Gamma \subseteq E : \mathbf{1}_\Gamma \in L^1(\mathcal{I};\mathbb{R})\big\},$$

the mapping

$$\Gamma \in \sigma\big(L^1(\mathcal{I};\mathbb{R})\big) \longmapsto \mu_I(\Gamma) \equiv \tilde{I}\big(\mathbf{1}_\Gamma\big) \in [0,\infty)$$

is a finite measure on $\big(E, \sigma(L^1(\mathcal{I};\mathbb{R}))\big)$, $\sigma\big(L^1(\mathcal{I};\mathbb{R})\big)$ is the completion of $\sigma(\mathbf{L})$ with respect to μ_I, $L^1(\mu_I;\mathbb{R}) = L^1(\mathcal{I};\mathbb{R})$, and

$$\tilde{I}(f) = \int_E f\, d\mu_I \quad \text{for } f \in L^1(\mathcal{I};\mathbb{R}).$$

Finally, if $\big(E, \mathcal{B}, \nu\big)$ is any finite measure space with the properties that $\mathbf{L} \subseteq L^1(\nu;\mathbb{R})$ and $I(f) = \int_E f\, d\nu$ for every $f \in \mathbf{L}$, then $\sigma(\mathbf{L}) \subseteq \mathcal{B}$ and ν coincides with μ_I on $\sigma(\mathbf{L})$.

Proof. Let $\mathcal{H}$ denote the collection of sets described on the right-hand side of the first equation. Using Theorem 8.2.4, one can easily show that $\mathcal{H}$ is a σ-algebra over E and that

$$\Gamma \in \mathcal{H} \longmapsto \mu_I(\Gamma) \equiv \tilde{I}\big(\mathbf{1}_\Gamma\big) \in [0,\infty)$$

defines a finite measure on $(E, \mathcal{H})$. Our first goal is to prove that

$$L^1(\mathcal{I};\mathbb{R}) = L^1(\mu_I;\mathbb{R}) \quad \text{and} \quad \tilde{I}(f) = \int_E f\, d\mu_I \text{ for all } f \in L^1(\mathcal{I};\mathbb{R}). \qquad (*)$$

To this end, for given $f : E \longrightarrow \mathbb{R}$ and $a \in \mathbb{R}$, consider the functions

$$g_n \equiv \big[n(f - f \wedge a)\big] \wedge \mathbf{1}, \quad n \in \mathbb{Z}^+.$$

If $f \in L^1(\mathcal{I};\mathbb{R})$, then each g_n is also an element of $L^1(\mathcal{I};\mathbb{R})$, $g_n \nearrow \mathbf{1}_{\{f>a\}}$ as $n \longrightarrow \infty$, and therefore $\mathbf{1}_{\{f>a\}} \in L^1(\mathcal{I};\mathbb{R})$. Thus we see that every $f \in L^1(\mathcal{I};\mathbb{R})$ is $\mathcal{H}$-measurable. Next, for given $f : E \longrightarrow [0,\infty)$, define

$$f_n = \sum_{k=0}^{4^n} \frac{k}{2^n} \mathbf{1}_{\{k<2^n f \le k+1\}} \quad \text{for} \quad n \in \mathbb{Z}^+.$$

If $f \in L^1(\mathcal{I};\mathbb{R}) \cup L^1(\mu_I;\mathbb{R})$, then (cf. the preceding and use linearity) $f_n \in L^1(\mathcal{I};\mathbb{R}) \cap L^1(\mu_I;\mathbb{R})$, $f_n \nearrow f$, and so $f \in L^1(\mathcal{I};\mathbb{R}) \cap L^1(\mu_I;\mathbb{R})$ and $\tilde{I}(f) = \int_E f\,d\mu_I$. Hence, we have now proved (∗).

Our next goal is to show that $\overline{\sigma(\mathbf{L})}^{\mu_I} = \mathcal{H} = \sigma\big(L^1(\mathcal{I};\mathbb{R})\big)$. Since $\mathbf{L} \subseteq L^1(\mathcal{I};\mathbb{R})$ and every element of $L^1(\mathcal{I};\mathbb{R})$ is $\mathcal{H}$-measurable, what we know so far is that $\sigma(\mathbf{L}) \subseteq \sigma\big(L^1(\mathcal{I};\mathbb{R})\big) \subseteq \mathcal{H}$. In addition, by the first part of Lemma 8.2.3, it is clear that $\mathcal{H} = \overline{\mathcal{H}}^{\mu_I}$. Thus, all we need to do is show that

$$\mathcal{H} \subseteq \overline{\sigma(\mathbf{L})}^{\mu_I}. \tag{$**$}$$

But if $\Gamma \in \mathcal{H}$, then we can choose $\{\varphi_n : n \ge 1\}$ and $\{\psi_n : n \ge 1\}$ from $\mathbf{L}_u \cap L^1(\mathcal{I};\mathbb{R})$ such that $-\varphi_n \le \mathbf{1}_\Gamma \le \psi_n$, $-I(\varphi_n) \nearrow \tilde{I}(\mathbf{1}_\Gamma)$, and $I(\psi_n) \searrow \tilde{I}(\mathbf{1}_\Gamma)$. Further, after replacing φ_n and ψ_n by $\varphi_1 \wedge \cdots \wedge \varphi_n$ and $\psi_1 \wedge \cdots \wedge \psi_n$, we may and will assume that each of these sequences is non-increasing. Next, take $\varphi = \lim_{n\to\infty} \varphi_n$ and $\psi = \lim_{n\to\infty} \psi_n$. Then φ and ψ are in $L^1(\mathcal{I};\mathbb{R})$, $-\varphi \le \mathbf{1}_\Gamma \le \psi$, and $-\tilde{I}(\varphi) = \tilde{I}(\psi)$. To complete the proof that $\Gamma \in \overline{\sigma(\mathbf{L})}^{\mu_I}$ from here, first observe that every element of $\mathbf{L}_u$ is $\sigma(\mathbf{L})$-measurable and therefore that φ and ψ are also. Hence, both the sets $C = \{\varphi < 0\}$ and $D = \{\psi \ge 1\}$ are elements of $\sigma(\mathbf{L})$. Finally, from $-\varphi \le \mathbf{1}_\Gamma \le \psi$ and $-\tilde{I}(\varphi) = \tilde{I}(\psi)$, it is easy to check that $-\varphi \le \mathbf{1}_C \le \mathbf{1}_\Gamma \le \mathbf{1}_D \le \psi$ and therefore that $\mu_I(D \setminus C) = \mu_I(D) - \mu_I(C) \le \tilde{I}(\psi) + \tilde{I}(\varphi) = 0$, which means that $\Gamma \in \overline{\sigma(\mathbf{L})}^{\mu_I}$. Thus, we have completed the proof of (∗∗) and therefore of the desired equalities.

We have now completed the proof of everything except the concluding assertion about uniqueness. But if $\mathbf{L} \subseteq L^1(\nu;\mathbb{R})$, then obviously $\sigma(\mathbf{L}) \subseteq \mathcal{B}$. Moreover, if $\int f\,d\nu = I(f)$ for all $f \in \mathbf{L}$, then one can prove that $\nu \restriction \sigma(\mathbf{L}) = \mu_I \restriction \sigma(\mathbf{L})$ as follows. It is clear that $\sigma(\mathbf{L})$ is generated by the Π-system of sets Γ of the form $\Gamma = \{f_1 > a_1, \dots, f_\ell > a_\ell\}$, where $\ell \in \mathbb{Z}^+$, $\{a_1, \dots, a_\ell\} \subseteq \mathbb{R}$, and $\{f_1, \dots, f_\ell\} \subseteq \mathbf{L}$. Thus, since $\nu(E) = \mu_I(E)$, Theorem 2.1.2 says that we need only check that ν agrees with μ_I on such sets Γ. To this end, define

$$g_n = \left[n \min_{1\le k\le \ell} (f_k - f_k \wedge a_k) \right] \wedge 1,$$

note that $\{g_n : n \ge 1\} \subseteq \mathbf{L}$ and $0 \le g_n \nearrow \mathbf{1}_\Gamma$, and conclude that

$$\nu(\Gamma) = \lim_{n\to\infty} \int g_n\,d\nu = \lim_{n\to\infty} \int g_n\,d\mu_I = \mu_I(\Gamma).$$

□

8.2.3 An Extension Theorem

With the preceding results at hand, I can now treat the situations described in examples (**ii**) and (**iii**) of Example 8.2.1 in the introduction to this section. In this subsection I will prove and apply the following extension criterion, which handles (**ii**).

Theorem 8.2.6 *Let $\mathcal{A}$ be an algebra of subsets of E, and suppose that $\mu : \mathcal{A} \longrightarrow [0,\infty)$ is a finitely additive measure with the property that (8.2.1) holds. Then there is a unique finite measure, again denoted by μ, on $\big(E, \sigma(\mathcal{A})\big)$ that is an extension of μ from $\mathcal{A}$.*

Proof. Define $\mathbf{L}(\mathcal{A})$ and I on $\mathbf{L}(\mathcal{A})$ as in (**ii**) of the Example 8.2.1. As was asserted there, I is an integral on $\mathbf{L}(\mathcal{A})$. In addition, it is an easy matter to see that $\sigma(\mathcal{A}) = \sigma\big(\mathbf{L}(\mathcal{A})\big)$. Hence the desired existence and uniqueness statements follow immediately from Theorem 8.2.5. □

To demonstrate how this criterion gets applied, I will use it to carry out a general construction that plays an important role in probability theory, where it is interpreted as guaranteeing the simultaneous existence of infinitely many independent random variables. The setting is as follows. For each i from a non-empty index set $\mathcal{I}$, $(E_i, \mathcal{B}_i, \mu_i)$ is a probability space. Given $\emptyset \neq S \subseteq \mathcal{I}$, set $E_S = \prod_{i\in S} E_i$, and use $\varPi_S$ to denote the natural projection map from $E \equiv E_{\mathcal{I}}$ onto E_S, the one that takes $x \in E_{\mathcal{I}}$ to $x \restriction S$. Finally, let $\mathfrak{F}$ stand for the set of all non-empty, finite subsets F of $\mathcal{I}$, and denote by $\mathcal{B}_{\mathcal{I}}$ the σ-algebra over E generated by sets of the form

$$\varGamma_F \equiv \varPi_F^{-1}\left(\prod_{i\in F} \varGamma_i\right), \quad F \in \mathfrak{F} \text{ and } \varGamma_i \in \mathcal{B}_i \text{ for } i \in F.$$

The goal is to show there is a unique probability measure $\mu \equiv \prod_{i\in\mathcal{I}} \mu_i$ on $(E, \mathcal{B}_{\mathcal{I}})$ with the property that

$$\mu(\varGamma_F) = \prod_{i\in F} \mu_i(\varGamma_i) \tag{8.2.6}$$

for all choices of $F \in \mathfrak{F}$ and $\varGamma_i \in \mathcal{B}_i$, $i \in F$.

Begin by observing that uniqueness is clear. Indeed, the collection that generates $\mathcal{B}_{\mathcal{I}}$ is obviously a $\varPi$-system that contains E, and therefore, by Theorem 2.1.2, the condition in (8.2.6) can be satisfied by at most one measure. Secondly, observe that there is no problem when $\mathcal{I}$ is finite. In fact, when $\mathcal{I}$ has only one element there is nothing to do at all. Moreover, if we know how to handle $\mathcal{I}$'s containing $n \in \mathbb{Z}^+$ elements and $\mathcal{I} = \{i_1, \dots, i_{n+1}\}$, then we can take $\prod_1^{n+1} \mu_{i_m}$ to be the image of

$$\left(\prod_1^n \mu_{i_m}\right) \times \mu_{i_{n+1}}$$

under the mapping

$$\big((x_{i_1},\dots,x_{i_n}),x_{i_{n+1}}\big) \in \left(\prod_1^n E_{i_m}\right) \times E_{i_{n+1}} \longmapsto \big(x_{i_1},\dots,x_{i_{n+1}}\big) \in \prod_1^{n+1} E_{i_m}.$$

Therefore, we know how to construct μ when $\mathcal{I}$ is finite.

Now assume that $\mathcal{I}$ is infinite. Given $F \in \mathfrak{F}$, use μ_F to denote $\prod_{i\in F}\mu_i$. In order to construct μ, introduce the algebra

$$\mathcal{A} \equiv \bigcup_{F\in\mathfrak{F}} \Pi_F^{-1}(\mathcal{B}_F), \quad \text{where } \mathcal{B}_F \equiv \prod_{i\in F}\mathcal{B}_i,$$

and note that $\mathcal{B}_{\mathcal{I}}$ is generated by $\mathcal{A}$. Next, observe that the map $\mu : \mathcal{A} \longrightarrow [0,1]$ given by

$$\mu(A) = \mu_F\big(\Pi_F^{-1}(\Gamma)\big) \quad \text{if } A = \Pi_F^{-1}(\Gamma) \text{ for some } F \in \mathfrak{F} \text{ and } \Gamma \in \mathcal{B}_F$$

is well-defined and finitely additive. To see the first of these, suppose that, for some $\Gamma \in \mathcal{B}_F$ and $\Gamma' \in \mathcal{B}_{F'}$, $\Pi_F^{-1}(\Gamma) = \Pi_{F'}^{-1}(\Gamma')$. If $F = F'$, then it is clear that $\Gamma = \Gamma'$ and that there is no problem. On the other hand, if $F \subsetneq F'$, then $\Gamma' = \Gamma \times E_{F'\setminus F}$ and therefore, since the μ_i's are probability measures, that

$$\mu_{F'}(\Gamma') = \mu_F(\Gamma) \prod_{i\in F'\setminus F} \mu_i(E_i) = \mu_F(\Gamma).$$

Thus, μ is well-defined on $\mathcal{A}$. Moreover, given disjoint elements A and A' of $\mathcal{A}$, choose an $F \in \mathcal{I}$ for which $A,\ A' \in \Pi_F^{-1}(\mathcal{B}_F)$, note that $\Gamma = \Pi_F(A)$ is disjoint from $\Gamma' = \Pi_F(A')$, and conclude that

$$\mu(A \cup A') = \mu_F(\Gamma \cup \Gamma') = \mu_F(\Gamma) + \mu_F(\Gamma') = \mu(A) + \mu(A').$$

In view of the preceding paragraph and Theorem 8.2.6, all that remains is to show that if $\{A_n : n \ge 1\}$ is a non-increasing sequence from $\mathcal{A}$ and $\bigcap_{n=1}^\infty A_n = \emptyset$, then $\mu(A_n) \searrow 0$. Equivalently, what we need to show is that if $\{A_n : n \ge 1\} \subseteq \mathcal{A}$ is non-increasing and, for all $n \in \mathbb{Z}^+$ and some $\epsilon > 0$, $\mu(A_n) \ge \epsilon$, then $\bigcap_{n=1}^\infty A_n \ne \emptyset$. Thus, suppose that such a sequence is given, and choose $\{F_n : n \ge 1\}$ such that $A_n \in \Pi_{F_n}^{-1}(\mathcal{B}_{F_n})$ for each $n \in \mathbb{Z}^+$. Without loss of generality, assume that there exists a sequence $\{i_n : n \ge 1\}$ of distinct elements of $\mathcal{I}$ such that $F_n = \{i_1,\dots,i_n\}$ for each $n \in \mathbb{Z}^+$. Thus, under the condition that $\mu(A_n) \ge \epsilon > 0$ for all $n \in \mathbb{Z}^+$, we must produce a sequence $\{a_{i_n} : n \ge 1\}$ such that $a_{i_n} \in E_{i_n}$ and $(a_{i_1},\dots,a_{i_n}) \in \Pi_{F_n}(A_n)$ for each $n \in \mathbb{Z}^+$. Indeed, given such a sequence $\{a_{i_n} : n \ge 1\}$, one can construct an element a of $\bigcap_{n=1}^{\infty} A_n$ by determining a by $\Pi_{F_n} a = (a_{i_1}, \dots, a_{i_n})$ for each $n \in \mathbb{Z}^+$ and, if $S \equiv \mathcal{I} \setminus \bigcup_{n=1}^{\infty} F_n \neq \emptyset$, $\Pi_S a$ to be any element of E_S.

To find the a_{i_m}'s, first choose and fix a reference point $e_i \in E_i$ for each $i \in \mathcal{I}$. Next, for each $n \in \mathbb{Z}^+$, define $\Phi_n : E_{F_n} \longrightarrow E$ so that

$$\Phi_n(x_{F_n})_i = \begin{cases} x_i & \text{if } i \in F_n \\ e_i & \text{if } i \in \mathcal{I} \setminus F_n, \end{cases}$$

and set $f_n = \mathbf{1}_{A_n} \circ \Phi_n$. Obviously,

$$\epsilon \leq \mu(A_n) = \int_{E_{F_n}} f_n(x_{F_n})\, \mu_{F_n}(dx_{F_n}).$$

Furthermore, if, for each $m \in \mathbb{Z}^+$, we define the sequence $\{g_{m,n} : n \geq m\}$ of functions on E_{F_m} by $g_{m,m}(x_{F_m}) = f_m(x_{F_m})$ and

$$g_{m,n}(x_{F_m}) = \int_{E_{F_n \setminus F_m}} f_n\big((x_{F_m}, y_{F_n \setminus F_m})\big)\, \mu_{F_n \setminus F_m}(dy_{F_n \setminus F_m})$$

when $n > m$, then $g_{m,n+1} \leq g_{m,n}$ and

$$g_{m,n}(x_{F_m}) = \int_{E_{i_{m+1}}} g_{m+1,n}\big((x_{F_m}, y_{i_{m+1}})\big)\, \mu_{\{i_{m+1}\}}(dy_{i_{m+1}})$$

for all $1 \leq m < n$. Hence, $g_m \equiv \lim_{n\to\infty} g_{m,n}$ exists and, by the monotone convergence theorem,

$$g_m(x_{F_m}) = \int_{E_{i_{m+1}}} g_{m+1}\big((x_{F_m}, y_{i_{m+1}})\big)\, \mu_{\{i_{m+1}\}}(dy_{i_{m+1}}). \tag{$*$}$$

Finally, since

$$\int_{E_{i_1}} g_1(x_{i_1})\, \mu_{\{i_1\}}(dx_{i_1}) = \lim_{n\to\infty} \int_{E_{F_n}} f_n(x_{F_n})\, \mu_{F_n}(dx_{F_n}) = \lim_{n\to\infty} \mu(A_n) \geq \epsilon,$$

there exists an $a_{i_1} \in E_{F_1}$ such that $g_1(a_{i_1}) \geq \epsilon$. In particular (since $g_1 \leq f_1$), this means that $a_{i_1} \in \Pi_{F_1}(A_1)$. In addition, from $(*)$ with $m = 1$ and $x_{F_1} = a_{i_1}$, it means that there exists an $a_{i_2} \in E_{i_2}$ for which $g_2(a_{i_1}, a_{i_2}) \geq \epsilon$, and therefore (since $g_2 \leq f_2$) $(a_{i_1}, a_{i_2}) \in \Pi_{F_2}(A_2)$. More generally, if $(a_{i_1}, \dots, a_{i_m}) \in E_{F_m}$ and $g_m(a_{i_1}, \dots, a_{i_m}) \geq \epsilon$, then (since $g_m \leq f_m$) $(a_{i_1}, \dots, a_{i_m}) \in \Pi_{F_m}(A_m)$, and, by $(*)$, there exists an $a_{i_{m+1}} \in E_{F_{i_{m+1}}}$ for which $g_{m+1}(a_{i_1}, \dots, a_{i_{m+1}}) \geq \epsilon$. Hence, by induction, we are done and have proved the following theorem.

Theorem 8.2.7 *Referring to the setting above, there exists a unique probability measure* $\mu = \prod_{i\in\mathcal{I}} \mu_i$ *on* $(E, \mathcal{B}_{\mathcal{I}})$ *for which* (8.2.6) *holds.*

8.2.4 Another Riesz Representation Theorem

Part (**iii**) of Example 8.2.1 is another of the representation theorems for which F. Riesz is famous, and it is the one for which he developed the ideas that led to Daniell's theory of integration. In order to prove it as an application of Theorem 8.2.4, I must first prove the lemma of Dini alluded to earlier.

Lemma 8.2.8 (Dini's Lemma) *Let $\{f_n : n \geq 1\}$ be a non-increasing sequence of non-negative, continuous functions on the topological space E. If $f_n \searrow 0$ pointwise, then $f_n \longrightarrow 0$ uniformly on each compact subset $K \subset\subset E$*

Proof. Without loss of generality, assume that E itself is compact.

Let $\epsilon > 0$ be given. By assumption, we can find for each $x \in E$ an $n(x) \in \mathbb{Z}^+$ and an open neighborhood $U(x)$ of x such that $f_n(y) \leq \epsilon$ for all $y \in U(x)$ and $n \geq n(x)$. At the same time, by the Heine-Borel Theorem, we can choose a finite set $\{x_1, \dots, x_L\} \subseteq E$ for which $E = \bigcup_{\ell=1}^L U(x_\ell)$. Thus, if $N(\epsilon) = n(x_1) \vee \cdots \vee n(x_L)$, then $f_n \leq \epsilon$ whenever $n \geq N(\epsilon)$. □

Given a topological space E, let $C_b(E;\mathbb{R})$ denote the space of bounded continuous functions on E, and turn $C_b(E;\mathbb{R})$ into a metric space by defining $\|g - f\|_u$ to be the distance between f and g. I will say that $\Lambda : C_b(E;\mathbb{R}) \longrightarrow \mathbb{R}$ is a **non-negative linear functional** if Λ is linear and $\Lambda(f) \geq 0$ for all $f \in C_b\big(E;[0,\infty)\big)$. Observe that, because $\|f\|_u \mathbf{1} \pm f \geq 0$, $|\Lambda(f)| \leq \|f\|_u \Lambda(\mathbf{1})$. Finally, if Λ is a non-negative linear functional on $C_b(E;\mathbb{R})$, say that Λ is **tight** if it has the property that, for every $\delta > 0$, there exist a compact $K_\delta \subset\subset E$ and an $A_\delta \in (0,\infty)$ such that

$$|\Lambda(f)| \leq A_\delta \sup_{x \in K_\delta} |f(x)| + \delta \|f\|_u \quad \text{for all } f \in C_b(E;\mathbb{R}).$$

Notice that when E is itself compact, every non-negative linear functional on $C_b(E;\mathbb{R})$ is tight.

Theorem 8.2.9 (Riesz Representation) *Let E be a topological space, set $\mathcal{B} = \sigma\big(C_b(E;\mathbb{R})\big)$, and suppose that $\Lambda : C_b(E;\mathbb{R}) \longrightarrow \mathbb{R}$ is a non-negative linear functional that is tight. Then there is a unique finite measure μ on $(E,\mathcal{B})$ with the property that $\Lambda(f) = \int_E f\,d\mu$ for all $f \in C_b(E;\mathbb{R})$.*

Proof. Clearly, all that we need to do is show that $\Lambda(f_n) \searrow 0$ whenever $\{f_n : n \geq 1\} \subseteq C_b(E;\mathbb{R})$ is a non-increasing sequence that tends (pointwise) to 0. To this end, let $\epsilon > 0$ be given, set $\delta = \frac{\epsilon}{1+2\|f_1\|_u}$, and use Dini's Lemma to choose an $N(\delta) \in \mathbb{Z}^+$ for which $|f(x)| \leq \frac{\epsilon}{2A_\delta}$ whenever $n \geq N(\delta)$ and $x \in K_\delta$, where K_δ and A_δ are the quantities appearing in the tightness condition for Λ. Then, $\big|\Lambda(f_n)\big| \leq \epsilon$ for $n \geq N(\delta)$. □

Remark 8.2.1 Although the preceding theorem contains the essence of what functional analysts call the Riesz Representation Theorem, it differs from the usual statement in two ways. In the functional analytic context, Riesz's theorem is thought of as a description of the dual space (cf. Exercise 8.1.5) $C(E;\mathbb{R})^*$ of the Banach space $C(E;\mathbb{R})$ when E is compact. Because the dual space contains all continuous, linear functionals, not just the non-negative ones, the usual statement is that, for compact E's, $C(E;\mathbb{R})^*$ can be identified with the space of finite signed measures on $\sigma\big(C(E;\mathbb{R})\big)$. Thus, to get this statement from Theorem 8.2.9, one has to show that every continuous linear functional on $C(E;\mathbb{R})$ can be decomposed into the difference of two non-negative ones. (Theorem 8.1.3 and Exercise 8.1.5 are closely related to a special case of this decomposition.) A second remark that should be made is that $\sigma\big(C(E;\mathbb{R})\big)$, which is called the **Baire σ-algebra**, does not always coincide with the Borel σ-algebra $\mathcal{B}_E$. It is always contained in the Borel σ-algebra, but it may be strictly smaller. On the other hand, if the topology on E admits a metric ρ, then it is an easy matter to show that the Baire and Borel σ-algebras coincide. Indeed, for any open set $G, G = \{x : \rho(x, G^{\complement}) > 0\}$ is a Baire set.

8.2.5 Lévy Continuity and Bochner's Theorems

The Fourier transform of a Borel probability measures on $\mathbb{R}^N$ was introduced in § 7.4, but, as yet, the question of which functions on $\mathbb{R}^N$ are the Fourier transform of such a measure is open. Here I will present S. Bochner's answer to that question, and, along the way, prove an important result due to P. Lévy.

The first step is taken in the following corollary of Theorem 8.2.9.

Lemma 8.2.10 *Let Λ be a linear functional on $C_c^\infty(\mathbb{R}^N;\mathbb{R})$ with the properties that it is non-negative (i.e., $\varphi \geq 0 \implies \Lambda(\varphi) \geq 0$) and $\Lambda(\varphi_n) \nearrow 1$ if $0 \leq \varphi_n \nearrow 1$. Then there is a unique Borel probability measure μ on $\mathbb{R}^N$ such that*

$$\Lambda(\varphi) = \int \varphi \, d\mu \text{ for all } \varphi \in C_c^\infty(\mathbb{R}^N;\mathbb{R}).$$

Proof. Begin by observing that $\Lambda(\varphi) \leq \Lambda(\psi)$ if $\varphi \leq \psi$.

Choose a function $\rho \in C^\infty\big(\mathbb{R}^N;[0,\infty)\big)$ with the properties that $\rho(x) = 0$ if $|x| \geq \frac{1}{2}$ and $\int \rho d\lambda_{\mathbb{R}^N} = 1$, and let $\eta \in C_c^\infty\big(\mathbb{R}^N;[0,1]\big)$ be the function given by

$$\eta(x) = \int_{|y|\leq\frac{3}{2}} \rho(x-y)\,dy.$$

Clearly, $\eta \in [0,1]$ everywhere, $\eta = 1$ on $\overline{B_{\mathbb{R}^N}(0,1)}$, and $\eta = 0$ off $B_{\mathbb{R}^N}(0,2)$. For $n \geq 1$, define ρ_n and η_n so that $\rho_n(x) = n\rho(nx)$ and $\eta_n(x) = \eta\big(2^{-n}x\big)$, and check that $\eta_n = 1$ on $\overline{B_{\mathbb{R}^N}(0,2^n)}$ and $\eta_n = 0$ off $B_{\mathbb{R}^N}(0,2^{n+1})$. In particular, this means that $\eta_n \nearrow 1$.

The first step is to extend Λ as a non-negative linear function on $C_c(\mathbb{R}^N;\mathbb{R})$. To this end, let $\varphi \in C_c(\mathbb{R}^N;\mathbb{R})$ be given, and set $\varphi_n = \rho_n * \varphi$. Then $\varphi_n \in C_c^\infty(\mathbb{R}^N;\mathbb{R})$, $\|\varphi_n\|_u \le \|\varphi\|_u$, and, by Theorem 6.3.7, $\|\varphi_n - \varphi\|_u \longrightarrow 0$. Thus

$$\begin{aligned}\sup_{n>m}|\Lambda(\varphi_n) - \Lambda(\varphi_m)| &= \sup_{n>m}|\Lambda(\varphi_n - \varphi_m)| \\ &\le \sup_{n>m}\|\varphi_n - \varphi_m\|_u \le 2\sup_{n\ge m}\|\varphi_n - \varphi\|_u \longrightarrow 0 \text{ as } m \to \infty.\end{aligned}$$

Therefore $\{\Lambda(\varphi_n) : n \ge 1\}$ converges to a limit $\Lambda(\varphi)$, and $|\Lambda(\varphi)| \le \|\varphi\|_u$. In addition, it should be obvious that this extension of Λ is both linear and non-negative.

The next step is to extend Λ as a non-negative, linear functional on $C_b(\mathbb{R}^N;\mathbb{R})$ with the property that $\Lambda(\mathbf{1}) = 1$ and therefore that $|\Lambda(\varphi)| \le \|\varphi\|_u$. Namely, given $\varphi \in C_b(\mathbb{R}^N;\mathbb{R})$, set $\varphi_n = \eta_n\varphi$. Then, for $1 \le m < n$,

$$|\Lambda(\varphi_n) - \Lambda(\varphi_m)| = |\Lambda(\varphi_n - \varphi_m)| \le \|\varphi\|_u\Lambda(\eta_n - \eta_m) \le \|\varphi\|_u\big(1 - \Lambda(\eta_m)\big).$$

Since $\Lambda(\eta_m) \nearrow 1$, this proves that $\Lambda(\varphi) \equiv \lim_{n\to\infty}\Lambda(\varphi_n)$ exists, and it is easy to check that $\varphi \in C_b(\mathbb{R}^N;\mathbb{R}) \longmapsto \Lambda(\varphi) \in \mathbb{R}$ is a non-negative linear functional for which $\Lambda(\mathbf{1}) = 1$.

The final step is to show that Λ is tight. But, given $\delta > 0$, choose $n \ge 1$ so that $1 - \Lambda(\eta_n) < \delta$, and set $K_n = \overline{B_{\mathbb{R}^N}(0, 2^{n+1})}$. Then

$$\begin{aligned}|\Lambda(\varphi)| &\le |\Lambda(\eta_n\varphi)| + \big|\Lambda\big((1 - \eta_n)\varphi\big)\big| \le \sup_{x\in K_n}|\varphi(x)| + \|\varphi\|_u\big(1 - \Lambda(\eta_n)\big) \\ &\le \sup_{x\in K}|\varphi(x)| + \delta\|\varphi\|_u.\end{aligned}$$

Knowing that Λ is tight, apply Theorem 8.2.9 to complete the proof. □

Lemma 8.2.11 *Let $\{\mu_n : n \ge 1\}$ be a sequence of Borel probability measures on $\mathbb{R}^N$, and assume that*

$$\Lambda(\varphi) \equiv \lim_{n\to\infty}\int \varphi\, d\mu_n$$

exists for all $\varphi \in C_c^\infty(\mathbb{R}^N;\mathbb{R})$. If in addition

$$\lim_{R\to\infty}\sup_{n\ge 1}\mu_n\big(B_{\mathbb{R}^N}(0,R)^\complement\big) = 0,$$

then there is a Borel probability measure μ to which $\{\mu_n : n \ge 1\}$ converges weakly.

Proof. By Lemma 7.4.2, it suffices to show that there is a μ for which $\Lambda(\varphi) = \int \varphi\, d\mu$. Since it is clear that Λ is a non-negative linear functional on $C_c^\infty(\mathbb{R}^N;\mathbb{R})$ satisfying $|\Lambda(\varphi)| \le \|\varphi\|_u$, Lemma 8.2.10 says that it suffices to show that $\Lambda(\varphi_k) \nearrow 1$ if $0 \le \varphi_k \nearrow 1$. But, by Lemma 8.2.8, if $\varphi_k \nearrow 1$, then it does so uniformly on compacts. Now let $\epsilon > 0$, and choose $R > 0$ so that $\sup_{n\ge1} \mu_n\big(B_{\mathbb{R}^N}(0,R)^{\complement}\big) < \epsilon$ and k so that $\sup\{1-\varphi_k(x) : x \in \overline{B_{\mathbb{R}^N}(0,R)}\} < \epsilon$. Then

$$1 - \int \varphi_\ell \, d\mu_n \le \int (1-\varphi_k)\, d\mu_n \le 2\epsilon$$

for all $\ell \ge k$ and $n \ge 1$, from which it follows that $\Lambda(\varphi_k) \nearrow 1$. □

The following lemma provides a connection between the second hypothesis in Lemma 8.2.11 and the Fourier transform.

Lemma 8.2.12 *If ν is a Borel probability measure on $\mathbb{R}^N$, then*

$$\nu\big(B_{\mathbb{R}^N}(0,r^{-1})^{\complement}\big) \le 2N \sup_{|\xi|\le N^{\frac12} r} |1-\hat\nu(\xi)|$$

for all $r > 0$.

Proof. Note that

$$|1-\hat\nu(\xi)| \ge \mathfrak{Re}\big(1-\hat\nu(\xi)\big) = \int \Big(1-\cos\big(2\pi(\xi,x)_{\mathbb{R}^N}\big)\Big)\,\nu(dx).$$

Thus, for any $\mathbf{e} \in \mathbb{S}^{N-1}$,

$$\begin{aligned}\sup_{|\xi|\le r} |1-\hat\nu(\xi)| &\ge \frac{1}{r}\int_{[0,r]} |1-\hat\nu(t\mathbf{e})|\,dt \\ &\ge \int \left(1-\frac{\sin\big(2\pi r(\mathbf{e},x)_{\mathbb{R}^N}\big)}{2\pi r(\mathbf{e},x)_{\mathbb{R}^N}}\right)\nu(dx) \ge \left(1-\frac{1}{2\pi}\right)\nu\big(\{x : |(\mathbf{e},x)_{\mathbb{R}^N}| \ge r^{-1}\}\big).\end{aligned}$$

Since

$$\nu\big(B_{\mathbb{R}^N}(0,r^{-1})^{\complement}\big) \le N \sup_{\mathbf{e}\in\mathbb{S}^{N-1}} \nu\big(\{x : |(\mathbf{e},x)_{\mathbb{R}^N}| \ge (N^{\frac12} r)^{-1}\}\big),$$

the required estimate follows. □

Theorem 8.2.13 (Lévy's Continuity Theorem) *Let $\{\mu_n : n \ge 1\}$ be a sequence of Borel probability measures on $\mathbb{R}^N$, and assume that $\widehat{\mu_n}(\xi) \longrightarrow f(\xi)$ for each $\xi \in \mathbb{R}^N$. Then $f = \hat\mu$ for some Borel probability measure μ if and only if $\widehat{\mu_n} \longrightarrow f$ uniformly on compacts, in which case $\mu_n \overset{\mathrm{w}}{\longrightarrow} \mu$.*

Proof. If $f = \hat\mu$ for some μ, Theorem 7.4.3 says that $\widehat{\mu_n} \longrightarrow f$ uniformly on compacts.

Now assume that $\widehat{\mu_n} \longrightarrow f$ uniformly on compacts. Then, for each $\epsilon > 0$, there is a $m \ge 1$ such that $\sup_{|\xi|\le1} |\widehat{\mu}_n(\xi) - \widehat{\mu}_m(\xi)| < \epsilon$ for all $n \ge m$. At the same time, there exists a $0 < \delta \le N^{-\frac12}$ such that $\sup_{|\xi|\le N^{\frac12}\delta} |1-\widehat{\mu_n}(\xi)| \le \epsilon$ for $1 \le n \le m$. Hence, by Lemma 8.2.12,

$$\mu_n\big(B_{\mathbb{R}^N}(0,\delta^{-1})\big) \le 2N\epsilon \text{ for all } n \ge 1,$$

and therefore

$$\lim_{R\to\infty}\sup_{n\ge 1}\mu_n\big(B_{\mathbb{R}^N}(0,R)^{\complement}\big) = 0.$$

By Lemma 8.2.11, this means that $f = \hat{\mu}$ for some μ and $\mu_n \xrightarrow{\text{w}} \mu$. □

Theorem 8.2.13 will play a role in the proof of Bochner's characterization of functions that are the Fourier transforms of probability measures. Before stating his result, I need to define the notion of a **non-negative definite function**. Namely, a function $f : \mathbb{R}^N \longrightarrow \mathbb{C}$ with the properties that $f(-\xi) = \overline{f(\xi)}$ and, for any $\ell \ge 2$, $\{\xi_1,\dots,\xi_\ell\} \subseteq \mathbb{R}^N$, the Hermitian matrix

$$\big(\big(f(\xi_j - \xi_k)\big)\big)_{1\le j,k\le \ell}$$

is said to be non-negative definite. Equivalently, for all $\{z_k : 1 \le k \le \ell\} \subseteq \mathbb{C}$,

$$\sum_{j,k=1}^{\ell} f(\xi_j - \xi_k) z_j \overline{z_k} \ge 0.$$

Theorem 8.2.14 (Bochner's Theorem) *Given a function $f : \mathbb{R}^N \longrightarrow \mathbb{C}$, $f = \hat{\mu}$ for some Borel probability measure μ if and only if $f(0) = 1$, f is non-negative definite, and f is continuous.*

Proof. The "only if" assertion is easy. Indeed, if $f = \hat{\mu}$, then certainly $f(0) = 1$, $f(-\xi) = \overline{f(\xi)}$, and f is continuous. In addition, if $\{\xi_k : 1 \le k \le \ell\} \subseteq \mathbb{R}^N$ and $\{z_k : 1 \le k \le \ell\} \subseteq \mathbb{C}$, then

$$\sum_{j,k=1}^{\ell} f(\xi_j - \xi_k) z_j \overline{z_k} = \sum_{j,k=1}^{\ell} \int \mathfrak{e}_{\xi_j - \xi_k}(x) z_j \overline{z_k}\, \mu(dx)$$
$$= \int \left(\sum_{j,k=1}^{\ell} \mathfrak{e}_{\xi_j}(x) z_j \overline{\mathfrak{e}_{\xi_k}(x) z_k}\right) \mu(dx) = \int \left|\sum_{j=1}^{\ell} \mathfrak{e}_{\xi_j}(x) z_j\right|^2 \mu(dx) \ge 0.$$

The "if" assertion is more challenging. First note that since

$$\begin{pmatrix} 1 & f(\xi) \\ \overline{f(\xi)} & 1 \end{pmatrix}$$

is non-negative definite, it's determinant $1 - |f(\xi)|^2$ is non-negative, and so $|f(\xi)| \le 1$.

Assume for the moment that $f \in L^1(\lambda_{\mathbb{R}^N};\mathbb{C})$, and set (cf. (7.3.2)) $\varphi = \check{f}$. Then $\varphi \in C_0(\mathbb{R}^N;\mathbb{C})$, and because $f \in L^1(\lambda_{\mathbb{R}^N};\mathbb{C}) \cap C_{\mathrm{b}}(\mathbb{R}^N;\mathbb{C} \subseteq L^2(\lambda_{\mathbb{R}^N};\mathbb{C})$, $\varphi \in L^2(\lambda_{\mathbb{R}^N};\mathbb{C})$ which means, by Theorem 7.3.8, $f = \mathfrak{F}\varphi$ (a.e.,$\lambda_{\mathbb{R}^N}$). The goal now is to first show that $\varphi \geq 0$ and then that $\int \varphi \, d\lambda_{\mathbb{R}^N} = 1$. Since φ is bounded and continuous, we will know that it is non-negative if

$$\int \varphi(x)\psi(x)^2 \, \lambda_{\mathbb{R}^N}(dx) \geq 0$$

for all $\psi \in C_c^\infty(\mathbb{R}^N;\mathbb{R})$. To this end, observe that $\hat{\psi}$ is a continuous and, by Theorem 7.3.4, $\hat{\psi} \in L^1(\lambda_{\mathbb{R}^N};\mathbb{C})$. Since ψ is real valued and therefore $\hat{\psi}(-\xi) = \overline{\hat{\psi}(\xi)}$, Theorem 7.3.3 says that

$$\psi(x) = \int \mathfrak{e}_x(-\xi)\hat{\psi}(\xi)\, d\xi = \int \mathfrak{e}_x(\xi)\overline{\hat{\psi}(\xi)}\, d\xi.$$

Hence, by Theorem 4.1.5,

$$\begin{aligned}
\int \varphi(x)\psi(x)^2\, dx &= \lim_{R\to\infty} \int_{B(0,R)} \varphi(x)\psi(x)^2\, dx \\
&= \lim_{R\to\infty} \int_{B_{\mathbb{R}^N}(0,R)} \varphi(x) \left(\iint_{\mathbb{R}^N\times\mathbb{R}^N} \mathfrak{e}_x(\eta-\xi)\hat{\psi}(\xi)\overline{\hat{\psi}(\eta)}\, d\xi\times d\eta \right) dx \\
&= \lim_{R\to\infty} \iint_{\mathbb{R}^N\times\mathbb{R}^N} \hat{\psi}(\xi)\overline{\hat{\psi}(\eta)} \left(\int_{B_{\mathbb{R}^N}(0,R)} \varphi(x)\mathfrak{e}_{\eta-\xi}(x)\, dx \right) d\xi\times d\eta \\
&= \iint_{\mathbb{R}^N\times\mathbb{R}^N} f(\eta-\xi)\hat{\psi}(\xi)\overline{\hat{\psi}(\eta)}\, d\xi\times d\eta.
\end{aligned}$$

Thus, what we need to show is that

$$\iint_{\mathbb{R}^N\times\mathbb{R}^N} f(\eta-\xi)\hat{\psi}(\xi)\overline{\hat{\psi}(\eta)}\, d\xi\times d\eta \geq 0.$$

Because $\hat{\psi} \in L^1(\lambda_{\mathbb{R}^N};\mathbb{C})$ and f is bounded,

$$\iint\limits_{\mathbb{R}^N\times\mathbb{R}^N} f(\eta-\xi)\hat{\psi}(\xi)\overline{\hat{\psi}(\eta)}\, d\xi\times d\eta = \lim_{R\to\infty} \iint\limits_{Q(R)\times Q(R)} f(\eta-\xi)\hat{\psi}(\xi)\overline{\hat{\psi}(\eta)}\, d\xi\times d\eta,$$

where $Q(R) = [-R,R]^N$. Further, because $\hat{\psi}$ and f are continuous,

$$\iint\limits_{Q(R)\times Q(R)} f(\eta-\xi)\hat{\psi}(\xi)\overline{\hat{\psi}(\eta)}\, d\xi\times d\eta = \lim_{\|\mathcal{C}\|\to 0} \sum_{I,J\in\mathcal{C}} f(\xi_J-\xi_I)\hat{\psi}(\xi_I)\overline{\hat{\psi}(\xi_J)} \geq 0,$$

where $\mathcal{C}$ is a finite, exact cover of $Q(R)$ by non-overlapping rectangles and $I \in \mathcal{C} \longmapsto \xi_I \in I$ is an associated choice function.

Now that we know $\varphi \geq 0$, what remains is the show that $\int \varphi\, d\lambda_{\mathbb{R}^N} = 1$, and then take $d\mu = \varphi\, d\lambda_{\mathbb{R}^N}$. For that purpose, recall the functions g_t in (7.3.4) and their Fourier transforms $h_t \equiv \widehat{g_t}$ in (7.3.6). Because h_t is bounded, even, and $\lambda_{\mathbb{R}^N}$-integrable, $\check{h}_t = \widehat{h}_t$, and so, Theorem 7.3.3 says that $\widehat{h_t} = g_t$. Thus, by (7.3.3),

$$\int h_t \varphi \, d\lambda_{\mathbb{R}^N} = \int g_t \bar{f} \, d\lambda_{\mathbb{R}^N}.$$

Since $0 \le h_t(x) \nearrow 1$ as $t \searrow 0$ and $\varphi \ge 0$,

$$\lim_{t \searrow 0} \int h_t \varphi \, d\lambda_{\mathbb{R}^N} = \int \varphi d\lambda_{\mathbb{R}^N}$$

by Theorems 3.2.2 and 6.3.7. At the same time, because f is continuous and $f(0) = 1$, $\lim_{t \searrow 0} \int g_t \bar{f} \, d\lambda_{\mathbb{R}^N} = 1$.

Having handled the case when $f \in L^1(\lambda_{\mathbb{R}^N}; \mathbb{C})$, I will use Theorem 8.2.13 to handle the general case. Namely, set $f_t = g_t f$. Obviously f_t is a bounded, continuous, $\lambda_{\mathbb{R}^N}$-integrable function for which $f_t(0) = 1$ and $f_t(-\xi) = \overline{f_t(\xi)}$. To see that it is non-negative definite, let $\{\xi_k : 1 \le k \le \ell\} \subseteq \mathbb{R}^N$ and $\{z_k : 1 \le k \le \ell\} \subseteq \mathbb{C}$ be given. Then

$$\sum_{j,k=1}^{\ell} f_t(\xi_j - \xi_k) z_j \overline{z_k} = \sum_{j,k=1}^{\ell} f(\xi_j - \xi_k) z_j \overline{z_k} \int \mathfrak{e}_{\xi_j - \xi_k} g_t \, d\lambda_{\mathbb{R}^N}$$

$$= \int g_t(x) \left(\sum_{j,k=1}^{\ell} f(\xi_j - \xi_k) \mathfrak{e}_x(\xi_j) z_j \overline{\mathfrak{e}_x(\xi_k) z_k} \right) dx \ge 0.$$

Hence, by the preceding, we know that, for each $t > 0$, there is a μ_t for which $f_t = \widehat{\mu_t}$, and, since $f_t \longrightarrow f$ uniformly on compacts as $t \searrow 0$, Theorem 8.2.13 guarantees that there is a μ for which $f = \hat{\mu}$. □

Remark 8.2.2 It is somewhat ironic that the more or less trivial part of Bochner's theorem, the one that says that $\hat{\mu}$ is non-negative definite, turns out to be more useful in practice than the part that requires some work. The reason is that, unless one already knows that it is the Fourier transform of a probability measure, it is extremely difficult to determine when a function is non-negative definite. For example, without knowing that it is the Fourier transform of $\frac{\delta_1 + \delta_1}{2}$, it would be hard to show that cos is non-negative definite. Similarly, the easiest way to check that $\xi \rightsquigarrow \frac{\sin(2\pi)\xi}{2\pi\xi}$ is non-negative definite is to recognize it as the Fourier transform of the measure with density is $\frac{1}{2}\mathbf{1}_{[-1,1]}$. The same is true of functions of the form $f_\alpha(\xi) = e^{-|2\pi\xi|^\alpha}$ for $\alpha \ge 0$. By the results in Exercise 7.3.2, one knows that f_α is non-negative definite if $\alpha \in \{0, \frac{1}{2}, 1, 2\}$, and it turns out (cf. § 3.3.2 in my probability book) to be non-negative definite if and only if $\alpha \in [0, 2]$. In Exercise 8.2.8 below, you will show that $e^{-|\xi|^4}$ fails to be non-negative definite.

8.2.6 Exercises for § 8.2

Exercise 8.2.1 Using the existence and properties of the Riemann integral developed in § 1.1, prove the existence of $\lambda_{\mathbb{R}^N}$ as a application of Theorem 8.2.9. Similarly, show that Theorem 8.2.9 can be used to derive Theorem 2.2.17 from the results in § 1.1.2.

Exercise 8.2.2 It should be clear that the Bernoulli measures β_p constructed in § 2.2.4 can also be obtained as the special case of Theorem 8.2.7 in which $\mathcal{I} = \mathbb{Z}^+$ and $E_i = \{0,1\}$ and $\mu_i(\{1\}) = p = 1 - \mu_i(\{0\})$ for all $i \in \mathbb{Z}^+$. The purpose of this exercise is to show that β_p can also be constructed as an application of Theorem 8.2.9. To this end, first recall that $\Omega = \{0,1\}^{\mathbb{Z}^+}$ is a compact metric space. Next, given $f \in C(\Omega;\mathbb{R})$ and $\epsilon > 0$, show that there is an n such that $|f(\omega') - f(\omega)| < \epsilon$ if $\omega'(i) = \omega(i)$ for $1 \le i \le n$, and use this to show that

$$\Lambda_p(f) \equiv \lim_{n\to\infty} \sum_{\eta \in \{0,1\}^n} f \circ \psi_n(\eta) p^{\sum_{i=1}^n \eta(i)} q^{n - \sum_{i=1}^n \eta(i)} \text{ exists,}$$

where, for each $n \ge 1$, $\psi_n : \{0,1\}^n \longrightarrow \Omega$ is defined so that $[\psi_n(\eta)](i)$ is either $\eta(i)$ or 0 according to whether $1 \le i \le n$ or $i > n$. Further, show that Λ_p is a non-negative linear functional on $C(\Omega;\mathbb{R})$ and that β_p is the associated measure in Theorem 8.2.9.

Exercise 8.2.3 Let E be a compact metric space and $\{\mu_n : n \ge 1\}$ a sequence of finite Borel measures on E with the property that $\sup_{n\ge1} \mu_n(E) < \infty$. The purpose of this exercise is to show that there exist a subsequence $\{\mu_{n_m} : m \ge 1\}$ and a finite Borel measure μ with the property that $\{\mu_{n_m} : m \ge 1\}$ converges to μ in the sense that $\int \varphi\, d\mu_{n_m} \longrightarrow \int \varphi\, d\mu$ for all $\varphi \in C(E;\mathbb{R})$. The first version of this result was proved when E is a compact interval in $\mathbb{R}$ and is known as the **Helly–Bray Theorem**.

A critical ingredient in the argument outlined below is the fact that $C(E;\mathbb{R})$ with the uniform topology is separable. That is, there is a sequence $\{\varphi_k : k \ge 1\} \subseteq C(E;\mathbb{R})$ that is dense in the sense that, for each $\varphi \in C(E;\mathbb{R})$, $\inf_{k\ge1} \|\varphi - \varphi_k\|_{\mathrm{u}} = 0$. For example, if E is a compact interval in $\mathbb{R}$, then the Weierstrass Approximation Theorem shows that the set of polynomials with rational coefficients will be dense in $C(E;\mathbb{R})$, and the general case can be reduced to this one.[4]

Now let $\{\mu_n : n \ge 1\}$ be given, and use Λ_n to denote the non-negative, continuous linear functional on $C(E;\mathbb{R})$ given by $\Lambda_n(\varphi) = \int \varphi\, d\mu_n$. Also, let $\{\varphi_k : k \ge 1\}$ be a dense sequence in $C(E;\mathbb{R})$.

(**i**) Using a diagonalization procedure, find a subsequence $\{\mu_{n_m} : m \ge 1\}$ with the property that $\lim_{m\to\infty} \Lambda_{n_m}(\varphi_k)$ exists for all $k \ge 1$.

---

[4] See, for example, Lemma 9.1.4 in the second edition of my book *Probability Theory, an Analytic View.*

(ii) Continuing **(i)**, define $\Lambda(\varphi_k) = \lim_{m\to\infty} \Lambda_{n_m}(\varphi_k)$, and use the density of $\{\varphi_k : k \geq 1\}$ in $C(E;\mathbb{R})$ to show that $\Lambda(\varphi) \equiv \lim_{m\to\infty} \Lambda_{n_m}(\varphi)$ exists for all $\varphi \in C(E;\mathbb{R})$.

(iii) Referring to **(ii)**, show that $\varphi \rightsquigarrow \Lambda(\varphi)$ is a non-negative, continuous linear functional on $C(E;\mathbb{R})$, apply Theorem 8.2.9 and the second half of Remark 8.2.1 to see that there exists a Borel measure μ for which $\Lambda(\varphi) = \int \varphi\, d\mu$, and conclude that $\{\mu_{n_m} : m \geq 1\}$ converges to this μ in the desired sense.

Exercise 8.2.4 Recall the space $\mathscr{S}(\mathbb{R};\mathbb{C})^*$ of tempered distributions introduced in § 7.3.4, and suppose Λ is a tempered distribution with the property that $\varphi \geq 0 \implies \Lambda(\varphi) \geq 0$. You are to show in this exercise that there is a Borel measure μ on $\mathbb{R}$ and a $k \in \mathbb{N}$ such that

$$\int (1+x^2)^{-\frac{k}{2}}\mu(x) < \infty \text{ and } \Lambda(\varphi) = \int \varphi\, d\mu \text{ for } \varphi \in \mathscr{S}(\mathbb{R};\mathbb{C}).$$

In particular, $\Lambda(\varphi) \in \mathbb{R}$ for φ in the space $\mathscr{S}(\mathbb{R};\mathbb{R})$ of real valued elements of $\mathscr{S}(\mathbb{R};\mathbb{C})$. Here are some steps that you might follow.

(i) Choose a function $\eta \in C^\infty\big(\mathbb{R};[0,1]\big)$ such that $\eta = 1$ on $[-1,1]$ and $\eta = 0$ off $(-2,2)$, and set $\eta_R(x) = \eta(R^{-1}x)$ for $R \geq 1$ and $x \in \mathbb{R}$. Next, define $\Lambda_R(\varphi) = \Lambda(\eta_R\varphi)$ for $\varphi \in \mathscr{S}(\mathbb{R};\mathbb{C})$, and show that $\big|\Lambda_R(\varphi)\big| \leq \|\varphi\|_{\mathrm{b}}\Lambda(\eta_R)$.

(ii) Using the result in **(i)**, show that Λ_R has a unique extension to $C_{\mathrm{b}}(\mathbb{R};\mathbb{C})$, and apply Theorem 8.2.9 to show that there is a Borel measure μ_R on $\mathbb{R}$ such that

$$\mu_R(\mathbb{R}) = \Lambda(\eta_R) \text{ and } \Lambda_R(\varphi) = \int \varphi\, d\mu_R \text{ for } \varphi \in C_{\mathrm{b}}(\mathbb{R};\mathbb{C}).$$

(iii) Show that there is a Borel measure μ on $\mathbb{R}$ such that $\mu = \mu_R$ on $\mathcal{B}_{\mathbb{R}}\big[(-R,R)\big]$ and therefore that $\mu\big((-R,R)\big) \leq \Lambda(\eta_R)$ for all $R \geq 1$. Now apply Theorem 7.3.12 to produce an $f \in L^2(\lambda_{\mathbb{R}};\mathbb{C})$ and an $m \in \mathbb{N}$ such that

$$\Lambda_R(\eta_R) = \big(\mathcal{H}^{\frac{m}{2}}\eta_R, \bar{f}\big)_{L^2(\lambda_{\mathbb{R}};\mathbb{C})},$$

use this to show that $\Lambda(\eta_R) \leq C(1+R^2)^{\frac{m+1}{2}}$ for some $C < \infty$. Finally, take $k > m+3$.

Exercise 8.2.5 When $p > 2$, the argument in part **(iii)** of Exercise 8.1.4 does not work. Nonetheless, at least when $\Lambda \in L^p(\mu;\mathbb{R})^*$ is non-negative, in the sense that $\Lambda(\varphi) \geq 0$ when $\varphi \geq 0$, a combination of the results in this and the previous sections does work.

(i) Let $(E,\mathcal{B},\mu)$ be a finite measure space, $p \in [1,\infty)$, and $\Lambda \in L^p(\mu;\mathbb{R})^*$. Assuming that Λ is non-negative, show that $\big(E, L^p(\mu;\mathbb{R}),\Lambda\big)$ is an integration theory, and use Theorem 8.2.5 to produce a finite measure ν such that $\Lambda(\varphi) = \int \varphi\, d\nu$ for all $\varphi \in L^p(\mu;\mathbb{R})$.

(ii) Continuing **(i)**, show that $\nu \ll \mu$, and set $f = \frac{d\nu}{d\mu}$. Using Theorem 6.2.4, show that f is an element of $L^{p'}(\mu;\mathbb{R})$ for which $\Lambda(\varphi) = \int \varphi f \, d\mu$ whenever $\varphi \in L^p(\mu;\mathbb{R})$.

(iii) Extend the preceding to cover μ's which are σ-finite.

Exercise 8.2.6 Suppose that $\{\mu_t : t \in (0,\infty)\}$ is a family of Borel probability measures on $\mathbb{R}^N$, and assume that it is a **convolution semigroup** in the sense that $\mu_{s+t} = \mu_s * \mu_t$. Assuming[5]

$$L(\xi) = \lim_{t \searrow 0} \frac{\widehat{\mu_t}(\xi) - 1}{t}$$

exists for each $\xi \in \mathbb{R}^N$, show that, $t \rightsquigarrow \widehat{\mu_t}(\xi)$ is a continuously differentiable function satisfying $\frac{d}{dt}\widehat{\mu_t}(\xi) = L(\xi)\widehat{\mu_t}(\xi)$. Conclude that $\widehat{\mu_t}(\xi) = e^{tL(\xi)}$, and therefore that $\xi \rightsquigarrow L(\xi)$ is continuous.

Exercise 8.2.7 The observation in Exercise 8.2.6 provides an alternate way of computing the Fourier transforms of measures which are elements of a convolution semigroup. In this exercise, you are to see how it applies to the measures whose densities are the functions in Exercise 6.3.3. The numbering here is the same as it was there.

(i) Let μ_t be the measure whose density is $\gamma_{t^{\frac{1}{2}}}$. Then, from Exercise 6.3.3, we know that $t \rightsquigarrow \mu_t$ is a weakly continuous convolution semigroup. To compute the associated function L, show that

$$\begin{aligned} \frac{\widehat{\mu_t}(\xi) - 1}{t} &= (2\pi)^{-\frac{N}{2}} \int_{\mathbb{R}^N} \frac{\cos\left(t^{\frac{1}{2}} 2\pi(\xi, x)_{\mathbb{R}^N}\right) - 1}{t} e^{-\frac{|x|^2}{2}} \, dx \\ &\longrightarrow -2\pi^2|\xi|^2 (2\pi)^{-\frac{N}{2}} \int_{\mathbb{R}^N} (\mathbf{e}, x)_{\mathbb{R}^N}^2 e^{\frac{-|x|^2}{2}} \, dx \end{aligned}$$

for every $\mathbf{e} \in \mathbb{S}^{N-1}$, and conclude that $L(\xi) = -2\pi^2|\xi|^2$ and therefore that $\widehat{\mu_t}(\xi) = e^{-t2\pi^2|\xi|^2}$. This provides a derivation of (7.3.6) that doesn't involve analytic continuation.

(ii) Let μ_t be the measure whose density is ν_{t^2}. Again $t \rightsquigarrow \mu_t$ is a weakly continuous convolution semigroup. Show that

$$\frac{\widehat{\mu_t}(\xi) - 1}{t} = \pi^{-\frac{1}{2}} \int_{(0,\infty)} \frac{e^{i2\pi\xi x} - 1}{x^{\frac{3}{2}}} e^{-\frac{t^2}{x}} \, dx \longrightarrow L(\xi) \equiv \pi^{-\frac{1}{2}} \int_{(0,\infty)} \frac{e^{i2\pi\xi x} - 1}{x^{\frac{3}{2}}} \, dx.$$

By comparing this result to the one in part **(ii)** of Exercise 7.3.2, show that

$$\int_{(0,\infty)} \frac{1 - \cos x}{x^{\frac{3}{2}}} \, dx = \int_{(0,\infty)} \frac{\sin x}{x^{\frac{3}{2}}} \, dx = (2\pi)^{\frac{1}{2}},$$

[5] It turns out that this assumption is unnecessary since a corollary of the Lévy–Khinchine formula, this limit always exists.

a calculation that is usually done using Cauchy's integral formula.

(iii) Let μ_t be the measure with density P_t. Using the same line of reasoning as in the preceding, show that

$$\frac{\widehat{\mu_t}(\xi)-1}{t} \longrightarrow L(\xi) \equiv \frac{4\pi|\xi|}{\omega_N}\int_{\mathbb{R}^N} \frac{\cos\big((\mathbf{e},x)_{\mathbb{R}^N}\big)-1}{|x|^{N+1}}\,dx,$$

where $\mathbf{e}$ is any element of $\mathbb{S}^{N-1}$. By comparing this to the result in part **(iii)** of Exercise 7.3.2, show that

$$\int_{\mathbb{R}^N} \frac{\cos\big((\mathbf{e},x)_{\mathbb{R}^N}\big)-1}{|x|^{N+1}}\,dx = \frac{\omega_N}{2}.$$

Next show that

$$\int_{\mathbb{R}^N} \frac{\cos\big((\mathbf{e},x)_{\mathbb{R}^N}\big)-1}{|x|^{N+1}}\,dx = \left(\int_{\mathbb{S}^{N-1}} |(\mathbf{e},\omega)_{\mathbb{R}^N}|\,\lambda_{\mathbb{S}^{N-1}}(d\omega)\right)\left(\int_{(0,\infty)} \frac{\cos r-1}{r^2}\,dr\right),$$

and using part **(iii)** of Exercise 5.1.3 and **(i)** of Exercise 5.2.8, show that

$$\int_{\mathbb{S}^{N-1}} |(\mathbf{e},\omega)_{\mathbb{R}^N}|\,\lambda_{\mathbb{S}^{N-1}}(d\omega) = \frac{2\omega_{N-2}}{N-1} = \frac{\omega_N}{\pi}.$$

Conclude from these that

$$\lim_{R\to\infty}\int_{(0,R]} \frac{\sin r}{r}\,dr = \int_{(0,\infty)} \frac{\cos r-1}{r^2}\,dr = \frac{\pi}{2},$$

another calculation that is usually done using Cauchy's integral formula.

(iv) Let μ_t be the measure with density G_t, and show that the corresponding function L is given by

$$L(\xi) = \int_{(0,\infty)} \frac{e^{i2\pi\xi x}-1}{x} e^{-x}\,dx,$$

and therefore

$$\widehat{\mu_t}(\xi) = e^{-t\log(1-i2\pi\xi)}.$$

Using this together with part **(vi)** of Exercise 7.3.2, conclude that

$$\int_{(0,\infty)} \frac{\cos\xi x-1}{x}e^{-x}\,dx = \frac{1}{2}\log(1+\xi^2) \;\&\; \int_{(0,\infty)} \frac{\sin\xi x}{x}e^{-x}\,dx = \arctan\xi,$$

calculations that can be done by more elementary means.

Exercise 8.2.8 Another application of the result in Exercise 8.2.6 provides a proof that $\xi \rightsquigarrow e^{-\xi^4}$ is not a non-negative definite function on $\mathbb{R}$. Indeed, suppose it were, and take μ_t to be the measure with $\widehat{\mu_t}(\xi) = e^{-t(2\pi\xi)^4}$. Using

Theorems 7.4.1 and 7.3.4, show that

$$\lim_{t\searrow 0} \frac{\int_{\mathbb{R}} \varphi \, d\mu_t - \varphi(0)}{t} = -(2\pi)^4 \int_{\mathbb{R}} \xi^4 \hat{\varphi}(\xi) \, d\xi = -\partial^4 \varphi(0)$$

for $\varphi \in C_c^\infty(\mathbb{R};\mathbb{R})$. Next show that if $\varphi(0) = 0 \le \varphi$, then

$$\lim_{t\searrow 0} \frac{\int_{\mathbb{R}} \varphi \, d\mu_t - \varphi(0)}{t} \ge 0.$$

Finally, let $\eta \in C_c^\infty(\mathbb{R};[0,1])$ equal 1 on $[-1,1]$, set $\varphi(x) = x^4\eta(x)$, and thereby arrive at the contradiction $0 \le -4!$.

8.3 Carathéodory's Method

As I said before, the theory in § 8.2 does not apply when the measure under construction lacks sufficient finiteness properties. In this section I will present a theory, due to C. Carathéodory, that does not share this deficiency.

8.3.1 Outer Measures and Measurability

As distinguished from **(ii)** in Example 8.2.1, where we assumed that we already had a finitely additive measure on an algebra of subsets of a set E, the starting point here will an **outer measure** which is defined on *all* subsets of E. To be precise, an outer measure is a function (cf. § 2.1.2) $\tilde{\mu} : \mathcal{P}(E) \longrightarrow [0,\infty]$ with the properties that $\tilde{\mu}(\emptyset) = 0$, $\tilde{\mu}$ is *monotone* in the sense that $A \subseteq B \implies \tilde{\mu}(A) \le \tilde{\mu}(B)$, and $\tilde{\mu}$ is *subadditive* in the sense that

$$\tilde{\mu}(\emptyset) = 0 \quad \text{and} \quad \tilde{\mu}\left(\bigcup_{n=1}^\infty \Gamma_n\right) \le \sum_{n=1}^\infty \tilde{\mu}(\Gamma_n) \tag{8.3.1}$$

for all $\{\Gamma_n : n \ge 1\} \subseteq \mathcal{P}(E)$. Such functions appeared in the construction procedure in § 2.2.1, and the way they arose there is typical (cf. Remark 8.3.1 and Exercise 8.3.1).

Just as was the case in § 2.2.1, the idea is to pass from an outer measure to a measure by extracting from $\mathcal{P}(E)$ a σ-algebra on which its restriction is a measure. The difference is that at this level of abstraction it is even less obvious than it was there how to go about that extraction. The answer was given by Carathéodory, who said that the σ-algebra should consist of those $A \subseteq E$ with the property that

$$\tilde{\mu}(\Gamma) = \tilde{\mu}(\Gamma \cap A) + \tilde{\mu}(\Gamma \cap A^\complement) \quad \text{for all } \Gamma \subseteq E,$$

and I will say that $A \subseteq E$ is **Carathéodory measurable** with respect to $\tilde{\mu}$, or, more economically, **C-measurable**, if it satisfies this condition. Observe that, because $\tilde{\mu}$ is subadditive, A is C-measurable if and only if

$$\tilde{\mu}(\Gamma) \geq \tilde{\mu}(\Gamma \cap A) + \tilde{\mu}(\Gamma \cap A^{\complement}) \quad \text{for all } \Gamma \subseteq E. \tag{8.3.2}$$

As we will see below, this definition is remarkably clever in that it always produces a σ-algebra on which $\tilde{\mu}$ is a measure.

Lemma 8.3.1 *Given an outer measure $\tilde{\mu}$ on $\mathcal{P}(E)$, let $\mathcal{B}_{\tilde{\mu}}$ denote the collection of all C-measurable subsets of E. Then $\mathcal{B}_{\tilde{\mu}}$ is an algebra and $\tilde{\mu}(A) = 0 \implies A \in \mathcal{B}_{\tilde{\mu}}$.*

Proof. We first show that $A \in \mathcal{B}_{\tilde{\mu}}$ if $\tilde{\mu}(A) = 0$. Indeed, since $\tilde{\mu}(\Gamma \cap A) = 0$ and $\tilde{\mu}(\Gamma \cap A^{\complement}) \leq \tilde{\mu}(\Gamma)$, (8.3.2) holds. In particular, this shows that $\emptyset \in \mathcal{B}_{\tilde{\mu}}$. Obviously, $A \in \mathcal{B}_{\tilde{\mu}} \implies A^{\complement} \in \mathcal{B}_{\tilde{\mu}}$. Thus, all that remains is to show that $\mathcal{B}_{\tilde{\mu}}$ is closed under finite unions, and clearly this reduces to showing that $A, B \in \mathcal{B}_{\tilde{\mu}} \implies A \cup B \in \mathcal{B}_{\tilde{\mu}}$. But $\tilde{\mu}(\Gamma) = \tilde{\mu}(\Gamma \cap A) + \tilde{\mu}(\Gamma \cap A^{\complement})$ and

$$\tilde{\mu}(\Gamma \cap A^{\complement}) = \tilde{\mu}\big((\Gamma \cap A^{\complement}) \cap B\big) + \tilde{\mu}\big((\Gamma \cap A^{\complement}) \cap B^{\complement}\big) = \tilde{\mu}(\Gamma \cap A^{\complement} \cap B) + \tilde{\mu}\big(\Gamma \cap (A \cup B)^{\complement}\big).$$

Thus,

$$\tilde{\mu}(\Gamma) = \tilde{\mu}(\Gamma \cap A) + \tilde{\mu}\big(\Gamma \cap A^{\complement} \cap B\big) + \tilde{\mu}\big(\Gamma \cap (A \cup B)^{\complement}\big).$$

Finally, $\Gamma \cap (A \cup B) = (\Gamma \cap A) \cup \big(\Gamma \cap A^{\complement} \cap B\big)$, and so, by subadditivity,

$$\tilde{\mu}\big(\Gamma \cap (A \cup B)\big) \leq \tilde{\mu}(\Gamma \cap A) + \tilde{\mu}\big(\Gamma \cap A^{\complement} \cap B\big).$$

Hence $\tilde{\mu}(\Gamma) \geq \tilde{\mu}\big(\Gamma \cap (A \cup B)\big) + \tilde{\mu}\big(\Gamma \cap (A \cup B)^{\complement}\big)$. □

Theorem 8.3.2 *If $\tilde{\mu}$ is an outer measure on $\mathcal{P}(E)$ and $\mathcal{B}_{\tilde{\mu}}$ is the collection of all subsets that are C-measurable with respect to $\tilde{\mu}$, then $\mathcal{B}_{\tilde{\mu}}$ is a σ-algebra. Moreover, if μ is the restriction of $\tilde{\mu}$ to $\mathcal{B}_{\tilde{\mu}}$, then μ is a measure and $(E, \mathcal{B}_{\tilde{\mu}}, \mu)$ is complete.*

Proof. Since $\mathcal{B}_{\tilde{\mu}}$ is an algebra, we will know that it is a σ-algebra once we have shown that it is closed under countable unions of mutually disjoint sets. Thus, suppose that $\{A_n : n \geq 1\}$ is a sequence of mutually disjoint elements of $\mathcal{B}_{\tilde{\mu}}$, and set $B_n = \bigcup_{m=1}^{n} A_m$ and $B = \bigcup_{m=1}^{\infty} A_m$.

By Lemma 8.3.1, we know that $B_n \in \mathcal{B}_{\tilde{\mu}}$ for each $n \geq 1$. We will now show that

$$\tilde{\mu}(\Gamma \cap B_n) = \sum_{m=1}^{n} \tilde{\mu}(\Gamma \cap A_m) \quad \text{for all } n \geq 1 \text{ and } \Gamma \subseteq E. \tag{$*$}$$

There is nothing to do when $n = 1$, and if $(*)$ holds for n then, because $B_n \in \mathcal{B}_{\tilde{\mu}}$, $(\Gamma \cap B_{n+1}) \cap B_n = \Gamma \cap B_n$, and $(\Gamma \cap B_{n+1}) \cap B_n^{\complement} = \Gamma \cap A_{n+1}$,

$$\tilde{\mu}(\Gamma \cap B_{n+1}) = \tilde{\mu}(\Gamma \cap B_n) + \tilde{\mu}(\Gamma \cap A_{n+1}) = \sum_{m=1}^{n} \tilde{\mu}(\Gamma \cap A_m) + \tilde{\mu}(\Gamma \cap A_{n+1}).$$

Given $(*)$, we know that

$$\tilde{\mu}(\Gamma \cap B) \geq \tilde{\mu}(\Gamma \cap B_n) = \sum_{m=1}^{n} \tilde{\mu}(\Gamma \cap A_m) \quad \text{for all } n \geq 1,$$

and so $\tilde{\mu}(\Gamma \cap B) \geq \sum_{m=1}^{\infty} \tilde{\mu}(\Gamma \cap A_m)$. Since the opposite inequality holds by subadditivity, we have shown that

$$\tilde{\mu}(\Gamma \cap B) = \sum_{m=1}^{\infty} \tilde{\mu}(\Gamma \cap A_m) \quad \text{for all } \Gamma \subseteq E. \tag{$**$}$$

Starting from $(*)$ and $(**)$, the proof that $B \in \mathcal{B}_{\tilde{\mu}}$ is easy. Simply use them to check that

$$\begin{aligned}\tilde{\mu}(\Gamma) = \tilde{\mu}(\Gamma \cap B_n) + \tilde{\mu}(\Gamma \cap B_n^{\complement}) &\geq \sum_{m=1}^{n} \tilde{\mu}(\Gamma \cap A_m) + \tilde{\mu}(\Gamma \cap B^{\complement}) \\ &\longrightarrow \tilde{\mu}(\Gamma \cap B) + \tilde{\mu}(\Gamma \cap B^{\complement}).\end{aligned}$$

Hence, we now know that $\mathcal{B}_{\tilde{\mu}}$ is a σ-algebra. In addition, from $(**)$ with $\Gamma = B$, we see that $\tilde{\mu} \upharpoonright \mathcal{B}_{\tilde{\mu}}$ is countably additive and therefore that $(E, \mathcal{B}_{\tilde{\mu}}, \mu)$ is a measure space.

Finally, to check completeness, suppose that $A \subseteq B \subseteq C$, where $A, C \in \mathcal{B}_{\tilde{\mu}}$ and $\mu(C \setminus A) = 0$. Then $\tilde{\mu}(B \setminus A) \leq \mu(C \setminus A) = 0$ and so

$$\tilde{\mu}(\Gamma \cap B) \leq \tilde{\mu}(\Gamma \cap A) + \tilde{\mu}\big(\Gamma \cap (B \setminus A)\big) = \tilde{\mu}(\Gamma \cap A) \quad \text{for any } \Gamma \subseteq E.$$

Hence,

$$\tilde{\mu}(\Gamma) = \tilde{\mu}(\Gamma \cap A) + \tilde{\mu}(\Gamma \cap A^{\complement}) \geq \tilde{\mu}(\Gamma \cap B) + \tilde{\mu}(\Gamma \cap B^{\complement})$$

for all $\Gamma \subseteq E$. □

Remark 8.3.1 It is important to understand the relationship between C-measurability and μ-measurability. Thus, let $(E, \mathcal{B}, \mu)$ be a measure space, and define $\tilde{\mu}$ by

$$\tilde{\mu}(\Gamma) = \inf\{\mu(A) : \Gamma \subseteq A \in \mathcal{B}\}.$$

It is an easy matter to check that $\tilde{\mu}$ satisfies (8.3.1) and that its restriction to $\mathcal{B}$ is μ. Furthermore, if $A \in \mathcal{B}$ and $\Gamma \subseteq E$, then, for every $\mathcal{B} \ni B \supseteq \Gamma$,

$$\mu(B) = \mu(B \cap A) + \mu(B \cap A^{\complement}) \geq \tilde{\mu}(\Gamma \cap A) + \tilde{\mu}(\Gamma \cap A^{\complement}),$$

and therefore $\tilde{\mu}(\Gamma) \geq \tilde{\mu}(\Gamma \cap A) + \tilde{\mu}(\Gamma \cap A^{\complement})$. Hence, $\mathcal{B} \subseteq \mathcal{B}_{\tilde{\mu}}$. In fact, when $(E, \mathcal{B}, \mu)$ is a complete, σ-finite measure space, this inclusion is an equality. To see this, first suppose that μ is finite. Given $B \in \mathcal{B}_{\tilde{\mu}}$, choose a non-increasing sequence $\{C_n : n \geq 1\} \subseteq \mathcal{B}$ such that $B \subseteq C_n$ and $\mu(C_n) \leq \tilde{\mu}(B) + \frac{1}{n}$ for each $n \geq 1$. Then $C \equiv \bigcap_{n=1}^{\infty} C_n \in \mathcal{B}$, $B \subseteq C$, and $\tilde{\mu}(B) = \mu(C)$. Hence,

$$\tilde{\mu}(B) = \mu(C) = \tilde{\mu}(C \cap B) + \tilde{\mu}(C \cap B^{\complement}) = \tilde{\mu}(B) + \tilde{\mu}(C \setminus B),$$

and so $\tilde{\mu}(C \setminus B) = 0$. Because $B^{\complement}$ is also C-measurable, one can work by complementation to produce an $A \in \mathcal{B}$ such that $A \subseteq B$ and $\tilde{\mu}(B \setminus A) = 0$. Combining these, one has that $\mu(C \setminus A) \leq \tilde{\mu}(C \setminus B) + \tilde{\mu}(B \setminus A) = 0$ and therefore that $B \in \mathcal{B}$. Now suppose that μ is infinite but σ-finite, and write $E = \bigcup_{n=1}^{\infty} E_n$, where $\{E_n : n \geq 1\}$ is a sequence of elements of $\mathcal{B}$ with finite μ-measure. Note that $\tilde{\mu}_n \equiv \tilde{\mu} \upharpoonright \mathcal{P}(E_n)$ bears the same relationship to $\mu_n \equiv \mu \upharpoonright \mathcal{B}[E_n]$ as $\tilde{\mu}$ does to μ. Further, observe that if A is C-measurable with respect to $\tilde{\mu}$, then $A_n \equiv A \cap E_n$ is C-measurable with respect to $\tilde{\mu}_n$. Thus, by the preceding, A_n is μ_n-measurable, which means that A_n is μ-measurable as well. Therefore, $A = \bigcup_{n=1}^{\infty} A_n$ is also μ-measurable.

8.3.2 Carathéodory's Criterion

The problem left open by the considerations in the preceding subsection is that of determining for a given outer measure what sets, besides the empty set and the whole space, are C-measurable. This subsection is devoted to Carathéodory's elegant solution to this problem.

Theorem 8.3.3 (**Carathéodory**) *Let (E, ρ) be a metric space and $\tilde{\mu}$ an outer measure on $\mathcal{P}(E)$. If, for all A, $B \subseteq E$,*

$$\rho(A, B) > 0 \implies \tilde{\mu}(A \cup B) = \tilde{\mu}(A) + \tilde{\mu}(B), \tag{8.3.3}$$

then $\mathcal{B}_E \subseteq \mathcal{B}_{\tilde{\mu}}$. In particular, (8.3.3) *implies that there is a Borel measure μ that is the restriction of $\tilde{\mu}$ to $\mathcal{B}_E$.*

The key step in proving this theorem is taken in the following lemma.

Lemma 8.3.4 *Assume that* (8.3.3) *holds, and let G be an open subset of E. Given $A \subseteq G$, set $A_n = \{x \in A : \rho(x, G^{\complement}) \geq \frac{1}{n}\}$ for $n \geq 1$. Then $\tilde{\mu}(A) = \lim_{n\to\infty} \tilde{\mu}(A_n)$.*

Proof. Obviously $A_n \subseteq A_{n+1} \subseteq A$, and therefore $\tilde{\mu}(A_n) \leq \tilde{\mu}(A_{n+1}) \leq \tilde{\mu}(A)$. Hence, the limit $\lim_{n\to\infty} \tilde{\mu}(A_n)$ exists, and all that we need to do is check that it dominates $\tilde{\mu}(A)$; and clearly we need to do so only when $K \equiv \sup_{n\geq 1} \tilde{\mu}(A_n) < \infty$. To this end, first observe that, because G is open, $A_n \nearrow A$. Next, set $D_n = A_{n+1} \setminus A_n$. Then

$$\begin{aligned}\tilde{\mu}(A) &= \tilde{\mu}\left(A_{2m} \cup \bigcup_{n\geq m} D_{2n} \cup \bigcup_{n\geq m} D_{2n+1}\right)\\ &\leq \tilde{\mu}(A_{2m}) + \sum_{n\geq m} \tilde{\mu}(D_{2n}) + \sum_{n\geq m} \tilde{\mu}(D_{2n+1})\end{aligned}$$

for all $m \geq 1$. Thus, we will be done if we show that both

$$\sum_{n=1}^{\infty} \tilde{\mu}(D_{2n}) \quad \text{and} \quad \sum_{n=1}^{\infty} \tilde{\mu}(D_{2n+1})$$

are convergent series. But $\rho(D_{2(n+1)}, D_{2n}) > 0$, and so, by induction, (8.3.3) implies that

$$K \geq \tilde{\mu}(A_{2M+1}) \geq \tilde{\mu}\left(\bigcup_{n=1}^{M} D_{2n}\right) = \sum_{n=1}^{M} \tilde{\mu}(D_{2n})$$

for all $M \geq 1$. Similarly, $\sum_{n=1}^{M} \tilde{\mu}(D_{2n+1}) \leq K$ for all $M \geq 1$. Hence, both series converge. □

Proof of Theorem 8.3.3. We must show that every open set G is C-measurable. To this end, let $\Gamma \subseteq E$ be given, set $A = \Gamma \cap G$, and take the sets A_n as in the statement of Lemma 8.3.4. Then, $\rho(A_n, G^{\complement}) \geq \frac{1}{n}$, and so, by that lemma and (8.3.3),

$$\tilde{\mu}(\Gamma) \geq \tilde{\mu}\big(A_n \cup (\Gamma \cap G^{\complement})\big) = \tilde{\mu}(A_n) + \tilde{\mu}(\Gamma \cap G^{\complement}) \longrightarrow \tilde{\mu}(\Gamma \cap G) + \tilde{\mu}(\Gamma \cap G^{\complement}).$$

□

8.3.3 Hausdorff Measures

Hausdorff measures provide an example of the sort of situation to which Carathéodory's method applies but Daniell's does not.

Recall from § 4.2.2 the definition given in (4.2.3) of $\mathbf{H}^N$ and the fact proved in Theorem 4.2.4 that $\mathbf{H}^N = \lambda_{\mathbb{R}^N}$ on $\mathcal{B}_{\mathbb{R}^N}$. One of advantage of Hausdorff's description is that it lends itself to generalizations that can detect sets that are invisible to Lebesgue measure. Namely, for $N \in \mathbb{N}$, let Ω_N be the volume of the unit ball in $\mathbb{R}^N$, and, with the understanding that $\Omega_0 = 1$, define[6]

$$\Omega_s = (N + 1 - s)\Omega_N + (s - N)\Omega_{N+1} \quad \text{for } s \in [N, N+1],$$

[6] When s is not an integer, there is no canonical choice of Ω_s and any positive number will serve.

and, for $\delta > 0$ and $s \in [0, \infty)$, set

$$\mathbf{H}^s(\Gamma) = \lim_{\delta \searrow 0} \mathbf{H}^{s,\delta}(\Gamma) \text{ where } \mathbf{H}^{s,\delta}(\Gamma) \text{ equals}$$

$$\inf\left\{\sum_{C \in \mathcal{C}} \Omega_s \operatorname{rad}(C)^s : \mathcal{C} \text{ a countable cover of } \Gamma \text{ with } \|\mathcal{C}\| \le \delta\right\}.$$

Just as in the case when $s = N$, $\mathbf{H}^{s,\delta}(\Gamma)$ is non-increasing as a function of $\delta > 0$, and so there is no doubt that this limit exists.

Theorem 8.3.5 *For each $s \in [0, \infty)$, the restriction of $\mathbf{H}^s$ to $\mathcal{B}_{\mathbb{R}^N}$ is a Borel measure.*

Proof. When $s = 0$, it is clear that $\mathbf{H}^0$ is the counting measure: $\mathbf{H}^0(\Gamma) = \operatorname{card}(\Gamma)$. Thus, we will assume that $s > 0$.

For each $\delta > 0$, it is easy to check that $\mathbf{H}^{s,\delta}$ is an outer measure, and so it obvious that $\mathbf{H}^s(\emptyset) = 0$ and that $\mathbf{H}^s$ is monotone. In addition, because $\mathbf{H}^{s,\delta}$ is non-increasing in $\delta > 0$, for any $\{\Gamma_n : n \ge 1\} \subseteq \mathcal{P}(\mathbb{R}^N)$,

$$\mathbf{H}^{s,\delta}\left(\bigcup_{n=1}^{\infty} \Gamma_n\right) \le \sum_{n=1}^{\infty} \mathbf{H}^s(\Gamma_n) \quad \text{for all } \delta > 0,$$

and so

$$\mathbf{H}^s\left(\bigcup_{n=1}^{\infty} \Gamma_n\right) \le \sum_{n=1}^{\infty} \mathbf{H}^s(\Gamma_n).$$

Thus $\mathbf{H}^s$ is also an outer measure, and therefore it suffices to prove that $\mathbf{H}^s(A \cup B) \ge \mathbf{H}^s(A) + \mathbf{H}^s(B)$ if $\operatorname{dist}(A, B) > 2\delta_0$ for some $\delta_0 > 0$. To this end, let $\mathcal{C}$ be a countable cover of $A \cup B$ with $\|\mathcal{C}\| \le \delta \le \delta_0$, and define $\mathcal{C}_A = \{C \in \mathcal{C} : C \cap A \ne \emptyset\}$ and $\mathcal{C}_B = \{C \in \mathcal{C} : C \cap B \ne \emptyset\}$. Then $\mathcal{C}_A$ and $\mathcal{C}_B$ are disjoint, $\mathcal{C}_A$ is a countable cover of A with $\|\mathcal{C}_A\| \le \delta$, and $\mathcal{C}_B$ is a countable cover of B with $\|\mathcal{C}_B\| \le \delta$. Hence

$$\sum_{C \in \mathcal{C}} \Omega_s \operatorname{rad}(C)^s \ge \sum_{C \in \mathcal{C}_A} \Omega_s \operatorname{rad}(C)^s + \sum_{C \in \mathcal{C}_B} \Omega_s \operatorname{rad}(C)^s \ge \mathbf{H}^{s,\delta}(A) + \mathbf{H}^{s,\delta}(B),$$

from which it follows that $\mathbf{H}^{s,\delta}(A \cup B) \ge \mathbf{H}^{s,\delta}(A) + \mathbf{H}^{s,\delta}(B)$ for all $0 < \delta \le \delta_0$. Now let $\delta \searrow 0$. □

Remark 8.3.2 It should be clear that the Borel measure $\mathbf{H}^s$ is translation invariant for all $s \in (0, \infty)$, but that $\mathbf{H}^N$ is the only $\mathbf{H}^s$ that assigns $[0, 1]^N$ a positive, finite measure and is therefore the only one to which Corollary 2.2.12 applies. Applying that corollary, one knows that $\mathbf{H}^N \restriction \mathcal{B}_{\mathbb{R}^N}$ is a positive multiple of $\lambda_{\mathbb{R}^N}$, and one might think that this a priori information would afford another derivation, one that does not require the isodiametric inequality, of the fact, proved in § 4.2.2, that $\mathbf{H}^N \restriction \mathcal{B}_{\mathbb{R}^N} = \lambda_{\mathbb{R}^N}$. Indeed, all that one has to show is that $\mathbf{H}^N\big(B_{\mathbb{R}^N}(0,1)\big) = \Omega_N$ in order to conclude that the positive factor is 1. However, when one attempts to prove this equality, one realizes that one is more or less forced to derive the isodiametric inequality.

It is obvious that if $\Gamma \subseteq \mathbb{R}^M$ and $F : \Gamma \longrightarrow \mathbb{R}^N$ is uniformly Lipschitz continuous with Lipschitz constant L, then

$$\mathbf{H}^s\big(F(\Gamma)\big) \le L^s\mathbf{H}^s(\Gamma) \quad \text{for all } s \in [0,\infty). \tag{8.3.4}$$

Equally obvious is the fact that the measures $\mathbf{H}^s$ corresponding to distinct s's are radically different. Indeed,

$$\mathbf{H}^{s_2,\delta}(\Gamma) \le \frac{\Omega_{s_2}\delta^{s_2-s_1}}{\Omega_{s_1}}\mathbf{H}^{s_1,\delta}(\Gamma) \quad \text{for } s_1 < s_2.$$

Hence,

$$\mathbf{H}^s(\Gamma) > 0 \implies \mathbf{H}^{s'}(\Gamma) = \infty \text{ for all } s' < s$$
$$\mathbf{H}^s(\Gamma) < \infty \implies \mathbf{H}^{s'}(\Gamma) = 0 \text{ for all } s' > s.$$

and so there is at most one s for which $\mathbf{H}^s(\Gamma) \in (0,\infty)$. More generally, for any $\Gamma \subseteq \mathbb{R}^N$, the **Hausdorff dimension** of Γ is the number

$$\mathrm{Hdim}(\Gamma) \equiv \inf\left\{s : \lim_{R\to\infty} \mathbf{H}^s\big(\Gamma \cap B(0,R)\big) = 0\right\}, \tag{8.3.5}$$

which is obviously a number in the interval $[0,N]$. Although $\mathbf{H}^s(\Gamma)$ will be 0 for $s > \mathrm{Hdim}(\Gamma)$ and ∞ for $s < \mathrm{Hdim}(\Gamma)$, $\mathbf{H}^s(\Gamma)$ can be zero, some finite positive number, or infinite when $s = \mathrm{Hdim}(\Gamma)$. Be that as it may, the reason for its name is that Hausdorff dimension is usually thought of as some sort of measure-theoretic analog of the topological notion of *dimension*. However, one has to be a little careful not to push too hard on this analogy with topological dimension. For example, topological dimension is a topological invariant (i.e., it is invariant under homeomorphisms) whereas, although (8.3.4) says that Hausdorff dimension is a bi-Lipschitz invariant (i.e., it is invariant under homeomorphisms that are locally Lipschitz continuous and have locally Lipschitz continuous inverse), it is not a topological invariant. Thus, topological dimension is more stable than Hausdorff dimension. On the other hand, none of the topological definitions of dimension has the resolving power possessed by Hausdorff's. See Exercise 8.3.5 for an example of its power.

8.3.4 Hausdorff Measure and Surface Measure

People who work in geometric measure theory[7] think of Hausdorff measure as providing a generalization of surface measure on submanifolds of $\mathbb{R}^N$. At least at a qualitative level, it is easy to understand what they have in mind. Namely, an n-dimensional submanifold of $\mathbb{R}^N$ is a set $M \subseteq \mathbb{R}^N$ with the property that, for each $p \in M$ there is an $r > 0$ and a diffeomorphism $\Psi \in C^1\big(B_{\mathbb{R}^n}(0,1); \mathbb{R}^N\big)$ taking $B_{\mathbb{R}^n}(0,1)$ onto $B_{\mathbb{R}^N}(p, 2r) \cap M$. Thus, by (8.3.4), the pushforward of $\lambda_{B_{\mathbb{R}^n}(0,1)}$ under Ψ is a measure that is bounded above and below by positive multiples of $\mathbf{H}^n \restriction \mathcal{B}_{B_{\mathbb{R}^N}(p,r)\cap M}$, and so, when M is an n-dimensional differentiable submanifold of $\mathbb{R}^N$, $\mathbf{H}^n \restriction \mathcal{B}_{B_{\mathbb{R}^N}(p,r)\cap M}$ is commensurate with the restriction to $\mathcal{B}_M$ of what differential geometers call the induced Riemannian measure λ_M on M.

In fact, it turns out that $\mathbf{H}^n \restriction M$ is actually equal to λ_M. In that I have discussed λ_M only for hypersurfaces, I will prove this result only in that case. That is, I will show that for hypersurfaces of the sort described in § 5.2.2, the surface measure λ_M constructed there (cf. (5.2.6)) is equal to the restriction of $\mathbf{H}^{N-1}$ to $\mathcal{B}_M$.

Since both λ_M and $\mathbf{H}^{N-1} \restriction \mathcal{B}_M$ are Borel measures and λ_M is finite on compacts, we will know that they are equal as soon as we know that for each point $p \in M$ they agree on all sufficiently small open neighborhoods of p. Further, by Lemma 5.2.6, each element of M lies in a bounded neighborhood V in M for which there is a continuously differentiable choice of $p \in V \longmapsto \mathbf{n}(p) \in \mathbb{S}^{N-1} \cap \mathbf{T}_p(M)^\perp$ and a twice continuously differentiable map[8] Ψ of an open U in $\mathbb{R}^{N-1}$ onto V for which the map

$$(u, \xi) \in \tilde{U} \equiv U \times (-\rho, \rho) \longrightarrow \tilde{\Psi}(u, \xi) \equiv \Psi(u) + \xi \mathbf{n}\big(\Psi(u)\big) \in \mathbb{R}^N$$

is a twice continuously differentiable diffeomorphism onto (cf. (5.2.5)) $V(\rho)$ for a sufficiently small $\rho > 0$. Hence, we need only show that $\lambda_M(V) = \mathbf{H}^{N-1}(V)$ for such a neighborhood V. In addition, without loss in generality, I will assume that U is convex, $\mathbf{n} \circ \Psi$ has bounded first derivatives, and that both $\tilde{\Psi}$ and $\tilde{\Psi}^{-1}$ have bounded first and second derivatives.

I begin by proving that $\lambda_M(V) \le \mathbf{H}^{N-1}(V)$, and for that purpose I will use the following lemma.

Lemma 8.3.6 *There exists a $K < \infty$ such that (cf. (5.2.5))*

$$\lambda_{\mathbb{R}^N}\big(\Gamma(\delta)\big) \le (1 + K\delta)^N 2\delta \Omega_{N-1} \mathrm{rad}(\Gamma)^{N-1}$$

---

[7] The bible of this subject is H. Federer's formidable *Geometric Measure Theory*, published by Springer-Verlag as volume 153 in their Grundlehren series. For a less daunting treatment, see L.C. Evans and R. Gariepy's *Measure Theory and Fine Properties of Functions*, published by the Studies in Advanced Math. Series of CRC Press.

[8] For those who know about normal coordinates, the proofs of the results that follow are simpler if Ψ is chosen so that the coordinate system that it determines is normal.

for all $0 < \delta < \rho$ *and* $\Gamma \in \mathcal{B}_V$ *with* $\operatorname{diam}(\Gamma) < \delta$.

Proof. Without loss in generality, we will assume that $\mathbf{0}_{\mathbb{R}^{N-1}} \in U$, $\Psi(\mathbf{0}_{\mathbb{R}^{N-1}}) = 0_{\mathbb{R}^N} \in \Gamma$, and $\mathbf{n}(\mathbf{0}_{\mathbb{R}^N}) = \mathbf{e}_N$, where $(\mathbf{e}_1, \dots, \mathbf{e}_N)$ is the standard, orthonormal basis for $\mathbb{R}^N$ and I have used the subscript on $\mathbf{0}$ to distinguish between the origin in $\mathbb{R}^{N-1}$ and the one in $\mathbb{R}^N$.

Define $\pi : \mathbb{R}^N \longrightarrow \mathbb{R}^{N-1} \times \{0\}$ to be the projection map given by $\pi(x) = x - (x, \mathbf{e}_N)_{\mathbb{R}^N}\mathbf{e}_N$. Given $q \in \Gamma$ and $|\xi| < \delta$, write

$$q + \xi\mathbf{n}(q) = \pi\big(q + \xi\mathbf{n}(q)\big) + \big(q + \xi\mathbf{n}(q), \mathbf{e}_N\big)_{\mathbb{R}^N}\mathbf{e}_N.$$

Because $\pi\big(q + \xi\mathbf{n}(q)\big) = \pi(q) + \xi\pi\big(\mathbf{n}(q)\big)$ and $\pi\big(\mathbf{n}(q)\big) = \pi\big(\mathbf{n}(q) - \mathbf{e}_N\big)$,

$$\big|\pi\big(q + \xi\mathbf{n}(q)\big) - \pi(q)\big| \le |\xi|\big|\mathbf{n}(q) - \mathbf{e}_N\big|,$$

and so we can find a $K < \infty$ such that (cf. (5.2.1))

$$\pi\big(q + \xi\mathbf{n}(q)\big) \in \pi(\Gamma)^{(K\delta\operatorname{rad}(\Gamma))} \cap (\mathbb{R}^{N-1} \times \{0\}).$$

At the same time, if $u(t) = \big(\Psi(t\Psi^{-1}(q)), \mathbf{e}_N\big)_{\mathbb{R}^N}$ for $t \in [0,1]$, then $u(0) = 0$, and, because

$$\frac{d}{dt}\Psi\big(t\Psi^{-1}(q)\big)\Big|_{t=0} \in \mathbf{T}_{\mathbf{0}_{\mathbb{R}^N}}(M),$$

$\dot{u}(0) = 0$. Thus,

$$(q, \mathbf{e}_N)_{\mathbb{R}^N} = u(1) = \int_{[0,1]} (1-t)\ddot{u}(t)\,dt,$$

and so, because $|\Psi^{-1}(q)|$ is dominated by a constant times $|q|$, we can adjust K so that $|(q, \mathbf{e}_N)_{\mathbb{R}^N}| \le K\delta^2$. Hence, after readjusting K again, we have

$$\big|\big(q + \xi\mathbf{n}(q), \mathbf{e}_N\big)_{\mathbb{R}^N}\big| \le (1 + K\delta)\delta.$$

Combining these, we now know that

$$\Gamma(\delta) \subseteq \big[\pi(\Gamma)^{(K\delta\operatorname{rad}(\Gamma))} \cap (\mathbb{R}^{N-1} \times \{0\})\big] \times \big[-(1+K\delta)\delta, (1+K\delta)\delta\big],$$

and therefore, by the isodiametric inequality for $\lambda_{\mathbb{R}^{N-1}}$, that

$$\lambda_{\mathbb{R}^N}\big(\Gamma(\delta)\big) \le \Omega_{N-1}\big((1+K\delta)\operatorname{rad}(\Gamma)\big)^{N-1} 2\delta(1+K\delta).$$

□

Given Lemma 8.3.6, it is easy to see that $\lambda_M(\Gamma) \le \mathbf{H}^{N-1}(\Gamma)$ for all $\Gamma \in \mathcal{B}_V$. Indeed, if $\Gamma \in \mathcal{B}_V$ and $0 < \delta < \rho$, let $\mathcal{C}$ be a countable cover of Γ with $\|\mathcal{C}\| < \delta$, and, (cf. Exercise 8.3.4) without loss in generality, assume that $\mathcal{C} \subseteq \mathcal{B}_V$. By Lemma 8.3.6, $\lambda_{\mathbb{R}^N}\big(C(\delta)\big) \le (1+K\delta)^N 2\delta\Omega_{N-1}\operatorname{rad}(C)^{N-1}$ for each $C \in \mathcal{C}$, and therefore

$$\frac{\lambda_{\mathbb{R}^N}(\Gamma(\delta))}{2\delta} \le \sum_{C\in\mathcal{C}} \frac{\lambda_{\mathbb{R}^N}(C(\delta))}{2\delta} \le (1+K\delta)^N \sum_{C\in\mathcal{C}} \Omega_{N-1}\mathrm{rad}(C)^{N-1}.$$

Hence $\frac{\lambda_{\mathbb{R}^N}(\Gamma(\delta))}{2\delta} \le (1+K\delta)^N \mathbf{H}^{N-1,\delta}(\Gamma)$, and so, after letting $\delta \searrow 0$ and using (5.2.6), we obtain $\lambda_M(\Gamma) \le \mathbf{H}^{N-1}(\Gamma)$. In view of the remarks with which I began this discussion, we now know that $\lambda_M \le \mathbf{H}^{N-1} \restriction \mathcal{B}_M$.

In order to prove the opposite inequality, I will use the following two lemmas.

Lemma 8.3.7 *For each open subset V' of V and $\delta \in (0,\rho)$, there exist sequences $\{p_n : n \ge 1\} \subseteq V'$ and $\{r_n : n \ge 1\} \subseteq (0,\delta)$ such that $\overline{B_{\mathbb{R}^N}(p_n,r_n)} \subseteq V'(\delta)$ for all $n \ge 1$, $\overline{B_{\mathbb{R}^N}(p_m,r_m)} \cap \overline{B_{\mathbb{R}^N}(p_n,r_n)} = \emptyset$ for all $m \ne n$, and*

$$\mathbf{H}^{N-1}\left(V' \setminus \bigcup_{n=1}^{\infty} \left(\overline{B_{\mathbb{R}^N}(p_n,r_n)} \cap V'\right)\right) = 0.$$

In particular

$$\mathbf{H}^{N-1}(V') = \sum_{n=1}^{\infty} \mathbf{H}^{N-1}\left(\overline{B_{\mathbb{R}^N}(p_n,r_n)} \cap V'\right).$$

Proof. The proof is modeled on the proof of Lemma 4.2.3.

We begin by showing that if $\Gamma \in \mathcal{B}_V$ and $\lambda_{\mathbb{R}^{N-1}}\big(\Psi^{-1}(\Gamma)\big) = 0$, then $\mathbf{H}^{N-1}(\Gamma) = 0$. To this end, for a given $\delta \in (0,\rho)$ and $\epsilon > 0$, choose a countable cover $\mathcal{C}$ of $\Psi^{-1}(\Gamma)$ by cubes Q in $\mathbb{R}^{N-1}$ such that $\|\mathcal{C}\| < \delta$, and $\sum_{Q\in\mathcal{C}} \lambda_{\mathbb{R}^{N-1}}(Q) < \epsilon$. Then the collection $\{\Psi(Q) : Q \in \mathcal{C}\}$ is a countable cover of Γ, and $\mathrm{rad}\big(\Psi(Q)\big) \le L\mathrm{rad}(Q)$, where $L = \mathrm{Lip}(\Psi)$ is the Lipschitz norm of Ψ. Hence,

$$\mathbf{H}^{N-1,L\delta}(\Gamma) \le L^{N-1} \sum_{Q\in\mathcal{C}} \Omega_{N-1}\mathrm{rad}(Q)^{N-1} \le \left(\frac{L\sqrt{N-1}}{2}\right)^{N-1} \Omega_{N-1}\epsilon.$$

In view of the preceding, it suffices to choose $\{p_n : n \ge 1\} \subseteq V$ and $\{r_n : n \ge 1\} \subseteq (0,\delta)$ so that the balls $B_n = \overline{B_{\mathbb{R}^N}(p_n,r_n)}$ are mutually disjoint, $B_n \subseteq V'(\delta)$ for all $n \ge 1$, and

$$\lambda_{\mathbb{R}^{N-1}}\left(U' \setminus \bigcup_{n=1}^{\infty} \left(\Psi^{-1}(B_n \cap V')\right)\right) = 0,$$

where $U' = \Psi^{-1}(V')$. Using Lemma 2.2.10, choose a countable cover $\mathcal{C}_0$ of U' by non-overlapping cubes Q of diameter less than δ. If $Q \in \mathcal{C}_0$ is centered at c and has side length 3ℓ, then

$$\overline{B_{\mathbb{R}^{N-1}}(c,\alpha^2\ell)} \subseteq \Psi^{-1}\Big(\overline{B_{\mathbb{R}^N}(\Psi(c),\alpha\ell)} \cap V'\Big) \subseteq \mathring{Q},$$

where $\alpha = \big(\mathrm{Lip}(\Psi)\wedge\mathrm{Lip}(\Psi^{-1})\big)^{-1}$. Thus, if $B_Q = \overline{B_{\mathbb{R}^N}\big(\Psi(c),\alpha\ell\big)}$, then $B_Q \subseteq V'(\delta)$ and $\lambda_{\mathbb{R}^{N-1}}\big(\Psi^{-1}(B_Q\cap V')\big) \geq 2\theta\lambda_{\mathbb{R}^{N-1}}(Q)$ for some $\theta \in \big(0,\frac{1}{2}\big)$ that depends only on N and α. Hence,

$$\lambda_{\mathbb{R}^N}\left(U' \setminus \bigcup_{Q\in\mathcal{C}_0} \Psi^{-1}(B_Q\cap V')\right) \leq (1-2\theta)\lambda_{\mathbb{R}^N}(U').$$

One now proceeds in exactly the same way as we did at the analogous place in the proof of Lemma 4.2.3. Namely, we can find a finite subset $\{Q_1,\dots,Q_{n_1}\} \subseteq \mathcal{C}_0$ for which

$$\lambda_{\mathbb{R}^{N-1}}\left(U' \setminus \bigcup_{m=1}^{n_1} \Psi^{-1}(B_{Q_m}\cap V')\right) \leq (1-\theta)\lambda_{\mathbb{R}^{N-1}}(U').$$

Applying the same argument to $U_1' \equiv U' \setminus \bigcup_{m=1}^{n_1}\Psi^{-1}(B_{Q_m}\cap V')$, one can remove a θth of the $\lambda_{\mathbb{R}^{N-1}}$-measure of $\mathcal{U}_1'$. Hence, proceeding by induction, at each step consuming a θth of the $\lambda_{\mathbb{R}^{N-1}}$-measure of what is left at the end of the step before, one can construct the required sequence. The details are left to the reader. □

Lemma 8.3.8 *There exists a $K<\infty$, depending only on Ψ, such that*

$$\lambda_M\big(B_{\mathbb{R}^N}(p,r)\cap V\big) \geq (1+Kr)^{-1}\Omega_{N-1}r^{N-1}$$

for $p\in V$ and $r<1\wedge \mathrm{dist}(p,V^{\complement})$.

Proof. Set $u_0 = \Psi^{-1}(p)$ and $A = \big(\frac{\partial\Psi}{\partial u}(u_0)\big)^{-1}$. There exists an $R\in(0,1]$, depending only on Ψ, such that $u_0 + A(u-u_0)\in U$ if $|u-u_0|<R$, and so we can define $\Psi_p(u) = \Psi\big(u_0 + A(u-u_0)\big)$ for $u\in B_{\mathbb{R}^{N-1}}(u_0,R)$.

Observe that $\frac{\partial\Psi_p}{\partial u}(u_0) = I_{\mathbb{R}^{N-1}}$, and therefore that there exists an $\alpha\in[0,1)$, depending only on Ψ, such that

$$\Psi_p\Big(B_{\mathbb{R}^{N-1}}\big(p,(1-\alpha r)\big)\Big) \subseteq B_{\mathbb{R}^N}(p,r)\cap V \quad \text{for } r\in(0,R]$$

and $J\Psi_p(u) \geq (1-\alpha r)$ for $u\in B_{\mathbb{R}^{N-1}}(u_0,r)$. Thus, by (5.2.10),

$$\lambda_M\big(B_{\mathbb{R}^N}(p,r)\big) \geq (1-\alpha r)^{N-1}\Omega_{N-1}r^{N-1}.$$

□

Given Lemmas 8.3.7 and 8.3.8, it is quite easy to show that $\lambda_M(V) \geq \mathbf{H}^{N-1}(V)$. Indeed, choose $\epsilon>0$ so that $V_\epsilon = \{x:\ \mathrm{dist}(x,V\complement)>\epsilon\}\neq\emptyset$. By Lemma 8.3.7, for arbitrarily small $\delta>0$, there exist $\{p_n:\ n\geq 1\}\subseteq V_\epsilon$ and

$\{r_n : n \geq 1\} \subseteq (0,\delta)$ such that the closed balls $\overline{B_{\mathbb{R}^N}(p_n, r_n)}$ are mutually disjoint and

$$\mathbf{H}^{N-1}(V_\epsilon) = \sum_{n=1}^{\infty} \mathbf{H}^{N-1}\big(\overline{\mathbf{B}_{\mathbb{R}^N}(p_n, r_n)}\big) \cap V_\epsilon\big),$$

and, by Lemma 8.3.8, there is a $K < \infty$ such that

$$\mathbf{H}^{N-1}\big(\overline{B_{\mathbb{R}^N}(p_n, r_n)} \cap V_\epsilon\big) \leq \Omega_{N-1} r_n^{N-1} \leq (1 + K\delta)\lambda_M\big(B_{\mathbb{R}^N}(p_n, r_n) \cap V\big).$$

Hence, since

$$\lambda_M(V) \geq \sum_{n=1}^{\infty} \lambda_M\big(B_{\mathbb{R}^N}(p_n, r_n) \cap V\big),$$

after letting $\delta \searrow 0$ we have that $\mathbf{H}^{N-1}(V_\epsilon) \leq \lambda_M(V)$, from which $\mathbf{H}^{N-1}(V) \leq \lambda_M(V)$ follows when $\epsilon \searrow 0$.

Putting this together with the earlier result, and taking into account the remarks preceding Lemma 8.3.6, we now have a proof of the following theorem.

Theorem 8.3.9 *If $N \geq 2$ and λ_M is the surface measure for a hypersurface M in $\mathbb{R}^N$ of the sort described in § 5.2.2, then λ_M is the restriction to $\mathcal{B}_M$ of $\mathbf{H}^{N-1}$.*

8.3.5 Exercises for § 8.3

Exercise 8.3.1 Here is a ubiquitous procedure, which abstracts the procedure used in § 2.2.1 for constructing outer measures. Let E be a non-empty set and $\mathcal{R}$ a collection of subsets of E with the properties that $\emptyset \in \mathcal{R}$ and for each $\Gamma \subseteq E$ there is a countable cover $\mathcal{C}$ of Γ by elements of $\mathcal{R}$. Further, let $V : \mathcal{R} \longrightarrow [0,\infty]$ with $V(\emptyset) = 0$, and define

$$\tilde{\mu}(\Gamma) = \inf\left\{\sum_{I \in \mathcal{C}} V(I) : \mathcal{C} \subseteq \mathcal{R} \text{ is a countable cover of } \Gamma\right\}.$$

Show that $\tilde{\mu}$ is an outer measure.

Exercise 8.3.2 In Theorem 4.2.4 we saw that $\mathbf{H}^{N,\delta} = \mathbf{H}^N$ for all $\delta > 0$. Show that the analogous equality is false when $s < N$. In addition, show that if $s < N$, then $\mathbf{H}^s$ is not (cf. § 2.1.2) regular.

Exercise 8.3.3 In connection with the preceding exercise, it should be recognized that the restriction of $\mathbf{H}^{s,\delta}$ to $\mathcal{B}_{\mathbb{R}^N}$ is not even a finitely additive measure when $s < N$ and $\delta > 0$. For example, suppose that $N \geq 2$ and

$\delta > 0$, and consider the sets $A = B_{\mathbb{R}^N}(0,\delta) \cap \{x \in \mathbb{R}^N : x_1 = 0\}$ and $B = B_{\mathbb{R}^N}(0,\delta) \cap \{x \in \mathbb{R}^N : x_1 \neq 0\}$. Obviously A and B are disjoint elements of $\mathcal{B}$. On the other hand, show that $\mathbf{H}^{1,\delta}(A \cup B) < \mathbf{H}^{1,\delta}(A) + \mathbf{H}^{1,\delta}(B)$.

Exercise 8.3.4 Just as in Exercise 4.2.2, show that for each $s > 0$ $\mathbf{H}^s$ is unchanged if one restricts to covers either by closed sets or by open sets. In addition, show that if $\mathbf{H}^s_{\mathrm{b}}$ is defined by restricting to covers by either open balls or closed balls, then $\mathbf{H}^s = \mathbf{H}^s_{\mathrm{b}}$ when $N = 1$ and $\mathbf{H}^s \leq \mathbf{H}^s_{\mathrm{b}} \leq 2^s \mathbf{H}^s$ when $N \geq 2$. Thus, in determining Hausdorff dimension, one can restrict to covers by either open or by closed balls.

Exercise 8.3.5 Recall the Cantor set C and its construction in Exercise 2.2.3, and set $c = \frac{\log 2}{\log 3}$. The goal of this exercise is to show that $2^{-c-2}\Omega_c \leq \mathbf{H}^c(C) \leq 2^{-c}\Omega_c$ and therefore that $\mathrm{Hdim}(C) = c$. It is known that $\mathbf{H}^c(C) = 2^{-c}\Omega_c$, but this requires more work. Be that as it may, the result here is already a significant refinement of the statement that C is an uncountable set of Lebesgue measure 0.

(**i**) Show that $\mathbf{H}^{c,\delta}(C_n) \leq 2^{-c}\Omega_c$ for every $n \geq 1$ and $\delta \geq 3^{-n}$, and conclude from this that $\mathbf{H}^c(C) \leq 2^{-c}\Omega_c$.

(**ii**) For each $n \in \mathbb{N}$, let $\mathcal{C}_n$ denote the collection of 2^n disjoint, closed intervals whose union is C_n. Given any open interval J and $n \geq 1$, show that $J \supseteq I$ for some $I \in \mathcal{C}_{n-1}$ if J has non-empty intersection with five or more elements of $\mathcal{C}_n$.

(**iii**) If J is an non-empty open interval with $\mathrm{rad}(J) < \frac{1}{2}$, show that, for all $n \geq 0$,

$$4\mathrm{rad}(J)^c \geq \sum_{\substack{I \in \mathcal{C}_n \\ I \subseteq J}} \mathrm{rad}(I)^c.$$

Hint: Assume that $n \geq 1$ and that $J \supseteq I$ for some $I \in \mathcal{C}_n$. Next, let $m \in \mathbb{N}$ be the smallest $k \in \mathbb{N}$ such that $J \supseteq I$ for some $I \in \mathcal{C}_k$. Because $\mathrm{rad}(J) < \frac{1}{2}$, $1 \leq m \leq n$. Now use (**ii**) to see that J can have non-empty intersection with at most four elements of $\mathcal{C}_m$, and conclude that

$$4\mathrm{rad}(J)^c \geq \sum_{\substack{I' \in \mathcal{C}_m \\ I' \cap J \neq \emptyset}} \mathrm{rad}(I')^c = \sum_{\substack{I' \in \mathcal{C}_m \\ I' \cap J \neq \emptyset}} \sum_{\substack{I \in \mathcal{C}_n \\ I \subseteq I'}} \mathrm{rad}(I)^c \geq \sum_{\substack{I \in \mathcal{C}_n \\ I \subseteq J}} \mathrm{rad}(I)^c.$$

(**iv**) Assume that $\mathcal{C}$ is a countable cover of C by open intervals J with $0 < \mathrm{rad}(J) < \frac{1}{2}$, and show that $\sum_{J \in \mathcal{C}} \mathrm{rad}(J)^c \geq 2^{-c-2}$. In doing this, first show that, without loss in generality, one may assume that $\mathcal{C}$ is finite and that there exists an $n \in \mathbb{Z}^+$ such that each $I \in \mathcal{C}_n$ is contained in some $J \in \mathcal{C}$. From this and the second part of Exercise 8.3.4, conclude that $\mathbf{H}^c(C) \geq 2^{-c-2}\Omega_c$.

Index

D. W. Stroock, *Essentials of Integration Theory for Analysis*, Graduate Texts in Mathematics 262, https://doi.org/10.1007/978-3-030-58478-8

Printed by Printforce, the Netherlands